THE ANATOMICAL BASIS OF MOUSE DEVELOPMENT

Front cover picture:
A 3D reconstruction of an E9 mouse embryo showing the following tissues: neural tube (yellow), somites (green), gut (pale blue), heart (atrium: purple; primitive ventricle: mauve), aortae (red), early vascular system (dark blue), optic vesicle (orange), and otic pit (brown).

Back cover picture:
A 3D reconstruction of the cardiovascular system of an E9 embryo within the translucent ectoderm. The tissues shown are the heart (atrium: purple; primitive ventricle: mauve; bulbus cordis: pale pink), aortae and arterial system (red) and early vascular system (pale blue).

(We thank Renske Brune and Richard Baldock for these pictures.)

THE ANATOMICAL BASIS OF MOUSE DEVELOPMENT

Matthew H. Kaufman &
Jonathan B.L. Bard

Department of Biomedical Sciences
Edinburgh University
Edinburgh EH8 9AG
United Kingdom

ACADEMIC PRESS
A Harcourt Science and Technology Company

San Diego San Francisco New York Boston London Sydney Tokyo

This book is printed on acid-free paper

Copyright © 1999 by Academic Press

All Rights Reserved
No part of this publication may be reproduced or transmitted in any form or by any means electronic or mechanical, including photocopy, recording, or any information storage and retrieval system, without permission in writing from the publisher.

Academic Press
A Harcourt Science and Technology Company
525 B Street, Suite 1900, San Diego, California 92101-4495, USA
http://www.academicpress.com

Academic Press
Harcourt Place, 32 Jamestown Road, London NW1 7BY, UK
http://www.academicpress.com

ISBN 0-12-402060-7

A catalogue record for this book is available from the British Library

Typeset by M Rules, London
Printed in United States by Maple-Vail Book Manufacturing Group

00 01 02 03 04 MV 9 8 7 6 5 4 3 2

Contents

1	**INTRODUCTION**	1
2	**THE EARLY STAGES**	**4**
2.1	From fertilization to implantation	4
	Maturation of the egg	4
	The immediate effects of fertilization (TS 1–2, 0–24 hours)	5
	Post-fertilization events (TS 2–3, 1–2 days)	9
	The preimplantation stages (TS 4–5, 3–4 days)	10
	Implantation (TS 6–8, 4.5–6 days)	11
2.2	The postimplantation embryo	15
	Staging early postimplantation mouse embryos	15
	The late egg cylinder (TS 9, 6.5 days)	16
	The primitive streak stage (TS 10, 6.5–7 days)	18
	Early neurulation (TS 11, 7.5 days)	18
	Early human embryonic development	19
2.3	Postimplantation extra-embryonic development	23
	Formation of amniotic, exocoelomic and ectoplacental cavities, and the fate of the amnion and chorion	23
	"Turning" and the reorganization of the membranes	26
	The yolk sac	26
	The umbilical cord and placenta	27
2.4	Early organogenesis	33
	The external appearance of the embryo	33
	Early neurulation	36
	Neural crest migration	36
	The early auditory system	36
	Mesoderm differentiation and somite formation	37
	The early cardiovascular system	37
	The primitive gut	39
	The septum transversum	39
	The diaphragm	40
	The intraembryonic cavities	42
3	**THE TRANSITIONAL TISSUES**	**47**
3.1	The neural crest	47
	Origin	48
	Cephalic neural crest	48
	The occipital region and spinal cord	49

3.2	Somites and their derivatives (muscles, dermis and vertebrae)		51
	Somite formation and breakdown		51
	The myotome and musculature		52
	The dermatome and dermis		55
	The sclerotomes and vertebral axis		55
3.3	The branchial arch system		60
	The branchial arch arteries		63
	The branchial arch cartilages		68
	The nerve supply of the branchial arches		69
	The branchial grooves (clefts)		70
	The branchial pouches		72
	Branchial arch derived muscles		76

4 THE MAJOR ORGAN SYSTEMS — 77

4.1	The heart and its associated vascular system		77
	The development of the heart		77
	The arterial and venous systems		87
	Post-birth changes		91
4.2	The limbs		93
	Normal development		93
	Bone formation		96
	Muscle formation		102
	The blood supply of the limbs		103
	The nerves		105
4.3	The urogenital system		109
	The renal system		109
	The early stages		111
	The metanephros		114
	Later development of the renal system		115
	The autonomic nerve supply to the kidney		117
	The genital system		117
	Primordial germ cells		118
	The early "indifferent" gonad stages		119
	The female gonads and reproductive duct system		121
	The male gonads and reproductive duct system		123
	The external genitalia		127
	Autonomic nerve supply to the gonads		128
	Implications for genetic analysis		128
4.4	The gut and its associated tissues		129
	Early gut development		129
	The foregut		132
	The midgut		134
	The hindgut		138
	Gut function		139
	Systems and organs associated with the gut		143
	The lower respiratory tract and lungs		143

	The derivatives of the septum transversum	147
	The liver	149
	The spleen	152
	The pancreas and gall bladder	152
4.5	The mouth and nose region	156
	The early stages	157
	The palate	158
	The nasal cavities, nasal septum and vomeronasal organ	161
	The external nose and nasal cartilages and bones	162
	The tongue	162
	The thyroid gland	165
	The sublingual, submandibular and parotid salivary glands	166
	The lips	167
	The vibrissae	169
	The teeth	169
4.6	The brain and spinal cord	171
	The basic subdivisions of the central nervous system	173
	Early neural tube development (TS 11–16)	178
	The forebrain	179
	The meninges and choroid plexus	184
	The midbrain	185
	The hindbrain	186
	Blood supply of the brain	188
	The spinal cord	189
	The peripheral nervous system	191
	Appendix: The pituitary	192
4.7	The eye and the ear	194
	The eye	194
	The early stages	194
	The lens	195
	The retina	195
	The cornea, iris and pupil	197
	The eyelids and other peripheral tissue	198
	The nerve supply	198
	The musculature	199
	The blood supply	199
	The ear	202
	The inner ear	202
	The middle ear	202
	The external ear	204
	The nerve supply	205
	The blood supply	206
	Appendix: Functions of the auditory/vestibular apparatus	206
	Hearing	206
	Vestibular function	207

4.8	The skull	209
	The evolutional context	209
	Membrane and dermal bones (the calvarium)	213
	The chondrocranium and cranial base	214
	The vault of the skull	215
	The facial skeleton (viscerocranium)	216
	The blood supply	217

5 THE INDEXES 218

5.1	Introduction	218
5.2	Table of mouse and rat developmental stages	220
5.3	The tissues present in each stage of mouse development	224
5.4	Index of first occurrences of tissues	255
5.5	Glossary	263
5.6	References	266
5.7	Author index	279
5.8	Tissue index	282
5.9	Subject index	288

THE WIRING (LINEAGE) DIAGRAMS

The general organization of the book (Figure 1.1)	2
Cell lineages in the early mouse embryo (Figure 2.2.2)	18
The branchial arches and their derivatives	61
The heart	78
The forelimb	94
The urogenital system	
The renal system	110
The female genital system	120
The male genital system	124
The gut and its associated organs	
The foregut	130
The foregut-midgut junction, midgut and hindgut	131
The respiratory system	142
The brain and spinal cord	
The forebrain and pituitary	174
The midbrain and hindbrain	175
The eye	196
The ear	201

Acknowledgements

We are particularly grateful to Duncan Davidson for our many enjoyable discussions about mouse developmental biology and for his help both in making the tissue lists and in adapting them for the databases, and to Richard Baldock and Christophe Dubreuil for programming the software that allows the anatomy database to be visualized over the internet, all of whom are at the MRC Human Genetics Unit in Edinburgh. It is also a pleasure to thank our colleagues Renske Brune of the Department of Biomedical Sciences, Edinburgh University, and Martin Ringwald at the Jackson Laboratory in Maine, USA. Finally, we would like to express our gratitude to Ian Lennox who drew all of the illustrations from informal drawings produced by MHK and to Sheila Hunter for her help in typing into the computer the many, many changes that were made to drafts of the manuscript.

Preface

The purpose of this book is to act as a resource on anatomical information for developmental biologists trying to elucidate the mechanisms underpinning mouse embryogenesis. To this end, it contains a series of essays describing the developmental anatomy of the major organ systems and their constituent tissues, together with indexes detailing when tissues first appear and which tissues are present in each stage of mouse embryogenesis. There are also diagrams showing developmental lineages for most of the major organ systems with sufficient explanatory text to make them comprehensible to those as yet unfamiliar with the richness of mouse developmental anatomy.

Our intention has been to make the majority of this book readable by someone with relatively little knowledge of mouse developmental anatomy, while also being helpful to the professional anatomist who will inevitably feel that some of the text moves a little slowly. We have however tried to include a fair amount of detail that may be of interest to biologists whose substantial knowledge of mouse development may not extend either to all parts of the embryo or to the later stages of development, and here we have particularly tried to emphasize those ways in which mouse embryogenesis differs from that of the human, and, in these parts, someone with limited background may find the text hard going. We thus appreciate that no reader will enjoy the whole book, but we hope that different readers will find different parts useful.

The origins of this book lie in an attempt to make an index for *The Atlas of Mouse Development* (Kaufman, 1994). That book described in a stage-by-stage way the development of the mouse using photomicrographs of serially sectioned material on which all the major tissues were marked. Indexing such a work was not a trivial matter, and it soon became clear that it would be necessary to determine the stage at which each of the more than a thousand principal organs, components of organs and tissues in the mouse embryo first appeared and to give this as the index item (see Kaufman, 1994). Of itself, however, this catalogue would still not be particularly useful, as an alphabetical list of tissues makes little anatomical sense. We therefore found that we had no choice but to group and categorize tissues into organ systems.

This done, the demands of the database project (see below) led us to make an index of all the constituent tissues in each of the 26 Theiler developmental stages (Theiler, 1972), and here we had, of course, to allow for the fact that many of the early tissues are transitional and break down as embryogenesis proceeds. At this point, it seemed sensible to use these lists longitudinally and put together in diagrammatic form the development of all the major organ systems and so detail their lineage. The next logical step was to supplement this information with textual information about mouse developmental anatomy and so produce this book.

This last step has however taken considerably longer than we expected as there is rather less published on the developmental anatomy of the mouse than we had expected and some of the descriptions in this book go beyond what can be found in the text part of the *Atlas* or, indeed, in the published literature. And here, we had the problem of deciding just how much of this literature to cite. As this book covers the whole of mouse developmental anatomy, there are thousands of articles which might have been mentioned. We have however taken a more parsimonious course and, rather than reference each and every anatomical description, have decided to give for each tissue a few key anatomical references that should be construed as guides to further reading. We have however tried to be more comprehensive where lineage and other experimental data are discussed, but, because the flood of data being published is so great and it is hard to know which work will have long-term importance, we have decided to exclude molecular data, except in a few cases where it directly impinges on developmental anatomy. We therefore apologize to those authors who may feel that mention of their work has been unreasonably omitted.

Anyone who picks up this book may well be surprised that it is not lavishly illustrated with micrographs, and contains only simple line drawings to illustrate key anatomical features. This is not because we wanted this book to be as unreadable as possible, but because it is meant as a reference work to be used in conjunction with *The Atlas of Mouse Development*; we also wanted to keep down its price. Although there are no explicit cross-citations between the two books, readers should have little difficulty in going from the essays, drawings and indices here to the micrographs of sectioned, stage-specific material there. It should also be mentioned that the legends to these drawings are in many cases extensive as they are intended to allow the reader to follow the details of tissue development without direct reference to the text.

A further feature of the book that we hope will be useful are the indexes: these include timing, staging and, in particular, long lists of the tissues present in the embryo at each stage, with tissues new to a particular stage being indicated in bold type. One role of these lists is to help mouse developmental biologists interpret molecular marker data and identify those genes involved in the ontogeny of those new structures, another is to provide a tissue check list for analysing mutant mice. Computer-literate readers may wonder whether these lists are available over the internet – and of course they are (at *http://genex.hgu.mrc.ac.uk/*) as this text complements a much larger project, the making of a database linking domains of gene expression with graphical and textual representations of mouse developmental anatomy, with access to this database being over the internet. Constructing this database is a collaboration between the scientists of the MRC Human Genetics Unit in Edinburgh, the Jackson Laboratory in Maine, USA, and the Department of Anatomy at Edinburgh University.

With the availability of internet access, the reader may wonder why we felt that it was worth publishing these lists here, and the reasons are simple. First, computer access is not always as user-friendly as it might appear, particularly as there are limitations on screen size and hence on the amount of material that can be displayed and easily browsed through. Second, there is the more technical problem that the material available on a screen is in a particularly detailed format designed for entering gene-expression data in the database rather than for browsing (in the database, the tissues are arranged in a hierarchy, with each only mentioned once, while the level of detail there [*e.g.* each somite and its derivatives are separately mentioned at all stages when they are present] goes beyond what is needed for most purposes), and both views have their uses; third, we wanted to include data on those later stages of development that are unlikely to be part of the database in the shorter term, and, finally, we thought that it would be sensible to link the lists to text descriptions of embryogenesis. In this way, we hoped to make the essence of mouse developmental anatomy available in a form that would be helpful both to experienced developmental anatomists and to those approaching the subject for the first time.

Finally, we should say that we are aware that, to many contemporary biologists, anatomy, even developmental anatomy, is a dead, completed subject of minimal academic interest which derives from data gleaned by our forefathers. We do not take that view; on the contrary, it seems to us that the stimulus of molecular genetics has made us realize just how little we know of tissues and the mechanisms underpinning their ontogeny (a decade ago, who had heard of rhombomeres?). And here, we have a reciprocal, even an inductive, interaction: developmental anatomy provides an assay system for much of molecular genetics, while the results of this work are not only illuminating the genetic basis of wide swathes of mammalian developmental anatomy, but are also providing the technology for probing more deeply into tissue morphogenesis than had hitherto been possible. If this book helps to lubricate that interaction, it will have served its purpose.

Matthew Kaufman
Jonathan Bard

1
INTRODUCTION

The first part of this book gives a detailed analysis of mouse developmental anatomy and is designed to fulfil two different roles. The first is to provide an illustrated text description of mouse embryogenesis, particularly for those lacking a detailed background in mammalian development. The second is to give a chronological description of how the major organ systems form and hence show the essential lineages that give rise to these systems. This latter information is presented in a form of "wiring" diagrams that show the time (or Theiler stage) when a particular component of a system first appears and, through links or wires, the "parents" and "children" of those components. The text and diagrams are together intended to provide a simple (perhaps oversimple for professional mouse embryologists) description of tissue development and so introduce anatomical terminology in a way that helps explain its meaning. The latter part of the book is a series of indexes designed to help the reader find out which tissues are to be found at each stage of development and when a particular tissue first appears. As an additional aid, a brief glossary of some common anatomical terms is given at the end of the book.

The systems that are described here are subdivided into three groups. The first covers the early development and lineage arrangements of the embryo up to about 9 days post coitum (or embryological day 9, normally abbreviated to E9, the convention that will be used here). The second describes the transitional systems that participate in the development of many later tissues (the neural crest, the somites and the branchial arches). The third includes the major organ systems together with a few that are of particular interest (e.g. the pituitary gland). The relationships between the various systems and tissues that are covered here are demonstrated in the wiring diagram (Figure 1.1). If there is a particular system that a reader wishes to know about and that is not covered here, a wiring diagram can usually be constructed from the data given in the various indexes.

A difficulty that may be encountered in the interpretation and use of these diagrams is that the lineage analysis is mainly based on anatomical data, under the traditional criterion that, if a tissue in a specific locale changes with time and an altered one appears in the same place, then the new tissue derives from the old. This is not a bad approach, partly because most cells (with some notable exceptions *e.g.* the neural crest, the somite derivatives, some neuronal cells and the primordial germ cells) do not move during development, once the essential geometry of the embryo is in place, and partly because intermediate stages can often be found. On the other hand, it is not completely satisfactory and we would of course have preferred to have based the diagrams on proper experimental data using lineage markers and transplantation studies. Unfortunately, such work has, in general, only been done for the preimplantation and early postimplantation stages of mouse development and, even here, the experimental story is incomplete. This is, however, an area in which rapid progress is being made through the use of cell markers and genetic manipulation, and, although we have tried to include recent studies based on such techniques, we appreciate that some will have been missed while work in progress will render parts of the lineage information out of date far sooner than we would like.

The other limitation of these wiring diagrams is that we have probably underestimated the sophistication of some of the interactions between tissues, and future research will certainly demonstrate sins of omission, but not, we hope, too many of those of commission. In particular, it is likely that early tissues will have more "parents" or "children" than are now known while we have almost certainly given too simple a view of the interactions that underpin organogenesis. The reader should therefore be aware of the limitations of this work and expect that minor changes can be expected as more experimental work is done using molecular and lineage markers. The authors would therefore appreciate readers informing them of any gaps, inadequacies and errors in the text.

The other problem that raises difficulties is the question of developmental age. In a litter of mouse embryos, there is inevitably some anatomical variation and a simple citation of days postconception (dpc) or its equivalent, embryological age (E), is not always helpful in predicting what tissues are present in a given embryo. Moreover, the velocity of development is low at the beginning and end of embryogenesis and rather faster between E9 and E13.

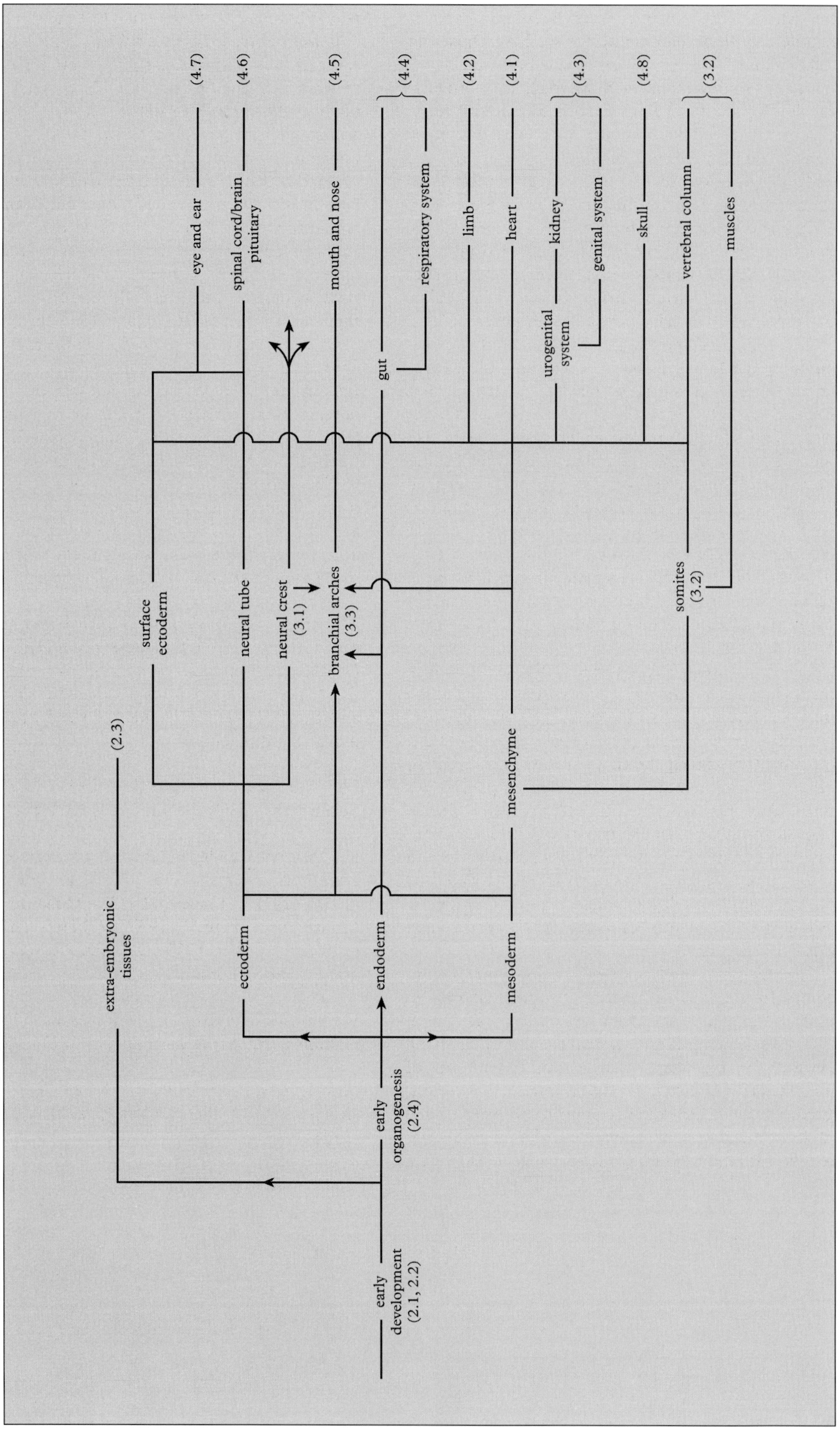

THE ANATOMICAL BASIS OF MOUSE DEVELOPMENT

Figure 1.1 The general organization of the book. The bracketed numbers associated with each tissue refers to the chapter in which it is discussed.

Several staging methods have been proposed to handle mouse development, but we have chosen to follow that of Theiler (1989) as it is based on developmental achievement, acknowledges this variation in developmental velocity and is also in fairly common use. We do, however, appreciate that many mouse developmental biologists prefer to use embryological age and have, therefore, included on each pair of pages a scale linking embryological age to Theiler stage (abbreviated to TS). As one of the indexes in Section 5, we include a detailed table for staging embryos based on the more obvious tissues that they have at a given Theiler stage, and that points out those that they have not yet developed.

As the authors' knowledge of mouse developmental anatomy and particularly developmental neuroanatomy is necessarily somewhat patchy, it has frequently been necessary to refer to the standard works and publications on rodent embryology. On many occasions, where insufficient information has been forthcoming even from these sources, a selected number of textbooks on *human* embryology as well as adult anatomy and physiology have also needed to be consulted. In places throughout the present text, these sources have been specifically cited, while in many other locations either general reviews or occasionally papers dealing with specific topics have been cited.

Another feature in a textbook of *mouse* developmental anatomy the value of which may not be immediately obvious to the reader, is the considerable number of "notes" that appear at intermittent intervals throughout the text. These mainly refer to *human* developmental anatomy, and emphasize for those that may wish to put certain topics into a clinical context, either the *similarities* but more commonly the *differences* between the morphogenesis and anatomical features of these two species. It is hoped that this may be useful, for example, to those that are evaluating the potential of mouse models for human clinical conditions.

One aspect of the book which caused the authors some anguish was the extent to which material is repeated as topics are often dealt with under more than one heading. Here, we had the choice of trying either to be terse, and so increasing the amount of cross-referencing, or to make chapters locally complete. By and large, we have chosen the latter course in the expectation that this would make the text easier to follow for most readers, but apologize to those who would have preferred the book to be shorter.

It will be immediately apparent from a perusal of some of the chapters that many of the references are to work published more than 10 years ago. Indeed, many of the items cited were published before 1950, and some were published before 1930. This should not cause too much distress, and is simply a reflection of the fact that many of the standard references are to laboriously and meticulously conducted analyses of large numbers of serially sectioned mammalian and (occasionally) avian embryos. Few researchers now have the expertise (or even the patience) to undertake such work, and those who do are unlikely to have their work published in the major journals.

Of the general works that have been of greatest assistance in the preparation of the text, specific mention should be made of the following, all of which have in their own way proved to be both instructive and often invaluable. While *mouse* (as well as mammalian) embryologists and developmental biologists will be familiar with most of the following reference texts, these may be less well-known to others, and are all highly recommended. Most are readily available for consultation in biomedical libraries, and many are themselves profusely referenced.

The standard recommended sources for the *mouse* are: Green (1966); Grüneberg (1963); Kaufman (1992, reprinted with index, 1994, 1997); Otis and Brent (1954); Rugh (1968, reprinted 1990); Snell and Stevens (1966); Theiler (1972, reprinted with minor changes, 1989).

For the *rat*, see: Christie (1964); Hebel and Stromberg (1986); Keibel (1937); Witschi (1962).

For the *human*, see: Frazer (1926); Gasser (1975); Hamilton and Mossman (1972); Keibel and Mall (1910, 1912); Larsen (1993); McMinn (1990); Moore and Persaud (1993); O'Rahilly and Müller (1987); Williams (1995), and the *macaque*, Heuser and Streeter (1941).

For the *chick*, see Bellairs and Osmond (1998) and Hamburger and Hamilton (1951).

For more general references, see Bard (1992), Butler and Juurlink (1987), Gilbert (1997) and Wolpert (1998).

2
THE EARLY STAGES

2.1 From fertilization to implantation

Introduction

This chapter covers the events surrounding the production, maturation and fertilization of the mouse egg, together with the early stages of its development, up to implantation and the formation of the proamniotic canal in the egg cylinder. In terms of time, this is a substantial period in the development of the mouse as some 60 hours elapse between the initial hormonal stimulation of the ovary and the fertilization of the resulting oocyte, while the development of the egg through cleavage, morula and inner cell mass stages through to implantation and the formation of the egg cylinder take another six or so days. In terms of embryogenesis, however, this nine-day period is almost irrelevant as, at its end, although the external extra-embryonic tissues are in place, the only embryonic tissues present are the initial derivatives of the inner cell mass, the primitive ectoderm and endoderm.

The focus of the first part of this chapter is therefore on the physiological processes that lead to oogenesis and fertilization together with the various post-fertilization events and blocks that ensure that a healthy egg is fertilized by a single sperm. The second part of the chapter describes the early stages of development and includes a discussion of morula morphogenesis, the formation of the early extra-embryonic tissues, the loss of the zona pellucida, implantation and the formation of the egg cylinder. Comparatively little space is given to the formation of the mouse embryo proper as its development in these early stages is restricted to the appearance of the inner cell mass and its partitioning into inner primitive ectoderm (or epiblast) and a layer of primitive endoderm that starts to line the blastocoelic cavity. It is not until the following chapter that we can begin a substantive analysis of mouse embryogenesis.

Maturation of the egg

Maturation starts within the mouse ovary when **ovarian follicles** are stimulated to undergo folliculogenesis by the pituitary gonadotrophin follicle-stimulating hormone (FSH). This induces them to progress from an immature state to form antral-stage (or mature Graafian) follicles (Figure 2.1.1); 40–48 h later, they are exposed to another pituitary gonadotrophin, luteinizing hormone (LH), one of whose roles includes initiating oocyte maturation (for review, see Gosden *et al.*, 1997). In spontaneously cycling females, this hormone allows the eggs within, on average, 6–12 antral-stage follicles to "mature" from the resting (or dictyate/dictyotene) state, an early stage of prophase, to complete the first meiotic (or "reduction") division. The first polar body is then extruded and these eggs then continue to differentiate until they "block" at the metaphase stage of the second meiotic division which occurs shortly before ovulation (which is also stimulated by LH, see Edwards and Gates, 1959). At this point, the egg is some 65 μm in diameter (Kaufman, 1983a; Figure 2.1.2).

Ovulated eggs are immediately surrounded by a collection of follicle cells termed the **corona radiata**, and collections of such units amalgamate together to

Figure 2.1.1. Diagram showing the typical features of an "antral stage" or mature "Graafian" follicle shortly before ovulation. Note that the oocyte is immediately surrounded by a loose collection of follicle cells, termed the *corona radiata*, and that this unit is only loosely attached to the cells that line the follicle by a narrow bridge of cells (termed the antrum). When ovulation occurs, the oocyte surrounded by its corona radiata is expelled from the follicle with a considerable volume of follicular fluid. The "empty" follicle then becomes converted into the corpus luteum of pregnancy (if fertilization and subsequently implantation occurs) and its component cells secrete progesterone which supports the pregnancy until its hormonal role is taken over by hormones secreted by the placenta.

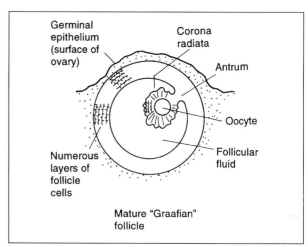

Figure 2.1.2. Diagram showing the principal features observed in a recently ovulated (or secondary) oocyte shortly after ovulation. Note that cellular processes from follicle cells of the corona radiata penetrate through the zona pellucida to make contact with the surface membrane of the egg. Within the perivitelline space, the first polar body is usually still intact at this time, but tends to fragment, often within a few hours of ovulation. Within the vitellus of the egg, membrane-bound cortical granules are seen at this stage in a subcortical location (though, with postovulatory ageing of the egg, these tend to drift away from the periphery towards the centre of the egg). The second meiotic spindle is also located towards the periphery of the egg, and rotates through 90° shortly after fertilization (or activation of the egg) to facilitate extrusion of the second polar body. Chromosomes at metaphase of the second meiotic division are present on the equator of the spindle apparatus.

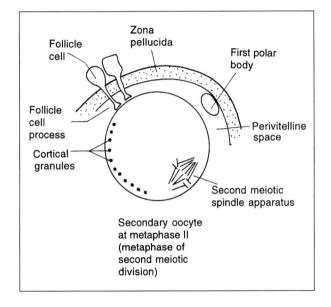

form the **cumulus mass**, and this is believed to play a critical role at the time of fertilization (Fowler et al., 1986). Follicular fluid produced by the follicle cells is also expelled from the antral follicles at the time of ovulation. The majority of the ovulated eggs pass into the opening at the proximal (rostral) end of the oviduct (or infundibulum) which is surrounded by a fringe of finger-like processes (this region of the oviduct is termed the **fimbriated os**) and the eggs are believed to be guided towards the os through what may be a chemotactic mechanism, although it is still not clearly understood (Sterin-Speziale et al., 1978).

Changes in the structure of the fimbria and the location of the infundibulum are associated with changing muscle activity in the supporting ligaments of the ovary, so that at the time of ovulation the fimbriae are drawn close to the surface of the ovary[1] (Pauerstein, 1974). The movements of the fimbriae and the increased activity of the mesentery which supports the oviduct at this time may be associated with the declining levels of oestrogen as ovulation approaches (Fredericks et al., 1977; Halbert and Conrad, 1975). The rhythm of contraction of the fimbriae also depends on levels of prostaglandin and norepinephrine (noradrenaline), and are impaired under conditions of progesterone dominance.

The eggs that enter the oviduct are guided in the direction of its **ampullary region** (where fertilization occurs, see below) by the unidirectional wafting of the cilia that line the oviduct. The cilia are also believed to play a role in transporting the sperm to the site of fertilization, assisted by muscular activity, as well as facilitating the movement of the fertilized embryo along the oviduct towards the uterus. The activity of these cilia and the muscular activity of the oviducts is discussed by Edwards (1980a; see also Woodruff and Pauerstein, 1969).

Ovulated eggs require a further stimulus to complete the second meiotic division, and this is usually provided by the fertilizing spermatozoon (Figure 2.1.3). The follicular cells which remain within the antral follicles from which the eggs were ovulated are then further stimulated by LH, so that this unit becomes converted into the **corpus luteum**. This secretes progesterone which maintains the pregnancy until its critical role is taken over by the hormone-producing tissues of the placenta. Should fertilization not occur, the corpus luteum regresses, and a new oestrus cycle begins.

The immediate effects of fertilization (Theiler stages 1–2, 0–24 hours)

Once the egg has been activated by the fertilizing sperm, usually in the ampullary region of the oviduct, various events occur in rapid succession (for review, see Gilbert, 1997). The universal activating stimulus (which has been observed in all species studied) is believed to be the release of intracellular-bound calcium ions (Steinhardt et al., 1974; Parrington et al., 1998), some of which are associated with membrane phospholipids. This release of calcium ions seems to be related (although the exact relationship is unclear) to the change that occurs in the resting potential of the vitelline membrane (the external cytoplasmic envelope of the egg) which is termed the *fast (or vitelline) block to polyspermy*, which temporarily inhibits the fertilization of the egg by supplementary

---

[1] In the human, the fimbriae partially enclose the ovaries, being arranged over the preovulatory follicles (Edwards and Steptoe, 1975). A similar arrangement is also seen in other species such as the rabbit, guinea pig and cat to form an effective ovarian bursa. In other species, such as in mice, rats and hamsters, where a periovarian sac surrounds the ovary, this assists the transport of oocytes through the fibriated os.

TS	1	2	3	4	5	6	7	8	9	10	11	12	13	14	15	16	17	18	19	20	21	22	23	24	25	26
E	0	1	2	3	4	4.5	5	5.5	6	6.5	7	7.5	8	8.5	9	9.5	10	10.5	11	11.5	12	12.5	13.5	14.5	15.5	16.5	17.5

Figure 2.1.3. Characteristic features seen in a mature mammalian spermatozoon. This is divided into three regions: the head, mid- or middle piece and the tail, from which projects a terminal axial filament. A plasma membrane covers the entire surface of the sperm. In the head region, the nucleus contains the DNA of the cell in a particularly condensed form, and its rostral part is covered by an acrosome cap. The latter contains proteolytic enzymes that are released following the "acrosome reaction", and these facilitate the movement of the sperm through the zona pellucida. A representative transverse section through the middle piece reveals the characteristic 9 + 2 arrangement present, with 2 axial filaments surrounded by an inner (smaller) and an outer (larger) ring of 9 subfibrils. Within the outer "shell" of the middle piece mitochondria are located, within which energy conversion necessary for sperm motility takes place. The 9 + 2 arrangement seen here is also observed in cilia and in flagellae, although it is believed to have developed independently in these three systems.

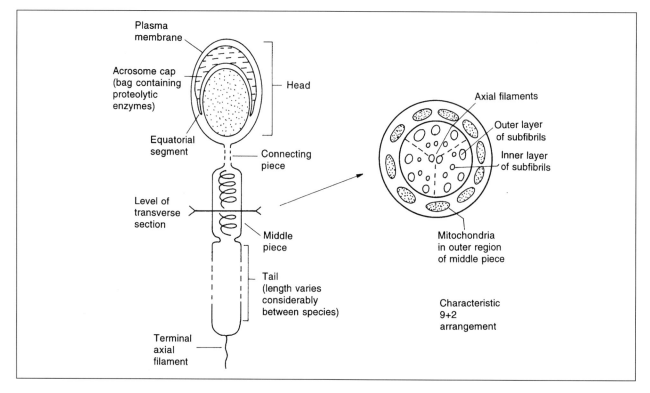

(perivitelline) spermatozoa. Shortly afterwards, the **cortical granules** that are located just beneath the vitelline membrane release their contents into the perivitelline space (Whitaker and Steinhardt, 1982; Kline, 1993; Yanagimachi, 1994). These consist of proteolytic enzymes which alter the biochemical and physical properties of the **zona pellucida** (the acellular glycoprotein membrane surrounding the egg which is now known to be produced by the oocyte (Wassarman, 1990), rather than by the follicle cells as was previously thought). This release of the contents of the cortical granules and its consequences is sometimes termed the *zona reaction*, and constitutes the *slow block to polyspermy*, since it has the effect of inhibiting supernumerary spermatozoa from attaching to the outer surface of the *zona*, and blocking the further progress of those sperm that are in the process of penetrating through the zona (Kaufman, 1983b).

The fertilizing sperm triggers a series of periodic increases in Ca^{2+} that occur at regular intervals, and may continue for several hours. These are termed Ca^{2+} oscillations (Cuthbertson *et al.*, 1981; Miyazaki and Igusa, 1981). It is now believed that it is these oscillations that in some way trigger cortical granule exocytosis (Kline and Kline, 1992). In the mouse, the oscillations are of relatively low frequency, but the *pattern* of Ca^{2+} release appears to be species-specific (Parrington *et al.*, 1998).

As the zona pellucida serves several functions, it is important to be aware of its complex chemical composition and physical properties. One of its most important functions is as an immunological barrier, and it is therefore responsible for species-specific gamete adhesion. Equally, it (usually) acts as a barrier to the fertilization of an egg by an unrelated species, though this may not be a completely efficient process if the two species are closely related. In the human, immunological incompatibility between partners may be an occasional cause of infertility: the zona may block the penetration of sperm from one individual, but allow the sperm from a genetically unrelated male to penetrate it and fertilize the egg.

Of particular importance is the fact that the zona is made up of a number of glycoproteins, termed ZP1, ZP2 and ZP3 (for review, see Wassarman, 1990; Gosden *et al.*, 1997). Each has a slightly different, but principally structural, role during the fertilization process. ZP3 is believed to be the most critical of the three, as it is the *primary sperm receptor,* and *acrosome reaction inducer*, while ZP2 is believed to serve as a *secondary receptor* during gamete adhesion. After the activation component of fertilization, both ZP2 and ZP3 are modified by the enzymes released from the cortical granules, and their activity constitutes the *slow block to polyspermy*. A sperm receptor (non-protein) oligosaccharide is also recognized by sperm during the process of gamete adhesion. In the mouse, the zona is about 7 μm in thickness, and is permeable to most macromolecules such as enzymes and even small viruses. All three of the principal zona glycoproteins are synthesized at the same time and make up about 7–8% of the total protein output of the egg.

The interaction between the sperm and the egg at fertilization is quite complex, and the process has a number of component steps.

1. *Attachment.* There is no evidence of species specificity between the sperm and the outer surface of the zona pellucida.
2. *Primary binding.* There is now some evidence of species specificity, and, once primary binding of the sperm to the surface of the zona has occurred, it is not possible to remove sperm of the appropriate species from the surface of the zona by gentle suction using a pipette. The zonal component to which the sperm attaches is termed the *primary sperm receptor.* Sperm will bind to the surface of ovulated eggs, but not to the surface of "immature" eggs isolated from the ovary. Substantial ultrastructural and histochemical changes take place in the zona pellucida of the mouse during the final stages of oocyte maturation prior to ovulation (Kaufman *et al.*, 1989). After ovulation, the zona shows evidence of maturation.
3. *Acrosome reaction.* This mainly occurs when sperm bind to the surface of the zona. The enzymes released from its *acrosome cap* allow the sperm to penetrate through the zona to reach first the vitelline membrane and then, following sperm penetration, the vitellus of the egg (Figure 2.1.4).
4. *Secondary binding.* This occurs after the acrosome reaction: when the *inner acrosome membrane* is exposed, *secondary receptors* in the zona then bind

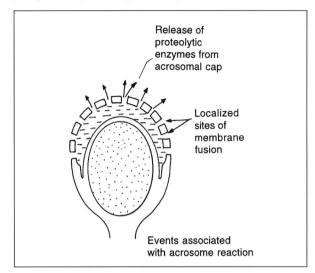

Figure 2.1.4. Diagram displaying the critical changes that occur during the "acrosome reaction". Localized sites of membrane fusion occur between the outer region of the acrosome cap and the overlying plasma membrane. Pores form through which the contents of the acrosome cap are released. The contents consist of a variety of proteolytic enzymes including acrosin, trypsin and chymotrypsin, and their action facilitates the passage of the sperm through the zona pellucida.

to proteins on this membrane.

5. *Penetration.* The sperm penetrates through the zona in an arc-like path (whose width is that of the sperm head) by digesting one or more of the glycoproteins of the zona with proteinase (possibly *acrosin*) associated with and presumably released from the inner acrosome membrane. The pathway through which the fertilizing sperm passes acts as a site of weakness and this is believed to play a critical role in facilitating the escape of the embryo from within the zona pellucida shortly before implantation (see below and Figure 2.1.5a–c).

Figure 2.1.5a–c. The arc-like pathway taken by the fertilizing spermatozoon as it passes through the zona pellucida is illustrated in (a). When the head of the sperm has entered the perivitelline space, it makes initial contact with the microvillous part of the surface of the egg (b), and eventually the plasma membrane of the sperm head fuses with the vitelline or plasma membrane of the egg to form the zygote. The region overlying the second meiotic spindle apparatus is relatively avillous, and gamete fusion cannot occur in this region. This is the site where the second polar body is extruded (c). The sperm nucleus gives rise to the male pronucleus, while the product of the second meiotic division that is retained within the egg gives rise to the female pronucleus. Both pronuclei have a haploid chromosome constitution.

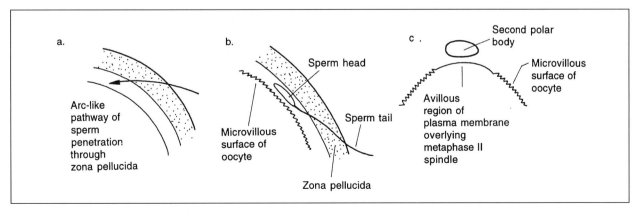

TS	1	2	3	4	5	6	7	8	9	10	11	12	13	14	15	16	17	18	19	20	21	22	23	24	25	26
E	0	1	2	3	4	4.5	5	6	6.5	7	7.5	8	8.5	9	9.5	10	10.5	11	11.5	12	12.5	13.5	14.5	15.5	16.5	17.5

6. *Fusion.* When the sperm reaches the *perivitelline space*, the plasma membrane of the sperm head can then fuse with the vitelline or plasma membrane of the egg to form the *zygote*. Membrane fusion only occurs at the villous surface of the egg, and not in the relatively avillous region overlying the second meiotic spindle.
7. *Fast (primary) block to polyspermy.* Within a few seconds of gamete fusion, supernumerary sperm are prevented from fusing with the vitellus of the egg because there is a rapid change in the electrical properties of the entire surface of the vitellus due to the rapid depolarization of the plasma membrane initiated at the site of fusion, but which rapidly spreads over the entire surface of the egg.
8. *Cortical reaction.* Cortical granules (lysosome-like organelles) located beneath the plasma membrane of the egg fuse with the egg plasma membrane (the cortical reaction) and release their contents (lytic enzymes) into the perivitelline space (Szöllösi, 1967; Figure 2.1.6).
9. *Zona reaction.* The *slow (secondary) block to polyspermy* is due to the modification of the physical and chemical properties of the zona. After fertilization, the zona rapidly "hardens" as a consequence of the chemical changes induced in the zona by the proteolytic enzymes released from the cortical granules. The zona is retained until shortly before implantation when the blastocyst escapes from the zona by the process of "hatching" (see below).
10. *Volume loss.* Following fertilization, there is a slight decrease in the volume of the vitellus.

It is of considerable importance that mating only occurs when the female mouse is in oestrus, because mistimed fertilization leads to chromosomal abnormalities (mostly polyploidy) and the efficiency of the various "blocks" decreases with increased postovulatory ageing of the egg. This is particularly evident in relation to the "slow block to polyspermy" which dramatically decreases in efficiency over a period of possibly 12 hours due to the central migration of the cortical granules from their original subcortical location towards the centre of the egg (Szöllösi, 1971, 1973). Similarly, as an additional consequence of postovulatory ageing of the egg, the second meiotic spindle tends to lose its subcortical attachment and also "drifts" towards the centre of the egg, and this almost certainly inhibits the extrusion of the second polar body. There is also ultrastructural evidence (Szöllösi, 1971) of deterioration of the components of the spindle apparatus which may also be postovulatory age-related, and may result in chromosome detachment during late metaphase/anaphase.

The fertilization of a postovulatory aged egg may lead to a wide variety of chromosomally abnormal (and genetically unbalanced) states, most of which die during the pre- or early postimplantation period (Kaufman, 1983a). The chromosomal constitution of these embryos depends to a considerable degree on whether or not the second polar body is extruded, and whether all, or only some of the female chromosomes condense and become incorporated onto the first cleavage spindle apparatus, so taking part in the first cleavage division. If only a proportion of the maternally-derived chromosomes condense, then the embryo may have a heteroploid (*i.e.* aneuploid) chromosome constitution, and its chance of survival to birth is extremely limited. If the second polar body fails to be extruded, perhaps because of the central location of the second meiotic spindle at the time of fertilization, then digynic triploidy results. These embryos also rarely survive beyond the early postimplantation period. If more than one sperm fertilizes an egg, then diandric triploidy may result.

Occasionally, none of the female chromosomes in an egg activated by the fertilizing sperm take part in embryonic development, and this leads to the development of an *androgenetic* embryo. Such embryos need to be contrasted with those that undergo *gynogenesis* where, following fertilization, only the female genome takes part in subsequent embryonic development. A further subgroup also needs to be distinguished from these two groups where, in both cases, the egg is activated by a spermatozoon, and that is those eggs which undergo *parthenogenetic development*. This is defined as the activation of an egg, and subsequent production of an embryo, with or without eventual development into an adult, from a female gamete in the absence of any contribution from a male gamete (Kaufman, 1981a, 1983a,

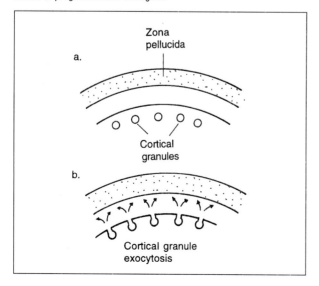

Figure 2.1.6. (a) This shows the normal relationship between the cortical granules and the vitellus of the egg. (b) Some minutes after activation of the egg, and the electrical changes associated with the "fast block to polyspermy", the cortical granules are drawn to the surface of the egg and release their contents into the perivitelline space. Their contents consist of proteolytic enzymes, and these interact with the glycoproteins that constitute a high proportion of the zona pellucida, and induce a "zona reaction". This alters both the physical and chemical properties of the zona and is said to constitute the "slow block to polyspermy", because following the zona reaction (i) sperm seem unable to attach to the outer surface of the zona, and (ii) sperm that are passing through the zona appear to be unable to progress further through it.

modified after Beatty, 1957). In these cases, the egg may be activated by a wide range of stimuli (particularly chemical ones), although in mammals this is, of course, almost exclusively an experimentally induced phenomenon.

Post-fertilization events (Theiler stages 2–3, 1–2 days[2])

The most obvious event that is associated with the completion of the second meiotic division is the extrusion of the second polar body, and this is usually seen within about 3–5 hours of activation. Shortly afterwards, the sperm nucleus decondenses to form the **male pronucleus**, while the product of the second meiotic division that remains within the egg differentiates into the **female pronucleus** (Austin, 1965). The two pronuclei are first evident at about 4–5 hours after activation and, despite the fact that both have a haploid chromosome constitution, the male pronucleus is always about twice the volume of its female counterpart. Both increase in volume throughout the first cell cycle, and achieve their maximum volume just before the first cleavage division. Shortly after the pronuclei are first observed, the DNA within them is doubled (during the S-phase of the first cell cycle).

As the fertilized egg is about to enter the first cleavage division, the outlines of the two pronuclei gradually disappear, and this coincides with the condensation of their maternally- and paternally-derived chromosomes. These initially associate (during prometaphase) on the equator of the first cleavage spindle; the paired chromatids then become apparent and these segregate (during anaphase) to the two poles of the spindle apparatus. All the events associated with the first cleavage division, from the time that the pronuclear outlines disappear to cytoplasmic cleavage (cytokinesis) with the formation of the 2-cell stage embryo, are completed in about 2 hours (Kaufman, 1973). For diagram showing the principal features of the 2-cell stage embryo see Figure 2.1.7.

The total duration of the first cell cycle, between fertilization and the first cleavage division, is normally about 16–18 hours. After a further 36 or so hours and two approximately synchronous divisions, the 8-cell stage is reached; subsequent divisions become increasingly asynchronous (for ultrastructural observations on the early cleavage stages in the mouse, with particular interest in the appearance of the nucleoli between the 2-cell stage and the morula,

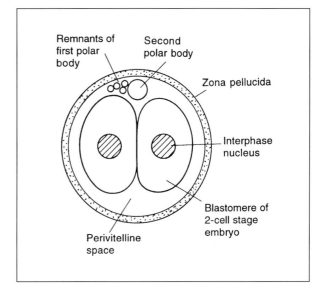

Figure 2.1.7. Diagram illustrating the typical features seen in a 2-cell stage fertilized embryo. Note that the two blastomeres each possess a diploid interphase nucleus. Within the perivitelline space, the most conspicuous feature is the second polar body that was extruded at the completion of the second meiotic division, shortly after activation. The second polar body may remain as an intact unit until the time of implantation. Also present are remnants of the first polar body. This was extruded at the completion of the first meiotic division, shortly before ovulation, and commonly fragments or degenerates shortly afterwards. The embryo is surrounded by the acellular zona pellucida. By this stage, it is unusual to see follicle cells adherent to the outer surface of the zona, either because their cell processes, that formerly passed through the zona, have been withdrawn, or because the follicle cells have degenerated and been discarded into the genital tract.

see Hillman and Tasca, 1969). Up to this 8-cell stage, it is possible to demonstrate that each blastomere is totipotential, being capable (under appropriate experimental conditions) of forming a normal viable fetus which can reach term (Tarkowski and Wróblewska, 1967; Kelly, 1975, 1979). By now, the embryo, each of whose blastomeres maintains a uniform microvillous distribution on their surface, has usually reached the utero-tubal junction and will shortly pass into the lumen of the uterus (Hunter, 1988).

Reproduction in the mouse is extremely efficient and most of the fertilized eggs resulting from natural oestrus cycles successfully develop to birth. If exogenous gonadotrophin hormones are injected to induce *superovulation*, up to a maximum of almost 100 eggs may be induced to ovulate (although 15–30 eggs is more usual) and be successfully fertilized. Because of crowding in the uterus following implantation, a high proportion of such large numbers of embryos usually fail to develop to birth.

---

[2] For detailed analyses of the ultrastructure of the gametes, and the events associated with fertilization and the preimplantation stages of mouse development, see Van Blerkom and Motta (1979, 1984).

TS	1	2	3	4	5	6	7	8	9	10	11	12	13	14	15	16	17	18	19	20	21	22	23	24	25	26
E	0	1	2	3	4.5	5	6	6.5	7	7.5	8	8.5	9	9.5	10	10.5	11	11.5	12	12.5	13.5	14.5	15.5	16.5	17.5

The preimplantation stages (Theiler stages 4–5, 3–4 days)

Perhaps the most important event to occur in this period is the process of *compaction* when the cleavage-stage embryo condenses and becomes a **morula** with the macroscopically observable individuality of the blastomeres being lost. About one third of the component cells now becomes surrounded by a second population of cells and, a little later, fluid starts to accumulate in the intercellular spaces within the morula. The former or inner group of cells is destined to form the **inner cell mass** (ICM), most of which will eventually give rise to the embryo proper, while the outer shell of cells forms the **trophectoderm** or **trophoblast cells** which mainly form such extra-embryonic tissues as the embryonic component of the placenta (Gardner and Papaioannou, 1975; Figure 2.1.8). The trophectoderm subdivides into two regions, according to whether it overlies the ICM (termed the *polar* trophectoderm or Rauber's layer) or whether it surrounds the blastocoele (termed the *mural* trophectoderm).[3]

The formation of the inner cell mass is the first occasion during embryogenesis when there are separate populations on the "inside" and "outside" of the embryo (the "inside–outside" hypothesis, see Tarkowski and Wróblewska, 1967). The mechanism underlying this segregation is both simple and elegant: at about the 8-cell stage, individual blastomeres divide to give two morphologically dissimilar products, the process being termed *polarization* (Johnson and Ziomek, 1981). Cells that will end up on the inside of the morula tend to be smaller, have fewer surface microvilli and tend to produce extracellular glycoproteins on their surface that make them stickier than their complementary division products which will form the outer shell of trophoblast cells. Once this division has occurred, the different fates of the two division products is said to be *determined*. Once formed, the trophoblast cells do not divide and are said to be "end" cells, though the DNA within the nucleus continues to replicate, a process termed *endoreduplication*.

The critical fact that determines when compaction/polarization occurs is not cell number but developmental age as represented by the number of DNA replications that have taken place since fertilization (Smith and Johnson, 1985). Thus, both isolated blastomeres and individual cells within aggregates of cleavage-stage embryos will each behave according to their inbuilt clocks (Smith and McLaren, 1977). Indeed, numerous aggregation and disaggregation studies have confirmed the relationship between developmental age and events such as compaction, polarization and blastocoelic cavity formation (for review, see Gilbert, 1997a).

As compaction occurs, *zonulae occludens* or gap junctions are first evident from about the 8-cell stage, and connect all of the cells of the embryo. Later, such cell communication becomes localized to small groups of cells, and these complexes are believed to play a critical role in the development of the embryo (Lo, 1984). Adhesive complexes (or desmosomes) subsequently form between the cells (Lo and Gilula, 1979; Lo, 1984). Slightly later, fluid accumulation takes place between the cells of the morula and, due to the almost watertight seals (tight junctional complexes) that form between the trophectoderm cells, these spaces expand and aggregate to form the blastocoele (or blastocoelic cavity) thus allowing the embryo to reach the blastocyst stage (Figure 2.1.8) (Enders, 1971; Nadijcka and Hillman, 1974;

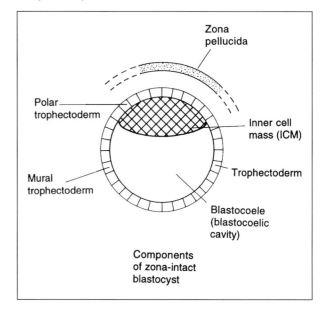

Figure 2.1.8. Diagram illustrating the typical features seen in a *zona-intact* blastocyst. The embryo is seen to contain two discrete components: the inner cell mass (ICM) and an outer shell of trophectoderm. The fluid-filled cavity largely bounded by the trophectoderm is the blastocoele (or blastocoelic or blastocystic) cavity. The latter has a microenvironment that has unique biochemical features, differing in many respects from the chemical composition of the plasma, and the secretions of the female genital tract. The fate of the ICM is different from that of the trophectoderm: the former gives rise to the embryo proper as well as some of the embryonic components of the extra-embryonic membranes (e.g. the ectodermal component of the amnion), while the trophectoderm (or trophoblast tissue) gives rise exclusively to extra-embryonic components.

---

[3] Following the escape of the embryo from within the zona pellucida, the fates of these two trophectoderm components differs: the polar trophectoderm cells are actively involved in invading into the maternal decidual cells, and eventually give rise to the *secondary* trophoblast giant cells. The mural trophoblast cells, in contrast, are end-cells, in that they do not divide following their differentiation to form trophectoderm cells, and they give rise to the *primary* giant cells. During the invasion of the maternal tissues, the polar trophectoderm cells closest to the ICM retain their discrete cell boundaries (termed cytotrophoblast), while those that migrate away from this region lose their cell boundaries to form the so-called syncytiotrophoblast. The different rôles of the cytotrophoblast and the syncytiotrophoblast have yet to be fully determined, but they are believed to play an important role in hormone production, for example, of chorionic gonadotrophin, and possibly also steroidogenesis.

Ducibella, 1977). The presence of the watertight seals isolates the blastocoelic fluid compartment from the outside of the embryo and enables it to become a special microenvironment (for the various types of junctional complexes seen at this stage, see Figure 2.1.9a–c).

The gradual accumulation of fluid between the cells of the morula, and their aggregation together to form a single cavity, defines the early blastocyst stage. With time, an increasing volume of fluid accumulates in this location, and the embryo increases in volume to form the *expanded* blastocyst stage. The gradual expansion in the diameter of the embryo at this time plays a part in the escape (or "hatching") of the blastocyst from the zona pellucida, an essential prerequisite before implantation can occur. At the same time, uterolysins (principally proteolytic enzymes) present in the uterine fluid act on, and gradually thin, the zona pellucida.

Dziadek (1979) has investigated the properties of isolated inner cell masses after isolating them by immunosurgery and allowing them to develop *in vitro*. Within a short period of time, a layer of endoderm cells formed on the outer surface of them. When this layer was removed, a second layer of endoderm formed consisting of predominantly alkaline phosphatase-positive cells (a biochemical "marker" characteristic of *visceral* endoderm cells). It was concluded that the first layer of endoderm cells to form consisted principally of primary or *parietal* endoderm cells, as these were largely alkaline phosphatase-negative.

Implantation (Theiler stages 6–8, 4.5–6 days)

Embryos will only develop beyond the blastocyst stage once they have implanted themselves in the endometrial lining of the uterus, an event that normally occurs after the 64-cell stage when about one third of the embryo's cells comprise the ICM, with the remainder forming the trophectoderm. For implantation to take place, the embryo has to escape from the *zona pellucida*, as it is a prerequisite for the normal *decidual response* (or *decidual reaction*) that the trophectoderm makes *direct* contact with the endometrial cells lining the uterus (Bell, 1985). The zona gradually thins and splits open due to a combination of factors, the most important of which, as mentioned above, is believed to be its exposure to proteolytic enzymes (proteases, termed *uterolysins* that are secreted by the endometrium, Fishel, 1985). This thinning of and eventual "hatching" from the *zona pellucida* is the consequence of the gradual expansion of the blastocyst due to the principally unidirectional flow of water into the blastocoelic cavity, combined with a weakening of the zona caused by the pathway created by the fertilizing sperm (TS 6) together with the effect of the uterolysins (Cole, 1967).

Once the blastocyst is zona-free, it can stimulate the endometrium to undergo a series of changes associated with the decidual response or reaction (TS 7). According to Dickson (1966), unimplanted blastocysts grow in size as they undergo giant cell transformation, as well as increasing in length. After the zona pellucida is lost, giant cell transformation starts at the abembryonic pole and spreads to reach the region of the ICM

Figure 2.1.9a–c. At the 8- to 16-cell stage in the mouse, *compaction* occurs, with the loss of cell individuality. The embryo at this stage is then termed a *morula*. Numerous sites of membrane adhesion, seen at the ultrastructural level to be due to the presence of desmosomes, differentiate between adjacent cells (a). Small region of mural trophectoderm at the abembryonic pole of an expanded blastocyst stage embryo (b), with a close-up view of the region between adjacent trophectoderm cells (c). At the early blastocyst stage, an additional type of junctional complex appears, termed a tight junction, in the peripheral region between adjacent trophectoderm cells.

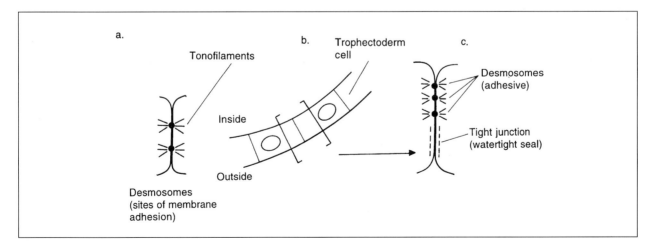

TS	1	2	3	4	5	6	7	8	9	10	11	12	13	14	15	16	17	18	19	20	21	22	23	24	25	26
E	0	1	2	3	4	4.5	5	6	6.5	7	7.5	8	8.5	9	9.5	10	10.5	11	11.5	12	12.5	13.5	14.5	15.5	16.5	17.5

about 12 h later. *In vivo*, implantation occurs at about this time. It is only the cells that constitute the polar region of the trophectoderm (the polar trophoblast cells; *i.e.* the trophectoderm cells that overlie the inner cell mass) and not the mural trophectoderm (the cells that surround the blastocoelic cavity) that play an active role in implantation, and they will shortly form the ICM-derived component of the **ectoplacental cone**[4] (Gardner and Papaioannou, 1975; Copp, 1978, 1979).

According to Copp (1978), the polar trophectoderm cells divide faster than the mural trophectoderm cells. He suggests that the ICM induces a high rate of proliferation on the polar trophectoderm and this consequently results in a shift of cells from the polar to the mural region during blastocyst development. Mural trophectoderm cells close to the ICM divide faster than those further away from the ICM. A considerable increase in cell death was also reported in the ICM, and possibly also in the trophectoderm, but it was speculated that this was unlikely to have a morphogenetic role. After implantation, the polar trophectoderm cells migrate into the blastocoelic cavity and contribute to the formation of the extra-embryonic ectoderm (Copp, 1979). When the ectoplacental cone cells migrate away from their original location, most are destined to form the **secondary trophoblast giant cells** that will eventually be located at the junctional zone between the embryonic and maternal components of the placenta. These cells proliferate and spread anti-mesometrially so as to surround the rest of the conceptus: these cells disappear before term. This population of cells contrasts with those trophoblast cells that line the blastocoelic cavity, termed the **mural trophoblast** that, following implantation, are destined to form the **primary trophoblast giant cells**. For observations on the development of the trophoblast in the mouse embryo, see Duval (1892), Jenkinson (1902), Snell and Stevens (1966) and for experimental studies in which the fate of these cells was investigated, see Gardner and Papaioannou (1975). The transformation of the mural trophectoderm cells to form the primary giant cells (see Alden, 1948; Dickson, 1966) has been confirmed experimentally following the analysis of trophectodermal vesicles in which development of the ICM has been suppressed (Ansell and Snow, 1975), while the origin of the secondary giant cells has been confirmed in studies in which the trophectoderm and ICM isolated from blastocysts with different alleles at the glucose phosphatase isomerase locus (*Gpi-I*) were recombined and the fate of the different components studied (Gardner *et al.*, 1973). At, or shortly before implantation (~ E4.5), the cells on the blastocoelic surface of the ICM delaminate to form the primary embryonic endoderm (see Dziadek, 1979).

The underlying mechanism involved in primitive endoderm formation has yet to be established, but it is now clear that these cells are derivatives of the ICM rather than the trophectoderm, as had previously been thought (Dalcq, 1957). It is unclear whether they form by a process analogous to polarization; it does seem more likely, however, that those totipotent cells exposed to the blastocoelic fluid differentiate to form primary or primitive endoderm, with the earliest manifestation of this process of delamination occurring shortly before implantation and while the blastocyst is still zona-intact. In contrast, the remaining ICM cells maintain their pluripotential properties, forming the **epiblast** component of the embryo. ICM cells can give rise to **embryonic stem cells** when blastocysts are maintained, under special conditions, in tissue culture (Evans and Kaufman, 1981).

Formation of the egg cylinder

The transition from the **blastocyst** to the **egg cylinder** takes place during the early postimplantation period (TS 6/7). Two important changes are observed in the configuration of the embryo at this time. First, there is the peripheral migration of the primitive endoderm from its initial site of delamination in the region overlying the blastocoelic surface of the ICM, so that it eventually lines the entire blastocoelic surface of the **mural trophoblast**, where it forms the extra-embryonic distal (or parietal) layer of endoderm. Following the completion of primitive endoderm migration, the blastocoelic cavity then changes its name, becoming the **primary yolk sac cavity** (Snell and Stevens, 1966). These events occur during TS 7.

Second, and at the same time, a substantial increase occurs in the number of ICM cells present, due to the high level of mitotic activity of these pluripotential cells, and this results in a concomitant increase in its volume. This increase in the growth of the ICM region is directed towards the abembryonic pole of the implanting blastocyst which, because of its changed shape, is now at the **egg cylinder** stage.

During TS 8, changes occur in the configuration of the egg cylinder, the most obvious being the first appearance of the **proamniotic cavity** which forms within the primitive ectoderm (or epiblast). Cavitation is initially observed at the most distal part of the egg cylinder (its future embryonic pole), but it progressively extends proximally towards what will shortly be recognized as its abembryonic pole, to form the **proamniotic canal**; cavitation therefore extends in the direction of the site formerly occupied by the polar trophoblast, towards the region of the ectoplacental cone (see Figure 2.2.1).

The outer surface of the egg cylinder is covered by a layer of primitive (or proximal) endoderm, while the cells which line the trophoblast shell constitute

---

[4] The polar trophectoderm cells overlying the abembryonic pole of the egg cylinder are initially cuboidal in shape, and protrude, in the form of a cap-like mass, into the maternally-derived decidual cells in this location. It is the further growth of these trophectoderm-derived cells that gives rise to the ectoplacental cone. The most peripheral of the trophectoderm cells migrate from the ectoplacental cone in an irregular fashion between the maternal decidual cells and invade the maternal blood vessels forming intercellular lacunae – the basis for the primitive placenta seen in the mouse.

the extra-embryonic parietal (or distal) layer of endoderm. It is at this stage that the first evidence of **Reichert's membrane** is seen. This consists of acellular material secreted by the parietal endoderm; the isolated units of acellular material fuse together to form a continuous layer (Reichert's membrane) which eventually completely separates the parietal endoderm layer of cells from the subjacent mural trophoblast cells, some of which show early evidence of giant cell formation. These will in due course form the **primary giant cells**, in contrast to the **secondary giant cells** which, as indicated earlier, are derived from the terminal differentiation products of the polar trophoblast cells, and even at this stage show early evidence of migration into the maternal decidual tissue. In addition to increasing in volume, giant cell DNA endoreduplicates (*i.e.* duplicates without undergoing cell division or increasing their chromosome constitution).

Implantation, and the formation of the ectoplacental cone

At implantation, changes occur in the shape of the maternal stromal fibroblast tissue surrounding the implanting blastocyst to give the so-called **decidual reaction**. The first evidence of this is the appearance of large polygonal cells that contain high levels of glycogen and lipid and extensive interdigitation then occurs between the microvilli of polar trophectoderm and maternally-derived epithelial cells (Enders and Schlafke, 1965). The decidual tissue is now invaded by maternal capillaries, and large quantities of amorphous extracellular material soon appear (Brökelmann and Biggers, 1979), while gap junctions form between the epithelial cells about 24 h after decidualization.

Shortly after the polar trophectoderm starts to proliferate, its cells interact with maternal decidual cells to form the ectoplacental cone (or *träger*) that caps the mesometrial surface of the developing egg cylinder (see below) and acts as a spearhead of tissue that penetrates the *decidua basalis* of the endometrium. At the peripheral margin of the polar trophectoderm closest to the ectoplacental cone, Reichert's membrane (see above) merges with the proximal or visceral endoderm. In this region, the endoderm is columnar and has a higher metabolic activity than the squamous visceral endoderm of the embryo (Gupta *et al.*, 1982).

Later, during the early somite period (TS 13), the embryonic vasculature of the ectoplacental cone is provided by the allantois, with the placenta developing in the region where the extending allantois contacts and then fuses with the **chorionic plate** (*i.e.* the site where the extra-embryonic mesoderm is in contact with polar trophectoderm, the future cytotrophoblast).

At the periphery, or junctional zone of the ectoplacental cone, the trophectoderm cells differentiate to form the syncytiotrophoblast (*i.e.* where the individuality – the distinct cell boundaries – of the cytotrophoblast cells is lost; Theiler, 1972). The peripheral trophoblast cells of the ectoplacental cone invade the newly formed decidual tissue and eventually penetrate the maternal blood spaces (Amoroso, 1952; Snell and Stevens, 1966; Zybina, 1970).

It has been demonstrated experimentally that the polar trophectoderm cells can undergo giant cell transformation irrespective of the presence of the ICM, but that the formation of the ectoplacental cone specifically requires the presence of the ICM (Gardner and Johnson, 1973; Ansell, 1975).

Differentiation of the ICM to form the egg cylinder

At about E 4.5, the cells at the anti-mesometrial pole of the ICM delaminate to form the embryonic or visceral endoderm. While the underlying mechanism is still unclear, it is probably relevant that ICMs isolated prior to endoderm formation and injected into empty zonae pellucidae will develop the ultrastructural features characteristic of endoderm after 1–2 days in the oviduct (Rossant, 1975). Similarly, whenever spontaneous differentiation of mouse embryonal carcinoma (or EC) cells is observed *in vitro*, primary (*i.e.* visceral) endoderm cells differentiate first (Evans and Martin, 1975): the product is a "simple" embryoid body, and this structure is equivalent to the endoderm-invested ICM of the E 4.5–5 mouse embryo.

The remainder of the ICM, the **epiblast**, retains its totipotentiality and, as its cells increase in number, the epiblast with its anti-mesometrial covering of visceral endoderm expands into the blastocoelic cavity to form the egg cylinder. The visceral endoderm cells gradually move away from the blastocoelic surface of the ICM until they eventually line the blastocoelic surface of the mural trophectoderm to form the parietal or distal endoderm with its characteristic flattened morphology. By this time (E5, TS 7), the blastocoelic cavity has diminished in size to a slit-like space, and is now termed the **primary yolk sac cavity**.

Within the egg cylinder, a number of clefts appear during TS 7 (E 5.5) and these gradually coalesce to form the **proamniotic cavity**. The mesometrial two-thirds of this cavity is lined by simple cuboidal epithelium (the **extra-embryonic ectoderm**), while the anti-mesometrial portion is lined by a layer of cuboidal/columnar cells (the **embryonic ectoderm**). At this stage, the distal pole of the egg cylinder is cup-shaped with a lining of ectoderm cells and a covering of endoderm. This is unlike the situation observed in most mammalian species where the

TS	1	2	3	4	5	6	7	8	9	10	11	12	13	14	15	16	17	18	19	20	21	22	23	24	25	26
E	0	1	2	3	4	4.5	5	6	6.5	7	7.5	8	8.5	9	9.5	10	10.5	11	11.5	12	12.5	13.5	14.5	15.5	16.5	17.5

outer region of the early postimplantation embryo is covered with ectoderm, and its inner part is lined by endoderm. The detailed topological morphology of the advanced egg cylinder, the process of gastrulation with the formation of the mesoderm (the third or intermediate germ layer) and the differentiation of the egg cylinder to form the primitive streak stage is discussed in Chapter 2.2. The process of turning (or entypy), whereby the germ layers are inverted, so that the arrangement in the mouse and related species becomes similar to that in other mammalian species is discussed in detail in Chapter 2.4.

2.2 The postimplantation embryo (TS 9–11)

Introduction

This chapter covers the development of the early embryo from the period of implantation through gastrulation, the primitive streak stage and early neurulation. This period of development is difficult to follow for two reasons: first, the embryo is very small indeed (less than 0.5 mm in length at neurulation) and, second, it is during this period that differences in the details of both their timing and their morphogenesis among the various mouse strains are most pronounced. The chapter therefore starts with a detailed discussion of this topic.

The chapter ends with a brief discussion of early human embryogenesis. Because the arrangements of the embryonic disc, the amnion and the yolk sac in the mouse and human are very different, the events that occur during their primitive streak stages and that result in the linking of the intra- and extra-embryonic coelomic cavities are also different. While most of later mouse development is typical of mammalian embryogenesis, it is worth spending a little time on early human development here as it represents the more normal mammalian arrangement.

Staging early postimplantation mouse embryos

Embryo staging during the period between the egg cylinder stage and the establishment of the neural folds that extends to gastrulation and early neurulation (TS 9–10, E6.5–7) is particularly difficult. This is partly because timing varies slightly among different strains of mice and partly because there is a considerable degree of variability in the time of first appearance of the different developmental landmarks. This applies both to embryos isolated from different litters at about the same time post-coitum, and to embryos isolated from a single litter. The problem is made worse by the fact that too few mouse strains have been examined in sufficient detail to determine whether there is a precise "normal" pattern and time scale of events or to indicate whether, for example, the F1 hybrids may be, in developmental terms, somewhat in advance of the norm, or whether some strains, *e.g.* the PO mice, may be somewhat retarded compared to such a norm (see later). It is therefore essential that researchers indicate in their publications which mouse strain(s) they are working with.

The exact reason for the degree of variability seen is presently unclear, though it has been suggested that a partial explanation may be that, in general, XY-bearing embryos, both in mice and in cattle, develop slightly faster than XX embryos during the pre-implantation period (mice: Tsunoda *et al.*, 1985; Burgoyne, 1993; cattle: Avery *et al.*, 1989; Xu *et al.*, 1992), and that this difference is carried through into the early postimplantation period, with XX embryos then catching up so that there is no obvious difference in their later rate of development (Seller and Perkins-Cole, 1987).

Any detailed discussion of these events, particularly any attempt to divide this period into substages is thus intrinsically complicated. Furthermore, the inconsistencies in terminology used by different authors in, for example, defining the primitive streak stage and describing the early events associated with neurulation have made meaningful comparisons between the different staging systems extremely difficult. Much useful light on this topic has, however, been shed with the studies by Fujinaga *et al.* (1992) on rodents of various unnamed strains and Downs and Davies (1993) on the PO strain (*i.e.* Pathology-Oxford strain), though it has to be emphasized that no universally applicable, and therefore entirely satisfactory staging system, has yet been produced.

The following account is based on the staging system recommended by Theiler (1972), and the descriptive account of the events observed generally

TS	1	2	3	4	5	6	7	8	9	10	11	12	13	14	15	16	17	18	19	20	21	22	23	24	25	26
E	0	1	2	3	4	4.5	5	6	6.5	7	7.5	8	8.5	9	9.5	10	10.5	11	11.5	12	12.5	13.5	14.5	15.5	16.5	17.5

follows that provided by Snell and Stevens (1966), but has been modified to take into account more recent work. The various substages used here are principally those recommended by Downs and Davies (1993), though modified to bring them into line with the standard Theiler staging scheme. Even with the scheme used here (Table 2.2.1), it is still only possible to give a general indication of the average embryonic age, and consequently the presence or otherwise of specific developmental landmarks, for specific Theiler stages and a considerable degree of variation is to be expected in a group of embryos removed from a single female, particularly in the earlier stages.

The late egg cylinder (Theiler stage 9, 6.5 days)

During TS 8, significant changes occurred in the configuration of the egg cylinder, the most obvious being the first appearance of the **proamniotic cavity** which forms within the primitive ectoderm (or epiblast). Cavitation is initially observed at the most distal part of the egg cylinder, but it progressively extends proximally (towards what will shortly be recognized as its abembryonic pole that lies within the extra-embryonic component of the primitive ectoderm) to form the more extensive **proamniotic canal**; cavitation therefore extends from the embryonic pole in the direction of the ectoplacental cone. (For details of the extra-embryonic development at this stage and the others in this chapter, see Chapter 2.3 on *Postimplantation extra-embryonic development.*)

During TS 9, the egg cylinder shows increasing evidence of differentiation, so that it is now possible to recognize that the ICM-derived cells form two morphologically distinct regions of the conceptus, a distal "embryonic" component and a proximal "extra-embryonic" component. It is appropriate to call this either the **late egg cylinder** or **pre-streak** stage. While both regions of the conceptus are covered by a single layer of primitive endoderm that had formed earlier (c. TS 5), the cells overlying the embryonic pole tend to show a mainly squamous morphology, while those overlying the extra-embryonic ectoderm have a largely cuboidal or columnar appearance. The extra-embryonic endoderm cells also characteristically possess apical vacuoles and have a microvillous or "brush" border. The transitional zone between the regions covered by cells with a squamous morphology and those with a cuboidal or columnar morphology does not exactly coincide with the conventional boundary between the embryonic and abembryonic components of the conceptus; the non-squamous cells extend in the direction of the embryonic pole on either side of the embryonic axis, but not in the region overlying it (M.H.K., unpublished observations).

The polar trophoblast-derived cells of the ectoplacental cone invade the maternal blood vessels, and consequently become bathed in maternal blood. This is thus the earliest stage in the formation of a specialized site where gaseous exchange between the embryo and the mother can take place, and it is here that the definitive placenta subsequently forms (see Chapter 2.3 *Postimplantation extra-embryonic development*).

At this early stage of development, the embryo, which is located at the embryonic pole of the late or advanced egg cylinder, is cup-shaped and bilayered with an outer hypoblast (or endoderm layer) and an inner ectodermal layer, the epiblast, with this latter sheet of cells facing the proamniotic canal and eventually forming the neural and surface ectoderm, together with the other tissues of the embryo (for more details, see later and Lawson *et al.*, 1991). Note that the rodents are the exception as compared to other mammals in that the organization of their germ layers at this stage is "inverted" as compared to the "normal" situation (*e.g.* in the human embryo) with which most mammalian embryologists are familiar. In order for the embryo to achieve its definitive arrangement (with an outer-facing ectoderm and, subjacent to this, the neural tube apposed to the amniotic cavity), rodents have to "invert" their germ layers (by changing the morphological organization of the embryo with respect to its extra-embryonic membranes) and this is done during the process of

Table 2.2.1 *Strain differences between the P0 strain and C57BL × CBA F1 hybrid embryos which complicate the definition of developmental stages (based principally on the work of Downs and Davies, 1993. For further details see http://genex.hgu.ac.uk/)*

		Staging system	
Stage	Age (days post-coitum)	Defining characteristic P0 mice	Defining characteristic C57BL×CBA F1 hybrid mice
TS 9	6.25–7.25	Late egg cylinder or pre-streak stage	Late egg cylinder or pre-streak stage
TS 9.5	6.25–7.25	Early streak	Early streak
TS 10	6.5–7.5	Primitive streak stage	Primitive streak stage
TS 10.5	6.5–7.5	Late streak stage	Late streak stage + allantoic bud
TS 11.0	7.25–8	Pre-allantoic bud stage	As TS 10.5
TS 11.25	7.25–8	Early allantoic bud stage	Allantoic bud stage
TS 11.5	7.25–8	Late allantoic bud stage	Late allantoic bud stage
TS 11.75	7.25–8	Early headfold stage	Early headfold stage
TS 12	7.5–8.75	1–7 pairs of somites	1–7 pairs of somites

"turning" which follows soon after gastrulation (see *Inversion of the germ layers and the process of turning* in Chapter 2.4 on *Early organogenesis*).

Towards the end of TS 9, and associated with an increase in the volume of the proamniotic canal that had formed during TS 8, both the embryonic and extra-embryonic regions become more clearly defined (Bonneville, 1950). While the ectodermal component of both tends to have a columnar morphology, those ectodermal cells at the embryonic pole tend to be even more elongated. In the most advanced embryos at this stage, it is possible to recognize early evidence of **gastrulation**, with the delamination of **intra-embryonic mesoderm** (Reinius, 1965; Poelmann, 1981a). During the process of gastrulation, these cells migrate from the embryonic ectoderm towards and through the primitive streak, and as they move away from the midline embryonic axis, they separate the overlying ectoderm from the subjacent endoderm. Once evidence of intra-embryonic mesoderm is seen, these embryos have reached the **early streak stage**, and it has been suggested that, as they represent a distinct substage of TS 9, they should now be designated as being at TS 9.5 (Downs and Davies, 1993). Sequential stages in the conversion of the egg cylinder into the primitive streak stage embryo are illustrated diagrammatically in Figure 2.2.1, while current views on the cell lineage relationships present in the early postimplantation mouse embryo are summarized in Figure 2.2.2.

Figure 2.2.1. Representative sagittal sections through a series of embryos showing stages in the conversion of the egg cylinder into the primitive streak embryo: (a) egg cylinder stage, (b) stage in which proamniotic canal present, (c) stage showing final evidence of closure of the proamniotic canal, (d) stage observed after the closure of the proamniotic canal. Note in particular that, at the egg cylinder stage, the embryo is principally bilayered, with an ectodermal core surrounded in all regions (apart from the site of formation of the ectoplacental cone) with a layer of endoderm. Only a small amount of mesoderm is present in the caudal region of the embryo at this stage. The visceral endoderm is seen to be in continuity with the parietal endoderm close to the site of the ectoplacental cone (a). Initially, the posterior amniotic fold, and slightly later the anterior amniotic fold forms; both of these folds are covered on their outer (medial) surface by extra-embryonic ectoderm, and on their inner surface by extra-embryonic mesoderm. In advanced embryos at this stage, the allantoic rudiment is seen within the posterior amniotic fold in the midline just caudal to the primitive streak; the proamniotic canal is consequently narrowed in the region between the anterior and posterior amniotic folds (b). With the fusion of the latter, the single cavity within the embryo becomes subdivided into three cavities: distally, the amniotic cavity; an intermediate cavity, the extra-embryonic coelomic (or exocoelomic) cavity; more proximally, the ectoplacental cavity. The intermediate membranes that subdivide the embryo at this stage are the amnion and chorion, respectively, both of which are bilayered. With progressive stages of embryonic development, the amount of intra- and extra-embryonic mesoderm present increases, and separates the overlying ectoderm from the subjacent endoderm. In all of these diagrams the future headfold region of the embryo is shown to the left, the primitive streak region is located to the right, and the cavity surrounding the embryo (with the exception of its most proximal part) is the yolk sac cavity (from Kaufman, 1990, with permission).

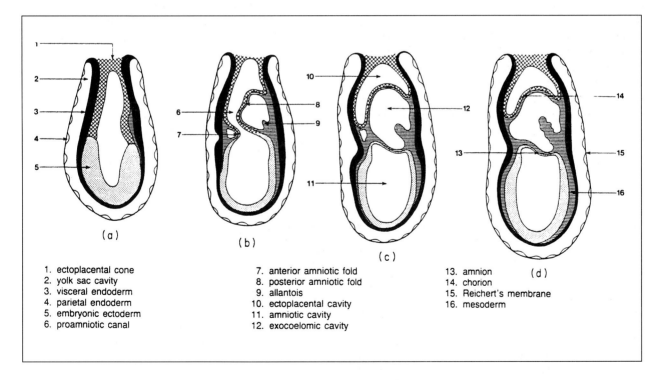

1. ectoplacental cone
2. yolk sac cavity
3. visceral endoderm
4. parietal endoderm
5. embryonic ectoderm
6. proamniotic canal
7. anterior amniotic fold
8. posterior amniotic fold
9. allantois
10. ectoplacental cavity
11. amniotic cavity
12. exocoelomic cavity
13. amnion
14. chorion
15. Reichert's membrane
16. mesoderm

TS	1	2	3	4	5	6	7	8	9	10	11	12	13	14	15	16	17	18	19	20	21	22	23	24	25	26
E	0	1	2	3	4	4.5	5	6	6.5	7	7.5	8	8.5	9	9.5	10	10.5	11	11.5	12	12.5	13.5	14.5	15.5	16.5	17.5

Figure 2.2.2. Cell lineages in the early mouse embryo (from Bard and Kaufman, 1994).

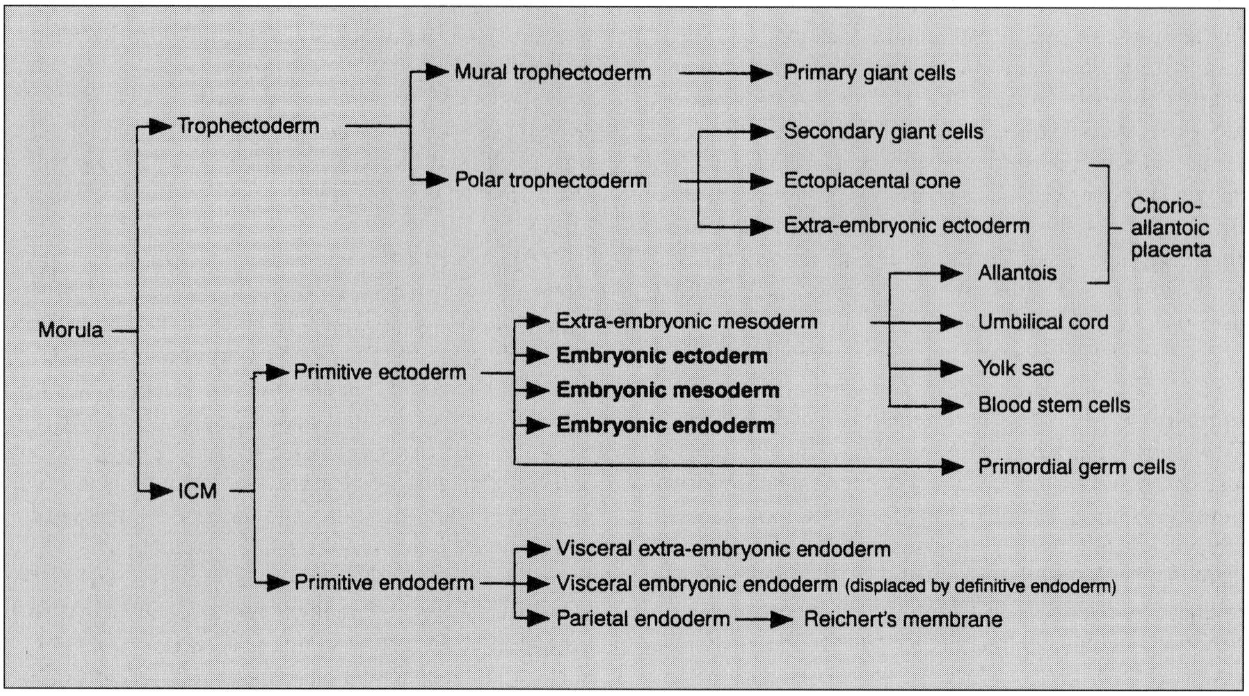

```
                          ┌─► Mural trophectoderm ───► Primary giant cells
         ┌─► Trophectoderm ┤                         ┌─► Secondary giant cells
         │                 └─► Polar trophectoderm ──┼─► Ectoplacental cone     ┐
         │                                           └─► Extra-embryonic ectoderm ┤ Chorio-
         │                                                                        ├ allantoic
         │                                                   ┌─► Allantois      ┘ placenta
         │                         ┌─► Extra-embryonic mesoderm ─► Umbilical cord
Morula ──┤        ┌─► Primitive ectoderm ─┼─► Embryonic ectoderm ─► Yolk sac
         │        │                       ├─► Embryonic mesoderm ─► Blood stem cells
         │        │                       └─► Embryonic endoderm
         └─► ICM ─┤                                                ─► Primordial germ cells
                  │                       ┌─► Visceral extra-embryonic endoderm
                  └─► Primitive endoderm ─┼─► Visceral embryonic endoderm (displaced by definitive endoderm)
                                          └─► Parietal endoderm ──► Reichert's membrane
```

The primitive streak stage (Theiler stage 10, 6.5–7 days)

It is during this stage that most of gastrulation takes place, although it is difficult to follow this process in the mouse in anywhere near the detail that is possible in non-mammalian (*e.g.* avian or amphibian) vertebrates. It seems that the primitive groove, which was first apparent at about TS 7 in the posterior/caudal region of the epiblast, extends anteriorly/rostrally towards the presumptive head region so providing the first morphological evidence of an embryonic axis. There, at the rostral tip of the primitive streak, a "node" forms and, while this is not as well differentiated and obvious in the mouse as is Henson's node in the chick, it probably plays a similar role in controlling the subsequent morphogenesis of the embryo.

At about TS 10, the cell movements that characterize gastrulation start and involve three co-ordinated activities. A proportion of cells within the epiblast layer move medially towards the groove and through the node which itself migrates caudally. Once through the node, these cells detach from the overlying ectoderm and, in ways that remain unclear, the most medially located of the early migrating cells become **definitive endoderm** by displacing primitive endoderm, while the later migrating cells are believed to form only embryonic mesoderm (Poelmann, 1981b; Tam *et al.*, 1993). Two small but nevertheless defined regions close to either end of the embryonic axis do not have an intervening layer of mesoderm cells to separate the surface ectoderm from the subjacent endoderm, and they will form the **buccopharyngeal** (TS 12) and **cloacal membranes** (TS 17) that will be located at the **stomatodaeum** (TS 12) and **proctodaeum** (TS 19) (*i.e.* at the rostral and caudal ends of the primitive gut tube, respectively). Some of the cells of the node migrate rostrally to form a **notochordal process**, but later separate from it to form the definitive **notochord**.

The process of gastrulation continues for some hours as the node moves caudally and eventually reaches the caudal end of the embryo. Neither the forces nor the directing cues that underpin gastrulation are yet understood, but careful marking of cells within the early epiblast has allowed the movements to be followed and a fate map of the gastrulating cells to be drawn (Lawson *et al*, 1991; Lawson and Pedersen, 1992).

Early neurulation (Theiler stage 11, 7.5 days)

Staging

This stage is characterized by the increasing definition of the neural plate and by the development of the extra-embryonic tissues and the amniotic, exocoelomic and ectoplacental cavities that then develop. Its analysis is, however, made difficult by the fact that the timing of the various events that occur during TS 11 is quite variable. Furthermore, it has recently been appreciated that there is a considerable degree of strain variability as to when, for example, the **allantoic bud** is formed as compared to the time when the amniotic, exocoelomic and ectoplacental cavities appear. (For details of the extra-embryonic development at this stage, see Chapter 2.3 on *Postimplantation extra-embryonic development.*)

To help establish some uniformity in the nomenclature of this stage, it has recently been suggested (Downs and Davies, 1993) that the original TS 11, as defined by Theiler (1972), should be divided into 4 substages (Table 2.2.1). In the first substage of TS 11, no evidence of allantoic bud formation is seen, and embryos are consequently said to be at the **pre-allantoic bud stage**; at TS 11.25, the **early allantoic bud stage**, there is minimal evidence of an allantoic bud, while at TS 11.5, a more advanced degree of allantoic bud formation is present, and these embryos are defined as being at the **late allantoic bud stage**. At TS 11.75 when the neural folds are elevating, embryos are now said to have reached the **early head-fold stage**.

As with all compromises, this revised nomenclature for TS 11 is only partly satisfactory. The use of substages for TS 11 is largely based on the findings of Downs and Davies (1993) who used PO strain mice and is entirely appropriate when considering their development. However, in the original Theiler scheme (Theiler, 1972), and as modified by Kaufman (1992), (C57BL×CBA) F1 hybrid mice were studied. In these mice, an early allantoic rudiment (or allantoic bud) is first evident shortly before the posterior amniotic fold meets and fuses with the anterior amniotic fold. Accordingly, in this strain of mice, and many others, one would not expect to see embryos at the first substage of TS 11 (amended), as the allantoic bud is already present in the majority of embryos at TS 10.5, and certainly in all mice by early TS 11 (see Table 2.2.1).

Embryonic development

Throughout all of the substages of TS 11, the embryonic ectoderm continues to differentiate. At the early part of this stage, the **neural plate** forms in the rostral part of the embryonic axis, becoming better defined in the mid-stage. A little later, and towards the end of the stage (at TS 11.75), the neural folds in the rostral part of the embryonic axis can be seen to have elevated slightly and are deemed to have formed the early **head-folds**. In the midline of the embryonic axis, the neural groove is located between the neural folds, though caudally it merges imperceptively into the primitive groove. Equally, while the neural plate is initially clearly defined rostrally and laterally along the embryonic axis, more posteriorly it gradually merges into the **primitive streak region**. Subjacent to the neural groove in the midline of the embryonic axis, it is possible to recognize the **notochordal plate**, though more caudally it becomes increasingly poorly defined as it merges into the transitional ectoderm/mesoderm tissue subjacent to the primitive streak.

At about this time, the primitive endoderm is gradually displaced laterally by the definitive endoderm, and in the midline by the **notochordal plate**, both of which are derived from the epiblast cells which migrate from the nodal region at the rostral extremity of the primitive streak. It has also been suggested (Tam and Beddington, 1992) that some of the overlying embryonic ectoderm cells, give rise directly to a small proportion of the definitive endoderm cells and to the cells of the notochord without their first having to migrate through the primitive streak.

A slight indentation gradually develops in the endoderm at the ventral midline towards the end of this stage, though it does not become well defined until the early part of TS 12. This represents an early stage in the differentiation of the **foregut pocket**, and all of the endoderm cells that line it are believed to be derived from the anterior end of the primitive streak (Tam and Beddington, 1992). This stage also coincides with the first appearance of the **intra-embryonic coelomic cavity** (or coelom). This forms from isolated spaces that appear within the intra-embryonic mesoderm and that subsequently amalgamate to give a rostrally-directed inverted-U-shaped channel which passes across the midline just rostral to the buccopharyngeal membrane (see Figure 2.4.5). By this means, continuity is achieved between the extra- and intra-embryonic coelomic cavities. The latter is lined with squamous mesothelial cells, some of which soon develop a columnar rather than a squamous morphology, and it is these cells within the prospective pericardial region of the coelom, across the midline, which will shortly differentiate (during the early part of TS 12) into the future **cardiogenic** or **myocardial plate**.

By the early part of TS 12, the head-folds have substantially increased in volume, while at the same time the formation of the first somites provides early evidence of differentiation in the paraxial component of the intra-embryonic mesoderm.

Early human embryonic development

In the human, implantation occurs during days E6–7. At about this time, it is possible to recognize that the blastocoelic surface of the inner cell mass (constituting the **epiblast**) is already covered by a single layer of endoderm, the **hypoblast**. By E8, the **amniotic cavity** is first seen within the epiblast, and is completely surrounded by epiblast cells a day later. At this stage, some of the more peripheral of the hypoblast cells have migrated away from the embryonic pole and now line all but the most abembryonic part of the **mural trophoblast**. These migrating hypoblast cells constitute the first elements of the **extra-embryonic endoderm**, and will shortly line

TS	1	2	3	4	5	6	7	8	9	10	11	12	13	14	15	16	17	18	19	20	21	22	23	24	25	26
E	0	1	2	3	4	4.5	5	6	6.5	7	7.5	8	8.5	9	9.5	10	10.5	11	11.5	12	12.5	13.5	14.5	15.5	16.5	17.5

the whole of the blastocoelic surface of the **blastocoelic cavity** which is then called the **primary yolk sac cavity** (Figure 2.2.3).

There has been much confusion over the years about how the extra-embryonic endoderm and mesoderm form. This is because so few well-fixed human embryos at this stage of development are available for histological analysis, and their features have been variously interpreted by different observers, although there now seems to be general acceptance of the fact that **Heuser's membrane** (as described by previous authors) is synonymous with the extra-embryonic endoderm (see, for example, Larsen, 1993). For a detailed overview of the histological events associated with blastogenesis, implantation and the establishment of the various extra-embryonic tissues and cavities in the human, see Vögler (1987).

On or about E10–11, the first evidence of extra-embryonic mesoderm is seen, and its cells can only be derived from the proliferation and differentiation of cells at the peripheral margin of the epiblast. These cells appear mesenchymatous, and spread out from the embryonic pole of the blastocyst, separating the extra-embryonic endoderm from the mural trophoblast as they migrate towards the extra-embryonic pole of the conceptus. It is, however, currently unclear whether the extra-embryonic mesoderm initially spreads out as a monolayer, or as two distinct layers, with one lining the mural trophoblast and the other covering the blastocoelic surface of the extra-embryonic endoderm. The limited histological evidence so far available suggests that the second possibility is more likely to be correct.

At E11–12, when the extra-embryonic mesoderm has completely lined the mural trophoblast, and what appears to be a second distinct layer has covered the extra-embryonic endoderm, the space between these two layers constitutes the **extra-embryonic coelomic** (or **exocoelomic**, or **chorionic**) **cavity**. The blastocoelic cavity is now called the **secondary yolk sac cavity**, while the wall of the **yolk sac** is composed of an inner endodermal surrounded by an outer mesodermal layer, both of which are extra-embryonic in origin[1] (see Figure 2.2.4).

While the exact sequence of events that occurs at

Figure 2.2.3. On or about the 8th day of human pregnancy, the polar trophectoderm cells of the zona-free blastocyst have eroded their way through the maternal endometrium into the subjacent tissues. The embryo is at the *bilaminar disc stage*, and the *amniotic cavity* has only recently appeared as a space within the *epiblast* (the collection of the totipotential cells derived from the inner cell mass, ICM). The layer of ICM-derived cells closest to the blastocoelic cavity (at this stage, the layer that is subjacent to the embryonic endoderm) now forms the embryonic ectoderm, and will give rise to the embryo proper. In contrast, the ICM-derived cells that line the polar trophectoderm will give rise to the ectodermal component of the amnion. A layer of embryonic endoderm has delaminated from the ICM cells that were exposed to the *blastocoelic cavity*, and these cells are seen to have initiated their migration away from their site of origin. When the parietal endoderm cells completely line the blastocoelic cavity, the latter is then termed the *primary yolk sac cavity*. The polar trophectoderm cells continue to divide, those closest to the ICM retain their individuality (and are therefore termed *cytotrophoblast*), while those that are more distant from the ICM lose their individuality (termed *syncytiotrophoblast*) and invade the maternal tissues in this location. The response of the maternal cells to the presence of the implanting blastocyst is termed the *decidual reaction*. The initial invasion of the host tissues by the invading blastocyst is indicated by the arrows at the top of the diagram.

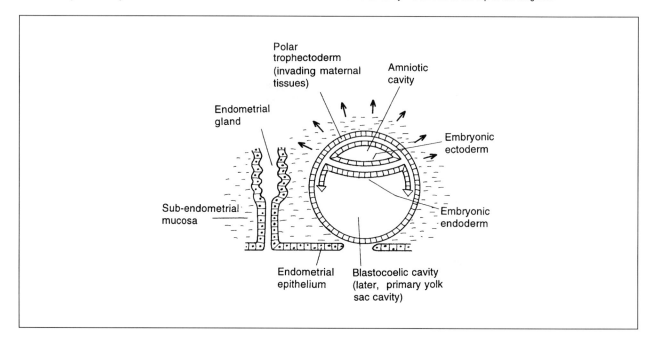

---

[1] In the human, the yolk sac mainly regresses during the first trimester (unlike the mouse where it is retained throughout gestation), although small remnants of the yolk sac having a maximum diameter of about 5 mm are commonly seen in human embryos of up to about 5 months, and may even be recognized in the full-term placenta.

During the period between 4–8 weeks of gestation, the amniotic cavity expands at the expense of the extra-embryonic coelomic cavity, so that at the end of this time, minimal evidence of the exocoelomic cavity is seen. The overall changes that take place with regard to the amniotic and extra-embryonic coelomic cavities and the yolk sac are shown in Figure 2.2.5a,b.

Figure 2.2.4. Diagrammatic representation of a section through the implantation site in the human on or about E 12/13. Note in particular that the embryo is still at the *bilaminar disc stage*, with a layer of embryonic ectoderm subjacent to which is a layer of embryonic endoderm (the four small arrows indicate the extent of the embryonic disc). The volume of the amniotic cavity has increased compared to that illustrated in Figure 2.2.3, and the yolk sac cavity is now seen to be completely lined by endoderm cells. Initially, a layer of *extra-embryonic mesoderm* intervenes between the outer shell of trophectoderm and the amniotic ectoderm and yolk sac endoderm, but slightly later (as illustrated here) this splits into two discrete layers: (i) that which lines the shell of trophectoderm, and (ii) that which covers the outer surface of the amnion and the yolk sac. Their only site of contact is at the *connecting stalk*, the future *umbilical cord*. The space that develops between the two layers of extra-embryonic mesoderm is termed the *extra-embryonic coelom* (or coelomic cavity). At the same time, once the latter has formed, the *primary* yolk sac cavity now becomes the *secondary* yolk sac cavity. The modified morphology of the polar trophectoderm cells is also illustrated in this diagram: those cells that were originally closest to the ICM are seen to retain their individuality (the cytotrophoblast), while those that are invading the maternal tissues are seen to have lost their individuality, representing the syncytiotrophoblast layer. Where the latter invades maternal *sinusoids* (a term given to the dilated maternal blood vessels in this region), this forms the basis for the uteroplacental circulation. The *primary stem villi* consist of a core of cytotrophoblast surrounded by a shell of syncytiotrophoblast, the latter being bathed in maternal blood. The cytotrophoblast (of polar trophectodermal origin) with its subjacent layer of extra-embryonic mesoderm is termed the *chorionic plate*. The decidual cells around the implanting conceptus become increasingly oedematous, while the capillaries in this area become "leaky", thus forming the basis for the "*Pontamine Blue Reaction*": when this dye is injected into the maternal vascular system in the mouse, it "leaks" into the maternal tissues around the implantation site. This is one of the earliest manifestations of pregnancy in this species.

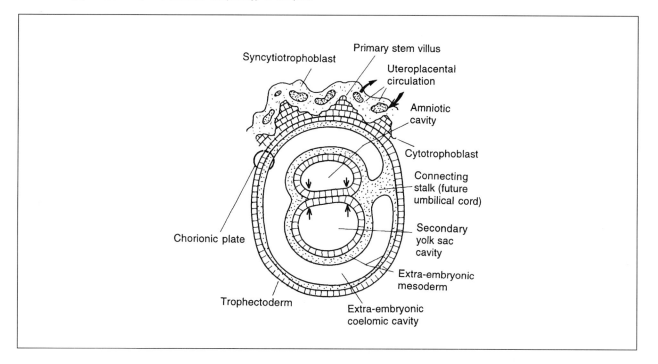

this stage is unclear, it has recently been proposed (Lawson and Pedersen, 1992) that there is a second wave of proliferative activity within the hypoblast on or about E12, which eventually results in the displacement of the first wave of extra-embryonic endoderm cells (the original lining of the primary yolk sac) by the cells which later form the lining of the **secondary** (or **definitive**) **yolk sac**. If this suggestion is correct, then the more recently formed of the extra-embryonic endodermal derivatives of the hypoblast then give rise exclusively to the endodermal component of the wall of the definitive yolk sac, while the cells that constitute the first wave of proliferative activity of the hypoblast degenerate and subsequently take no further part in development.

At about E14–15, the volume of the extra-embryonic coelomic cavity expands dramatically, while the conceptus, which now consists of a **bilaminar embryonic disc** (with its amniotic and secondary yolk sac cavities) is suspended from the **chorionic plate** by the mesenchymal **connecting stalk** which has now formed, there being no exocoelomic cavity here. It is also on E14 that the **primitive streak**, the first evidence of an embryonic axis, is seen, with the **primitive** (or **Hensen's**) **node** being located at its caudal end.

Gastrulation occurs on or about E16 in the human. During this process, epiblast cells migrate towards and enter the primitive streak and separate the overlying epiblast (which now constitutes the **ectoderm**) from the subjacent hypoblast (which now constitutes the **endoderm**), the intermediate layer thus formed constitutes the third of the definitive germ layers, the **intra-embryonic mesoderm**. Recent mouse

TS	1	2	3	4	5	6	7	8	9	10	11	12	13	14	15	16	17	18	19	20	21	22	23	24	25	26
E	0	1	2	3	4	4.5	5	5.5	6	6.5	7	7.5	8	8.5	9	9.5	10	10.5	11	11.5	12	12.5	13.5	14.5	15.5	16.5	17.5

evidence (Lawson *et al.*, 1991; Lawson and Pedersen, 1992; Tam *et al.*, 1993) suggests that this picture may be oversimplified, in that the first epiblast cells entering the primitive streak may in fact displace the original layer of hypoblast that is subjacent to the primitive streak, and form the definitive embryonic endoderm. The epiblast cells that later migrate into the primitive streak are those that will give rise to the definitive mesoderm.

By E16–17, the embryo is now at the **trilaminar disc stage**, possessing its three definitive germ layers (ectoderm, endoderm and mesoderm). Even at this early stage, however, it is possible to see that the mesoderm shows early evidence of differentiation into three columns of tissue on either side of the midline, the **paraxial**, **intermediate** and **lateral plates**. At about the same time, a tubular mass of tissue migrates rostrally from the primitive node to form the **notochordal process**.

Shortly after the establishment of the three primary germ layers, a small cleft appears on either side of the lateral plate mesoderm, at the peripheral margin of the embryonic disc, approximately mid-way along the embryonic axis. It is at these sites that continuity becomes established between the **intra-embryonic coelomic cavity** (which forms following the splitting of the intra-embryonic mesoderm), and the extra-embryonic coelomic cavity which is already present, but will eventually be lost (see above and Figure 2.2.5a,b). The mesenchyme cells which line both cavities differentiate to form mesothelium. Within the embryo, this subsequently differentiates to form the lining of the pleural, pericardial and peritoneal cavities, as well as the coverings of the organs that differentiate within them (the visceral pleura, pericardium and peritoneum, respectively).

Figure 2.2.5a, b. Two stages in the resorption of the *human* extra-embryonic coelomic cavity, associated with the proportionate increase in the volume of the amniotic cavity. In the first stage illustrated (a), at about 3–4 weeks of pregnancy, a representative section through the trunk in the region of the midgut shows that in this region, at this stage, the midgut is in direct continuity with the yolk sac. With the gradual expansion in the volume of the amniotic cavity, the volume of the extra-embryonic coelom correspondingly diminishes. The layer of extra-embryonic mesoderm that is continuous with the mesodermal (i.e. outer or abembryonic) component of the amnion is termed the *somatopleure*, while that which forms the outer "wall" of the midgut (and is later associated with the differentiation of its various muscle layers) and yolk sac is termed the *splanchnopleure*. By about 8 weeks of pregnancy, a representative transverse section through a similar region of the trunk to that illustrated in (a), clearly shows that the extra-embryonic coelom has all but diminished, and only small discrete mesothelial-lined cavities containing extra-embryonic coelomic fluid are now present within the extra-embryonic mesoderm (b). Equally, the *connecting stalk* has both lengthened and proportionately diminished in diameter, and may now be termed the *umbilical cord*. The volume enclosed by the yolk sac is now extremely small, and it is connected to the distal part of the midgut loop by the *vitelline* (or vitello-intestinal) *duct*. The yolk sac subsequently regresses further, and usually completely disappears, as does the vitelline duct. The arrangement shown in both (a) and (b) should indicate that the umbilical cord inserts into the chorionic plate in the region of the developing placenta, rather than as (for simplicity) is displayed here.

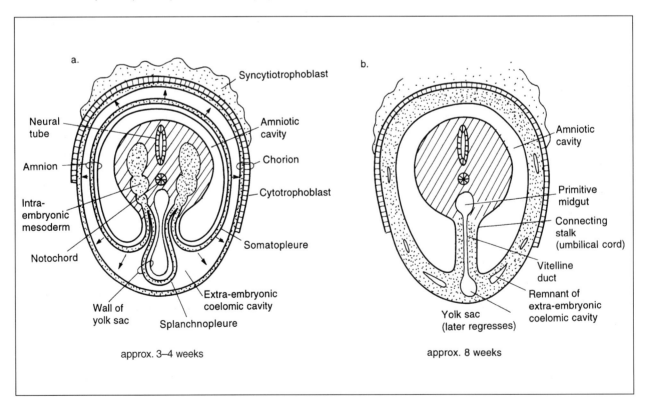

2.3 Postimplantation extra-embryonic development

Introduction

This chapter starts by reviewing the events that take place as the extra-embryonic membranes form, develop and differentiate, and considers the various cavities, such as that of the yolk sac and the extra-embryonic coelom, that appear and are lost as development proceeds. It then goes on to consider umbilical cord and placenta formation in some detail. From the point of view of embryogenesis, these tissues might seem of secondary importance as their prime role is to provide the physical, physiological and nutritional environment within which the embryo can develop normally. Such a view is, however, inadequate, particularly for the case of the yolk sac, as its constituents are the source of most of the red blood cells and all of the primordial germ cells. There is thus a considerable emphasis throughout this chapter on the role of these extra-embryonic tissues in mediating normal development.

Formation of amniotic, exocoelomic and ectoplacental cavities, and the fate of the amnion and chorion

The early postimplantation embryo (the egg cylinder) has an external trophoblast layer[1] whose polar cells at the ectoplacental cone (abembryonic pole) invade the maternal uterine lining and anastomose with its blood vessels (see *Chorioallantoic placenta* below), and an inner cell mass which is partitioning into an embryonic (distal) and an extra-embryonic (proximal) component (at the advanced egg cylinder stage), a process marked by the appearance of the **proamniotic cavity** which forms within the primitive ectoderm and extends from the embryonic pole towards the ectoplacental cone (TS 8–9). At this stage, both the embryonic and extra-embryonic components of the egg cylinder are bilayered with an inner ectodermal and an outer (*visceral*) endodermal layer. The *parietal* endoderm cells constitute a continuous layer which forms the outer boundary of the **primary yolk-sac cavity** (see previous chapter). The visceral endoderm (which is continuous with the parietal endoderm at the proximal extremity of the yolk-sac cavity) forms a continuous outer layer which covers the extra-embryonic and the embryonic ectodermal cells, with the primitive streak now forming in the latter (see Chapter 2.2 *The postimplantation embryo* for further details). Here, we focus on the development of the extra-embryonic tissues and the events that lead to the formation of the amniotic, yolk-sac and extra-embryonic cavities.

During the early part of TS 10, the **posterior amniotic fold** forms from the growth of a mass of extra-embryonic ectoderm cells located in the dorsal midline just proximal to the boundary between the embryonic and extra-embryonic ectoderm, and just caudal to the primitive streak region. Shortly afterwards, a cavity forms within this fold, that will in due course form a component of the **extra-embryonic coelomic (or exocoelomic) cavity**, and is lined by a layer of extra-embryonic mesoderm cells that had delaminated a stage or so earlier from the ectoderm. The formation of the exocoelomic cavity splits the extra-embryonic mesoderm, so that the distal, proximal and lateral components of its inner layer respectively form the outer part of the walls of the amniotic, ectoplacental and eventually the yolk-sac cavities (see Figure 2.2.1, and below).

Once the posterior amniotic fold is well established, a second cellular outgrowth (the **anterior amniotic**

---

[1] It has been suggested that the immunological privilege of the mouse trophoblast is due in part to an investing layer of fibrinoid that is probably secreted by the trophoblast and is believed to prevent the escape of fetal transplantation antigens (Kirby *et al.*, 1964). Other studies have confirmed that the fine structure of the mouse placenta is in all respects identical to that in the rat (Jollie, 1964; see also Enders, 1965).

TS	1	2	3	4	5	6	7	8	9	10	11	12	13	14	15	16	17	18	19	20	21	22	23	24	25	26
E	0	1	2	3	4	4.5	5	6	6.5	7	7.5	8	8.5	9	9.5	10	10.5	11	11.5	12	12.5	13.5	14.5	15.5	16.5	17.5

fold) develops in the midline on the opposite side of the conceptus, in the extra-embryonic ectoderm directly proximal to the presumptive head-fold region of neural ectoderm. At this stage, the **proamniotic canal** has a narrowed central region, but this will soon be obliterated once the posterior and anterior amniotic folds meet, and then join as their fused membranes break down, so forming the **exocoelomic cavity**. This event occurs towards the end of TS 10, and such embryos are designated as being at the **late streak stage**. (It has been suggested that this represents a distinct substage of TS 10, and embryos at this stage should now be designated as being at TS 10.5, see Downs and Davies (1993) and Table 2.2.1.)

By the end of TS 11, the advanced primitive-streak-stage conceptus thus has three discrete fluid-filled compartments (from proximal to distal), the ectoplacental, the exocoelomic (or extra-embryonic coelomic) and the amniotic cavities[2] that are respectively separated by the chorion and amnion. There is also a fourth compartment, the yolk-sac cavity which surrounds the egg cylinder and is bounded medially by the visceral endoderm and laterally by the parietal endoderm (see Figure 2.2.1).

The **amnion** in the mouse is generally considered to be a bilayered structure with an extra-embryonic ectodermal component that is continuous with the outer (*i.e.* ectodermal) surface of the embryo at its peripheral margin. It initially forms the distal (*i.e.* embryonic) boundaries to the posterior and anterior amniotic folds, only being completed when these fuse together and the intervening walls between them break down (see above). When this occurs the amnion becomes apposed to and subsequently adherent to the subjacent region of extra-embryonic mesoderm lining the exocoelomic cavity. After the embryo has *turned* (see below), it is completely surrounded by the amnion[3], except at the future site of insertion of the umbilical cord into the anterior abdominal wall of the embryo.

The **chorion** is also a bilayered structure, but it is formed entirely from the apposition and fusion of the abembryonic (extra-embryonic) mesodermal component of the posterior and anterior amniotic folds and, after these amalgamate and the intervening walls break down, the chorion is seen to form the continuous sheet of cells which separates the ectoplacental from the exocoelomic cavities. For observations on the development of the chorion in the rat, see Ellington (1987).

As the amniotic and exocoelomic cavities expand, the volume of the ectoplacental cavity gradually diminishes. Eventually its two inner walls come into contact and fuse together to form the ectoplacental (or chorionic) plate, and its lumen is then obliterated (TS 12/13, E8–8.5), although the two layers are still recognizable at its periphery for a short period. Meanwhile, during TS 11.25–11.5, the allantois increases in volume and grows towards, and eventually fuses with the surface cells at the peripheral margin of the chorion (TS 12/13). At this stage, the allantois, which has a very loose structure with cavities lined by endothelial cells, contains primitive nucleated red blood cells, due to the amalgamation of the allantoic vasculature with the yolk sac circulation (see below). At later stages, the volume of the amniotic cavity increases at the expense of the exocoelomic cavity which becomes completely obliterated when the anterior abdominal wall closes and the umbilical cord forms (see below).

During TS 11.25–11.5, the **allantois** increases in volume and its tip eventually fuses with the peripheral margin of the chorion. This allows its vasculature to amalgamate with that of the yolk sac and eventually with that of the embryo (towards the end of TS 12) to form the basis of the chorioallantoic (*i.e.* the primitive placental) circulation (for a detailed review of these events, see Downs, 1998). According to this author, the sequence of events is (i) allantoic bud formation, (ii) allantoic growth, (iii) chorioallantoic fusion, and (iv) overt vascularization of the allantois.

The mouse allantois gives rise to the future umbilical component of the chorioallantoic placenta, and its principal function, once it fuses with the chorion, is to serve as a connection between the embryo and the mother for the interchange of nutrients, gases and metabolic waste products. Although the development of the allantois varies considerably among species (Steven and Morriss, 1975), the mouse allantois is an entirely extra-embryonic mesodermal structure (Snell and Stevens, 1966).

The timing of the first appearance of the allantoic bud varies slightly both within and between mouse strains (Downs and Davies, 1993; Downs and Gardner, 1995; and see previous chapter), but its morphological features are believed to be similar in all

---

[2] The amniotic cavity is filled with amniotic fluid thought to have an initial composition similar to that of blood plasma, and to be transported into the cavity by the cells of the amnion through filtration or secretion. At about E13.5–14 (TS 22), after the mesonephros has regressed, and its excretory role taken over by the metanephros, urine starts to provide the major component of the fluid filling this compartment. Once the renal system and the gut are established and functioning, the amniotic fluid and excretory products are constantly recycled; being swallowed by the embryo and absorbed through the epithelial lining of its gut. Excess fluid and excretory products are then transferred to the maternal circulation via the placenta for subsequent removal through the mother's kidneys.

[3] In early human pregnancy, the amnion consists of epithelium and extra-embryonic somatopleure only. As pregnancy proceeds, there is said to be differentiation into 5 distinct layers (Bourne, 1962). In a more detailed analysis by Saunders and Rhodes (1973) the layers are described in the following terms. The first, the epithelial layer, is composed of a single layer of non-ciliated cuboidal cells. Subjacent to this is a basement membrane. Next, there is a compact fibroblastic layer, the cells of which may be actively phagocytic. Finally, there is a spongy layer of loose connective tissue. Contrary to former suggestions (Keiffer, 1926), the amnion at term is said not to have a blood supply.

(for its appearance in the rat, see Ellington, 1985). In most strains of mice, contact between the tip of the allantois and the chorion occurs in embryos with about 6–7 pairs of somites (TS 12), followed, shortly afterwards, by fusion of these two structures (Ellington, 1987; Downs, 1998). Indirect evidence from the staining of isolated fragments of allantois maintained in tissue culture for the presence of benzidine, which is indicative of the presence of haemoglobin, suggests that under certain circumstances this tissue may be capable of erythropoiesis (Downs and Harmann, 1997).

The **yolk sac**, the other important extra-embryonic membrane, also appears at the advanced egg cylinder/early primitive streak stage (TS 10), once the conceptus has become subdivided by the amnion and chorion. The yolk sac is located at the peripheral margin of the embryonic plate and forms the lateral, but not the proximal or distal, boundary of the exocoelomic cavity (Figure 2.3.1). It is also exclusively extra-embryonic in origin and is bilayered, deriving from the apposition of the *extra-embryonic endoderm* that forms the outer boundary of the middle part of the conceptus with the *extra-embryonic mesodermal* layer that lines the lateral margin of the exocoelomic cavity. As noted previously, the yolk sac cavity is located between the visceral and parietal layers of endoderm cells.

Reichert's membrane, which usually ruptures shortly before birth, is first recognized at the primitive streak stage (E6.5–7, TS 9), and is produced as an extracellular (acellular) layer by the extra-embryonic mesothelial (mesodermally-derived) component of the amnion. Similarly, the mesothelial cells that cover the gut in this region (i.e. the visceral peritoneal cells) are in continuity with the mesothelial (mesodermally-derived) cells of the yolk sac, and the surface ectoderm cells are in continuity with the ectodermal component of the amnion. In (b), the arrows indicate the location of the "umbilical ring".

Figure 2.3.1a, b. Two diagrams that show the site of communication between the intra- and extra-embryonic coelomic cavities (arrows, a) close to where the midgut communicates with the yolk sac, and the situation immediately after their closure (b) with the formation of two discrete cavities. In (a), it should be noted that the sheet of mesothelial cells that lines the intra-embryonic coelomic cavity in the prospective peritoneal region (i.e. the parietal peritoneal cells) is in continuity with the

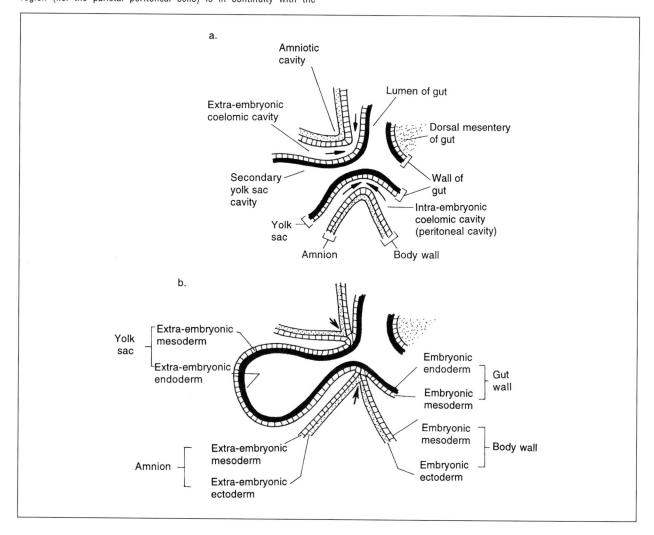

TS	1	2	3	4	5	6	7	8	9	10	11	12	13	14	15	16	17	18	19	20	21	22	23	24	25	26
E	0	1	2	3	4	4.5	5	6	6.5	7	7.5	8	8.5	9	9.5	10	10.5	11	11.5	12	12.5	13.5	14.5	15.5	16.5	17.5

(parietal) endoderm cells, and gradually thickens throughout gestation, although its thickness varies quite considerably over the membrane. While it separates the parietal endoderm cells (on its inner aspect) from the layer of primary giant cells (on its outer aspect) and forms a continuous peripheral boundary in all areas apart from the site occupied by the placenta, its exact role has yet to be determined. No equivalent structure forms during human embryonic development.

"Turning" and the reorganization of the membranes

From the primitive streak stage, when the first evidence of an embryonic axis is seen, until about the end of TS 12 when the embryo possesses about 6–8 pairs of somites, the configuration of the mouse embryo is quite different from that of the human embryo at comparable stages of development. The dorsal ectodermal surface of the mouse embryonic disc is deeply concave in all directions, and the subjacent endodermal surface correspondingly convex, with the embryonic axis conforming to this configurational arrangement and being U-shaped. The surface ectoderm and neural tube thus face dorsally into the amniotic cavity, while the exposed future midgut region faces ventrally and abuts the parietal endoderm, with only the yolk sac cavity intervening between them.

This situation is reversed when, by a mechanism that has yet to be fully understood, the embryo "turns", a process that is initiated during the advanced part of TS 12 and completed during TS 13/14 (E8–9.5, for details, see Chapter 2.4 *Early organogenesis*). Here, the embryo effectively "rolls" or rotates through 180° about the midpoint of its body axis (see Figure 2.4.2) and, as a result of this "turning", wraps itself in its amnion and yolk sac and reverses the orientations of its dorsal and ventral surfaces. The exposed surface of the midgut that was formerly on the convex surface of the embryonic axis thus becomes located within the concavity of the embryo, while the dorsal surface relocates to its convex surface. As a result of these changes, the mouse embryo adopts the normal so-called *fetal position* seen in the human at comparable stages of development.

Once the process of turning has been completed, the embryo becomes surrounded by its extra-embryonic membranes, with the amnion being immediately next to the embryo and separated from it by the amniotic fluid. The yolk sac is now located between the outer region of the amnion and the parietal layer of extra-embryonic endoderm that lines the mural trophectoderm, the two being separated by the exocoelomic and yolk-sac cavities, respectively (see Figure 2.4.1).

As well as becoming surrounded by its extra-embryonic membranes, the size of the peripheral boundary of the body wall where the embryonic endoderm is initially exposed (at the umbilical "ring"), diminishes, so that, once the embryo has turned and adopted the normal fetal position, the opening in the ventral part of the body wall (in the future umbilical region) has almost completely closed. By TS 16/17, the **umbilical cord** has become well defined (see below) and the link between the intra- and extra-embryonic coelomic cavities is lost. During TS 15–17, the outer layer of the amnion becomes increasingly adherent to (but never actually fuses with) the inner layer of the yolk sac and, by the end of this period, the extra-embryonic coelomic cavity is much diminished, while the intra-embryonic coelomic cavity enlarges and becomes compartmentalized to form the various body cavities (see *Organogenesis*).

The yolk sac

The yolk sac is retained in the mouse where it surrounds the amnion and continues to play a critical, though changing, role throughout gestation.[4] The persistence of the yolk sac in the mouse embryo means that the exocoelomic cavity is retained, although, because it is located between the amnion and the yolk sac whose membranes are closely adherent, its volume is small and diminishes to such an extent that it becomes a *virtual* rather than a *real* space. In the human, the link between the intra- and extra-embryonic coelomic cavities becomes closed off at the umbilicus; the exocoelomic cavity gradually becomes resorbed, disappearing completely by the end of the first trimester as the two layers of the extra-embryonic mesoderm come together. The human intra-embryonic coelom develops much as that in the mouse. (For further details, see *Early organogenesis*.)

At the primitive-streak/early-headfold stage (TS 10,11), the yolk-sac vasculature, which develops exclusively within its inner mesodermal component, first consists of scattered and isolated blood islands, but these rapidly amalgamate together to form the yolk-sac circulation. This eventually (TS 13, 8–12 pairs of somites, see below) fuses with the embryonic vasculature (which develops within the embryo as an independent system), so that the yolk sac-derived primitive nucleated red blood cells, but, and possibly more importantly, their haemopoietic/haematopoietic stem cells (see below) can then gain access to the embryonic circulation.

---

[4] It is thus very different from the human where the yolk sac disappears, leaving only a remnant that is sometimes retained as a small sac close to the site of insertion of the umbilical cord just subjacent to the fetal surface of the placenta (Figure 2.2.5b), and is connected to the atrophied vitelline (or vitello-intestinal) duct which is located at the midpoint of the primitive embryonic midgut, associated with its antimesenteric border.

While it was formerly believed by some authorities, and is still stated in some textbooks, that the primitive nucleated red blood cells come exclusively from the delamination of the endothelial cells lining the yolk sac blood islands, it is now clear from the histological analysis of the human yolk sac and the ultrastructural analysis of the yolk sac from advanced primitive streak and early somite-stage mouse embryos that the red blood cell precursors differentiate within the mesodermal component of the yolk sac (Kaufman, 1991). These cells then either, and most commonly, become surrounded by endothelial cells (see Streeter, 1920; Ingalls, 1921; Bloom and Bartelmez, 1940; Gladstone and Hamilton, 1941), or, less commonly, migrate towards the presumptive blood islands and then enter them through large discontinuities that form between cells in the endothelial wall, and that presumably close off directly afterwards (Kaufman, 1991). A small proportion of these nucleated red cells that join the embryonic circulation can divide, and probably represent the haemopoietic (or erythropoietic) stem cells that will later populate the septum transversum, although most of the cells of the septum transversum differentiate to form the parenchyma of the liver. Once the liver has formed, it immediately takes over the haemopoietic function of the yolk sac, and its stem cells are believed to be the progenitors of both the red and white cell lineages. While the liver retains this role throughout gestation, it is gradually supplemented by the haemopoietic activity of the spleen and slightly later by that in the bone marrow.

The yolk sac has another key role at the primitive-streak stage: it is the first place where the **primordial germ cells** (pgc) are seen. These can be readily recognized by their characteristic morphology (they are large rounded cells with a considerably greater diameter than any other cells in the yolk sac) and by their high level of intracellular alkaline phosphatase enzyme activity that can be demonstrated by appropriate histochemical staining of serially sectioned embryonic/extra-embryonic material. Their exact cellular origin is still a matter of debate, but it must be from within a small pool of epiblast cells, probably at the egg cylinder stage (TS 7–9), but there are currently no histochemical or molecular markers to identify them at these earlier stages.

During the advanced primitive streak stage, the pgc that have migrated from their site of origin in the wall of the yolk sac reach the base of the allantois. This region is then gradually incorporated into the caudal end of the primitive streak. The pgc then migrate via the wall of the hindgut and its dorsal mesentery to the medial (future gonadal) part of the urogenital ridge (Chiquoine, 1954; Ozdzeński, 1967). Once there, they differentiate and play a key role in gonad differentiation. (For further details, see *Primordial germ cells* in Chapter 4.3, *The urogenital system*.)

The endodermal layer of the yolk sac consists principally of columnar cells with a microvillous border which faces into the yolk sac cavity, and their main role at the primitive-streak stage and up to the time of placenta formation is to absorb nutrients from the yolk sac cavity and transfer them into the embryonic compartment. It is unclear whether they also serve an excretory role by removing metabolic waste products from within the embryonic compartment and transferring them into the yolk sac cavity.

The umbilical cord and placenta

The umbilical cord

In advanced unturned embryos where the endodermal lining of the future midgut is fully exposed to the yolk sac cavity, the ventral part of the body wall is almost completely absent in the region of the future midgut. With the process of turning, not only does the embryo finally adopt the characteristic "fetal" position (see previously), but it also becomes surrounded by the amnion and yolk sac, and the exposed midgut region (formerly on the outer part of the convexity of the U-shaped embryonic axis) becomes relocated within the concavity of the embryonic axis. At the same time, the peripheral boundary of the so-called **umbilical ring** diminishes, so that, even by TS 15, the region where the intra- and extra-embryonic coelomic (exocoelomic) cavities still communicate (around the **vitelline duct**) is far smaller than before; and, as the umbilical cord elongates, this continuity between the future peritoneal cavity and the exocoelom gradually disappears.

As this happens, the anterior abdominal wall in the periumbilical region forms, although the exact mechanism by which this is achieved is unclear. Both the amnion and yolk sac take origin from (or, possibly more correctly, insert into the peripheral boundary of) the umbilical ring. The endoderm of the yolk sac is continuous (at this junctional site) with the endoderm that forms the lining of the gut, while the ectoderm of the amnion at this site is continuous with the surface ectoderm of the embryo. These two membranes are both lined by a mesothelial layer of extra-embryonic mesoderm, and are separated by what remains of the exocoelomic cavity (Figure 2.3.1a,b).

Once the embryo has turned, the primitive **umbilical cord** is established, and contains the common umbilical artery (formed by the fusion at the base of the umbilical cord of the right and left umbilical arteries) and the common umbilical vein. These vessels are initially in continuity with and probably derived from

TS	1	2	3	4	5	6	7	8	9	10	11	12	13	14	15	16	17	18	19	20	21	22	23	24	25	26
E	0	1	2	3	4	4.5	5	6	6.5	7	7.5	8	8.5	9	9.5	10	10.5	11	11.5	12	12.5	13.5	14.5	15.5	16.5	17.5

the allantoic vessels, since the allantois becomes one of the principal components of the umbilical cord. The allantois with its associated vessels then fuses (TS 15/16) with the chorionic plate to form the embryonic/fetal portion of the **chorioallantoic placenta**.

On the decidual side of the conceptus, at the periphery of the ectoplacental cone, numerous clefts appear which become filled with maternal blood, and it is here that the embryonic and maternal circulations subsequently come into their closest contact. At this boundary zone, a single or, occasionally, several layers of scattered trophoblast giant cells are found, and these form an intermittent layer with the trophoblast giant cells that surround the conceptus; the former group are called the **secondary giant cells**, and are derived from the polar trophoblast cells, while the latter that are formed from the mural trophoblast are termed the **primary giant cells** (for detailed description of giant cell transformation, see Kaufman, 1983b, and previously).

The umbilical cord also contains the vitelline (vitello-intestinal) duct, a diverticulum which extends (in an antimesenteric direction) into the cord from the mid-point of the **midgut loop**, with its associated vitelline (and later mesenteric) vessels. The arterial supply to the vitelline duct is initially from the vitelline artery, and subsequently from the superior mesenteric artery, while its vitelline venous plexus drains into the portal vein. When the midgut loop herniates into the **"physiological" umbilical hernia**[5] (during TS 17), the hernial sac represents the proximal part of the original communication between the intra-embryonic coelom (now the peritoneal cavity) and the proximal part of the exocoelom. The allantoic mesodermal tissue from which the umbilical cord is largely formed gradually de-differentiates into an amorphous, largely acellular and translucent, substance termed **Wharton's jelly**. Finally, the cord is surrounded by the amniotic membrane.

The fully differentiated umbilical cord in the mouse thus contains two vessels of substantial diameter, the **common umbilical artery** and **common umbilical vein** (with only the former pulsating) that are embedded in the amorphous, mesenchyme-derived Wharton's jelly. Nerves may also be present, associated with the blood vessels (Spivack, 1943). The outside of the cord is covered by a layer of amnion, and this also covers the surface of the chorionic plate within the region bounded by the yolk sac.[6]

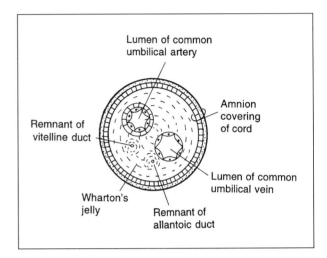

Figure 2.3.2. Representative transverse section through the distal part of the umbilical cord. Note that the majority of the cord consists of acellular mesenchyme-derived material termed Wharton's jelly, and that the outer covering of the cord is formed by the bilayered amnion. Its mesodermally-derived component is apposed to the tissue of the cord, while its ectodermally-derived component forms its outer covering. Within the cord, the two most obvious structures are the common umbilical artery whose endothelial lining, in advanced stages of pregnancy, is surrounded by a wall of smooth muscle, and the common umbilical vein.

In the case of the common umbilical vein, an outer wall of smooth muscle is also present, but is substantially thinner than that of the artery. Other structures that may also be observed in the cord are the remnants of the allantoic and vitelline ducts which extend distally from the urachus and mid-point of the antimesenteric wall of the midgut loop, respectively.

From about E15 (TS 23/24), increasing amounts of mesenchyme condense around the endothelial lining of the cord blood vessels, and this gradually differentiates to form the smooth muscle of their wall; this is particularly clearly seen in the wall of the artery (Figure 2.3.2).

While the midgut loop is within the physiological umbilical hernia (Figure 2.3.3), the gut is associated with two vessels of relatively similar diameter (though they are only about half the diameter of the umbilical vessels), the vitelline artery and vein. Of these two vessels, only the artery (which has a slightly larger diameter than the corresponding vein) pulsates, and it contains oxygenated blood. These vessels always run as a pair, but may be intertwined around the umbilical vessels during the first part of their course before they are separately directed towards the inner (mesoderm-derived) wall of the yolk sac where they branch

---

[5] The term "physiological" umbilical hernia is used to contrast this with the various pathological hernias that are encountered in the early newborn period in the periumbilical region (such as the "congenital" type of umbilical hernia, which is usually relatively small, and the often large deficiency of the anterior abdominal wall in the periumbilical region termed an "omphalocoele") because the herniation of the midgut loop into the proximal part of the umbilical cord is a "normal" or "physiological" event. Herniation of the midgut loop occurs because there is insufficient space in the peritoneal cavity to allow the elongation of the midgut to occur – an essential part of its differentiation. When space subsequently becomes available, largely because of the regression of the mesonephroi and their replacement by the considerably smaller metanephroi (or definitive kidneys), the overall increase in the volume of the peritoneal cavity and, but to a lesser extent, the proportionate decrease in the volume of the liver, the midgut loop can then return to the peritoneal cavity. (For further details, see "The gut and its associated glands").

[6] In the human, there are normally 3 vessels present in the cord, the 2 umbilical arteries, and the single umbilical vein. In about 1–2% of infants, the cord only contains 2 vessels, one artery accompanied by a single vein, and in about 50% of these cases, this is associated with abnormalities of the cardiovascular system.

Figure 2.3.3. Diagrammatic section through the proximal part of the umbilical cord in the region of the "physiological" umbilical hernia. Note in particular that the hernial sac contains a relatively small volume of *extra-embryonic coelom*, and that this is in continuity with the intra-embryonic coelomic cavity (i.e. the peritoneal cavity) in the region of the umbilicus. The two vessels of largest diameter present in this region of the cord are the *common umbilical artery* and *common umbilical vein*. The former carries deoxygenated blood to the placenta, while the latter carries oxygenated blood from the placenta to the right atrium via the ductus venosus and post-hepatic part of the inferior vena cava. The two umbilical vessels are characteristically located in the ventral part of the cord, while the *vitelline vessels*, that are associated with the midgut loop, have a considerably smaller diameter. The vitelline artery, a branch of the abdominal aorta, which carries oxygenated blood to the midgut, has a slightly greater diameter than that of the vein and pulsates, while the vitelline vein (which does not pulsate) drains to the portal vein. For simplicity in this diagram, neither the vitelline nor the allantoic ducts are shown. Note also that the mesothelial lining of the hernial sac is in continuity with the mesothelial lining of the peritoneal cavity.

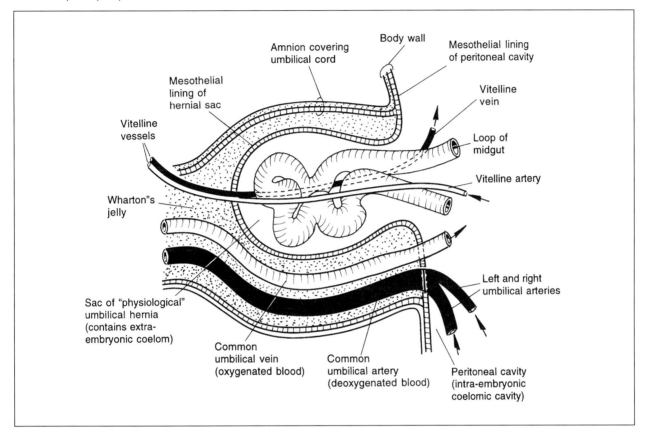

and become continuous with the yolk sac vasculature. Apart from the first part of their course, they do not run within the umbilical cord, but are invariably more rostrally located than the umbilical vessels and cord, and this is consistent with their relationship to the midgut (Figure 2.3.4).

The umbilical cord, which has a length of about one third of that of the crown–rump length of the embryo, passes directly towards the centre of the chorionic plate on the embryonic surface of the placenta. Here, the umbilical vessels break up into a series of large branches which course over a circular region which occupies the middle one third to one half of the embryonic surface of the placenta before entering it, and this region is bounded by a large peripheral blood vessel. More peripherally, the yolk sac is closely adherent to the chorionic plate, although it may readily be separated from it by gentle dissection.

The location of this large peripheral vessel closely coincides with the site of attachment of the yolk sac to the embryonic surface of the placenta. In the peripheral part of the placenta, the chorionic plate is separated for a short distance from the subjacent labyrinthine zone of the placenta; this is termed the **interplacental space** and is lined by a layer of cuboidal cells (on the side of the chorionic plate), and by a layer of flattened epithelial cells supported by a thick basement membrane (on the labyrinthine side). This space is continuous with the lumen of the yolk sac cavity, and is first seen at about E11 (TS 18).

Once the midgut loop has returned into the peritoneal cavity (towards the end of TS 24), the space within the cord formerly occupied by the gut is resorbed, and the abdominal wall reforms in this location. This is nevertheless still a potential site of weakness, because of the passage of the umbilical

TS	1	2	3	4	5	6	7	8	9	10	11	12	13	14	15	16	17	18	19	20	21	22	23	24	25	26
E	0	1	2	3	4	4.5	5	6	6.5	7	7.5	8	8.5	9	9.5	10	10.5	11	11.5	12	12.5	13.5	14.5	15.5	16.5	17.5

Figure 2.3.4. Diagram showing a more extensive view of the umbilical cord than that illustrated in Figure 2.3.3. Note in particular that the only part of the course of the vitelline vessels that is *within* the cord is close to the region of the "physiological" umbilical hernia. More distally, the vitelline vessels leave the cord and run, as a pair, towards the wall of the yolk sac where they branch and become continuous with the yolk sac vasculature. Once outwith the cord, these vessels are frequently entwined around it.

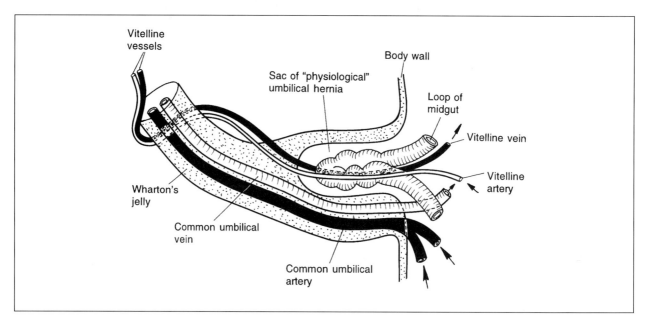

vessels through it. The vitelline duct[7] and the allantoic duct (the remnant of the allantois) generally regress as does the exocoelom within the cord, though the allantois may persist within the cord until term as a discontinuous epithelial strand. This portion of the allantois becomes the **urachus**, and passes distally from the apex of the bladder; after birth, it usually fibroses, and the peritoneal ridge overlying it (which passes from the apex of the bladder to the umbilicus) is known as the **median umbilical ligament** (Figure 2.3.5). For a detailed description of the human umbilical cord, see Chacko and Reynolds (1954), and for that of the rat, see Leeson and Leeson (1965).

The source and volume of the amniotic fluid

The relationship between the fetus and the amniotic fluid changes continuously throughout pregnancy, and the relative contribution of the maternal, fetal and placental membranes in the maintenance of amniotic fluid volume are subjects of considerable research interest (Fairweather and Eskes, 1973). Tracer studies (Vosburgh *et al.*, 1948) have demonstrated that the pool of amniotic fluid is not a static reservoir, but that there is a continuous interchange between the amniotic sac and the maternal and fetal circulations, with the fluid and its chemical constituents in a constant state of flux. Early in gestation, the fetal gastrointestinal tract contributes little to the elaboration or control of amniotic fluid volume, whereas at term (in the human, at least) fetal swallowing may be an important regulating function, although its true role in the control of amniotic fluid volume is far from established (Abramovich, 1973). There is some evidence that transudation through the walls of the fetal lung contributes to the supply of the amniotic fluid (Adams *et al.*, 1963; Fujiwara *et al.*, 1964).

Adams *et al.* (1963) described a fetal lamb that had no eyes, nose or mouth, and thus no communication between the trachea and the amniotic cavity. The tracheal fluid in this lamb was similar to that in the trachea of normal lambs, lending support to the hypothesis of its origin from the lung rather than from the amniotic cavity.

Originally, the amniotic fluid was believed to be a transudate (or filtration product) of the maternal plasma. Other studies, however, have concluded that the differences in composition between amniotic fluid and maternal plasma as correlated by Hutchinson *et al.* (1955) and Wirtschafter and Williams (1957) did not lend support to this theory. Most agree that later in gestation it is largely a product of fetal micturition. This certainly occurs from about 11 to 14 weeks in humans (Abramovich, 1968).

Pritchard (1966) determined that, in the human at term, about 450 ml of amniotic fluid were swallowed daily in the presence of a mean amniotic fluid volume

---

[7] Clinically, various problems may be encountered in relation to the vitelline duct. The proximal part of its course may remain patent, rather than completely regressing, as normally occurs, but the blind end does not retain its connection with the umbilicus. If the diverticulum is blind-ending, it is termed a *Meckel's diverticulum*. This may contain ectopic gastric, endometrial or other tissue, and may perforate rather like an appendix, giving rise to peritonitis. Occasionally, it may regress to form a fibrous band, but the latter remains attached to the umbilicus. This can act as an "axis of rotation", and can give rise to a *volvulus*, a condition in which the bowel twists, cutting off the blood supply to the region involved. Bowel obstruction associated with ischaemia and gangrene of the wall of the bowel often results. Both the vitelline duct and urachus may remain patent, leaving a fistula which opens in the region of the umbilicus, and through which either bowel contents or urine respectively may leak out.

Figure 2.3.5. Diagram of the umbilical region after the withdrawal of the midgut loop into the peritoneal cavity, and the resorption of the "physiological" umbilical hernial sac. Note that the largest structures present in the cord are the common umbilical artery and common umbilical vein. Two additional structures are also present, the *urachus*, being the rostral extremity of the urogenital sinus (bladder) that extends distally into the cord, and the *vitelline* (or vitello-intestinal) *duct*, being another narrow diameter diverticulum that in this case extends distally from the antimesenteric region at the mid-point of the midgut. Both of these ducts usually have only a relatively short course within the cord where their lumen is still patent. More distally, both are usually replaced by either a continuous fibrous band or intermittent segments of fibrous tissue. Clinical problems that may be encountered when either the vitelline duct or urachus fail to regress normally, are discussed in the main text. For simplicity in this diagram, the vitelline vessels are not shown.

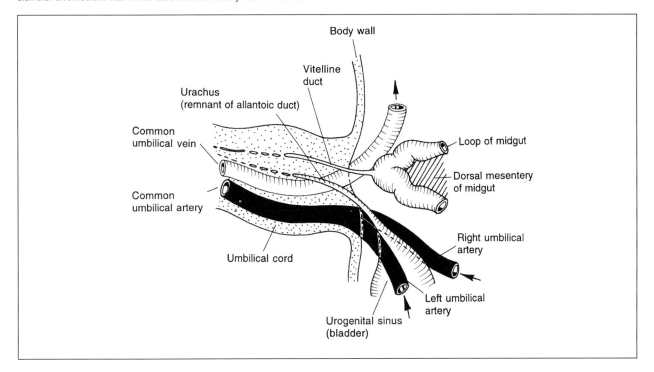

of about 850 ml. Increased amniotic fluid volume often accompanies those pregnancies in which fetal swallowing is impaired. According to Lloyd and Clatworthy (1958), 14% of pregnancies with oesophageal atresia and 66% of those with upper intestinal atresia have polyhydramnios. This is commonly seen in association with anencephalic pregnancies where the swallowing reflex is believed to be impaired, though this is not invariably seen as some anencephalics are capable of swallowing, and these cases are associated with a normal volume of amniotic fluid (Gadd, 1966). Rarely, normal volumes, and even more rarely increased volumes of amniotic fluid may be present in association with renal agenesis (Jeffcoate and Scott, 1959), and this was believed to be due to the presence of other congenital abnormalities such as iniencephaly[8] in which the normal pathway of removal of amniotic fluid by deglutition was probably absent.

The placenta

As early as E8–9 (TS 12/13), the maternally-derived decidual reaction to the presence of the conceptus shows evidence of increased vascularization. During E9–9.5 (TS 14/15), numerous clefts appear in the ectoplacental tissue close to the decidual cells. As maternal blood vessels become eroded, these spaces (or lacunae) become filled with maternal blood and form the basis of the **labyrinthine part of the placenta**; at this stage, only maternal blood circulates within the primitive placenta. More peripherally, the labyrinthine zone is bounded by an intermittent layer of **trophoblastic giant cells**. By E10 (TS 16), the allantoic vessels invade the chorionic plate, and the ectoplacental plate is fully converted into the labyrinthine part of the definitive placenta, with only a few layers of cells now separating the embryonic from the maternal blood (see below). In the region

---

[8] This is a rare congenital abnormality in which the foramen magnum (the large hole in the occipital bone through which the spinal cord passes) is enlarged and is associated with the absence of the spinous processes of the cervical, thoracic and often the lumbar vertebrae, so that the brain and much of the spinal cord occupies a single cavity. In the human, these pregnancies often go to term, and are frequently complicated by polyhydramnios. Such a condition has been induced in rat fetuses exposed to streptonigrin and other teratogenic substances (Warkany and Takacs, 1965; for review, see Warkany, 1971).

TS	1	2	3	4	5	6	7	8	9	10	11	12	13	14	15	16	17	18	19	20	21	22	23	24	25	26
E	0	1	2	3	4	4.5	5	6	6.5	7	7.5	8	8.5	9	9.5	10	10.5	11	11.5	12	12.5	13.5	14.5	15.5	16.5	17.5

close to the uterine insertion of the mesometrium (the mesentery from which the uterus is suspended from the wall of the pelvis), numbers of large, usually binucleated lipid-bearing decidual cells penetrate the connective tissue between the myometrial cells to form what will become the **metrial gland**.

During days E11–12 (TS 18–20), the placenta can be divided into a number of reasonably well-defined zones. That closest to the conceptus is the **chorionic plate** into which the yolk sac and allantoic vasculatures insert and where they amalgamate. The next region is the **labyrinthine zone**, and outside this is the **spongy zone** (or **spongiotrophoblast**), and this is separated from the maternally-derived decidual tissue by a dispersed layer of trophoblast giant cells. Towards the periphery of the placenta, there may be several layers of giant cells, whereas towards the central part of the placenta such giant cells are often only present as a monolayer and are usually considerably more widely dispersed. The (maternally-derived) decidual tissue is itself divided into a compact layer which is the region closest to the spongy layer, and a more peripheral zone where the decidual cells are much more loosely related to each other.

From about E10 (TS 16), the placental site is supplied by a single large branch of the uterine artery, and this usually passes towards the central region of the placenta directly to the chorionic plate. In developmentally more advanced placentas, it is possible to see that this vessel breaks up into a relatively small number of branches before its contents become subdivided and dispersed into a huge network of vascular spaces, each of which is filled with maternal red blood cells (rbc). Interspersed between these spaces are relatively small numbers of embryo-derived **trabecular cords**. These are surrounded by a layer of chorion-derived (cytotrophoblast) tissue and contain embryonic blood vessels within their (extra-embryonic mesodermally-derived) central core. These may be recognized because their contents consist exclusively of **primitive nucleated rbc**, large spherical cells that have a characteristic morphology of an enlarged and darkly-staining nucleus; they also have a larger overall diameter (and a volume that is about 4–5 times greater) than maternal rbc.

By E11 (TS 18), the trabecular cords extend from the embryonic surface of the placenta towards the embryonic/maternal junctional zone where they merge with the cytotrophoblast elements forming the trophoblastic part of the placental barrier. Transmission electron microscopy of the near-term mouse placenta has shown that the interspace between the allantoic (embryonic) capillary lumen and the lumen of the maternal blood sinuses consists of a trilaminar trophoblastic epithelium: a superficial discontinuous cellular layer consisting of cytotrophoblast, and two deeper syncytial cell layers. Two additional layers are also present, the endothelium that lines the embryonic capillary wall, and an amorphous basement membrane believed to be of chorion-derived mesenchymal origin (Kirby and Bradbury, 1965).

While the function of the labyrinthine zone of the placenta is clear (it is the principal site of gaseous exchange between the deoxygenated embryonic blood and the oxygenated maternal blood), the role of the spongy zone is still unknown. This region of the placenta only contains maternal blood vessels, and these open into the large venous sinusoids which are located within the decidual region of the placenta and that subsequently drain into the uterine venous system. Throughout gestation, the proportion of the placenta occupied by the labyrinthine zone gradually increases, so that by about E19 it occupies more than half of its entire volume.

The gross and histological morphology of the full-term placenta is very little different from that seen at about E14 (TS 22), and this represents only a slight change from that seen at about E11.5–12 (TS 19/20). Virtually all of the changes that occur during this period are due to the increase in the proportion of the placenta occupied by its labyrinthine zone.

For observations on the morphogenesis of the placenta and extra-embryonic membranes in a wide range of species, see Mossman (1937), and for a comprehensive account of *human* placentation, see Boyd and Hamilton (1970). The presence or absence of nerves associated with the umbilical blood vessels is still a contentious issue, and this topic is dicussed by Spivack (1943). A number of studies have confirmed that the fine structure of the mouse placenta is identical to that of the rat (Jollie, 1964; Enders, 1965).

2.4 Early organogenesis

Introduction

The events that take place at around the ninth day of development may well be the most interesting and important of mouse development. At the end of TS 11 (E7.5), gastrulation has occurred, the three germ layers are in place and the neural plate is beginning to differentiate. A day later, the development of each of the major organ systems is being initiated, the mesoderm is segmenting, the branchial arches are forming and the neural crest is just starting to migrate. Up to this point, it is practical to consider the embryo as a whole; another day later (TS 15, around E9.5) it is just too complex.

During TS 12 and 13, the embryo is thus a particularly busy place and the greater part of this chapter briefly documents this activity, but we use the main part of the text as an opportunity to point the reader towards the chapters in the remainder of the book where the various organ systems are discussed individually, and in more detail. The final part of this chapter considers two components of the embryo whose origins lie at this stage, but whose later development does not fit very easily elsewhere in the book. The first is the **septum transversum** and we consider both its origin and the tissues to which it gives rise, particularly the diaphragm. The second is the set of **body cavities** and we discuss both their formation and later partitioning. It should also be pointed out that it is around E9 when the heart and vascular system start to form and shortly after to function; their morphogenesis is discussed briefly here and more extensively in the chapter on the heart.

The external appearance of the embryo

The overall changes in the external features that are seen between advanced TS 11 and early TS 12 embryos are slight. At the end of TS 11, the most obvious feature is the presence of the early headfolds which reflect the growth of the neural folds in the prospective cephalic region and rostral part of the neural axis. Within the embryonic compartment, little else is clearly distinguishable either when freshly isolated embryos are examined under a dissecting microscope or when serially sectioned and stained material is examined histologically. At the beginning of the next stage, the characteristic features of the maturing embryo start to appear.

One of the first important landmarks to be seen are the early somites. These start to differentiate within the paraxial mesoderm during the early part of TS 12, and help to define this stage of development (during TS 12, embryos develop up to 7 pairs of somites, while TS 13-stage embryos possess 8–12 pairs). Just before the first evidence of somite formation is seen, however, the intra-embryonic mesoderm on either side of the primitive streak segregates into three initially poorly-defined longitudinally-running columns of tissue which appear in a craniocaudal sequence. Adjacent and lateral to the neural tube is the paraxial mesoderm; slightly lateral to this is the less well-defined intermediate-mesoderm, and beyond this is the lateral plate mesoderm. The other well-defined landmarks that form in the (TS 12) "unturned" embryo are the first and second branchial arches, the early heart, the otic placode and the tail.

An additional event initiated at about this time is the *turning* of the embryo when it reverses its orientation so that its gut, instead of facing outwards and in the direction of the trophoblast wall faces inwards so that it eventually becomes enclosed within the embryo. This reversal of the configuration of the embryo also closes up the peripheral margin of the body wall in the region of the midgut opening (the umbilical "ring") and helps define the internal cavities of the embryo which become partitioned as the heart and septum transversum form. A short time later, as the heart starts beating, the gut extends

TS	1	2	3	4	5	6	7	8	9	10	11	12	13	14	15	16	17	18	19	20	21	22	23	24	25	26
E	0	1	2	3	4	4.5	5	6	6.5	7	7.5	8	8.5	9	9.5	10	10.5	11	11.5	12	12.5	13.5	14.5	15.5	16.5	17.5

further rostrally and caudally, and the neural tube, neural crest and the associated sense organs start to differentiate.

Inversion of the germ layers, and the process of turning

The normal or characteristic "fetal" position is adopted by the human embryo and those of most other mammals at all stages of their embryogenesis. Here, the dorsal axis (or ectodermal surface) of the embryo is outermost, and forms its convex aspect, while its ventral surface, which includes the exposed part of the future midgut region, is located on the concavity of the embryo. The orientation of the early post-implantation mouse embryo, up to the end of the primitive streak stage, is the reverse of the normal mammalian arrangement and the embryo is said to be *unturned*. The reasons for this way of doing things are not fully understood, but may be a particularly economical use of space consistent with the development of 6–12 or more conceptuses within the relatively confined space in the uterine horns (see Snell and Stevens, 1966). This configuration, which is also seen in the rat, rabbit, guinea pig and their close relatives which also have large numbers of embryos, is referred to as *inversion of the germ layers* and it is reversed during the process of *turning*, after which the mouse embryo adopts the same configuration as the human embryo.

Turning (or axial rotation) occurs during a relatively restricted period between about E8–8.5 (advanced TS 12, when the embryo possesses about 4–6 pairs of somites) and E9 (TS 14, when the embryo possesses about 14–16 pairs of somites) after which the process is completed, and the embryo has adopted the normal

Figure 2.4.1. Diagrammatic sequence to illustrate changes in the conformation of the embryo and the way in which the extra-embryonic membranes surround it as it undergoes the process of "turning". The following stages are illustrated: (a) presomite headfold stage, E7.5–8; (b) embryo with 8–10 pairs of somites, about E8.5; (c) embryo with 10–12 pairs of somites, about E8.75; (d) embryo with 12–14 pairs of somites, about E9; (e) embryo with 15–20 pairs of somites, about E9.5; (f) the embryonic layers, extra-embryonic tissues, and cavities encountered in the embryos illustrated in (e). (From Kaufman, 1990, with permission.)

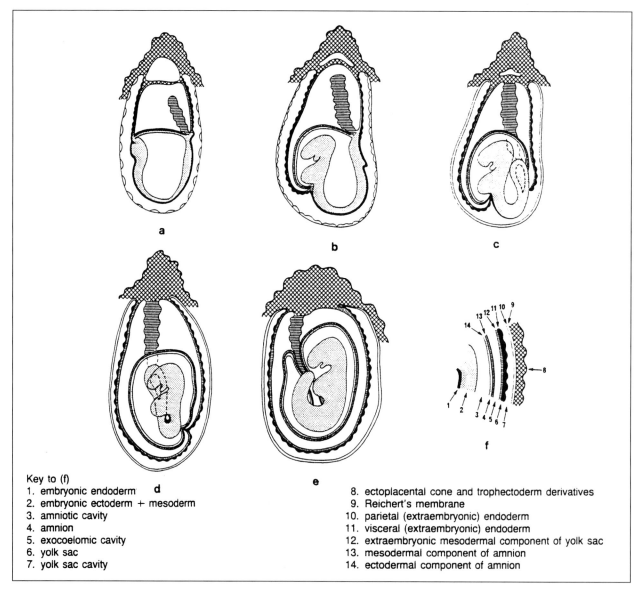

Key to (f)
1. embryonic endoderm
2. embryonic ectoderm + mesoderm
3. amniotic cavity
4. amnion
5. exocoelomic cavity
6. yolk sac
7. yolk sac cavity
8. ectoplacental cone and trophectoderm derivatives
9. Reichert's membrane
10. parietal (extraembryonic) endoderm
11. visceral (extraembryonic) endoderm
12. extraembryonic mesodermal component of yolk sac
13. mesodermal component of amnion
14. ectodermal component of amnion

fetal position. While the events that occur during the *turning* sequence are relatively straightforward, the geometric changes in the configuration of the extra-embryonic membranes are more difficult to envisage, and are most easily represented diagrammatically (see Figure 2.4.1). The embryo effectively *rolls* into its extra-embryonic membranes and becomes completely surrounded by both the amnion and, more peripherally, by its yolk sac. This change in the relationship between the embryo and its extra-embryonic membranes is a consequence of the fact that the latter are attached along the peripheral boundary of the body wall (the future site of attachment of the umbilical cord to the embryo), so that, if the embryo rotates, the membranes inevitably have to follow (for a detailed description of this process, see Kaufman, 1992).

If a transverse section through the midgut region, the middle part of the embryonic axis, is viewed from above, and in the direction of the tail, it is seen that the trunk region of the embryo rotates through 180° in an anticlockwise direction (see Figure 2.4.2). This simple mechanism enables the embryo to adopt the fetal position, while at the same time allowing it to become surrounded by its extra-embryonic membranes. While it is clearly of advantage to the mammalian embryo to adopt the fetal position, the turning mechanism that facilitates this change in its configuration has yet to be explained.

Figure 2.4.2. Diagrammatic sequence illustrating the changes in the relationship between the embryo and its extra-embryonic membranes when sequential transverse sections through the mid-trunk region are viewed during the five representative stages of the "turning" sequence illustrated in Figure 2.4.1. The transverse section illustrated in (a) is through the mid-trunk region of an "unturned" embryo, while the section illustrated in (e) is from an embryo that has completed the "turning" sequence. Sections (b)–(d) are from embryos that are progressing through the process of "turning". Note in particular that in (a) the dorsal surface of the embryo "faces" the region of the ectoplacental cone, while its ventral surface "faces" in the opposite direction. This contrasts with the situation observed in the turned embryo (e), where the dorsal surface of the embryo is now seen to be at 180° to the situation observed in (a). Equally, during the "turning" sequence, the embryo is now seen to be surrounded not only by its amnion, but also by its visceral yolk sac; a primitive umbilical cord (or body stalk) is also established in the region of the primitive yolk sac (the future midgut region). (From Kaufman, 1992, with permission.)

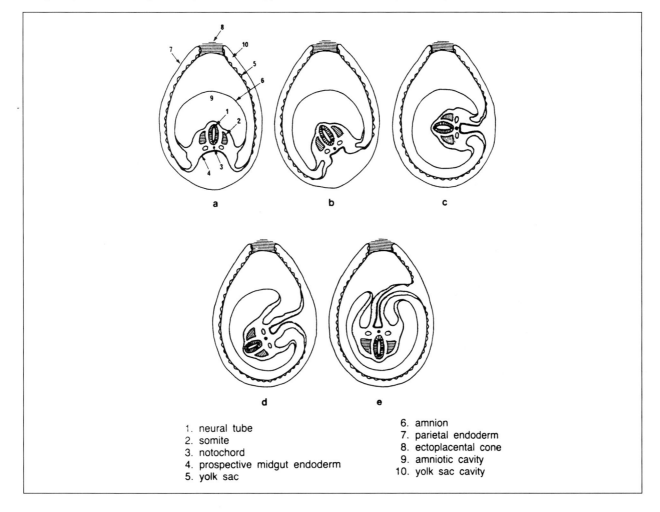

1. neural tube
2. somite
3. notochord
4. prospective midgut endoderm
5. yolk sac
6. amnion
7. parietal endoderm
8. ectoplacental cone
9. amniotic cavity
10. yolk sac cavity

TS	1	2	3	4	5	6	7	8	9	10	11	12	13	14	15	16	17	18	19	20	21	22	23	24	25	26
E	0	1	2	3	4	4.5	5	6	6.5	7	7.5	8	8.5	9	9.5	10	10.5	11	11.5	12	12.5	13.5	14.5	15.5	16.5	17.5

Early neurulation

The external appearance of the embryo changes dramatically during the early part of E8 (TS 12), and this roughly coincides with the onset of somitogenesis. Prior to TS 12, the embryo is *unturned*, has elevated, but only modestly developed neural folds in the cephalic region and relatively poorly developed neural folds elsewhere along its neural axis. This is particularly so beyond the mid-lordotic region (the most concave U-shaped part of the embryonic axis) of the embryo where the neural ectoderm is continuous caudally with that of the primitive streak region – there being no clear line of demarcation between the caudal part of the neural axis, and the most rostral part of the primitive streak region. Equally, the paraxial mesoderm has yet to condense to form discrete somites in the region caudal to the future cervical region of the embryonic axis.

During the early part of TS 12, the neural folds become increasingly prominent in the cephalic region, while the boundary zone separating the neural ectoderm from the surface ectoderm becomes progressively clearer along the embryonic axis (Adelmann, 1925). Apposition of the neural folds (the process of neurulation) first occurs at about the level of the fourth and fifth somites, and extends rostrally and caudally from this site, with fusion being initiated during the second half of TS 12. By TS 13, the presumptive diencephalon can be recognized in the forebrain region, while the first rhombomeres (A and B) are apparent in the hindbrain. (For a detailed description of the early events associated with neurulation, see *The brain* and *The spinal cord*.)

Optic "placodes" (evident as localized "thickened" regions of the neural ectoderm where the cells have a columnar rather than a cuboidal or squamous morphology) are first evident in histological sections through the lateral part of the prospective diencephalic region of the forebrain during the early part of TS 12, and these become increasingly indented in their central regions to form the **optic pits**. Towards the end of this stage, and during the early part of TS 13, the considerable depth of the optic evaginations (thus termed as they represent outgrowths of the future forebrain) results in the development of the optic eminences that are located on the outer (*i.e.* surface ectodermal) aspect of the cephalic neural folds in the future forebrain region. (For further details of eye development, see *The eye*.)

Neural crest migration

The other main event associated with the differentiation of the neural tube at TS 11/12 is the initiation of neural crest migration in the head region. Some of these cells begin to migrate towards the branchial arches which are now forming in the cephalic region, while others migrate into the peri-optic region and other regions which will in due course be associated with the facial primordia (Hall, 1988a). The maxillary component of the first pair of branchial arches, in particular, is becoming increasingly prominent at this time, though the mandibular component, which develops in the more ventral part of the first branchial arch, is not readily recognized until about E9–9.5 (TS 14) when the embryo has about 17–20 pairs of somites (for further details, see *The branchial arch system*).

The increase in the volume of the maxillary component of the first branchial arch that occurs at this time is principally due to the mass of trigeminal (V) neural crest cells that migrate from the junctional zone between the surface and neural ectoderm in the future mid-pontine region of the hindbrain into this arch. Despite the fact that neural crest cells migrate and intermingle with the branchial arch mesoderm, it is often still possible to distinguish between these two cell types because each has a characteristic cellular morphology. While this can be most readily done in sections histochemically stained to demonstrate the presence of (usually) migrating neural crest cells, these two cell types are also often distinguishable in conventionally-stained paraffin sections, although more unambiguous results are obtained through lineage labelling (Trainor and Tam, 1994). The cell bodies of branchial arch mesoderm (as well as those of cephalic mesenchyme) tend to have a spindle-shaped (fusiform) morphology that is associated with elongated cytoplasmic processes, and are surrounded by extensive intercellular spaces, whereas the cell bodies of the early migrating neural crest cells in particular tend to have a more ovoid shape, and are usually associated with short cytoplasmic processes with minimal intercellular spaces between them. (For further details, see *The neural crest*.)

The early auditory system

The earliest evidence of the differentiation of the inner ear apparatus (TS 13) is represented by a slight thickening of the surface ectoderm which appears on either side of the midline at the level of the prospective hindbrain region, opposite rhombomere B. This circumscribed region of initially cuboidal and later columnar epithelium represents the first appearance of the **otic placode**, and it differentiates in a region that has already shown some slight evidence of flattening. The placodal region is situated just lateral to the neural/surface ectodermal junction and is closely associated with the migration of neural crest cells from the facio-acoustic (VII–VIII) neural crest complex. These neural crest cells emerge from the future caudal region of the pons, and are directed towards the second branchial arch, where they will intermingle with the mesenchyme there. (For further details, see *The ear*.)

Mesoderm differentiation and somite formation

At about TS 11 and following gastrulation, the process is initiated whereby epiblast cells enter the primitive streak and separate the overlying epiblast from the subjacent hypoblast. The intra-embryonic mesoderm thus formed spreads laterally on either side of the primitive streak and differentiates into three initially poorly-defined columns of tissue which appear in a cranio-caudal sequence on either side of the neural tube and notochord (which at TS 13 extends rostrally to the point where it overlies the anterior tip of the foregut diverticulum, just subjacent to the midbrain region). Adjacent to the neural tube is the paraxial mesoderm which forms the somites in the body region and the somitomeres in the head region (at E14, there are 13–20 somites and 7 somitomeres). Slightly lateral to this is the less well-defined intermediate-mesoderm: this is exclusively involved in the differentiation of the urogenital system and, by TS 13 is starting to produce an identifiable **nephrogenic cord** (for details, see *Development of the gonads* and *The definitive kidney*).

Beyond this is the lateral plate mesoderm which is split (TS 12) into two components by the intra-embryonic coelom, an outer derivative (the somatopleure) which principally forms the body wall musculature, and an inner component (the splanchnopleure) which surrounds the primitive gut tube and differentiates to form the various layers of musculature that are associated with the wall of the gut (see Figure 2.4.3). They are also, in all likelihood, the origin of the tissues that will form the mesenteries (see below). These are mesodermally derived tissues with a mesothelial covering that support internal organs; their morphogenesis is, in general, still unclear (see below for discussion on the mesenteries of the early gut).

The early cardiovascular system

In the prospective thoracic region, a slight surface prominence across the ventral midline at late TS 11 is the first indication of the location of the pericardial component of the intra-embryonic coelom (the primitive pericardial cavity) within which the earliest stages of heart differentiation are occurring. During mid- to late-TS 12, the coelomic epithelial cells which are destined to form the **cardiogenic** (or **myocardial**) **plate** start to differentiate within the cells that line the ventral part of the wall of the future pericardial region of the intra-embryonic coelom. At around the same time, **endocardial cells**, a specialized endothelium that will form the lining of the primitive heart tube, are seen to be proliferating subjacent to the region where the cardiogenic plate is differentiating (see Figure 2.4.6).

Within the pericardial region of the embryo, these endothelial elements initially develop as two separate entities, one on either side of the midline, but they soon coalesce to form a single squat tubular structure and become continuous across the ventral midline. This tube extends caudally from the pericardial region just subjacent to the walls that line the right and left intra-embryonic coelomic channels (the future pericardio-peritoneal channels), where it bifurcates to form the endothelial lining of the two horns of the sinus venosus, the inflow tracts to the heart. Meanwhile, the myocardial plate is differentiating in the region of the cavity lining that overlies the endothelial cells, and soon surrounds them. The intervening space between the inner (endocardial/endothelial) and outer (myocardial) layers of the primitive heart is seen to be filled by **cardiac jelly** (mid-to-late TS 12). The reticular appearance of this very loose "mesenchyme" suggests that it is largely extracellular, and, although its exact origin is unclear, it may well be produced by the endothelial cells of the heart tube.

Elsewhere in the embryo, clusters of endothelial elements (sometimes termed "angiogenetic cell clusters") may be seen that soon amalgamate together to form the lining of the embryonic vasculature (Figure 2.4.3). These will amalgamate with the corresponding endothelial elements of the yolk sac vasculature (formerly the yolk sac blood islands) to form a single unified system (TS13), although there is as yet no evidence of the mesenchymal-cell-derived condensations which will in due course form the additional (principally muscular) components of the walls of these vessels. At this stage (mid- to late-TS 12), the embryonic vasculature consists of little more than the primitive outflow tract from the heart (the future aortic sac) from which emerges the paired first branchial (pharyngeal) arch arteries that are directed dorsally around the primitive foregut and the paired dorsal aortae, and the elements that will make up the venous return to the two horns of the sinus venosus.

It is, however, just possible to see that the primitive vascular elements also extend into the cephalic region via vessels that will subsequently differentiate into the internal carotid arteries (these represent the cephalic extensions of the dorsal aortae, and are first recognized as definitive vessels during TS 16) and their branches. The deoxygenated blood drains back to the heart via the primary head veins, which amalgamate to form the anterior cardinal and then the common cardinal venous system, elements of which are first recognized during TS 12/13. Throughout

TS	1	2	3	4	5	6	7	8	9	10	11	12	13	14	15	16	17	18	19	20	21	22	23	24	25	26
E	0	1	2	3	4	4.5	5	6	6.5	7	7.5	8	8.5	9	9.5	10	10.5	11	11.5	12	12.5	13.5	14.5	15.5	16.5	17.5

Figure 2.4.3. Diagram illustrating the relationship between the intra- and extra-embryonic coelomic cavities at about TS 12. In a transverse section through the middle part of the early embryonic disc, the arrow to the left of the diagram indicates the entrance to the intra-embryonic coelom from the extra-embryonic coelom. The former is seen to be an inverted U-shaped split which forms in the intra-embryonic mesoderm, and passes rostral to the buccopharyngeal membrane. The most rostral part of the intra-embryonic coelom is the location of its future pericardial region in which the primitive heart forms from the differentiation of mesothelial cells lining its ventral wall. The angiogenetic cell clusters amalgamate to form the components of the vascular system. At this stage, the intra-embryonic mesoderm is beginning to form into three columns: from medially to laterally, the paraxial mesoderm; the intermediate plate and lateral plate mesoderm.

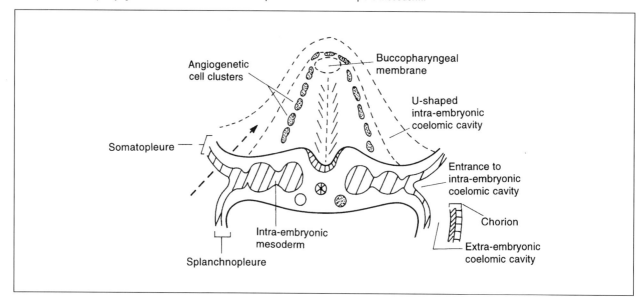

the early part of this stage, the embryonic vasculature consists of apparently empty vessels, for it is only when the embryonic and yolk sac (extra-embryonic) vascular systems amalgamate, usually during TS 13, in embryos with about 8–10 pairs of somites, that primitive nucleated red blood cells are first seen in the embryonic circulation. These primitive red blood cells are first recognized in the blood islands of the yolk sac (for further details, see *the yolk sac* in *Post-implantation extra-embryonic development*).

Towards the end of TS 12, the heart differentiates to form a substantial broad median tubular mass which is suspended from the dorsal wall of the pericardial cavity by a wide dorsal mesentery (the dorsal mesocardium). With the extensive growth and differentiation of the headfolds, the primitive heart gradually rotates on a transverse axis through almost 180°, so that it comes to lie ventral to the foregut (see Figure 2.4.6). Because of the longitudinal folding of the embryonic axis that occurs at this time, the septum transversum comes to lie (TS 13) between the pericardial cavity and the wall of the yolk sac (Hamilton and Mossman, 1972). The outer wall of the primitive heart tube, formed by the differentiation of the cardiogenic (or myocardial) plate initially bulges on either side of a deep ventral median sulcus or groove. It is at this stage that the first irregular, but subsequently more regular contractions of the heart are observed, and this coincides with the first ultrastructural evidence of myofibrillogenesis (Challice and Virágh, 1973; Virágh and Challice, 1973; Kaufman and Navaratnam, 1981; Navaratnam *et al.*, 1986). When the embryo increases beyond a certain critical mass, it is no longer possible for nutrients to reach all of its component parts by diffusion alone. It is for this reason that the heart becomes the first organ to function, while the circulatory system is the first functional unit to develop within the embryo. (For further details of the differentiation of the heart and vascular system, see *The heart and its inflow and outflow tracts*, and the *Branchial arch arteries* section of *The branchial arch system*.)

Blood cell formation

Red blood cell formation (erythropoiesis) initially occurs in association with the blood islands which form within the mesodermal component of the wall of the visceral yolk sac during the primitive-streak stage and also in association with the lateral mesoderm of the splanchnopleure (Godin *et al.*, 1993).

The yolk sac has another critical role at the primitive streak stage of development: it is the first place where the primordial germ cells (pgc) are seen and they can be recognized by their characteristic morphology and intracellular alkaline-phosphatase enzyme activity which may be readily demonstrated with appropriate histochemical stains (see Kaufman, 1994, and previously). During the advanced primitive streak stage (TS 10/11), the pgc become incorporated into the base of the allantois, and subsequently migrate via the wall of the hindgut and its dorsal mesentery to the urogenital ridge (Chiquoine, 1954; Ozdzeński, 1967). Once there, they differentiate and play a key role in gonad differentiation. (For further details, see section on *Primordial germ cells* in *Development of the gonads, internal duct system and external genitalia*.)

The primitive gut

The lining of the primitive gut and its associated glands mainly develops from the definitive endoderm which is located on the ventral aspect of the embryo. The first sign of the future gut is a slight indentation just caudal to the site of the future pericardial cavity which will become the **foregut diverticulum** (TS 11); this extends rostrally from a wide entrance pocket (or portal) into the future cephalic region, while caudally it merges with the future midgut region which itself extends caudally until it imperceptibly becomes the future hindgut region. By TS 12, the most rostral part of the hindgut is also represented by a wide entrance portal, though, as at the foregut–midgut junction, there is no clear line of demarcation between these two regions of the future gut. A stage later, an indentation associated with a thickening in the ventral surface of the foregut diverticulum (in due course located in the midline on the floor of the pharynx between the first and second branchial arches, termed the **foramen caecum**) provides the first evidence of the **thyroid primordium**.

During this early stage of gut differentiation, the primitive foregut is a blind-ending endodermally-lined diverticulum, the most rostral part of which makes contact (during TS 13/14) with an indentation of surface ectoderm termed the **stomatodaeum** (or **mouth pit**), the two being separated by a bilayered membrane, the **buccopharyngeal membrane**; this usually breaks down during TS 14 allowing continuity between the amniotic cavity that surrounds the embryo and the lumen of the gut tube. In the roof of the stomatodaeum, a slight indentation represents the site of **Rathke's pouch**, the source of the anterior pituitary. (For further details, see the Appendix to *The brain and spinal cord*.)

The future **hindgut diverticulum** differentiates slightly later than its equivalent in the foregut region, but it is possible to see that it extends into the proximal part of the tail region by TS 12. It is not, however, until about TS 17 that an ectodermal invagination (approximately comparable to the mouth pit) is recognized and the bilayered membrane separating it from the wall of the endodermally-lined **cloaca** is termed the **cloacal membrane**. This subsequently becomes subdivided by the downgrowth of the **uro-rectal septum** (TS 17/18) into the **anal membrane** (dorsally) and the **urogenital membrane** (ventrally), the two being eventually separated by the **perineal "body"** (about TS 21). The anal membrane generally breaks down during TS 25/26. (For further details, see *The gut, and its associated glands*.)

The early development of the caudal part of the embryonic hindgut region is more complex than that of the foregut region, because there is initially continuity between the **primitive hindgut** and the **urogenital sinus**, this overall region being termed the cloaca. The urogenital sinus, in due course, gives rise to the bladder (TS 22) and most of the urethra and, in the female, it seems that it also gives rise to the lower two-thirds or so of the vagina; the upper part of the vagina, as well as the rest of the reproductive duct system in the female coming from the **paramesonephric ducts** (for further details, see *The development of the gonads*).

A number of important gut-derived glandular structures differentiate from diverticula that emerge at the foregut-midgut junction, most notably the **biliary diverticulum** (or **bud**) (from which differentiates the intra- and extra-hepatic parts of the biliary system), and the ventral and dorsal **pancreatic diverticula** (which later amalgamate to form the definitive pancreas). The development of the liver and pancreas are considered in detail in the appropriate sections of *The gut and its associated glands*, though it is appropriate to consider the development of the septum transversum here, and, in doing so, we briefly consider liver development.

Much of the early gut tube is soon suspended from the posterior (dorsal) wall of the trunk by a midline dorsal mesentery, and, its most rostral part, to the anterior (ventral) wall of the trunk by a midline ventral mesentery. In fact, the majority of the foregut (the future pharynx and oesophagus) is not supported by a mesentery, but the rest of the foregut (the stomach and the rostral half of the duodenum) are suspended by both a dorsal and a ventral mesentery.

The septum transversum

The septum transversum forms from an aggregation of mesenchyme tissue that develops within the caudal part of the ventral mesentery of the foregut (TS 13). At the time of its first appearance, the septum transversum is at about the level of the third, fourth and fifth cervical somites. Over the next few days, the septum transversum becomes a particularly complex tissue as it is invaded by several other tissues that include the arborizations of the biliary tree (which form the intra-hepatic part of the biliary system) and the umbilical and vitelline venous systems (which are involved in the formation of the hepatic sinusoids and ductus venosus), while, at the early limb bud stage, its rostral surface is invaded by groups of somite-derived myoblasts. These myoblasts are believed to migrate into the cranial part of the septum, carrying their nerve supply from the cervical region of the spinal cord with them, and are later thought to migrate into the pleuro-peritoneal folds which then differentiate

TS	1	2	3	4	5	6	7	8	9	10	11	12	13	14	15	16	17	18	19	20	21	22	23	24	25	26
E	0	1	2	3	4	4.5	5	6	6.5	7	7.5	8	8.5	9	9.5	10	10.5	11	11.5	12	12.5	13.5	14.5	15.5	16.5	17.5

into the main muscle mass of the diaphragm (see below).

For these reasons, it is difficult to view the septum transversum in isolation, without taking into account the critical importance of its derivatives, the liver, components of the diaphragm and fibrous pericardium, as well as its influence in fashioning the vascular arrangement in the region of the foregut–midgut junction. This section thus discusses, in fairly general terms, both the events taking place during TS 13 and the later development of the tissues that derive from the septum.

The liver and its blood supply

Soon after it forms, some of the cells of the septum transversum form the parenchyma of the liver (TS 13, for further details, see *The gut and its associated glands*). A little later, this region is invaded by the intra-hepatic part of the biliary system which then repeatedly bifurcates, and it is also colonized by haematopoietic (or haemopoietic) stem cells (of yolk sac origin); the liver thus becomes the initial, and indeed the most important site of haematopoiesis during the prenatal period, its role in this process being supplemented by haematopoietic cells which later form in the spleen and bone marrow. (See *The yolk sac* in *Postimplantation extra-embryonic development* for observations on the source of these stem cells.)

Two sets of blood vessels, the **vitelline veins** and the **umbilical venous systems** enter and ramify within the septum transversum, and within the liver when it differentiates, and these have an important effect on the pattern of blood supply to the right atrium. The vitelline venous system will in due course play a critical role in the venous drainage of the gut, while the extra-hepatic part of the **right vitelline vein** eventually differentiates into the **portal vein**. This will drain the superior mesenteric, inferior mesenteric and splenic veins. Although the extra-hepatic part of the left vitelline vein regresses, the vitelline venous system arborizes within the liver to form the **hepatic sinusoids** which eventually drain into the right and left hepatic veins. These then drain into the post-hepatic part of the inferior vena cava, which then drains into the caudal part of the right atrium.

The **left umbilical vein** differentiates within the liver to form a conduit (the **ductus venosus**) which allows the oxygenated blood from the placenta to pass directly (via the inferior vena cava) to the right atrium (the intra- and extra-hepatic derivatives of the right umbilical vein largely regress). This intra-hepatic channel has a substantial diameter, and the oxygenated blood passes through it directly into the post-hepatic part of the inferior vena cava without being deflected into the hepatic sinusoids. Beyond the liver, the ductus joins the post-hepatic part of the inferior vena cava, which then enters the right atrium. After birth, when the ductus venosus no longer serves any useful function, it also regresses and fibroses shortly after birth to form the **ligamentum venosum**.

As the liver differentiates within the ventral mesentery of the foregut, this mesentery becomes subdivided (TS 21) into a **lesser omentum** (dorsally) and the **falciform ligament** (ventrally) within whose lower border is located the left umbilical vein during the prenatal period; after birth, this vein becomes obliterated and eventually fibroses, forming the **ligamentum teres** (or round ligament) which is continuous with the **ligamentum venosum**. (For a more detailed account of the liver, see the appropriate section in *The gut and its associated glands*; for observations on the fate of the ductus venosus, see *The heart and its inflow and outflow tracts*.)

The diaphragm

The **diaphragm** forms from the amalgamation of a number of structures in and around the septum transversum whose origins are quite diverse, but that each form at TS 21. Its four principal components are a central anterior portion (a derivative of the septum transversum that forms the **central tendon**), a dorsal, unpaired, midline portion (principally from the dorsal and ventral mesentery of the oesophagus), two dorso-lateral shelves of tissue (from the pleuro-peritoneal membranes, see Figure 2.4.4a–c) and a peripheral component that derives from the body wall. While the principal function of the diaphragm is to facilitate breathing, it also serves to separate the two **pleural cavities** (rostrally) from the **peritoneal cavity** (caudally).

The **central tendon** differentiates from the rostral and superior part of the septum transversum (TS 21) and is immediately below the fibrous pericardium with which it is partially blended. During TS 21/22, the fibrous pericardium and the central tendon appear to be completely adherent, but they gradually separate (TS 23–5), with the adhesion being largely lost by TS 26. That part of the superior surface of the liver which adheres to the central tendon of the diaphragm has no peritoneal covering, and is consequently termed the "**bare area of the liver**", and is bounded by the **coronary ligaments**. These are the sites of the peritoneal "reflections" where the "parietal" layer of peritoneum of the wall of the peritoneal cavity becomes continuous with the visceral layer of peritoneum that covers the majority of the liver.

The dorsal, unpaired, midline portion of the future diaphragm is mainly derived from the dorsal mesentery of the oesophagus, although some is also derived from a small portion of its ventral mesentery. The post-hepatic part of the inferior vena cava is also located within the ventral mesentery of the oesophagus, occupying the region between the oesophagus and the dorsal border of the septum transversum.

Two dorso-lateral shelves of tissue that originate in the body wall give rise to the **pleuro-peritoneal**

Figure 2.4.4. Three stages in the formation of the pleural cavities and the thoracic diaphragm. (a) A side view of the thoracic cavity at about TS 20 indicates the location of the early pleuro-pericardial membrane: this shelf grows medially from the side wall of the pericardio-peritoneal canal towards the ventral mesentery of the oesophagus. (b) and (c) views from above. At a relatively early stage (b, TS 20), much of the ventral part of the diaphragm is seen to be derived from the rostral part of the septum transversum (this will form the central tendon of the definitive diaphragm). A small part of the dorsal mesentery and a larger part of the ventral mesentery of the foregut acts as a "strut" towards which other components of the future diaphragm, such as the pleuro-peritoneal membranes (or shelves), grow. When these membranes complete the formation of the dorso-lateral part of the diaphragm, the caudal parts of the two pleuro-peritoneal canals are closed off. At a slightly later stage (c, TS 21), a body wall-derived component increases the surface area of the diaphragm by adding an additional component to its peripheral margin.

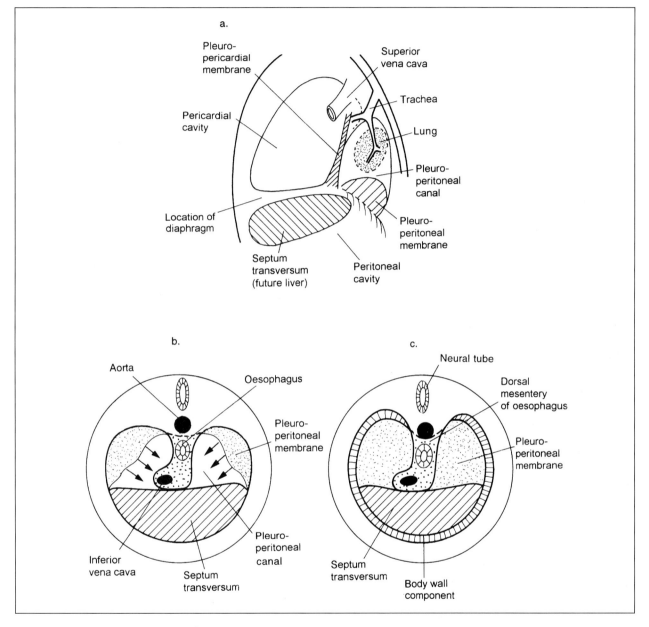

membranes, and these eventually fuse ventrally with the central tendon and with the oesophageal mesentery in the dorsal midline. The pleuro-peritoneal membranes are subsequently invaded by myoblast cells that migrate from mid-cervical somites. Their accompanying innervation (from the phrenic nerves) is thus from the corresponding ventral primary rami of the third to fifth cervical nerves. While the phrenic nerve is principally motor, it also contains a small sensory component, which has a widespread distribution.

The body wall component of the diaphragm forms as a narrow rim around most of the periphery of the above three structures (except for its dorsal, unpaired part), one of its functions being to accommodate the considerable expansion of the lungs that occurs at this time. The peripheral portion of the diaphragm is innervated by the lower 6 or 7 intercostal nerves (sensory alone). In addition, the **crura** of the diaphragm

TS	1	2	3	4	5	6	7	8	9	10	11	12	13	14	15	16	17	18	19	20	21	22	23	24	25	26
E	0	1	2	3	4	4.5	5	6	6.5	7	7.5	8	8.5	9	9.5	10	10.5	11	11.5	12	12.5	13.5	14.5	15.5	16.5	17.5

form during TS 21; these two fleshy processes attach the diaphragm to the spinal column, and form from condensations of tendinous tissue. These originate at the anterior part of the upper 3 lumbar vertebral bodies on the right, and upper 2 on the left, and their corresponding intervertebral discs, and insert into the dorsal part of the diaphragm.

The pleuro-peritoneal membranes are invaded by myoblast cells derived from cervical somites, and later innervated by spinal nerves from the same segmental levels, from the phrenic nerves, with root values of C3–C5, but principally from C4. These carry motor and sensory elements, motor to the dome of the diaphragm and sensory to the parietal pleura of the mediastinum (*i.e.* all of the tissues between the two pleural sacs), the parietal pericardium and the central parts of the diaphragmatic pleura and peritoneum. By contrast, the body wall-derived component at the peripheral margin of the diaphragm is innervated by proprioceptive branches from the lower intercostal nerves; their innervation is exclusively sensory.

Due to the differential growth of the caudal as compared to the rostral half of the embryo that occurs during the second half of gestation, the diaphragm seems to migrate caudally (or "descend") to its definitive location, at the level of the lower thoracic/upper lumbar vertebral bodies. The sequential changes that occur leading to the formation of the definitive diaphragm, with an indication of its various derivatives, are illustrated diagrammatically in Figures 2.4.4a–c (for additional information, see Wells, 1954).

The intraembryonic cavities

In order to determine how the various body cavities are established, it is necessary to understand both the initial configuration of the intra-embryonic coelomic cavity when it forms soon after gastrulation (TS 11), and the way in which it communicates with the extra-embryonic coelomic cavity.

The fate of the intra-embryonic coelom

During the late primitive-streak stage (TS 11), the middle part of the rostrally directed U-shaped cleft that forms within the intra-embryonic mesoderm passes across the ventral midline just rostral to the future site of the buccopharyngeal membrane (Figure 2.4.5). As the embryo folds along its longitudinal axis, mainly because of the extensive growth of the cephalic neural folds now occurring, the middle part of the "U", which is initially anterior (*i.e.* ventral) to the foregut, is carried ventrally and caudally to form the prospective pericardial cavity (see Figure 2.4.6). This rapidly expands (TS 12/13) to accommodate the growth and differentiation of the primitive heart, while the caudal parts of the two "arms" also expand, and, probably as a result of the lateral folding of the embryo, then meet and amalgamate to form the prospective peritoneal cavity. The rostral parts of the two "arms" which connect the pericardial cavity with the newly-forming peritoneal cavity do, however, remain separated, forming the two pericardio-peritoneal canals (or channels) that will later become the pleural cavities. These complex configurational changes, together with the location of the septum transversum in relation to the primitive foregut, are illustrated in Figure 2.4.7a–e.

As the configuration of the primitive heart tube changes within the pericardial cavity (see above), the heart becomes suspended by a wide dorsal mesentery from the dorsal wall of the pericardial cavity (the **dorsal mesocardium**). At this stage (TS 13/14), the most caudal parts of the early heart (the primitive

Figure 2.4.5. Very diagrammatic representation to show the location of an early stage in the development of the U-shaped intra-embryonic coelomic cavity, and its relationship to the primitive gut. Note in particular that the rostral part of the "U", where it crosses the ventral midline is closely associated with the buccopharyngeal membrane – this relationship is more clearly shown in Figure 2.4.6. The two "arms" of the "U" pass on either side of the primitive midgut, and will eventually amalgamate caudal to this region to form the primitive peritoneal cavity. At this early stage, the two arms of the intra-embryonic coelomic cavity form the pericardio-peritoneal channels (or canals), while the most rostral part of the "U" will shortly give rise to the pericardial component of the intra-embryonic coelomic cavity.

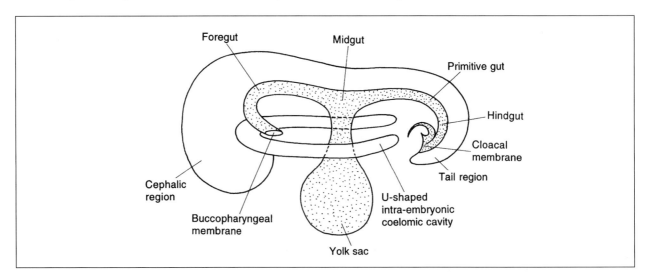

Figure 2.4.6. Diagrammatic sequence (a–c) representing 3 stages in the relocation of the pericardial component of the intra-embryonic coelomic cavity from immediately rostral to the buccopharyngeal membrane (a) to its definitive ventral "thoracic" location (c). This change in position is believed to be largely brought about by the considerable growth of the cephalic neural folds (the headfolds) that occurs during this period. Cellular differentiation to form the primitive myocardial (or cardiogenic) plate is initially associated with the mesothelial cells associated with the ventral wall of the intra-embryonic coelom in this location. With the relocation of this region, the myocardial plate later becomes associated with the dorsal wall of the pericardial cavity, and the primitive heart is eventually suspended from the dorsal wall by a broad mesentery (the dorsal mesocardium). In the lower sequence (d–f), the relationship between the cardiogenic plate and the subjacent endothelial tube is shown, with cardiac jelly interposed between the two, in a series of representative median sections through the pericardial cavity. The definitive location of the primitive heart tube is shown in a side-view (g), being suspended within the pericardial cavity by its dorsal mesentery. The location of its inflow and outflow regions is also shown.

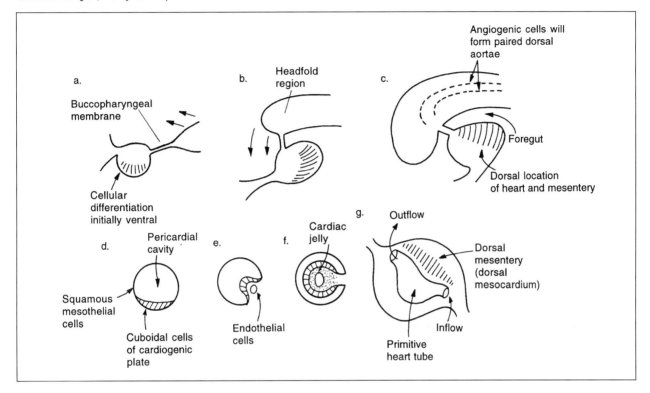

common atrial chamber and, even more caudally, the right and left horns of the sinus venosus) remain partly embedded within the postero-lateral part of the septum transversum (see above, and Chapter 3.1) which is itself differentiating within the ventral mesentery of the foregut. As this has expanded to accommodate the increasing volume of the septum transversum, it can easily be seen in sections and contrasts with the dorsal mesentery of the foregut which is only seen in relation to the most caudal region of the oesophagus and stomach and the first proximal part of the duodenum (the majority of the oesophagus has neither a dorsal nor a ventral mesentery).

Mesenchymal cells bordering the intra-embryonic coelomic cavities differentiate to form their epithelial lining, the mesothelium, at about TS 13. The first of these mesothelia to be seen is the parietal layer that lines the walls of the intra-embryonic coelom; mesothelia also provide a covering for the viscera, or the contents of these cavities (the visceral layer). In the pericardial region, both the **parietal** and **visceral pericardium** (as the mesothelial coverings are known) are first recognized during TS 18/19, while the parietal and visceral layers within the pleural and peritoneal cavities are not apparent until TS 22/23.

At about the level of the foregut–midgut junction, the two pericardio-peritoneal channels enlarge considerably at about TS 16 as the lung buds that had formed at about TS 15 expand laterally into them and, once they are sealed off by the closure of the pleuro-pericardial membranes medially (TS 21), and the pleuro-peritoneal membranes caudally, the lungs become confined within the right and left pleural cavities. It is also at about this stage of development that the midgut loop will herniate out of the peritoneal cavity into the "physiological" umbilical hernia (during TS 17). Slightly caudal to the umbilical region, and as a result of the lateral folding of the embryo, the two halves of the peritoneal cavity amalgamate across the ventral midline because there is no intervening ventral mesentery in this location (see Figure 2.4.7a and section D).

As the two lung buds expand laterally, two shelves of tissue, the **pleuro-pericardial membranes** (or

TS	1	2	3	4	5	6	7	8	9	10	11	12	13	14	15	16	17	18	19	20	21	22	23	24	25	26
E	0	1	2	3	4	4.5	5	6	6.5	7	7.5	8	8.5	9	9.5	10	10.5	11	11.5	12	12.5	13.5	14.5	15.5	16.5	17.5

Figure 2.4.7a–e. Representative transverse sections through various regions of the early embryo to illustrate the location of the intra-embryonic coelom and its relationship to the gut tube and heart. The location of the various transverse sections is illustrated in (a). These are seen to section the embryo in the pericardial region (b,A), through the septum transversum in the region of the developing liver (c,B), in the middle of the midgut region at the site of the primitive yolk sac, and later the site of the physiological umbilical hernia (d,C), and in a region caudal to the primitive yolk sac (e,D). The particular features to note are as follows: in the section through the pericardial region, the heart is suspended by its dorsal mesentery (dorsal mesocardium) from the dorsal wall of (at this stage of development) the pericardial component of the intra-embryonic coelomic cavity. The latter is located across the ventral midline in the future thoracic region of the embryo. Immediately dorsal to the pericardial component of the intra-embryonic coelom is the pharyngeal region of the foregut. The coelomic cavity in this region is lined by parietal pericardium and the myocardial plate is formed by the differentiation of a component of the wall of the coelomic cavity in this region. In section B, note that the transverse section passes through the two arms of the "U" of the intra-embryonic coelom (see Figure 2.4.8 for a clearer view), these represent the pericardio-peritoneal canals (or channels). The dorsal mesentery of the foregut in this region is more theoretical than real, whereas the liver is forming within the septum transversum which occupies most of the ventral mesentery of the foregut in this location. In section C, the peritoneal component of the intra-embryonic coelomic cavity is seen to be in two parts, one on either side of the primitive yolk sac. In section D, with the absence of a ventral mesentery of the gut beyond the foregut–midgut junction, the two sides of the peritoneal component of the coelomic cavity are in continuity across the ventral midline. In the pericardio-peritoneal canals and in the peritoneal region of the intra-embryonic coelom, the squamous mesothelial lining is continuous with that in the pericardial component of the intra-embryonic coelom, and in sections C and D, the gut is seen to be covered by a layer of visceral peritoneum of coelomic mesothelial origin.

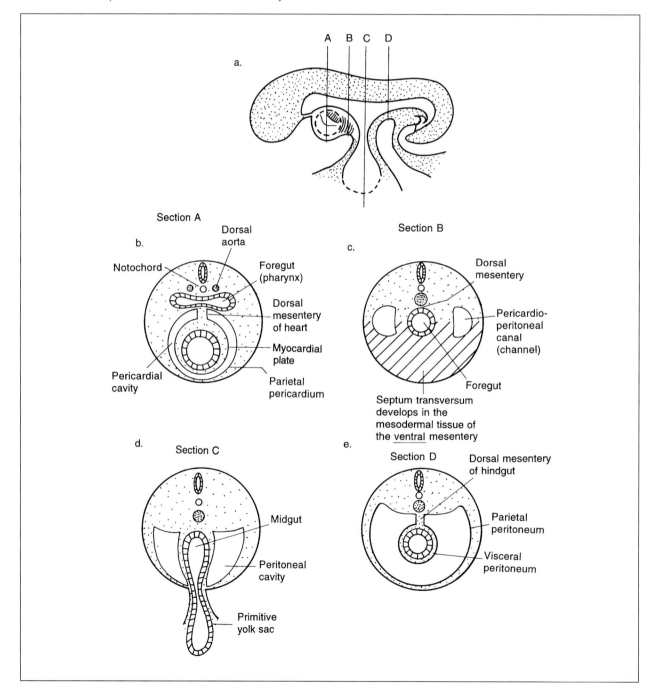

Figure 2.4.8a–c. Diagrammatic sequence showing three stages in the development of the lungs, the establishment of the pleural cavities and the fibrous pericardium. (a) A transverse section through an embryo at an early stage of lung bud development (TS 20/21). Note that at this stage there is continuity between the future pericardial component of the intra-embryonic coelomic cavity and the peritoneal cavity via the pericardio-peritoneal canals (or channels), the two arms of the U-shaped intra-embryonic coelomic cavity (see Figure 2.4.7). The primitive lung buds expand laterally into these canals. Two shelves of tissue (the pleuro-pericardial membranes) expand medially towards the dorsal mesentery of the heart, and, in the leading edges of these membranes are the common cardinal veins, just lateral to which are the phrenic nerves. (b) A transverse section through a later embryo (TS 23) in which the two pleuro-pericardial shelves have fused across the midline. This section is at the level of the transverse pericardial sinus, so that no dorsal mesocardium is seen. The two lung buds have now enlarged, and are surrounded (at this level) by the upper parts of the pleural cavities. (c) The lungs have expanded further on either side of the heart (which is now surrounded by the fibrous pericardium) in the direction of the ventral midline and are surrounded (except in their hilar region) by the left and right pleural cavities. The superior vena cava is now (TS 24) located close to the midline, at the level of the hilum of the lung buds. Only after the pleuro-peritoneal membranes have grown medially and closed off the pleuro-peritoneal canals, and thus completed the formation of the diaphragm, will the pleural cavities become two discrete units (Figure 2.4.4a–c).

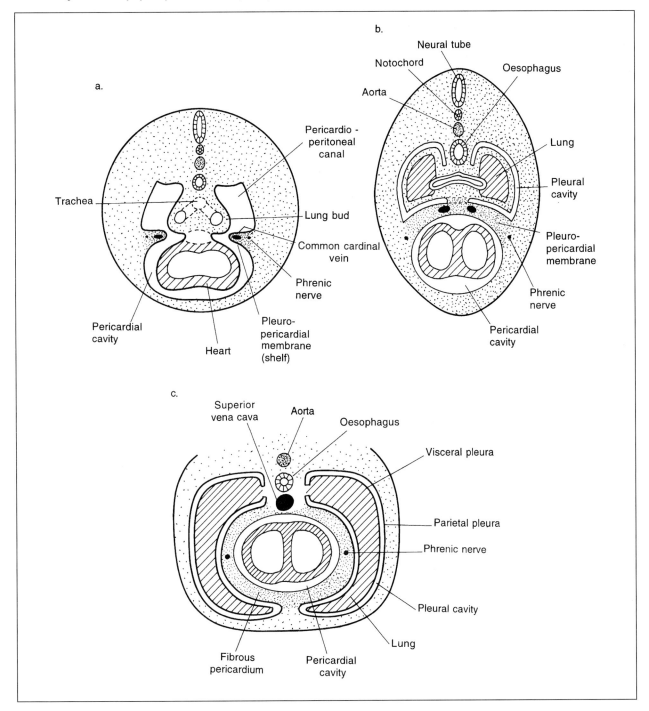

TS	1	2	3	4	5	6	7	8	9	10	11	12	13	14	15	16	17	18	19	20	21	22	23	24	25	26
E	0	1	2	3	4	4.5	5	6	6.5	7	7.5	8	8.5	9	9.5	10	10.5	11	11.5	12	12.5	13.5	14.5	15.5	16.5	17.5

folds), grow out from the lateral part of the body wall (TS 21; see Figure 2.4.8a–c). They are directed antero-medially and, when they meet close to the midline and fuse with the dorsal and ventral mesentery of the foregut (TS 23), they will separate the two pleural cavities from the **pericardial cavity** (see Figure 2.4.8b). As the two lung buds expand, they do so, initially, into the pericardio-peritoneal canals. With the medial growth and apposition of the two pleuro-pericardial shelves (or membranes), the rostral part of the pleural cavities becomes separated from the pericardial cavity. Subsequently, as the lungs expand ventrally, they almost completely surround the heart, being separated from it by the fibrous pericardium, the derivative of the pleuro-pericardial membrane (see Figure 2.4.8c). At their caudal poles, the two pleural cavities are initially in continuity with the peritoneal cavity. However, with the ventro-medial growth of the two pleuro-peritoneal membranes, the pleural cavities eventually become discrete units, completely separated from the more caudally located peritoneal cavity (for further details of this sequence, see Figure 2.4.4). It is close to the leading edges of the two pleuro-pericardial folds that the **common cardinal veins** are located. These form the two main vascular channels that drain the venous blood, principally into the right horn of the **sinus venosus**.

With the gradual regression of the left horn of the sinus venosus (its principal derivative is the **coronary sinus** and the oblique vein of the left atrium), the right horn receives venous blood from the right and left superior and the inferior venae cavae, and is gradually incorporated into the posterior wall of the right atrium.[1]

During TS 15/16, the lung buds begin to differentiate, and eventually expand considerably on either side of the heart (TS 23–25), It should be noted, however, that the relationship between the heart and the lungs is quite different during the early stages of lung differentiation as compared to that seen later in gestation. The reason for this is straightforward, and is related to the differential growth of the caudal as compared to the rostral half of the embryo during the second half of the postimplantation period, particularly during the period between E15.5 (TS 25) and term. Up to about TS 20/21, the primitive lung buds are almost exclusively caudal to the heart and during TS 22/23 are still mainly posterior and inferior to the heart. During TS 24, however, their volume increases and they expand rostrally, so that their apices now rise almost to the level of the outflow vessels of the heart, towards the upper extremity of the thoracic cavity. It is only when the first breaths are taken shortly after birth that the lungs expand fully and occupy all of the space available within the pleural cavities. Because the right lung, with its four lobes, is so much larger than the left, with its single lobe, it occupies the majority of the available space within the thoracic cavity.

The two pleuro-pericardial membranes later form the **fibrous layer of pericardium** within which are located the **phrenic nerves**. The membrane on the right side usually closes off its pleuro-pericardial channel shortly before the left, possibly because the right common cardinal vein is somewhat larger than that on the left side even at this early stage, and the membrane overlying it is correspondingly enlarged.

During TS 22, the peritoneal and the caudal part of the two pleural cavities are still linked via the two **pleuro-peritoneal canals** (or **channels**, also termed the **foramina of Bochdalek**). When these eventually close (TS 23) as a result of the growth and subsequent fusion of the two **pleuro-peritoneal membranes** with the **mesentery of the caudal region of the oesophagus**, the pleural and the peritoneal cavities become separated. This closure takes place in the region that is dorsal and lateral to the future central tendon of the **diaphragm**, and just rostral to the future gastro-oesophageal junction.[2]

---

[1] The situation in the mouse and human are slightly different: in humans, the venous drainage of the left side of the head and neck and left upper limb drain across the midline (through the left brachiocephalic vein) to unite with the superior vena cava, with the latter draining into the superior part of the right atrium. In the mouse, no left brachiocephalic vein forms, and the venous drainage of the left side of the head and neck and left upper limb is into the left superior vena cava. This vessel drains into the inferior vena cava just caudal to where this vessel enters the right atrium.

[2] In the human, congenital diaphragmatic hernias of the abdominal viscera into the pleural cavity are four to eight times more commonly encountered on the left side than on the right side and are almost invariably associated with failure of differentiation of the lung on that side. (For detailed discussion on the aetiology of congenital diaphragmatic hernias, see Warkany, 1971, pp. 751–757.)

3
THE TRANSITIONAL TISSUES

3.1 The neural crest

Introduction

It took well over 100 years from 1868, when Wilhelm His discovered the **neural crest** *(Zwischenstrang)* until direct information could easily be obtained about its role in mouse development. In the 1940s, research demonstrated that amphibian embryos such as *Ambystoma mexicanum* (the Axolotl) used neural crest to form pigment cells, neural tissue such as spinal ganglia, and parts of the head skeleton (Horstadius and Sellman, 1941, 1946; de Beer, 1937, 1947), and this has now been shown to be the case in all vertebrates. Since then, the main focus has been on the use of the chick–quail chimera model system (Le Douarin, 1974, 1976, 1982) to investigate the many cell types and tissues to which the neural crest contributes during avian embryogenesis (for an extensive recent review, see Hall, 1988a). In mammals, experimental transplantation and ablation studies have been carried out on the rat (*e.g.* Tan and Morriss-Kay, 1986), and it is gratifying that these observations largely appear to match those from the chick.

It should thus be emphasized that much of the information presently available on the origin and fate of neural crest in the mouse has not in fact been obtained from this species, but has been inferred on the basis of experimental studies involving avian embryos (*e.g.* Le Lièvre, 1978; Bee and Thorogood, 1980; Noden, 1983). Indeed, it is only with the recent availability of molecular markers and genetic technology that there has been substantial *direct* information about the role of the neural crest in mouse development. Nevertheless, in the belief that the

Table 3.1.1. *Well-known derivatives of the neural crest*

Cell types	Tissues or organs with neural crest cells
adipocytes	adipose tissue
adrenergic neurons	adrenal medulla
angioblasts	blood vessels
calcitonin-producing (C-) (also termed parafollicular) cells	brain
	connective tissue
cardiac mesenchyme	craniofacial bone (viscerocranium)
cholinergic neurons	craniofacial (pharyngeal arch) cartilages
chondroblasts, chondrocytes	dermis
chromaffin cells	eye (cornea, choroid, sclera, pupillary and ciliary muscles)
fibroblasts	
glial cells	gland connective tissue (parathyroid, thyroid, thymus, pituitary, lacrimal)
melanocytes	
odontoblasts	heart (aortic arch, spiral septum, trunco-conal septum)
osteoblasts, osteocytes	
Rohon–Beard cells	parasympathetic nervous system
satellite cells	peripheral nervous system
Schwann cells	spinal ganglia
sensory neurons	sympathetic nervous system
certain striated and smooth myoblasts *e.g.* intraocular muscles	thyroid gland (C-cells)
	tooth papilla, dentine
	ultimobranchial body

TS	1	2	3	4	5	6	7	8	9	10	11	12	13	14	15	16	17	18	19	20	21	22	23	24	25	26
E	0	1	2	3	4	4.5	5	6	6.5	7	7.5	8	8.5	9	9.5	10	10.5	11	11.5	12	12.5	13.5	14.5	15.5	16.5	17.5

neural crest has a similar fate in all advanced vertebrates including humans (see Larsen, 1993), the general consensus about the role of neural crest in mouse development is given here (Table 3.1.1), although the future use of genetic markers may expand our knowledge.

Origin

Neural crest cells arise along the lateral margins of the neural folds at the boundary between the surface and neural ectoderm. During the process of neurulation, these cells detach from the periphery of the neural plate and migrate throughout the body (Figure 3.1.1) where they differentiate into a particularly wide range of cell types present in many tissues and structures (Table 3.1.1, data from Hall, 1988). The first elements of the neural crest to be recognized (TS 12) arise in the mesencephalic region (the trigeminal (V) neural crest) when the cephalic neural folds in this region are still widely separated. Crest cells subsequently detach from other more rostral and caudal parts of the cephalic region. Along the neural axis, the crest cells usually detach in a cranio-caudal direction at or shortly before the fusion of the neural folds to form the neural tube, though some crest cells may detach shortly after neurulation is completed (see below). Crest cell migration is believed to have both active and passive components, with migration being guided by extracellular matrix molecules such as glycosaminoglycans and fibronectin as well as by other less clear mechanisms (Bronner-Fraser, 1982).

In the trunk region, neural crest cells may take either a ventral pathway through the anterior part of the sclerotome region of the somite or a dorsolateral route between the epidermis and the dermomyotome region of the somite. The former cells include those involved in the peripheral nervous system, while the latter include melanocytes (for review, see Gilbert, 1997).

Cephalic neural crest

Crest cells from the mesencephalon and caudal part of the prosencephalon give rise to the parasympathetic ganglia of the oculomotor (III) cranial nerve as well as the cephalic mesenchyme rostral to the level of the mesencephalon and the perioptic mesenchyme that surrounds the globe of the eye (TS 15). This later differentiates into two layers, an inner pigmented, vascular layer (the choroid, TS 21) and an outer fibrous layer (the sclera, TS 24) that are respectively believed to be homologous with the crest-derived pia-arachnoid that invests the brain and the dura mater that is thought to originate from paraxial mesoderm (the scleral bones in the chick embryo are said to be neural crest-derived; see Fyfe and Hall, 1983). The crest cells also differentiate to form the viscerocranium of the face together with the wall of the blood vessels in this region, as well as the pigment cells and neural elements (see below) that are seen throughout the body. In the facial region, mesenchymal cells proliferate and condense to form compact nodes. Some of these cells develop into capillaries while others change their shape and form osteoblasts and later secrete bone matrix (Gilbert, 1997). A similar arrangement is seen in relation to the formation of the flat bones of the skull, where neural crest-derived mesenchyme cells interact with extracellular matrix of the head epithelial cells to form bone (Hall, 1988b). It is generally believed that the head musculature derives from the mesoderm of the somitomeres, but the intraocular muscles that are associated with the pupil (the sphincter and dilator pupillae) and the ciliary body are probably

Figure 3.1.1. Diagram indicating the initial site of differentiation of neural crest, along the neural axis, at the junction between surface ectoderm and neural ectoderm (a). With the elevation and fusion of the neural folds across the dorsal midline, associated with the process of neurulation, the neural crest cells migrate away from their site of origin at the periphery of the neural plate to form a wide range of derivatives (detailed in Table 3.1.1).

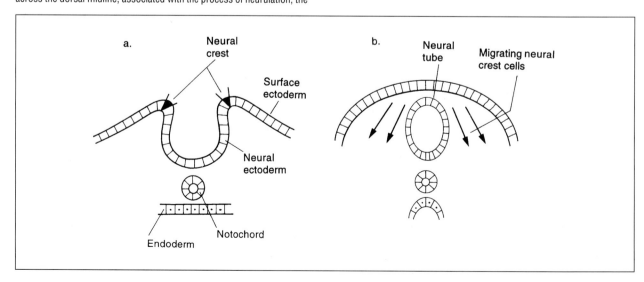

derived from the neural crest (Williams, 1995), being of neuroectodermal origin.

From the mesencephalic and rhombencephalic regions of the early brain, crest cells migrate into the pharyngeal arches (TS 12–16), giving rise, for example, to the cartilaginous precursors of the middle ear ossicles. Mandibular mesenchyme is also crest-derived, and its inductive interaction with mandibular epithelium is a prerequisite for the deposition of the membrane bone of the mandible (TS 21) (Hall, 1988a). These crest cells also form the dermis and smooth muscle of the face and ventral part of the neck, and the odontoblasts of the developing teeth (TS 21). Crest cells from the most caudal part of the rhombencephalon migrate into the ventral part of the fourth pharyngeal pouches where they will form the ultimobranchial bodies, one component of which, the C-cells, first amalgamate with the thyroid gland, and then become disseminated within it (see also *the branchial arches*).

Rhombencephalic crest cells also contribute to some of the cranial nerve ganglia, as well as neurons and glial cells in the parasympathetic ganglia of cranial nerves VII, IX and X, and to some neurons and all glial cells in the sensory ganglia of cranial nerves V, VII, VIII, IX and X (TS 13–15). By direct extrapolation from observations in avian embryos, it is believed that a proportion of the sensory neurons from some of these cranial nerves probably arise from surface ectodermal placodes, such as from the otic placode, their origin accordingly being similar to the sensory neurons of the olfactory system.

It is now clear that the crest cells that migrate into each pharyngeal arch are derived from distinct rhombomeres of the hindbrain neural tube (for review, see Gilbert, 1997), and that their fate is determined prior to migration. This situation apparently contrasts with that in the trunk region where the fate of the neural crest cells is mainly determined by the local conditions that they encounter during their migration and in their final destination.

The occipital region and spinal cord

Here, the neural crest gives rise to peripheral sensory neurons and their associated glia; the cell bodies of the dorsal root ganglia, and the cell bodies of the sympathetic and parasympathetic motor neurons located in their corresponding ganglia.

In the occipital region, it is believed that the neural crest cells lay down a connective tissue substratum during their migration which serves to guide the somitomere-derived myoblast cells when, at a slightly later stage, they migrate into the pharyngeal arches (see Larsen, 1993). Similarly, the fate of the myoblast cells is apparently not predetermined, but is dependent on the neural crest cells that they encounter during their migration, and at their final destination (see Larsen, 1993).

The sensory ganglionic cells within the dorsal root ganglia transmit impulses received from the periphery towards the spinal cord from end organs in the viscera, body wall and extremities. A pair of dorsal root ganglia develop at each segmental level along the spinal axis except (usually) at the first cervical and at the second and third coccygeal levels (and more caudally). The primordia of the most rostral pair of cervical dorsal root ganglia form at about TS 15, and the ganglia themselves are first seen by TS 16. The other ganglia form in a cranio-caudal succession, with the majority being present by about TS 18/19.

Teillet and Le Douarin (1983) have carried out excision and transplantation studies using labelled neural crest cells in chick–quail chimeras to investigate the influence of the notochord and neural tube on the development of the dorsal root ganglia. They found that both dorsal root ganglia and sympathetic ganglia depended on the survival and differentiation of somite-derived structures: in the absence of the notochord and neural tube, the somite-derived cells died rapidly, as did the neural crest cells that were present in the somitic mesoderm at that time. By contrast, those neural crest cells that reached the mesenchymal wall of the aorta, suprarenal glands and the gut survived and developed normally into neural and paraganglion cells.

In the cephalic region, neural crest-derived and placode-derived sensory ganglia of the head develop as cephalic mesenchyme and are not affected by the removal of the notochord or neural tube. This suggests that the peripheral ganglia are differentially sensitive to the presence of the notochord and neural tube. The presence of the neural tube allows both spinal and sympathetic ganglia to develop in the absence of the notochord. If the neural tube is removed, and the notochord left intact, spinal ganglia fail to develop, whereas sympathetic ganglia can develop.

The innervation of the enteric nervous system of the gut derives from two distinct neural-crest populations (Durbec *et al.*, 1996). A pool of crest cells from post-otic hindbrain give rise to most of the enteric nervous system (and the superior cervical ganglion), while a second lineage from trunk neural crest forms the more posterior sympathetic ganglia and also contributes to the foregut enteric nervous system. These observations thus supplement the earlier work of Pomeranz *et al.* (1991) who showed that crest cells originating in the occipito-cervical (vagal) and sacral regions migrate into the wall of the gut to innervate

TS	1	2	3	4	5	6	7	8	9	10	11	12	13	14	15	16	17	18	19	20	21	22	23	24	25	26
E	0	1	2	3	4	4.5	5	6	6.5	7	7.5	8	8.5	9	9.5	10	10.5	11	11.5	12	12.5	13.5	14.5	15.5	16.5	17.5

the entire length of the gastro-intestinal tract, forming the parasympathetic ganglia (TS 22) and supplying parasympathetic motor innervation to the viscera (TS 22).

The sympathetic autonomic ganglia form as a pair of ganglia associated with each of the thoracic and the upper three lumbar segmental units. In the cervical region, only three pairs of sympathetic ganglia are usually observed, though the lower cervical and first thoracic ganglia may be amalgamated together to form the cervico-thoracic or stellate ganglion (TS 21), while, in the thoracic region, a single pair of ganglia form at each segmental level. This arrangement is termed the thoraco-lumbar outflow from the spinal cord. The first of the thoraco-spinal sympathetic ganglia are seen at about TS 17, and the cervical sympathetic ganglia at about TS 21.

Additional peripheral sympathetic ganglia from the thoraco-lumbar outflow are located close to the major branches of the dorsal (abdominal) aorta in association, for example, with the origin of the coeliac trunk and the superior mesenteric, renal and inferior mesenteric arteries (at about TS 25).

The spinal neural crest also gives rise to non-neural elements, these include the pia and arachnoid coverings of the spinal cord (TS 21), as well as the glial cells associated with the various ganglia indicated above. Some neural crest differentiates into the Schwann cells which form the neurolemma, the thin membranous outer covering of the peripheral nerves. Neural crest also differentiates to form the neurosecretory chromaffin cells of the adrenal medulla (TS 22), and the neurosecretory cells associated with the heart and lungs. These cells also differentiate into the melanocytes (pigment cells) of the skin, and components of the aortico-pulmonary spiral septum of the outflow tract of the heart (TS 16) (see the relevant section of *The Heart and its Inflow and Outflow Tracts*).

3.2 Somites and their derivatives (muscles, dermis and vertebrae)

Introduction

At about TS 11 and following gastrulation, the paraxial mesoderm on either side of the embryo segregates from the rest of the mesoderm in a position lateral to the neural tube and the axial notochord and medial to the intermediate and lateral mesoderm. In the head region, this mesoderm forms 7 poorly defined **somitomeres**. A stage later, the first of what will be some 60 well-defined **somites** starts to form in the body and develop in a cranio-caudal sequence. This process continues until about TS 22 or TS 23 with the somites extending from what will be the base of the skull to the tip of the tail. The head somitomeres give rise to the musculature of the head and branchial arches, while the somites differentiate to give rise to the segmentally arranged **sclerotomes** and the associated **dermomyotomes**. The sclerotomes eventually form the vertebral axis while the latter separates into **myotomes** that will provide almost all the body and limb muscles and **dermatomes** that will form the dermis of the skin.

One interesting aspect of the somites is that they represent an evolutionary segmental pattern along the entire post-cranial part of the embryonic axis, although its only obvious remnants at birth are the vertebral elements and the segmented spinal nerves and the ribs which are (usually) confined to the thoracic region [very occasionally, "accessory" ribs may be found associated with the lowest of the cervical vertebrae]. It should be noted that there is also some evidence of segmentation in the anterior abdominal musculature, while the cutaneous nerve distribution (i.e. to the skin) clearly displays evidence of its segmental origin in the form of the dermatome pattern (see *Nerves* section of *The Limbs* chapter).

In this chapter, we first descibe the segmentation of the paraxial mesoderm, paying attention to its progressive differentiation caudally down the embryo, and then consider separately the development of the dermis, the musculature and the vertebral axis.

Somite formation and breakdown

Segmentation within the paraxial mesoderm starts in the head region and proceeds caudally as the anterior region of the unsegmented paraxial mesoderm in the posterior region of the embryo forms new somites every two hours or so, at least initially. The first sign of this segmentation is the appearance of **somitomeres** in the most rostral region of the embryo. These are loose aggregates of mesenchyme which are not easy to visualize, particularly in mouse embryos, and can be viewed as the first stage in the formation of the segmented somites (Tam and Meier, 1982; Tam, 1986).

In mice, somitomeres start to form on either side of the notochordal plate at the late primitive streak stage (TS 11), and extend caudally as a series of discrete units. At the late primitive streak stage, four somitomeres are present in the paraxial mesoderm on either side of the embryonic axis. These give rise to the cranial segments and head mesenchyme of neurulating embryos (Meier and Tam, 1982). In all species studied, the first seven pairs of somitomeres that form from head mesoderm do not go on to form discrete somites, but instead give rise to the striated muscles of the face, and those associated with the jaw and the pharyngeal arches (such as the extrinsic ocular muscles, the muscles of facial expression, the musculature of the tongue and the muscles of mastication).

In the region caudal to the head and initially in the prospective occipital and upper-cervical region (TS 12), the somitomeres soon form well segmented somites. The first group of the definitive somites to differentiate are thus believed to derive from the eighth, ninth, tenth and succeeding somitomeres, and appear in a strict cranio-caudal sequence every 2 hours or so, although the thoracic somites form a little faster and the tail ones a little slower (see Table 3.2.1). Somite formation continues until about E14.5 (TS 22/23), when a maximum of about 60 pairs will have been formed, extending caudally along the entire length of the post-cranial part of the

TS	1	2	3	4	5	6	7	8	9	10	11	12	13	14	15	16	17	18	19	20	21	22	23	24	25	26
E	0	1	2	3	4	4.5	5	5	6	6.5	7	7.5	8	8.5	9	9.5	10	10.5	11	11.5	12	12.5	13.5	14.5	15.5	16.5	17.5

Table 3.2.1. *Typical somite numbers*[1]

Theiler stage	Occipital	Cervical	Thoracic	Lumbar	Sacral	Tail[2]
TS 12 (E8)	4					
TS 13 (E8.5)	4	6				
TS 14 (E9)	4	7[3]	6			
TS 15 (E9.5)		7	13	1		
TS 16 (E10)		7	13	6	2	
TS 17 (E10.5)		7	13	6	4	4
TS 18 (E11)		PCC[4]	13	6	4	8
TS 19 (E11.5)		CC	PCC	6	4	12
TS 20 (E12)		cartilage	CC	PCC	PCC	16

[1] The total number of somites varies slightly even among embryos of the same developmental stage, although an individual is unlikely to have a number of somites that differs from that given here by more than two. The figures given here include the presence of dermomyotomes.

[2] Tail region numbers include somites whose sclerotomes will form the coccyx. Counting in this region can only be approximate. It is not clear how long somitogenesis continues in the tail.

[3] Somites start to break up into dermomyotomes and sclerotomes at about TS 14, or some 20 hours after they have first formed.

[4] In the cervical region, sclerotome cells condense to give precartilage condensations (PCC) that chondrify (CC) and then ossify, so giving, for example, the axis and atlas vertebrae in the neck region.

embryonic axis on either side of the neural tube and notochord. For additional observations on the differentiation of the somites, see Bellairs (1963), Tam (1981), Tam *et al.* (1982) and Meier and Tam (1982), and for an overview of this topic, see Bellairs *et al.* (1986), while their relationship to *Hox* coding and to adjacent tissues is discussed in Holland and Hogan (1988).

Shortly after it has formed, each somite undergoes a mesenchyme-to-epithelial transition and a central cavity, the myocoele, appears within the epithelial ball, and within which there may be a few cells. Under the inductive influence of neighbouring tissues (for review, see Gilbert, 1997a), the epithelialized somite becomes partitioned into presumptive sclerotome (ventromedial), myotome (dorsomedial) and dermotome (lateral) regions. The sclerotome cells then revert to being mesenchymal, breaking away from the epithelialized somite and migrating around the neural tube where they will form vertebrae. The myocoele is then eliminated and it is of interest that, in the chick at least, its cells then give rise to the ribs and the peripheral parts of the vertebral disc (Huang *et al.*, 1994, 1996). The remaining two components, the dermatome and the myotome, then form an epithelial bilayer, the dermomyotome. The cells on the medial aspect of the dermomyotome are tightly packed and give rise to the **myotome** (or muscle plate), while the more laterally located cells give rise to the **dermatome** (or skin plate). The myotome cells actively migrate and form most of the striated muscle of the body including that within the limbs while the cells of the dermotome will form the dermis of the skin. The time between the initial formation of a somite and its partitioning into a sclerotome and a dermomyotome is about 20 hours (see Figure 3.2.1 a–c).

The myotome and musculature

The cells of the **dermomyotome** form the dermis and body wall musculature, with the **myotome** cells of the lower cervical and upper thoracic region migrating into the forelimb bud to form the limb and pectoral girdle musculature. Similarly, myotome-derived cells from the lower lumbar and upper sacral regions migrate into the hindlimb bud to form the hindlimb and pelvic girdle musculature. In the mid-cervical region, some somite-derived myoblast cells migrate into the pleuroperitoneal membranes which then differentiate into the majority of the musculature of the diaphragm (see above), and accordingly receive their innervation from C3–5 (the phrenic nerves). The more caudal (sacral and coccygeal) myotome derivatives form the musculature of the tail.

The epithelial cells of the myotome soon revert to being mesenchymatous in character and, from about TS 18, the myotome cell masses become divided into a dorsal *epaxial* portion and a ventro-lateral *hypaxial* portion (or hypomere). The cells then migrate either dorsally, where they become closely related to the neural tube forming the precursors of the medial and lateral dorsal column muscle masses, or ventrally where they migrate into the body wall to form its associated layers of musculature (in the trunk region, for example, they form the characteristic inner, intermediate and outer layers of body wall musculature, and more ventrally the rectus muscles). The stages in their sequential cytodifferentiation from pre-myoblast (or presumptive myoblast) cells into myoblasts, then myotubes and eventually into fully differentiated muscle fibres is illustrated in Figure 4.2.7.

The mesenchyme between the epaxial and hypaxial muscles forms an intermuscular septum, which is attached to the transverse processes of the developing vertebrae, and effectively separates the two muscle masses completely. Once the transverse processes have formed, rib elements are laid down in the

Figure 3.2.1a–c. The differentiation of the paraxial mesoderm. Initially, paraxial mesoderm segregates to give discrete somites (a.1) within which a cavity, the myocoele, subsequently forms (a.2), and this subdivides them into an outer part that will become the dermomyotome, and an inner part that will become the sclerotome. The former gives the dermal structures and postural and body-wall muscle, while the cells of the latter migrate medially and surround the neural tube and notochord (a.3). (b) The myotome gives rise to two principal muscle masses: the most dorsally located give the paravertebral or epaxial muscle masses which are principally involved in the maintenance of posture, while the more ventral part differentiates to form the body wall musculature (the hypaxial muscles). The sclerotome-derived mesenchyme cells initally form the mesenchyme model and then the cartilage model for the vertebra (see Figure 3.2.2). Later, ossification centres form at various sites within these cartilage models that will amalgamate toether to form the definitive vertebra, with its body, neural arch and spine, transverse processes, etc. The dermal derivatives of the dermomyotome are not shown here.

Note that each vertebra has a basic organization that derives from the relationship between the somites and the spinal cord. Each somite, and hence a sclerotome, is associated with a "segmental unit" of the neural tube, from which emerges a pair of mixed (*i.e.* with motor and sensory roots) spinal nerves, with each having a ganglion associated with its dorsal root (c). This primitive arrangement is modified when the rostral half of each sclerotome mass fuses with the caudal half of the preceding sclerotome mass. This arrangement allows (i) the segmental spinal nerves to emerge between the definitive vertebral bodies (at the level of the intervertebral discs), and (ii) the somite-derived muscle masses, that are subsequently attached to the vertebral bodies, to act across the intervertebral joints, and thus facilitate vertebral movement. (see Figure 3.2.2a,b).

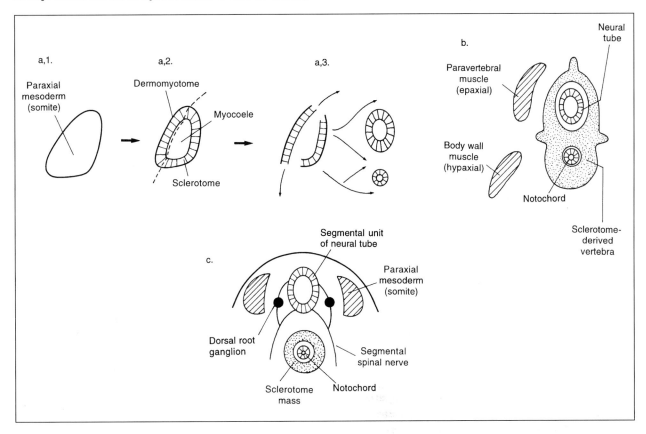

sclerotomic tissue (TS 19), and these extend ventrally into the region of the intersomitic clefts. At about this time, the epaxial muscles give rise to the extensors of the vertebral column. Further subdivision into medial and lateral groups occurs, giving rise to the short oblique and longer muscles of the back, respectively. The hypaxial group in the thoracic and lumbar regions give rise to the flexors of the vertebral column. In the thoracic region, the intercostal muscles form between the ribs, while in the abdominal region the muscles are present in the form of sheets. The most ventral component of this muscle mass forms the rectus group of muscles (for a detailed discussion of the fate of the myotomes in the human, see Hamilton and Mossman, 1972, where it is argued that the innervation of muscles in the adult provides clues to their embryonic origin, since the adult muscles retain their innervation regardless of how far they migrate during development).

The somite-derived mesenchyme cells that migrate into the limb girdles and limb buds differentiate *in situ*, and, at the myotube stage, they are segmentally innervated by spinal nerves. The nerve supply to the dorsal epaxial portion and ventro-lateral hypaxial portion (or hypomere) is from the dorsal (posterior) primary ramus and ventral (anterior) primary ramus, respectively. (For observations on the *principles* underlying the pattern of innervation of the muscles within the limbs, see the relevant section of *The limbs*.)

TS	1	2	3	4	5	6	7	8	9	10	11	12	13	14	15	16	17	18	19	20	21	22	23	24	25	26
E	0	1	2	3	4	4.5	5	6	6.5	7	7.5	8	8.5	9	9.5	10	10.5	11	11.5	12	12.5	13.5	14.5	15.5	16.5	17.5

The cranial and cervical musculature

Any consideration of the embryonic origin of the musculature in the cranial and neck regions must take into account information on the existence and fate of the ill-defined *somitomeres* which derive from rostral paraxial mesoderm as well as on the myogenic potential of the prechordal plate. It is now generally accepted that the main source of cranial musculature is from this paraxial mesoderm, and that some of the cells that give rise to the extrinsic ocular muscles pass through the prechordal plate. It has been suggested (Carlson, 1996) that the eventual fate of the cranial and cervical muscles depends on whether the myogenic cells in this region migrate through mesoderm-derived or neural-crest-derived mesenchyme. The origins of the various cranial and cervical muscles, and their innervation are presented in Table 3.2.2. Much of the information on the origin of the muscles in this region is necessarily *indirect*, and based on their innervation in the adult.

Origin of the non-somite-derived musculature

The non-somite-derived musculature derives from the *lateral plate mesoderm*. Shortly after gastrulation, intra-embryonic mesoderm is compartmentalized into paraxial, intermediate and lateral plate components, with the latter splitting to form the splanchnopleure and the somatopleure. The latter is itself subdivided into two components (*i.e.* into extra-embryonic and embryonic parts). That part beyond the margins of the presumptive embryo is continuous with the extra-embryonic mesoderm and covers both the amnion and yolk sac and, at the advanced egg cylinder stage, forms the lining of the extra-embryonic coelomic cavity – it plays no part in the formation of the embryo. The embryonic component of the somatopleure, however, gives rise to the inner lining of the body wall and to the vasculature of the limbs (Hamilton and Mossman, 1972).

Where the lateral plate splits, the space extends forwards within the embryonic mesoderm to form the intra-embryonic coelom. This eventually becomes horse-shoe shaped when the spaces from the two sides meet across the midline, rostral to the bucco-pharyngeal membrane, in the future pericardial region (see Figure 2.4.3). Within the embryo, the cavity is lined by the intra-embryonic parietal, or somatopleuric, mesoderm where it is in contact with the overlying ectoderm, and by the intra-embryonic visceral, or splanchnopleuric, mesoderm where it is in contact with the subjacent endoderm, and both can give rise to muscular tissue. Differentiation of the splanchnopleuric mesoderm in the pericardial region results in the formation of the *cardiogenic plate*, and subjacent to this the endothelial heart tubes soon appear. The cells of the cardiogenic plate differentiate during the early somite period into cardiac muscle, a special form of skeletal muscle whose characteristic features are that it is branched and in the form of a syncytium in that there is continuity between the sarcomeres through **intercalated discs**, an arrangement that facilitates the organized propagation of the cardiac contraction (Borysenko and Beringer, 1989). The endothelial heart tubes are continuous with the rest of the vascular system, and the smooth muscle that surrounds them and forms part of their wall, is also of somatopleuric origin (see Chapter 4.1, *The heart*).

All portions of the primitive gut (including the lower respiratory tract) and the yolk sac consist of a lining of endoderm which is covered by a layer of splanchnopleuric mesoderm which also covers any mesentery present. The intra-embryonic component of the latter gives rise to the longitudinal and circular

Table 3.2.2. *Origin of the cranial and cervical muscles*

Mesodermal origin	Muscles	Innervation (motor)
Somitomeres 1–3 and/or prechordal plate	All extrinsic ocular muscles except lateral rectus (see below)	Cranial nerves III, IV
Somitomere 4	Jaw-closing muscles (muscles of mastication)	Cranial nerve V (mandibular division)
Somitomere 5	Extrinsic ocular muscle – lateral rectus	Cranial nerve VI
Somitomere 6	Jaw-opening muscles + other second pharyngeal arch muscles (muscles of facial expression)	Cranial nerve VII
Somitomere 7	Third pharyngeal arch muscle (stylopharyngeus)	Cranial nerve IX
Occipital somites 1 and 2	Intrinsic laryngeal muscles	Cranial nerve X (recurrent laryngeal branch)
Occipital somites 2–4	Constrictors of pharynx	Cranial nerve X (superior laryngeal branch)
Occipital somites	Intrinsic muscles of tongue	Cranial nerve XII
Occipital somites	Muscles of neck region (sternomastoid, trapezius)	Cranial nerve XI (motor) and cervical plexus (C2–C4, sensory)

muscular and fibrous coats of the gut and to its covering of visceral coelomic mesothelium (visceral pleura and peritoneum). The muscle associated with the gut is of the smooth (or involuntary) variety, as is the **trachealis muscle** that bridges posteriorly the gap in the C-shaped cartilaginous tracheal "rings". Although the origin of the smooth muscle of the embryonic and adult urogenital tract (*e.g.* the myometrium, the wall of the ureter) is still unclear, most of the smooth muscle of the viscera develops *in situ* from the mesoderm-derived mesenchyme that surrounds these various organs.

There are a few exceptions: one may be the smooth muscle of the iris (particularly the *sphincter pupillae*) which is probably of neural crest (*i.e.* ectodermal) origin, although the *ciliary muscle* of the eye develops from the mesoderm surrounding the optic cup whose origin has both neural-crest and head-somitomere components. Another may be the myo-epithelial cells of the ducts of the sweat glands which are also believed to have an ectodermal origin. For further details of the differentiation of the lateral plate mesoderm, see any standard text-book of embryology; for observations on the histogenesis of the various types of muscle and fasciae, see Hamilton and Mossman (1972); and for more general observations on the role of the lateral plate mesoderm in the differentiation of the heart and vasculature, see Gilbert (1997).

The dermotome and dermis

Following the differentiation of the myotome, the cells of the **dermatome** lose their epithelioid character, become mesenchymatous and intermix with the somatopleuric mesenchyme and with other cell types to form the dermis and subcutaneous tissue (they also produce collagenous and elastic fibres). Interactions between dermal and epidermal cells allow epidermal appendages, such as hairs, to form. The muscles associated with the hair follicles (*arrectores pilorum*) are of mesenchymatous (dermatomal) origin and develop independently in the dermis, only later making contact with the root sheath of the hair follicle. Neural crest-derived cells also invade the epidermal/dermal region and become widely distributed within it. These cells develop extensive processes and are often termed *dendritic cells*. They appear in the hair roots during the early postnatal period.

The sclerotomes and vertebral axis

The cells of the **sclerotome** migrate medially and surround the neural tube and notochord and subsequently become partitioned into cranial and caudal halves, with the caudal part of one sclerotomal unit joining with the rostral part of the next to form individual vertebrae (see below). The first four pairs of somites are believed to form the occipital part of the base of the skull alone, while those of the next seven pairs of somites differentiate to give the atlas, the axis and the cervical vertebrae. In the thoracic region, the sclerotomal units form the vertebral bodies, and lateral extensions of their transverse process give rise to the ribs and costal cartilages with there being a possible role here for any cells in the myocoele (see above).

Each vertebra has a basic organization that derives from the relationship between the somites and the spinal cord. Each somite is associated with a "segmental unit" of the neural tube, from which emerges a pair of mixed (*i.e.* with motor and sensory roots) spinal nerves, with each having a ganglion associated with its dorsal root. Equally, a corresponding sclerotome mass is present at each segmental level. This primitive arrangement becomes modified when the rostral half of each sclerotome mass fuses with the caudal half of the preceding sclerotome mass. This arrangement allows (i) the segmental spinal nerves to emerge between the definitive vertebral bodies (at the level of the intervertebral discs), and (ii) the somite-derived muscle masses, that are subsequently attached to the vertebral bodies, to act across the intervertebral joints, and thus facilitate vertebral movement (see Figure 3.2.2a, b).

The upper cervical vertebrae

As mentioned, the first four somites do not form dermomyotomes and sclerotomes, and are lost just a day or so after they form (Table 3.2.1). Their cells are believed to contribute to the cartilaginous precursor elements of the **occipital bone**, a component of the base of the skull (see Figure 4.8.3 for a diagram of the elements that make up the occipital bone). The rostral half of the sclerotome of the fifth somite is thought to form the first cervical vertebra (C1, the **atlas**). The anterior arch of C1 consists of the **ossified central part of the triradiate ligament** (or **hypochordal bow**), but this vertebra lacks a centrum (or vertebral body) because it has always been believed that the primitive centrum of C1 becomes a discrete unit (the **odontoid process** of C2), whose caudal surface (normally) fuses with the rostral surface of the centrum of C2. Recent 3-D reconstructions of the human craniovertebral junction of embryos/fetuses between 45–77 d reveal that the odontoid process appeared to develop from a short cartilaginous eminence of the axis (David *et al.*, 1998). This finding suggests that, although the body of C2 incorporates the centra of C1

TS	1	2	3	4	5	6	7	8	9	10	11	12	13	14	15	16	17	18	19	20	21	22	23	24	25	26
E	0	1	2	3	4	4.5	5	6	6.5	7	7.5	8	8.5	9	9.5	10	10.5	11	11.5	12	12.5	13.5	14.5	15.5	16.5	17.5

Figure 3.2.3a, b. (a) Superior view of the three "centres" of ossification present in a "typical" vertebra at around birth. Note that one is located in each half of the neural arch, and one in the centrum of the vertebra, the site of the future vertebral body. A single large cartilaginous element is present at the site of the future spinous process that unites (dorsally) the two halves of the neural arches, and two smaller cartilaginous elements that are located (ventrally) between the pedicles of the neural arches and the equivalent dorso-lateral parts of the vertebral body. The arrows indicate the direction of spreading of ossification during the post-natal period. (b) Superior view of a "typical" lumbar vertebra with its various parts named.

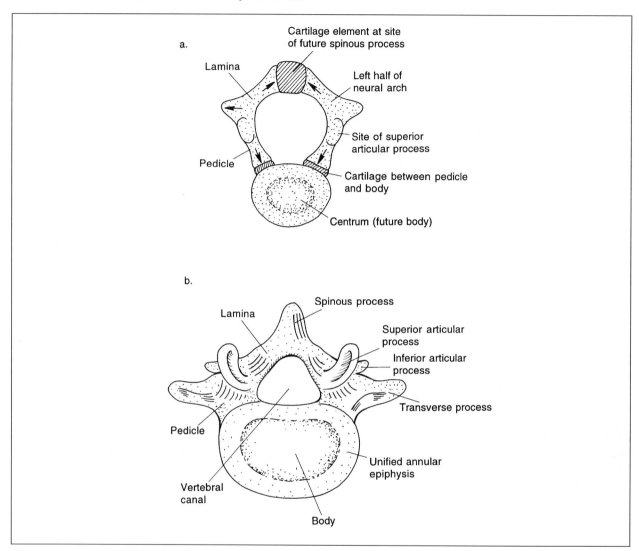

although the articulations between the primordia of the costal cartilages and the sternum, and the cartilage primordia of the sternal elements are not easily seen before about TS 22/23.

Ossification proceeds laterally and ventrally until the entire rib primordium becomes ossified. By about E14.5 (TS 22), only the first 11 ribs usually show evidence of ossification, with ossification within the 12th and 13th ribs occurring about a day later (during TS 23). The most ventral part of the cartilage primordia of the ribs do not ossify, but differentiate to form the **costal cartilages**.

Differentiation of the sternum

In the thoracic region, where the ribs and costal cartilages form as lateral extensions of the transverse processes of the thoracic vertebrae, the growing ends of the costal elements make contact with the sternal bars (or bands) which, at this stage, consist of mesenchymal condensations. The latter are in contact at the rostral end of the thorax, where the body wall is complete, at the site of the future manubrial condensation. More caudally, the sternal bands are increasingly widely separated from each other. Fusion of the two sternal bands subsequently occurs in a cranio-caudal sequence. [When isolated and maintained in tissue culture, the two sternal bands move towards and fuse with each other (Chen, 1952a, b)].

The most rostral elements fuse to form a midline unpaired rudiment, the pre-cartilage primordium of the **manubrium**. As fusion occurs more caudally, pre-cartilage primordia of the individual **sternebrae**, and finally the **xiphoid process** (or xiphisternum) are seen. The costal cartilages articulate with the **sternum** in the inter-sternebral regions.

From about TS 22/23, each of the sternebrae are seen to possess two cartilaginous precursor elements, but by TS 24, these fuse across the midline to produce individual sternebral cartilaginous units, and it is within each of the latter that ossification centres are seen at about E17. An ossification centre is usually present within the fourth sternebra by about E 18. At the time of birth, the manubrium is usually seen to contain a single enlarged ossification centre (although Grüneberg (1963) indicates that the manubrium may possess two cartilaginous primordia). There are three "square-shaped" sternebral ossification centres located between the medial insertions of the 2nd and 3rd, 3rd and 4th, and 4th and 5th costal cartilages, while a smaller centre is present between the medial insertions of the 5th and 6th costal cartilages. A single large ossification centre is usually present which occupies the upper two-thirds or more of the xiphoid process, although occasionally two symmetrical centres may be present, one located on either side of the midline.

If the pattern of ossification of the sternum is abnormal (*e.g.* the ossification of one half of a sternebral centre is out-of-step with its partner), then the corresponding costal cartilages invariably insert into the sternum at different levels on the two sides, thus indicating that the regular segmental differentiation pattern of the sternal elements is in some way disrupted. Accordingly, both the ossification pattern and the costal-cartilages' insertion sites of the two sides become asymmetrical, representing an exaggerated version of the "crankshaft-sternum" described by Theiler (1989). This situation is particularly clearly observed in *Engrailed-1* "knockout" embryos (Wurst *et al.*, 1994), and after experimental amniotic-sac puncture carried out during E12.5–13.5 (Chang *et al.*, 1996). It seems likely that the latter procedure, by altering the intra-uterine equilibrium within the amniotic fluid compartment, has a disruptive effect on the normal pattern of segmentation of the sternal elements that causes a similar pattern of ossification to that seen in the *Engrailed-1* knockout mice. Whether the underlying mechanism is similar in these two situations has yet to be established.

Variations in the total number of ribs present may occasionally be encountered. The extra ribs are usually associated with the lowest of the cervical vertebrae, which then take on the characteristic features of thoracic vertebrae with articulations for the head and neck regions of their respective ribs. When this situation arises, the supernumerary ribs may be either unilateral or bilateral. It is very rare to see more than two cervical (supernumerary) ribs, and, when present, they either resemble the first thoracic rib or are even more diminutive than this rib. More frequently, bifid or fused ribs are encountered, and these are sometimes seen in association with an abnormal pattern of sternal ossification. For a detailed discussion of the incidence of abnormalities of vertebral, rib and sternal ossification in mice, see Grüneberg (1963).

The intersegmental vertebral vascular system

When 3-D reconstructions are made of the endothelial lining of the entire vascular system, it is possible to recognize the presence of the intersegmental arterial branches from the dorsal aortae from as early as about TS 14 (Kaufman *et al.*, 1997). At this stage, they extend caudally from the mid-trunk region, but are mostly restricted to the level of the forelimb bud. At slightly later stages, intersegmental arterial branches are seen to have extended both rostrally and caudally. These intersegmental vessels give off dorsal branches which vascularize both the developing neural tube and the somite-derived paravertebral or epaxial muscle masses (which will subsequently differentiate to form the deep muscles of the back). Cutaneous branches from these vessels supply the skin of the back, while the more substantial ventral branches supply the somite-derived hypaxial muscle masses.

In the thoracic region, the intersegmental vessels differentiate to form the intercostal arteries. More caudally, they form the segmental lumbar and sacral arteries, while, more rostrally, they form a series of vessels that branch from the left and right vertebral arteries, being directed on either side of the vertebral bodies. The two vertebral arteries anastomose to form the basilar artery, and branches from this vessel anastomose with rostral branches from the dorsal aorta that give rise to the internal carotid artery and its principal branches that supply the developing brain (for further details, see Chapter 4.6, *The brain and spinal cord*). In the sacral region, the intersegmental arteries become considerably modified with the linking of the 5th lumbar intersegmental arteries to the umbilical arteries. The principal branches from these vessels become the internal and external iliac vessels. The most caudal part of the dorsal aorta also becomes much reduced in size, forming the median sacral artery.

Where the limb buds form, the intersegmental vessels also become much modified and enlarged, giving rise to the principal axial vessels to the forelimb and hindlimb. These are derived from the 7th cervical and 5th lumbar intersegmental arteries, respectively (for further details, see Chapter 4.2, *The Limbs*).

The venous drainage of the vertebral axis is initially via the anterior and posterior cardinal veins, which later differentiate to form the components of the systemic venous system. The most obvious segmental remnants of the vertebral vascular system are seen in the thoracic region, namely the intercostal vessels (for further details, see Chapter 4.1, *The heart and inflow and outflow tracts*).

TS	1	2	3	4	5	6	7	8	9	10	11	12	13	14	15	16	17	18	19	20	21	22	23	24	25	26
E	0	1	2	3	4	4.5	5	6	6.5	7	7.5	8	8.5	9	9.5	10	10.5	11	11.5	12	12.5	13.5	14.5	15.5	16.5	17.5

3.3 The branchial arch system

Introduction

Here, we consider the development of various structures associated with the lower half of the face and neck region, which derive from the transitional structures collectively known as the **branchial** (also termed the **pharyngeal, aortic** or **visceral**) **arch system**. In all vertebrates with jaws (Gnathostomata), the first and second branchial arches are involved in the development of the upper and lower jaw and also give rise to components of the middle ear apparatus, while the third and succeeding arches were originally associated with the gill system in fish. The derivatives of this part of the branchial arch apparatus were formerly principally associated with gaseous exchange between the individual and its aqueous environment and were segmentally arranged on either side of the primitive pharynx. Then, as now, each pair of gills contained a cartilaginous bar whose main function was to provide support for the gill. In addition, each gill arch contained striated muscle masses that were segmentally innervated by a specific (cranial) nerve, as well as having its own arterial supply and venous drainage.

In higher vertebrates, where the branchial arch system no longer serves a respiratory function, evidence of its original segmental arrangement is only seen during early embryogenesis as the gill derivatives have become substantially modified to serve very different and often quite specialized roles. This is not altogether surprising, since, even in the early jawed fishes, the caudal part of the oropharyngeal region served to supply oxygenated water to the gills, while its most rostral part was adapted for the purpose of feeding. For this, the muscles of the first arch became specialized to form the muscles of mastication and played an essential role in the initial processing of the food. This early evolutionary modification has been maintained in higher vertebrates, and these animals also display the effects of specialization/adaptation with regard to the differentiation of all the more caudally located arches. The muscles of the second arch, for example, form the muscles of facial expression, while the first to fourth and sixth arches contribute to the formation of the tongue and larynx respectively, with the tongue muscles playing a critical role in deglutition, and both groups facilitating vocalization in some species (there is little evidence of development of the fifth branchial arch in higher vertebrates). As the tongue and thyroid gland develop in the floor of the mouth, the former from derivatives of the first to fourth arches and the occipital myotomes, and the latter in close association with the first arch, they are considered in detail elsewhere (see the *Mouth and nose region*).

All branchial arches have similar component structures, and this can be seen in the pair of diagrams that display a simplified lateral view together with a transverse section through the cephalic region of an early limb bud stage embryo which illustrates the general arrangement of the first branchial arch and its relationship to the oropharynx (Figure 3.3.1a,b). The general arrangement of the floor of the pharynx seen in an embryo with 4 pairs of pharyngeal arches is shown in Figure 3.3.1c,d. A branchial arch comprises an outgrowth of the surface (*i.e.* external) epithelium which is defined both rostrally and caudally by lateral grooves (or clefts) on the outer aspect of the embryo, complemented by a branchial pouch which is a laterally-directed, endodermally-lined extension (or outpouching) of the oropharynx. Mesenchyme is absent between the two epithelial layers in the site of the branchial membrane (where the branchial pouch and groove are apposed). Within the arches, mesenchyme is present, that within the first three arches being partly derived from neural crest (see *The neural crest*) with all of the arches containing additional mesenchymatous derivatives of paraxial mesodermal origin. Some of the latter are derived from head somitomeres, while others are derived from occipital somites (*e.g.* the future intrinsic muscles of the tongue). The branchial arch blood vessels are located within the mesenchyme, while a specific cranial nerve is associated with each arch.

The branchial arches are first readily recognized between 8 and 11 days of development, although their final differentiation to form their specialized derivatives is not complete until shortly before birth. The **maxillary component** of the first branchial arch, for example, is first evident during TS 12, the **mandibular component** of the first arch and the second (or **hyoid**) arch develop during TS 13, the third arch during TS 14, while the fourth and sixth arches only

THE BRANCHIAL ARCH SYSTEM

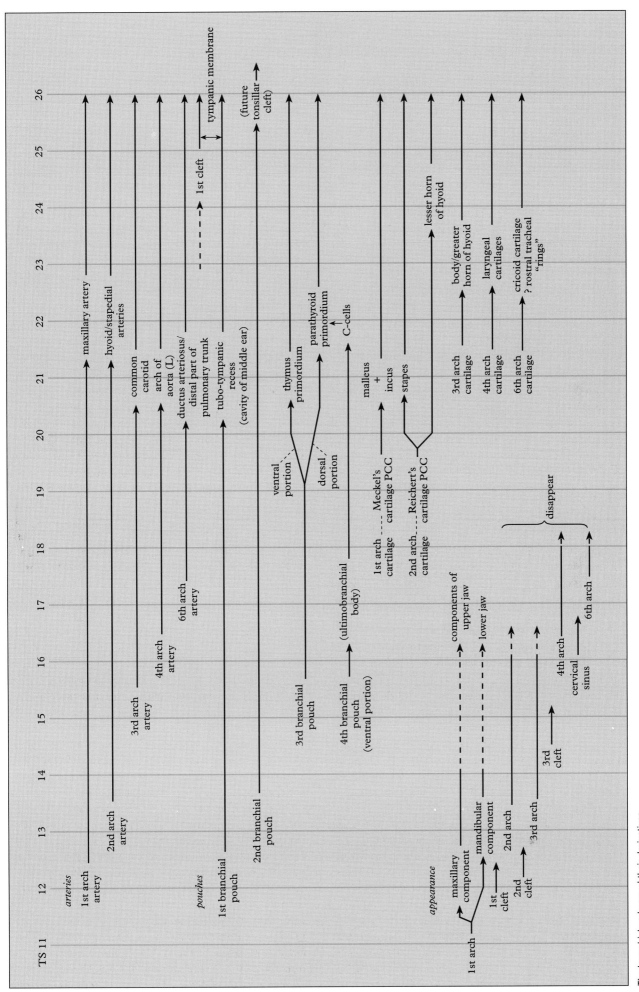

The branchial arches and their derivatives.

Figure 3.3.1 (a) Diagrammatic side view of the cephalic and thoracic regions of an embryo with the first two branchial arches present. The dotted lines indicate the region in which the cephalic neural folds are still widely open. At this stage, only the maxillary component of the first arch has developed. Two important landmarks present at this stage are the *optic placode*, a localized "thickened" area (*i.e.* showing early evidence of differentiation) of neural ectoderm on the inner aspect of the neural folds in the future diencephalic part of the future forebrain, and the *otic placode*. The latter is also a circumscribed region of "thickened" (*i.e.* cuboidal) cells, but is derived from *surface* rather than from *neural* ectoderm, and is located approximately opposite the junction between the future metencephalon (site of formation of the pons and cerebellum) and the myelencephalon (site of formation of the medulla oblongata). The horizontal line across the middle of this figure gives the approximate site of the transverse section in (b). Here, the neural groove in the future hindbrain region is clearly seen being bounded on either side by the neural-surface ectoderm boundary. The location of the otic placodes are also delineated, while the notochord is in the midline between the pharynx and the floor-plate of the neural groove. Within each branchial arch, only one of which is shown, may be found several structures: the first to appear is the arch artery, and slightly later its venous drainage. Only considerably later does the nerve supply enter the arch and even later does the cartilaginous bar differentiate within the mesenchyme.

(c) A diagrammatic side view of an embryo with four branchial arches, again showing as dotted circles the *optic vesicle* (the successor of the optic placode) and the *otocyst* (the successor of the otic placode). The horizontal line (arrowed at each end) gives the plane of (d). This view of the ventral surface of the foregut shows the arrangement of the branchial arches (note that the branchial arch cartilage would not appear until considerably later). On the outer aspect of the pharyngeal region of the embryo, indentations (the grooves or clefts) are present between each of the arches. On the pharyngeal surface, these indentations represent the location of the primitive branchial pouches. "Bulges" develop due to the proliferation of mesenchyme in the floor of each of the branchial arches, and these will in due course give rise to the various components of the tongue (see Figure 4.5.3). Only the pair of lateral lingual swellings deriving from the first arch are shown. As the pharynx narrows caudally, it becomes continuous with the oesophagus.

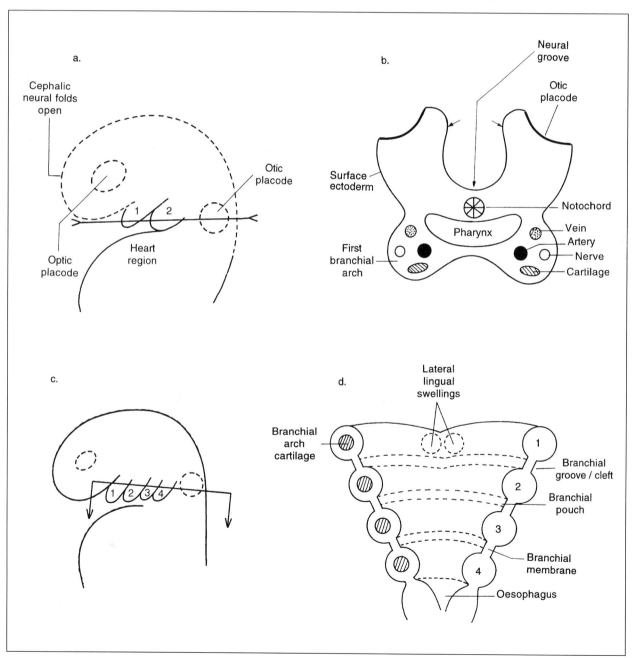

develop during TS 16/17. The first and second arches are separated externally by a deep ectodermally-lined groove (the first cleft), while the second groove develops between the second and third arches. The fourth and sixth arches are less well developed in higher vertebrates, and their corresponding externally-located grooves only poorly defined. The arteries are the earliest and most prominent structures seen in each arch, followed somewhat later by the corresponding veins; while the arteries are always large and clearly defined vessels, the tissues within the arches are drained by a relatively poorly defined venous network which only peripherally aggregate together to form larger vessels which drain into the anterior cardinal venous system. The nerves, pre-cartilage precursors and subsequently the cartilage "bars" themselves are not seen until considerably later in development. In the second arch, which is first seen at TS 13, the cartilage differentiates at TS 20, while the nerves do not appear until TS 20/21. The arch itself finally differentiates into its terminal structures at about TS 24.

The transformation of the various components of the primitive branchial arch system into the definitive arrangement seen in the adult is relatively complex and not well understood, and it is worth pointing out that the branchial arch components occasionally give rise to anomalous structures, many of which have been particularly noted in humans. For example, the branchial cleft (or groove) located externally between the second and subsequent arches may fail to close completely, giving rise to fistulae or sinuses, while anomalies may also arise due to the failure of migration or aberrant migration of cellular components. If cells fail to migrate from their initial site of origin, they may differentiate there (*e.g.* a "lingual" thyroid which may be found embedded in the dorsum of the tongue). If migration is incomplete, "rests" of cells may differentiate into ectopic organs or as cysts along their normal migratory pathway (*e.g.* a thyroglossal cyst may be found along the normal line of "descent" of the thyroglossal duct between the foramen caecum on the dorsum of the tongue and the normal location of its derivative, the thyroid gland). Alternatively, progenitor cells may migrate to an aberrant location (if, for example, the parathyroid glands fail to separate from the thymus as the latter descends into the chest, parathyroid tissue may be found in an aberrant site, namely in the superior part of the thoracic cavity (termed the superior mediastinum), in association with the thymus gland).

Here we consider in turn the derivatives of the branchial arch arteries, cartilages, nerve supply, grooves and pouches as well as the muscles which arise from myotome cells within the branchial arch mesenchyme (for summary, see Table 3.3.1).

The branchial arch arteries

The branchial component of the vascular system of the primitive embryo essentially consists of a bilaterally symmetrical system of arterial channels which develop in a cranio-caudal sequence, with each forming shortly after its corresponding branchial arch first appears. The heart is located in the thoracic region in the ventral midline and, at the early stages when the arches form, a single outflow vessel (the **ventral aorta**) is present. The dilated terminal portion of this vessel (formerly called the **aortic sac**) gives rise to the series of **branchial arch arteries** which are symmetrically arranged around the primitive oropharynx/foregut and join up on either side of the latter with a pair of **dorsal aortae**. The primitive symmetrical arrangement of aortic/pharyngeal arch arteries is displayed in Figure 3.3.2a. Figure 3.3.2b displays the arrangement seen at birth, while the definitive adult arrangement is displayed in Figure 3.3.2c. Since the first and second of the branchial arches were modified early in vertebrate evolution to form the basic elements of the mouth and components of the middle ear apparatus of jawed fish, their arch arteries were not, unlike the third and succeeding arch arteries, distributed to the true gills. In such lower vertebrates, these remaining more caudally located arch arteries divide into an afferent portion which carries blood to the fine capillary network within the gills, and an efferent vessel which carries the oxygenated blood to the dorsal aorta.

The development of the arches is very different in the higher vertebrates. First, these vessels pass through the branchial arches without breaking up into the fine capillary network characteristically seen in the gills. Second, the first one or two pairs, the most rostral of the branchial arch arteries, largely disappear by the time that the most caudally-located vessels have fully differentiated. Third, the fifth pair of branchial arches is never well developed in higher vertebrates and any arch arteries are transient structures. During vertebrate evolution, the remaining fourth and sixth vessels were extensively modified, and now form the large arteries of the neck and thorax. As it is difficult to isolate the differentiation of these branchial arch arteries from that of the arterial supply of the head and neck region, it seems sensible to consider the system as a whole here. It should also be noted here that the axial arteries to the forelimbs "ascend" during embryogenesis and become closely associated with the branchial arch system (though in developmental terms, their origin is from a quite separate source).

Before briefly describing the fate of the various branchial arch arteries, it should be noted that the common outflow tract from the primitive heart becomes subdivided by the **aortico-pulmonary spiral septum** into the **ascending aorta** and

TS	1	2	3	4	5	6	7	8	9	10	11	12	13	14	15	16	17	18	19	20	21	22	23	24	25	26
E	0	1	2	3	4	4.5	5	6	6.5	7	7.5	8	8.5	9	9.5	10	10.5	11	11.5	12	12.5	13.5	14.5	15.5	16.5	17.5

Table 3.3.1. *The branchial arch derivatives*

Tissue	1st arch	2nd arch	3rd arch	4th arch	6th arch
Nerve	Maxillary and mandibular divisions of trigeminal (V)	Facial nerve (VII)	Glosso-pharyngeal (IX)	Superior laryngeal branch of vagus (X)	Rec. laryngeal branch of vagus (X)
Artery	Terminal part of maxillary	Cortico-tympanic (*in adult*), stapedial (*in embryo*)	Common carotid	Arch of aorta, brachiocephalic	Ductus arteriosus, pulmonary trunk
N. crest cartilage	NCC maxillary: alisphenoid, mandibular: Meckel's + malleus and incus. Dermal mesenchyme: maxillary, zygomatic, squamous part of temporal, mandible	NCC stapes, styloid, stylohyoid lig., lesser horns and upper upper rim of body of hyoid	Lower rim and greater horns of hyoid	Laryngeal cartilage	Laryngeal cartilage
N. crest ganglia	Trigeminal	Geniculate, vestibular and accessory	Glosso-pharyngeal	Vagus	
Groove	External acoustic meatus and external ear				
Pouch	Middle ear, tympanic membrane, Eustachian tube	Tonsillar cleft and crypts of palatine tonsil	Parathyroid gland, thymus gland	Ultimobranchial bodies (para-follicular cells)	
Head mesoderm derivative	Muscles of mastication, myelohyoid, ant. belly of digastric, tensor tympani, tensor veli palatini (somitomere 4)	Muscles of facial expression, stapedius, post. belly of digastric (somitomere 6)	Stylopharyngeus (somites 7)	Muscles of pharynx, cricothyroid, Lev. veli palatini (Occipital somites 2–4)	Muscles. of larynx (Occipital somites 1 and 2)

pulmonary trunk (TS 16), respectively. While the former vessel differentiates to form the main outflow vessel from the left ventricle, the pulmonary trunk forms the main outflow vessel from the right ventricle. The arch arteries branch out from the distal part of the ventral aorta from a site formerly termed the aortic sac and now referred to as the distal or rostral part of the outflow tract (Moffat, 1959; see also Congdon, 1922).

The **first arch artery** largely disappears by about the time that the third arch arteries are fully differentiated (TS 16/17), and its only well-defined derivative is believed to be the **maxillary artery** (TS 19).

The **second arch** (or **hyoid**) **artery** also largely disappears by about the time that the fourth arch artery enlarges, and its only derivative is the **stapedial artery** (TS 22) although the ventral portions of both the first and second arch arteries may contribute to the development of the **external carotid artery** which is believed to be a sprout of the third arch artery (see Larsen, 1993).

The **third arch arteries** are of greater interest in that these vessels probably give rise to the distal part of the **common carotid arteries** (TS 21). In addition, they also give rise to the proximal part of the internal and external carotid arteries at the **bifurcation of the common carotid arteries** in the region of the **carotid sinus** and **carotid body**. The nerve supply to the latter region is from the glossopharyngeal (IX) nerve, the cranial nerve which specifically supplies the derivatives of the third branchial arch. The terminal branches of the third arch arteries become intracranial, and form the **ophthalmic**, the **anterior** and the **middle cerebral arteries** (TS 22).

The derivatives of the **fourth arch arteries** are also important as the **left branchial arch artery** gives rise to the **aortic arch** (TS 21) which extends distally from the rostral part of the outflow tract, the successor to the primitive ventral aorta, to the site of entry of the ductus arteriosus (this is termed the "preductal" part of the thoracic aorta). More distal to the site of entry of the ductus arteriosus, the "postductal" part of the "descending" thoracic/abdominal aorta is derived from the **left dorsal aorta**.

The **right fourth arch artery** by contrast forms the **brachiocephalic trunk** and possibly also the proximal part of the **right subclavian artery** (TS 20), though there may be a small contribution from the right dorsal aorta. As on the left side, the definitive subclavian artery is a derivative of the seventh (cervical) intersegmental artery which forms the principal (axial) arterial supply to the primitive forelimb.

The **fifth arch arteries**, if they develop, are only transient structures in higher vertebrates.

The **sixth arch arteries** give rise to the **right and left pulmonary arteries** (TS 19), while a distal

Figure 3.3.2a, b. These figures show how the primitive symmetrical arrangement of the aortic (pharyngeal) arch arteries seen in the hypothetical early mammalian embryo gradually evolves into the adult arrangement. (a) In the primitive symmetrical arrangement, the outflow tract from the heart is a single vessel, the ventral aorta, that gives a series of symmetrically arranged aortic arch arteries which pass dorsally on either side around the primitive pharynx, to unite (on either side) with a pair of dorsal aortae. In the hypothetical arrangement, 5 pairs of aortic arch arteries (1–4, then 6) are successively formed (the fifth pair seem not to form in higher vertebrates). Caudal to the heart, the paired dorsal aortae unite to form a single median (midline) dorsal aorta. The direction of blood flow in each of these vessels is shown in the diagram (with arrows), and the arterial blood supply to the forelimbs is via the *axial arteries*. The location of the two vagal trunks with their respective recurrent laryngeal branches is also shown.

During embryogenesis, some components of the aortic arch arterial system regress, disappear and reorganize. Those that disappear are indicated in (b) as dotted lines, and this is the arrangement present up to the time of birth. These include virtually all of the first and second arch arteries, leaving the *maxillary artery* and the *stapedial artery* as the only *named* derivatives of the first and second arch arteries, respectively. The third arch arteries give rise to the *common carotid arteries* including the region where these vessels branch to give the *internal* and *external carotid arteries*. The *carotid body* and *carotid sinus* are located here, and receive their nerve supply from the glossopharyngeal (IX) cranial nerve (the cranial nerve supply of the third arch). The rostral part of the internal carotid artery is derived from the dorsal aorta. The portion of the dorsal aorta between where it is joined by the third and fourth arch arteries, termed the *ductus caroticus*, disappears. The fourth arch artery *on the left* gives rise to the distal part of the ascending aorta, and the *arch of the aorta*, while that *on the right* gives rise to the *brachiocephalic* (innominate) *artery*, and possibly also the origin of the right subclavian artery. The sixth arch artery on the left gives rise to the *pulmonary trunk* (proximally) and, more distally, the *ductus arteriosus*. On the right side, the sixth arch artery largely disappears. Two small branches, one from each of the original sixth arch arteries, differentiate to form the left and right pulmonary arteries, that of the right side eventually taking origin from the pulmonary trunk. A segment of the right dorsal aorta between the site of insertion of the sixth arch artery and where it unites with the left dorsal aorta, also regresses and disappears. The two axial arteries, one to each of the forelimbs (originally the seventh pair of cervical intersegmental arteries) "ascend" to their definitive position with the rostral "ascent" of the forelimbs, and eventually form the subclavian arteries. These are not, however, components of the aortic arch system. The definitive relationship between the recurrent laryngeal branches of the vagal trunks on the two sides is now also seen to be asymmetrical; on the left side it hooks around the ductus arteriosus, while on the right side it hooks around the origin of the right subclavian artery.

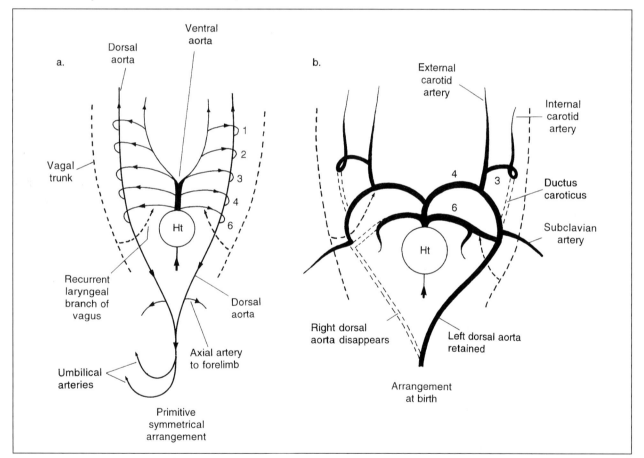

extension of the left arch artery additionally forms the **ductus arteriosus** (TS 21). It should be noted that the **pulmonary trunk**, which forms the main channel that drains the right ventricle after the outflow track of the heart has been subdivided by the **aortico-pulmonary spiral septum** (into the ascending aorta and pulmonary trunk), communicates with the distal part of the concavity of the arch of the aorta where it is continuous with the descending thoracic aorta, the derivative of left dorsal aorta

TS	1	2	3	4	5	6	7	8	9	10	11	12	13	14	15	16	17	18	19	20	21	22	23	24	25	26
E	0	1	2	3	4	4.5	5	6	6.5	7	7.5	8	8.5	9	9.5	10	10.5	11	11.5	12	12.5	13.5	14.5	15.5	16.5	17.5

Figure 3.3.2c. The principal features associated with the *adult* arrangement of the derivatives of the aortic arch arteries, following the changes that take place around birth. Note that the ascending aorta gives rise to the arch of the aorta from which a number of branches emerge, the most important of which are: the brachiocephalic (innominate) artery (which will give rise to the right subclavian and the right common carotid arteries), the left common carotid artery and the left subclavian artery. The right and left pulmonary arteries arise from the pulmonary trunk. A fibrous band, the *ligamentum arteriosum* now replaces the ductus arteriosus which closes off shortly after birth and subsequently fibroses. Note also the relationship between the recurrent laryngeal branches of the left and right vagus nerves (dotted lines). On the left side, the recurrent branch passes dorsally as a close relation of the ligamentum arteriosum, while on the right side it passes dorsally in the region of the junction between the brachiocephalic trunk and right subclavian artery.

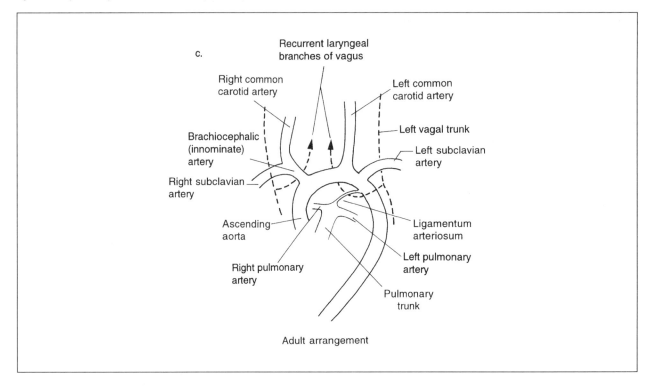

via the **ductus arteriosus**. This vessel is the terminal part of the left sixth arch artery beyond the origin of the arterial branch to the left lung bud, which subsequently differentiates to form the **left pulmonary artery** (TS 17). The ductus persists as a wide-calibred vessel until what is probably only a matter of minutes after birth, when it becomes completely occluded, and is eventually replaced by a fibrous band termed the **ligamentum arteriosum** (for discussion of the underlying mechanism(s) involved in the closure of the ductus arteriosus, see *The heart and its inflow and outflow tracts*).

On the right side, the communication between the sixth arch artery and the right dorsal aorta usually disappears beyond the origin of the arterial branch to the right lung bud, and this vessel subsequently differentiates to form the **right pulmonary artery**.

While the post-ductal part of the descending aorta is formed (as indicated above) from the left dorsal aorta, the right dorsal aorta normally completely disappears beyond its communication with the right fourth arch artery. The portion of both dorsal aortae between the points at which they are joined by the third and fourth arch arteries, termed the **ductus caroticus** (TS 17), also normally disappears shortly afterwards. The sequential changes that occur during the differentiation of the aortic (pharyngeal) arch arteries in the mouse between E9.5–15.5 are illustrated in Figure 3.3.3, and have been described in detail in the human embryo by Congdon (1922).

A segment of the right horn of the rostral part of the outflow tract persists and elongates to form the **brachiocephalic trunk**. This vessel terminates by giving origin to the right third and fourth arch arteries. The former vessel becomes the right **common carotid artery** and the commencement of the right **internal** and **external carotid arteries**, and (as indicated above) the stem of origin of the right subclavian artery. On both sides, the remainder of the internal carotid artery is derived from the rostral terminal part of the dorsal aorta. More caudally, the right dorsal aorta usually completely disappears.

The various abnormalities of the branchial arch arterial system that are occasionally encountered in the human, and may occasionally be induced experimentally in the mouse, (*e.g.* the presence of a double or right-sided aortic arch, a retro-oesophageal arch of the aorta and the situation where the right subclavian artery originates from the left side of the arch of the aorta beyond the origin of the left subclavian artery [see, for example, Kaufman, 1992]) can nearly all be explained in terms of the retention of components of the branchial arch arterial system that normally disappear, or the disappearance of vessels that are normally retained. The presence of these aberrant vessels may be associated with abnormalities of the outflow tract of the heart, and/or abnormalities of cardiac septation. Some of these aberrant vessels may

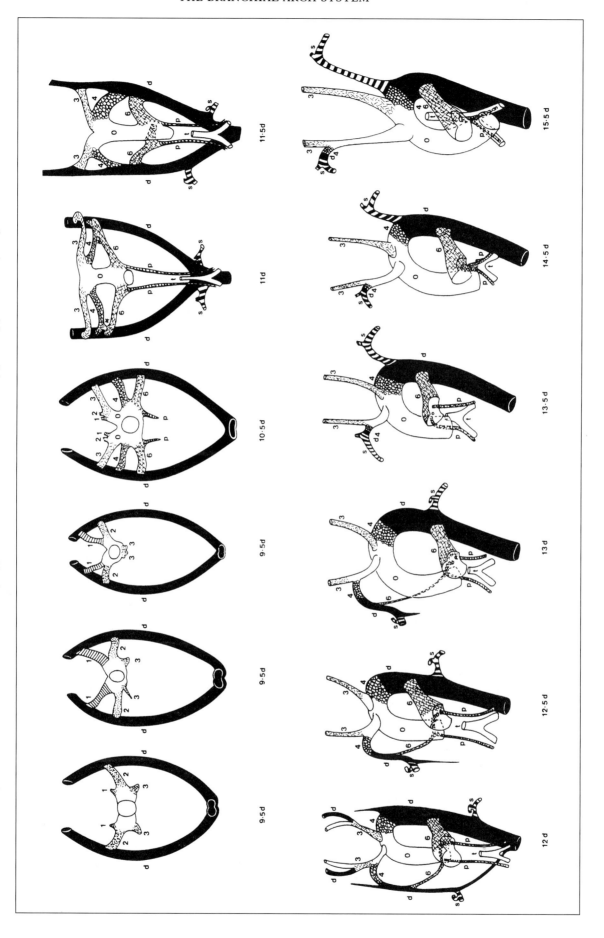

Figure 3.3.3. Sequential stages observed during the differentiation of the aortic (pharyngeal) arch arteries between E9.5 and E15.5. Key: 1. first aortic arch artery; 2. second aortic arch artery; 3. third aortic arch artery; 4. fourth aortic arch artery; 6. sixth aortic arch artery; d. dorsal aorta; o. derivative of ventral aorta (ascending part of arch of the aorta); p. pulmonary artery; s. subclavian artery (previously, lateral branch of seventh cervical intersegmental artery); t. trachea leading to the two main bronchi; *. (11d) possible location of the fifth aortic arch artery (transient) (from Kaufman, 1992).

TS	1	2	3	4	5	6	7	8	9	10	11	12	13	14	15	16	17	18	19	20	21	22	23	24	25	26
E	0	1	2	3	4	4.5	5	6	6.5	7	7.5	8	8.5	9	9.5	10	10.5	11	11.5	12	12.5	13.5	14.5	15.5	16.5	17.5

be explained as being due to the formation of either a partial or complete mirror image of the normal vascular arrangement.[1]

The branchial arch cartilages

Mesenchyme of neural crest origin initially condenses to form the pre-cartilage and subsequently the single cartilaginous elements which develop within each branchial arch, with the segmentation pattern observed in this area being generally associated with the corresponding pattern of homeotic (or Hox) gene expression in the rhombomeres within the hindbrain (see *The neural crest*).

The bony elements which differentiate to give rise to the **mandible** (TS 21) form by the process of intramembranous ossification both around but mainly lateral to the **first arch cartilage** (normally known as **Meckel's cartilage** (TS 19)). This is a transient structure over most of its length and, like the bones of the cranial vault (the calvarium, or skull cap), was originally a "dermal" bone in the primitive vertebrates. A second (caudally-directed) derivative of the first arch cartilage is the **sphenomandibular ligament** which passes from the **spine of the sphenoid bone** to the **lingula,** a small bony prominence located on the inner aspect of the mandible just anterior to the **mandibular foramen** through which pass the inferior alveolar vessels and nerve.

The **malleus** and **incus**, two of the **middle ear ossicles**, are also derived from the first arch cartilage (TS 19). The malleus represents the most dorsal extremity of Meckel's cartilage, while the incus develops from the chondrocranial element of the primitive upper jaw termed the **palatoquadrate** or **pterygoquadrate cartilage** which, though present in fish, has become much reduced during vertebrate evolution. It has also been suggested that the most anterior part of the pterygoquadrate cartilage may have become incorporated into the **greater wing of the sphenoid bone** (Hamilton and Mossman, 1972).

Within the mandibular process, two "membrane" bones (*i.e.* those formed by the process of intramembranous ossification) are laid down on either side of Meckel's cartilage using this element as a type of "model". The anterior and more laterally-located of these differentiates, as already mentioned, to form the mandible (TS 21), while the second, more medially-located "centre of ossification" differentiates to form the **tympanic plate** and subsequently fuses with the **squamous part of the temporal bone** and the cartilaginous **otic capsule**. In the upper part of the ramus of the mandible, secondary cartilage elements form, and their appearance precedes ossification; this region is associated with the formation of the **temporo-mandibular joint** (TS 25).

There is also a set of first-arch-derived skeletal elements located within the maxillary and mandibular processes that are formed by the direct ossification of membranous precursors derived from "dermal" mesenchyme of neural crest origin; these include the maxilla, premaxilla, the zygomatic and squamous portions of the temporal bone, the zygomatic, pterygoid, nasal, lacrimal and tympanic bones, the vomer and, as noted previously, the mandible. Together, these comprise the **membranous viscerocranium** of the skull (see *The development of the skull*).

The **second arch** (or **Reichert's**) **cartilage** (TS 20) gives rise to the third of the middle ear ossicles, the **stapes** (TS 20) as well as the **styloid process of the temporal bone** (TS 23), the **lesser horn** (or cornu) **of the hyoid bone** (TS 24) and the **stylohyoid ligament,** which connects the latter two structures. It also forms the **upper portion of the body of the hyoid bone** (TS 23).

The **third arch cartilages** differentiate to form the **greater horns** and the **lower portion of the body of the hyoid bone** (TS 23).

The derivatives of the **fourth** and **sixth arch cartilages** are the **laryngeal cartilages** (TS 23), consisting of the **thyroid, cricoid, arytenoid, corniculate** and **cuneiform cartilages**. While the first three branchial arch cartilages are of neural crest origin, those of the fourth and sixth arches are believed to originate from lateral plate mesoderm, although the literature is still ambiguous about their developmental history (for observations on the fate of the lateral plate mesoderm, see previous chapter). The **tracheal rings** (TS 23), and other components of the wall of the trachea probably derive from the splanchnic mesoderm which differentiates close to the ventral surface of the foregut.

The arrangements of the arch cartilage derivatives in the mouse are illustrated diagrammatically in Figure 3.3.4. In one of the earliest *avian* studies aimed at investigating the specification of skeletal tissues associated with each of the branchial arches, Noden (1983) undertook a series of excision and heterotopic transplantation studies to investigate neural

---

[1] In birds, it is the left side of the branchial arch system that undergoes partial regression rather than components of the right side as occurs in mammals, and as described in the main text. In the chick, six pairs of aortic arches are initially present, though these are not all present simultaneously because they appear in succession and some are only transitory. Initially, two paired dorsal aortae are recognized, though only portions of the aortae connecting the dorsal ends of the first three pairs of aortic arches persist as the dorsal carotid arteries, though the first two pairs of aortic arches largely disappear. Slightly later, on the left side, the entire segment of aorta between the third and sixth arches atrophies along with the left fourth arch: on the right side, however, the aorta degenerates only between the third and fourth aortic arches, both of which are retained. A continuation of both the fourth and sixth arch arteries of the right side persists, and these are involved in the formation of the arch of the aorta. The situation in the chick is clearly complex, and the arrangement apparently varies between different avian species (see Romanoff, 1960).

In reptiles, both dorsal aortae persist, but that of the left side is never as well developed as that of the right (Goodrich, 1930).

Figure 3.3.4. Diagrammatic outline of the head of a mouse to display the location of the principal derivatives of the branchial (pharyngeal) arch cartilages. Only those derivatives of the first four arches are normally described, and these are believed to be neural crest-derived.

The most important derivative of the first arch cartilage is *Meckel's cartilage*. This forms the "skeletal" element around which the membranous bone of the mandible develops; ossification occurs around, but principally lateral, to Meckel's cartilage. The intermediate part of the cartilage is replaced by the *anterior malleolar* and *spheno-mandibular ligaments*, and the *lingula*, a tongue-like protruberance on the inner aspect of the mandible that "guards" the entrance to the alveolar foramen (canal) through which passes the inferior alveolar nerve and blood vessels. From the lingula to the spine of the sphenoid bone passes the *spheno-mandibular ligament* (see above), whose most rostral fibres form part of the *anterior ligament of the malleus*. The two other named derivatives of the first arch cartilage are the *malleus* and *incus*, two of the middle ear ossicles. The named derivatives of the second arch cartilage (or *Reichert's cartilage*) are the *stapes* (the third of the middle ear ossicles), the *styloid process of the temporal bone*, the *stylo-hyoid ligament* (normally present as a fibrous band, but may ossify in some individuals (in the human, this is a feature of Eagle's syndrome)) and the *lesser horn* (or cornu) and upper part of the body of the hyoid bone. The *lower part of the body* and *greater horn* of the hyoid bone are the named derivatives of the third arch cartilage, while the *thyroid* and *cricoid cartilages* are the named derivatives of the fourth arch cartilage (although the cricoid cartilage may be of sixth arch origin; see Hamilton and Mossman, 1972).

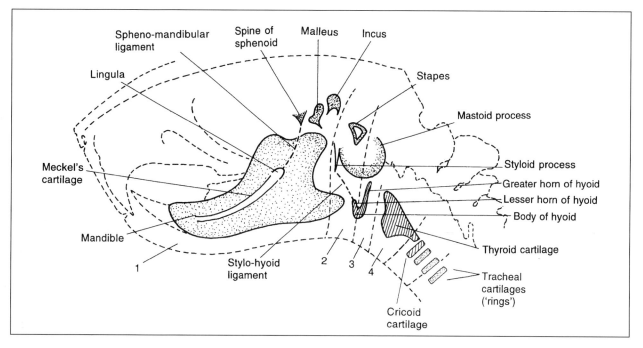

crest-derived connective tissues and mesodermal muscles. He concluded that the basis for patterning of the branchial arch skeleton and connective tissues resides with the neural crest population *prior to its emigration* from the neural epithelium, and is *not* determined within the pharynx or pharyngeal pouches. He was also of the view that the pattern of myogenesis from mesenchymal populations derived from paraxial mesoderm was also dependent upon properties inherent in the neural crest, a result more recently confirmed in the mouse (Trainor and Tam, 1994).

The nerve supply of the branchial arches

Each branchial arch has its own specific cranial nerve supply with both motor (or efferent) and sensory (or afferent) components. The motor nerves supplying the branchial muscles which, as mentioned earlier, derive from myotome cells that originate within the head somitomeres and the occipital somites are collectively termed **branchial efferent nerves**, while the sensory fibres from the dermis and mucous membranes of the craniofacial component of the head and neck region that largely derive from branchial mesenchyme pass back to the central nervous system via **branchial afferent nerves**.

The sensory nerve supply from the derivatives of the **first branchial arch** is from the **maxillary** and **mandibular divisions of the trigeminal (V) cranial nerve** (TS 20). Branches from these nerves supply the majority of the skin of the face as well as the teeth, the mucous membrane of the nasal cavities and paranasal sinuses, the palate and the mouth. They also mediate general sensation (*i.e.* touch) rather than special sensation (*i.e.* taste) from the anterior two-thirds of the surface of the tongue; the latter is mediated via the chorda tympani branch of the facial (VII) cranial nerve (see below). The skin and underlying dermis in the region of the forehead and most of the nose is of fronto-nasal rather than branchial arch origin, and is consequently supplied

TS	1	2	3	4	5	6	7	8	9	10	11	12	13	14	15	16	17	18	19	20	21	22	23	24	25	26
E	0	1	2	3	4	4.5	5	6	6.5	7	7.5	8	8.5	9	9.5	10	10.5	11	11.5	12	12.5	13.5	14.5	15.5	16.5	17.5

via the various branches of the ophthalmic (or first) division of the trigeminal (V) nerve.

The motor supply to the myoblast derivatives of the first branchial arch is via branches of the mandibular division of the trigeminal (V) nerve, and these innervate the **muscles of mastication** (temporalis, masseter, the medial and lateral pterygoids, mylohyoid, the anterior belly of the digastric, tensor tympani and tensor veli palatini). The tensor veli palatini is the only muscle to the soft palate supplied by the mandibular division of the trigeminal nerve, the rest are supplied by fibres from the accessory (XI) nerve which reach the pharyngeal plexus via the vagus (X) nerve.

The nerve supply to the derivatives of the **second branchial arch** is via branches of the **facial (VII) cranial nerve** (TS 21), which also has both sensory and motor components. The sensory component mediates special sensation (*i.e.* taste) and is distributed to taste buds located in the anterior two-thirds of the tongue. This modality of sensation is mediated via the **chorda tympani branch** of the facial (VII) nerve (TS 22) which travels with the lingual branches of the mandibular division of the trigeminal (V) nerve.

The motor supply of the facial (VII) nerve is to the **muscles of facial expression** (buccinator, orbicularis oris and oculi, auricularis, frontalis, risorius and platysma), as well as to the posterior belly of digastric, stylohyoid and stapedius muscles. The facial (VII) nerve also has a small visceral efferent component which terminates in some of the autonomic ganglia in the head region. Nerve fibres from these ganglia are distributed to the submandibular, sublingual, nasal and lacrimal glands.

The nerve supply to the derivatives of the **third branchial arch** is via branches of the **glossopharyngeal (IX) cranial nerve** (TS 20), and this also has both sensory and motor components. The sensory component supplies both general sensation and special sensation (*i.e.* taste) to the posterior one-third of the tongue, in the latter case being distributed to the **taste buds** (and **median vallate papilla**) here. It should be noted that there are substantial species differences in the *number* of circumvallate papillae present: in the mouse, only a **single** (median) **circumvallate papilla** is present, being located in the midline on the dorsum of the tongue, at the junction between the anterior two-thirds, and posterior one-third of the tongue. It has a *bilateral* innervation from both the right and left glossopharyngeal (IX) nerves. (For further details, see *Differentiation of the tongue* in *The mouth and nose region*.) The motor distribution of the glossopharyngeal (IX) nerve is exclusively to the stylopharyngeus muscle. Visceral efferent fibres from the glossopharyngeal (IX) nerve are distributed to the **otic ganglion** from which post-ganglionic fibres pass to the parotid and posterior lingual glands.

The nerve supply to the derivatives of the **fourth** and **sixth branchial arches** is via branches of the **vagus (X) cranial nerve** (TS 19). The derivatives of the fourth branchial arch (*e.g.* the pharyngeal constrictor muscles, cricothyroid and levator veli palatini muscles) are supplied by its **superior laryngeal branches**, while the derivatives of the sixth arch (*e.g.* the intrinsic muscles of the larynx) are supplied by its **recurrent laryngeal branches**. The sensory distribution of the vagus (X) nerve in the mouth region is to the mucous membrane of the epiglottis and root of the tongue via the internal laryngeal branches of the superior laryngeal nerve. The arrangements seen in the primitive jawed fishes and in early mammalian embryos are shown in Figure 3.3.5a,b.

It is interesting that the branches of the left and right vagus (X) nerves take slightly different courses on the two sides of the sixth arch. The left recurrent laryngeal branch arises from the vagus (X) nerve on the left side of the arch of the aorta and then winds below the arch immediately behind the attachment of the **ligamentum arteriosum** to the concavity of the arch of the aorta, ascending to supply the various muscles indicated above. In contrast, the right recurrent branch arises from the vagus (X) nerve in front of the first part of the subclavian artery, and then ascends to supply the intrinsic muscles of the larynx (see Figure 3.3.2a,b).

This difference between the two sides is readily explained by the developmental changes that occur to the branchial arch arterial system. In the initial symmetrical arrangement, the recurrent laryngeal nerve on each side loops around the sixth arch artery. On the left side this vessel later becomes the ductus arteriosus (TS 21), and, in the post-natal period, the ligamentum arteriosum, and this route is followed by the nerve. On the right side, however, the connection between the right sixth arch artery and the right dorsal aorta disappears.[2] The right recurrent laryngeal nerve then loops around the right subclavian artery (see Figure 3.3.2a,b).

The branchial grooves (clefts)

This series of dorso-ventrally directed grooves is present on the outer aspect of the neck/pharyngeal region of the early embryo, and each is located between successive branchial arches and is covered by surface ectoderm. By about TS 16/17, the first two pairs are clearly evident, while the third and fourth pairs, though present, are less clearly delineated.

The **first cleft**, between the first and second arches, is of considerable importance, since it persists as the epithelial lining of the **external acoustic meatus** – or external auditory canal (which only fully forms after birth). A series of about six irregularly-

---

[2] The disappearance of the right sixth arch artery approximately coincides with the most rostral degree of ascent of the forelimb bud, when its axial artery (the right seventh cervical intersegmental artery, which subsequently becomes the right subclavian artery) achieves its definitive position in relation to the brachiocephalic trunk (derived from the right horn of the aortic sac).

Figure 3.3.5a,b The organization of the cranial nerve supply and arches in the head regions of the ancestral vertebrate (a, based on Morriss-Kay and Tan, 1987) and the early mouse embryo (b).

In (a), the head region, the eye (e) and nerve supply to the vestibular system (ve) are outgrowths of the brain and spinal axis (lightly dotted area). The olfactory epithelium (o) receives its nerve supply from the most rostral part of the primitive brain; the eye is also an outgrowth from the rostral part of the brain. A series of cranial nerve ganglia are involved in the innervation of muscles (m) associated with each of the gill arches (2–6), and receive sensory information from the surface of head and gill arch regions. Two components are envisaged that will amalgamate to form the trigeminal (V) cranial nerve, which supplies the territory of the first arch; the facial (VII) cranial nerve supplies the territory of the second arch, while the territory of the third arch is supplied by the glossopharyngeal (IX) cranial nerve. The fourth and succeeding arches are supplied by the vagus (X) cranial nerve and its various branches. The first arch cartilage is already seen to have differentiated into two components, one of which is involved in forming part of the facial skeleton (the maxillary cartilage, mx) while the other forms the cartilaginous basis for the mandible (md). The gill arches each possess a discrete cartilaginous bar (2–6), and each, as well as the head region, possesses its own arterial blood supply (from the pharyngeal or aortic arch arteries, aa). These connect the ventral aorta (va) to the dorsal aorta (da) on either side of the pharyngeal region, their arterial supply coming from the primitive heart (ht).

(b) In the early mouse embryo, the eye is shown by a dotted circle, and its extrinsic ocular muscles (that control eye movements) are supplied by the oculomotor (III), trochlear (IV) and abducens (VI) cranial nerves. The cranial nerve supply to the first pharyngeal (aortic) arch is from the trigeminal (V) nerve, to the second arch is from the facial (VII) nerve, to the third arch is from the glossopharyngeal (IX) nerve and to the fourth and succeeding arches from the vagus (X) nerve. The derivatives of the occipital somites (that differentiate to form the intrinsic muscles of the tongue, and most of the extrinsic muscles of the tongue) are supplied by the hypoglossal (XII) cranial nerve.

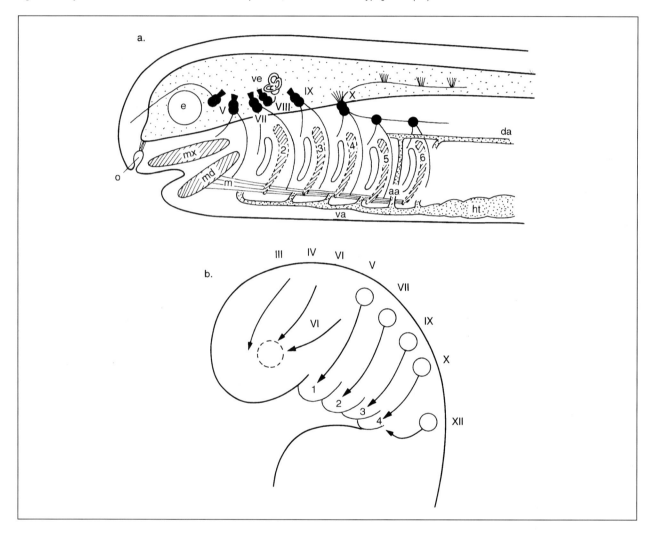

shaped swellings (the **auditory hillocks** or **tubercles**) develop on either side of the first groove during TS 19, and these eventually aggregate together to form the external ear.

As the second arch overgrows the third and fourth arches, the **second**, **third** and **fourth clefts** usually disappear leaving only a small depression overlying the region between the third and fourth arches that is termed the **cervical sinus** (TS 17, Figure 3.3.6a,b). This is usually lost, although it occasionally remains and is noted in humans as a persistent sinus and constant focus of infection from which mucus or pus may be discharged. Occasionally, pathological communications may be present between the pharynx and the

TS	1	2	3	4	5	6	7	8	9	10	11	12	13	14	15	16	17	18	19	20	21	22	23	24	25	26
E	0	1	2	3	4	4.5	5	6	6.5	7	7.5	8	8.5	9	9.5	10	10.5	11	11.5	12	12.5	13.5	14.5	15.5	16.5	17.5

Figure 3.3.6a–c. The formation and development of the branchial grooves and pouches. The initial symmetrical arrangement (a) is soon changed as the second arch overgrows the third and fourth arches, leaving an indentation on the surface, termed the *cervical sinus* (b). The later development of the first cleft and the first pouch is shown in (c): it forms the *external acoustic* (or auditory) *meatus* while the latter gives rise to the cavity of the middle ear, or *tubo-tympanic recess* within which are located the three ossicles: malleus, incus and stapes. The first pouch is connected to the oropharynx by the pharyngo-tympanic (or Eustachian) tube. The location of the tympanic membrane is also shown, as is its relationship to the malleus.

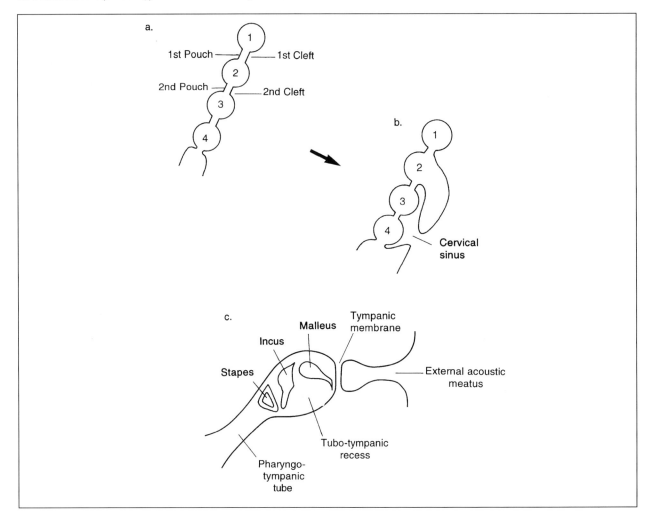

surface of the neck, and are most commonly due to the persistence of an abnormal communication between the second branchial cleft and pouch; these are termed **branchial fistulae**. If a deep indentation persists in a region formerly occupied by a branchial groove, it may form a **branchial cyst**, a site predisposed to infection. The tubular communication present between such a cyst and the surface is termed a persistent **branchial** (or **cervical** – because it is present in the neck region) **sinus**.

The branchial pouches

The most cranial part of the primitive foregut forms the pharynx which is particularly wide at its rostral extremity, but narrows caudally where it joins the oesophagus. On either side of the early pharynx are four well-defined pairs of **branchial pouches** (it is generally believed that no fifth pouch forms in the higher vertebrates). Each is located between branchial arches, so that the first pouch is between the first and second arches, the second pouch between the second and third arches, etc. As with the branchial arches, the pouches form and their endodermal linings differentiate in a cranio-caudal sequence. Subjacent to the endodermal lining of the pouch, this tissue is supported by branchial arch mesenchyme.

In all vertebrates, the endodermal epithelium lining each pouch contacts the surface ectoderm of the clefts to form a series of bilayered **branchial membranes**, that break down in fish to form the gill openings. In the terrestrial vertebrates, the bilayered branchial membrane is only kept in the first arch where it gives rise to the **tympanic membrane** or **ear drum** which only develops after birth, while the first and, more particularly, the succeeding pouches are substantially modified, giving rise to the tympanic (or middle ear) cavity (initially the tubo-tympanic recess, Figure 3.3.6c), the tonsillar cleft, the thymus (TS 21) and parathyroid (TS 22) glands and the ultimobranchial bodies (see below).

The **first pouch** is slightly unusual, in that its

ventral part is obliterated by the development of the tongue, while its dorsal portion forms a diverticulum termed the **tubo-tympanic recess** (TS 21). The lateral wall of this recess retains its close association with the ectoderm lining the first cleft, differentiating (after birth) into the pharyngeal component of the **tympanic membrane**, while medially the connection with the pharynx elongates to form the **pharyngo-tympanic** (or **Eustachian**) **tube** (TS 21). The tubo-tympanic recess, possibly with a small component from the dorsal part of the second pouch, will form the **middle ear** (or **tympanic**) **cavity** and this expands to envelop the **middle ear** (or **auditory**) **ossicles**, which (as indicated above) derive from dorsal components of the first and second arch cartilages (see Figure 3.3.6a–c). The middle ear cavity also expands dorsally in the postnatal period to connect with the air cells within the **mastoid process**, the bony protruberance that is located on the side of the skull just below the external ear. The mastoid process is not present at the time of birth in the human, but becomes increasingly prominent shortly afterwards – believed to be due to the "pull" of the sterno-cleido-mastoid muscle which inserts into it – and is involved in the maintenance of head posture during the postnatal period. Large numbers of "air cells" develop within the mastoid process (forming the **mastoid antrum**) which freely

Figure 3.3.7. Diagrammatic representation of the floor of the oropharynx to illustrate the derivatives of the pharyngeal pouches (modified after Larsen, 1993). The first pouch gives rise to the *tubo-tympanic recess* (the cavity of the middle ear, in which are located the three ossicles), and this is bounded laterally by the tympanic membrane. Medially, the first pouch is connected to the oropharynx by the pharyngo-tympanic (or Eustachian) tube. The endoderm of the second pouch (or *tonsillar cleft*) grows into the surrounding mesenchyme, and the pouch endothelium becomes the lining of the crypts of the *palatine tonsil*. The underlying mesenchyme becomes invaded by lymphoid tissue which subsequently differentiates into lymphoid nodules. The third pouch has a dorsal and a ventral component. The former gives rise to parathyroid tissue which subsequently migrates caudally with the derivative of the ventral portion of the third pouch, which forms the *thymus gland*. Along the path of descent of the thymus towards the superior mediastinum, the parathyroid tissue separates and becomes associated with the postero-medial surface of the thyroid gland to form the *parathyroid gland* (or parathyroid III). The fourth pouch in the mouse has a ventral but no dorsal derivative, and this gives rise to the *ultimobranchial body*. This fuses with the thyroid gland and the tissue disseminates within it, giving rise to the *parafollicular* (or *C-*) *cells* (for further details, see text).

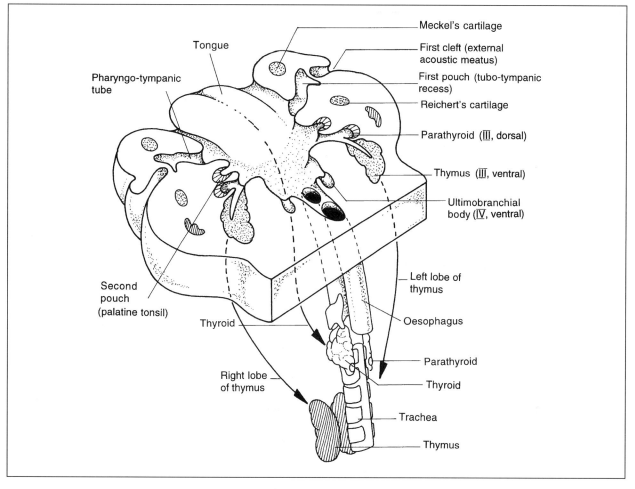

TS	1	2	3	4	5	6	7	8	9	10	11	12	13	14	15	16	17	18	19	20	21	22	23	24	25	26
E	0	1	2	3	4	4.5	5	6	6.5	7	7.5	8	8.5	9	9.5	10	10.5	11	11.5	12	12.5	13.5	14.5	15.5	16.5	17.5

communicate with the cavity of the middle ear.

Much of the ventral component of the **second pouch** is also obliterated by the developing tongue, although a small part of its dorsal portion may contribute to the tubo-tympanic recess (see above). The ventral part of the pouch loses its connection with the ectoderm of the second cleft, and the small part that remains develops into the **tonsillar cleft**. Within this, the endoderm proliferates and grows into the surrounding subjacent mesenchyme, while the pouch epithelium forms the lining of the **crypts of the palatine** tonsil. Lymphoid cells from elsewhere migrate into the underlying mesenchyme via the blood stream, and subsequently develop into the **lymphoid nodules** of the **tonsillar** bed. Similar lymphoid aggregations form in association with the first pouch (the **tubal tonsils**), in the dorsum of the tongue (the **lingual tonsils**) and in the dorsal pharyngeal wall (the **pharyngeal tonsils**).

The **third pouch** has a dorsal and a ventral component. The former expands and differentiates into a solid bar of tissue, while the ventral portion initially has a tubular structure. By about TS 22, the *dorsal* bars differentiate into the single pair of **parathyroid glands** (TS 22) that develop in the mouse (Figure 3.3.7). The epithelium of the *ventral* portion of the third pouch, possibly with an ectodermal component from the third ectodermal cleft (see Cordier and Haumoont, 1980), elongates to form two symmetrical tube-like structures and these migrate medially and eventually (by about TS 21) differentiate into the primordia of the two lobes of the **thymus gland**. These subsequently lose their connection with the wall of the pharynx and migrate caudally in close association with the parathyroid glands. At about this time, increased cellular activity is evident within the walls of the thymic rudiments, and this eventually leads to the obliteration of their lumens. It should be noted that, while *two* particularly large thymic lobes are formed in the mouse, these do not fuse to form a single median unit as occurs in humans. In the mouse, Cordier and Haumoont (1980) suggest that the primitive thymus has a central endodermal region surrounded by a peripheral ectodermally-derived region; in the adult, the medullary cells are said to be endodermally-derived, while the cortical cells are ectodermally-derived. According to Smith (1965), in a study of mouse embryos from E10 to birth and of newborn mice, there was already evidence of penetration of endodermal epithelium of third pouch origin by E10 by basophilic lymphoblasts. It was suggested that these cells may be the source of the humoral factor(s) associated with the thymus gland.

The origin of the myoid cells noted in the avian thymus were studied by Seifert and Christ (1990) using the chick–quail model. Cranial somites and prechordal mesoderm were grafted from quail to chick embryos. After somite transplantation, the host thymus did not contain graft-derived cells. However, after the implantation of prechordal mesoderm, graft-derived cells were found in the central cores of all of the visceral arches, as well as in the thymus anlage. These authors concluded that thymus myoid cells are derived from the axially located prechordal head mesoderm.

In a mouse study by Smith and Clifford (1962), the presence of *accessory* parathyroid tissue was frequently associated with the thymus. The thymus glands of embryos between E14–19, young mice and old mice between 8–14 months of age were analysed, and it was suggested that their location in the thymus was correlated with their similar pharyngeal arch origin.

The parathyroid glands later lose their connection with the thymus gland (TS 22/23) and become associated with the postero-medial surface of the thyroid gland. The thymus continues to differentiate after birth, and plays an important role in the development of the immune response. It usually descends into the upper part of the thoracic cavity, but may occasionally retain its connection with the parathyroids,[3] dragging them down into the thorax with it (Figure 3.3.8).

Each of the **fourth pouches** also has a ventral elongated portion which soon loses its connection with the wall of the pharynx and probably gives rise to the **ultimobranchial bodies** (Figure 3.3.8). These amalgamate with the thyroid gland, and their cellular components disseminate within it, ultimately giving rise to the **parafollicular** or **C-cells** which produce **calcitonin** (or thyrocalcitonin), a hormone associated with the regulation of calcium homeostasis. It should be said that there has been some discussion (see below) about the origin of these cells and it is believed that they may be of neural crest origin, deriving from crest cells that migrate into the fourth branchial pouches shortly after their formation. The situation, however, remains unclear. It has also been suggested that the ultimobranchial bodies may be derived from the transient **fifth branchial pouches** or, more controversially, that they may be derived from an **epibranchial placode** located next to the fourth branchial cleft (see Larsen, 1993). As already noted, and unlike the situation in humans, no dorsal derivatives of the *fourth* pouch form in the mouse: in

---

[3] The situation in the mouse also contrasts with that observed in humans in relation to the differentiation of the parathyroid glands: in humans, *two* pairs of parathyroid glands form, one pair, as in the mouse, develops from the dorsal part of the *third* pouch and is termed the parathyroid III; or "inferior" parathyroids because they descend further caudally than the other pair of parathyroid glands that develop from the dorsal part of the fourth branchial pouch. This second pair of parathyroid glands is called the parathyroid IVs or "superior" parathyroids, because they are normally found in a more superior location than the inferior parathyroids. For descriptive accounts of the development of the thyroid, parathyroids and thymus glands in humans, see Weller (1933) and Norris (1937).

In relation to the onset of functioning of the parathyroid glands in the human, Norris (1946) noted that the parathyroids derived from the fourth arch were first evident during week 5 (9 mm embryo), but that their rapid growth was only evident after about the 75 mm stage. Norris (1946) was also of the view that the parathyroid glands were active during both the embryonic and fetal periods.

Figure 3.3.8a–d. Four stages in the differentiation of the derivatives of the third and fourth pharyngeal pouches as viewed from the front of the embryo, indicating how their fate is closely related to that of the thyroid gland (modified after Larsen, 1993). These figures complement Figure 3.3.7. (a) Soon after their differentiation, the third pouch derivatives (the parathyroid III and thymus primordia) descend caudally. While the two thymus primordia migrate towards the midline (b) and eventually meet (c), they continue to descend until they reach their definitive location in the superior mediastinum (d). In the mouse, the thymus glands are usually present as two discrete lobes, whereas in the human they usually fuse across the midline. The parathyroids lose their contact with the thymus, and become closely associated with the postero-medial surface of the thyroid gland (c,d). The ultimobranchial bodies, of fourth pouch origin, descend to a limited extent until they make contact with, and eventually fuse with the thyroid gland, their cells eventually becoming disseminated within it (d). For further details, see main text.

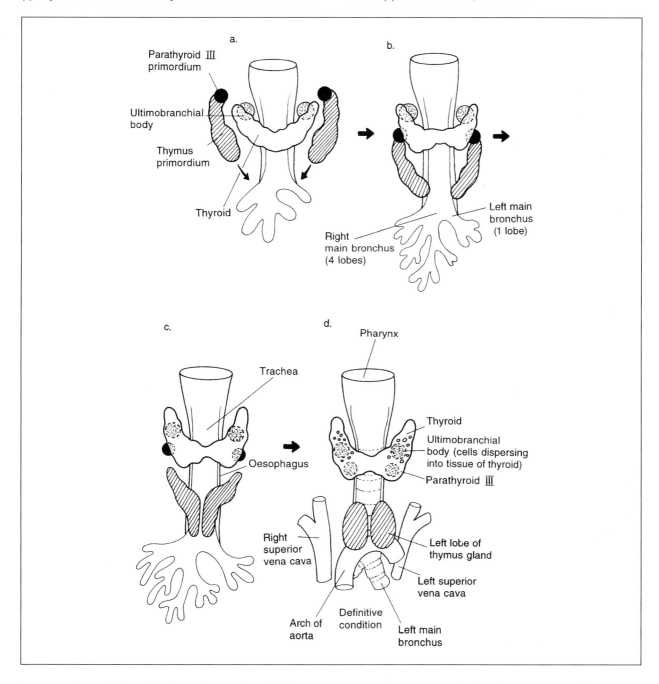

humans, these differentiate into the parathyroid IV, or superior pair of parathyroid glands.

Mérida-Velasco *et al.* (1989), from an analysis of human embryos at Carnegie Stages 14–15, suggest that the ultimobranchial bodies are derived from the *fifth* endodermal pharyngeal pouch, though its existence is controversial and not accepted by many embryologists (see Larsen, 1993). This is said to be colonized during Carnegie Stage 14 by cellular material of ectodermal placodal origin that originates in the most caudal part of the epibranchial placode, apparently with no participation of any other endo-

TS	1	2	3	4	5	6	7	8	9	10	11	12	13	14	15	16	17	18	19	20	21	22	23	24	25	26
E	0	1	2	3	4	4.5	5	6	6.5	7	7.5	8	8.5	9	9.5	10	10.5	11	11.5	12	12.5	13.5	14.5	15.5	16.5	17.5

dermal structure. These organs are said to arise independently until they eventually integrate into the thyroid gland. In mice, Fontaine (1979) demonstrated that the precursor cells of the thyroid C-cells were first present in the mesenchymal components of the *fourth* branchial arch until embryos possess about 28 pairs of somites (TS 16), after which time they begin to invade the endoderm of the fifth pouch. Cordier and Haumoont (1980) believed that the C-cells were already incorporated into the thyroid gland by E14 (see also observations of Munson *et al.* (1968) on the fourth pouch origin of the thyrocalcitonin-producing cells of the thyroid gland).

An overview of the fate and derivatives of the pharyngeal pouches is illustrated in Figure 3.3.7, while the fate of the third pouch derivatives in the mouse (the thymus and parathyroid glands) and their relationship to the thyroid and superior mediastinum is illustrated in Figure 3.3.8a–d. The fate of the fourth pouch derivative (the ultimobranchial bodies) is also shown. Occasionally, aberrant branchial arch derivatives are seen in the neck region of children, and have to be distinguished from more sinister structures[4].

The branchial arch derived muscles

Many of the muscles in the face and neck derive from arch tissue and transplant experiments in chicks and other experimental work has shown that the myoblasts giving rise to these muscles originate in the head somitomeres and occipital somites (*e.g.* Larsen, 1993; Noden, 1988). The myoblast cells that migrate into the first three branchial arches from **somitomeres 4–7**, are believed to be guided by connective tissue substrata previously laid down by the migrating neural crest cells (see Larsen, 1993) and to have their differentiation patterned by the neural crest cells that they encounter (Trainor *et al.*, 1994).

Myoblasts from somitomere 4 colonize the first branchial arch and give rise (TS 23–25) to the **muscles of mastication** (temporalis, masseter, the medial and lateral pterygoids, mylohyoid, the anterior belly of the digastric, tensor tympani and tensor veli palatini). Accordingly, these receive their nerve supply from the mandibular division of the trigeminal (V) nerve.

The second branchial arch muscles derive from somitomere 6 and give rise (from TS 22 onward) to the **muscles of facial expression** (buccinator, orbicularis oris and oculi, auricularis, frontalis, risorius and platysma), as well as the posterior belly of the digastric, the stylohyoid and the stapedius, and receive their nerve supply from the facial (VII) nerve.

The main muscle from the third branchial arch is the stylopharyngeus and its myotomes originate in somitomere 7. The muscles from the fourth branchial arch (the pharyngeal constrictor muscles, cricothyroid and levator veli palatini muscles) are from occipital somites 2–4, while those of the sixth (the intrinsic muscles of the larynx) derive from occipital somites 1 and 2. Accordingly, those muscles that are of third arch origin receive their nerve supply from the glossopharyngeal (IX) nerve, while those from the fourth and sixth arches are supplied by the vagus (X) nerve. As mentioned earlier, a summary of the fate of the pharyngeal arch arteries, skeletal elements, muscles and their associated cranial nerve supply is provided in Table 3.3.1.

---

[4] *The embryological origin of masses in the neck region in young children:* except for infections and enlarged lymph glands (inflammatory adenopathies), neck masses in children are uncommon, unlike those in adults, and seldom represent ominous disease states. These mostly consist of congenital cysts and sinuses of pharyngeal (branchial) arch origin, vascular malformations, salivary gland and thyroid anomalies and several varieties of benign and malignant neoplasms usually of non-squamous origin. A brief summary of the findings reported by Friedberg (1989) is provided here.

The commoner benign lesions are (i) pharyngeal sinuses and cysts, (ii) congenital pre-auricular sinuses, (iii) first pharyngeal cleft anomalies; these are uncommon, and may present either as first cleft stenosis, atresia or reduplication of the canal – leading to first cleft sinus or cervical/aural fistula, (iv) second pharyngeal cleft anomalies; these are very rare, and may present as persistent second pharyngeal cleft remnants, such as a small dimple or pin-hole located close to the anterior border of the sterno-mastoid muscle, (v) third and fourth pharyngeal cleft anomalies; these are the rarest of the group – the thymus may leave remnants along its course of descent, giving rise to paratracheal cysts – these are usually very mobile, and are often located just above the sterno-clavicular joint, (vi) mobile sinuses and cysts – most midline cysts are related to the thyroid gland, such as persistent thyroglossal ducts with/without associated cysts, (vii) lymphangiomas and cystic hygromas – these may be very extensive and spread across tissue boundaries, and are often particularly difficult to treat, (viii) salivary gland masses such as parotid tumours are only rarely encountered in children, (ix) capillary haemangiomas usually develop during the postnatal period, and their growth may be extremely rapid, (x) enlarged lymph nodes would have to be distinguished from solid lateral neck masses – while uncommon, they are usually neoplasms of neural origin, such as schwannomas, neurofibromas or neuroblastomas.

4
THE MAJOR ORGAN SYSTEMS

4.1 The heart and its associated vascular system

Introduction and overview

The heart starts to develop in the anterior part of the ventral region of the embryo soon after gastrulation is complete (E7.5–8). Two endothelial tubes form from a plexus of endothelial cells believed to be of lateral plate (splanchnopleuric) mesodermal origin which straddle the midline in the presumptive pericardial region of the **intraembryonic coelomic cavity** and these then aggregate to produce a single heart tube that is soon surrounded by a myocardial 'mantle' layer. This comes from the **cardiogenic plate** that differentiated from the lining of the ventral part of the intraembryonic coelomic cavity.

Between the endocardium and the myocardium is a layer of **cardiac jelly** (from TS 12) that consists of very loose mesenchyme or reticulum; its presence during early cardiogenesis is transient, and its exact origin and role are unclear, although it probably plays a part in valve formation. By E8.5, the bilayered heart tube is suspended from the more dorsally located foregut by a broad mesentery, the **dorsal mesocardium**, that gradually narrows.

Although the heart is just a simple tube at TS 12, a complex process of folding and remodelling turns it into a four-chambered organ that, by about TS 20, links the arterial and venous blood systems through the pulmonary and systemic circuits. By then, the primitive heart has three main cell layers: the inner **endocardium**, consisting of endothelial cells that are continuous with the lining cells of the blood vessels entering and leaving the heart, the intermediate **myocardial mantle** tissue that differentiates into **myocardial muscle**, and mesothelial tissue that is believed to migrate from the region of the sinus venosus. The latter covers the entire surface of the myocardium (TS 18) and differentiates into the outer "visceral" layer of **serous pericardium** or **epicardium** (late TS 18) that is continuous with the mesothelial lining of the pericardial cavity (*i.e.* the "parietal" pericardium). Although the essential structure of the heart has largely formed by about TS 20 (E12.5), the lungs do not function *in utero* and the pulmonary circuit is largely by-passed, its haemodynamics are therefore not finalized until shortly after birth.

In this chapter, we examine these events in some detail and also describe the development of the early vascular system. We also consider the changes that take place to the heart and mature vascular system after birth when the newborn mouse loses its umbilical link and maternally-derived oxygen supply and is thus forced to start breathing.

For a functional analysis of cardiac development, see Clark (1990), for the timing and sequence of events in mouse and human cardiogenesis, see Davis (1927), O'Rahilly (1971), Patten (1960), Vuillemin and Pexieder (1989a, 1989b), and for an overview of cardiac morphogenesis, see Kirby (1987) and Sweeny (1988). The classic descriptive accounts of early cardiogenesis in the human embryo by Davis (1927), and the development of the aortic arch arteries by Congdon (1922) provide baseline accounts from which all subsequent studies derive.

The development of the heart

The early stages (TS 11–13)

While some of the details of the early morphogenesis of the heart still remain unclear, reconstructions of serially sectioned embryos and scanning electron microscope (SEM) analyses suggest that the first sign of heart development is a thickening of the cells (TS 10/11) located in the ventral part of the lining of the future pericardial region of the U-shaped **intraembryonic coelomic cavity** that forms within embryonic mesoderm (see Figure 2.4.3), separating the splanchnopleuric and somatopleuric components at the lateral extremity of the embryo. These thickened cells then form the **cardiogenic plate** tissue

TS	1	2	3	4	5	6	7	8	9	10	11	12	13	14	15	16	17	18	19	20	21	22	23	24	25	26
E	0	1	2	3	4	4.5	5	6	6.5	7	7.5	8	8.5	9	9.5	10	10.5	11	11.5	12	12.5	13.5	14.5	15.5	16.5	17.5

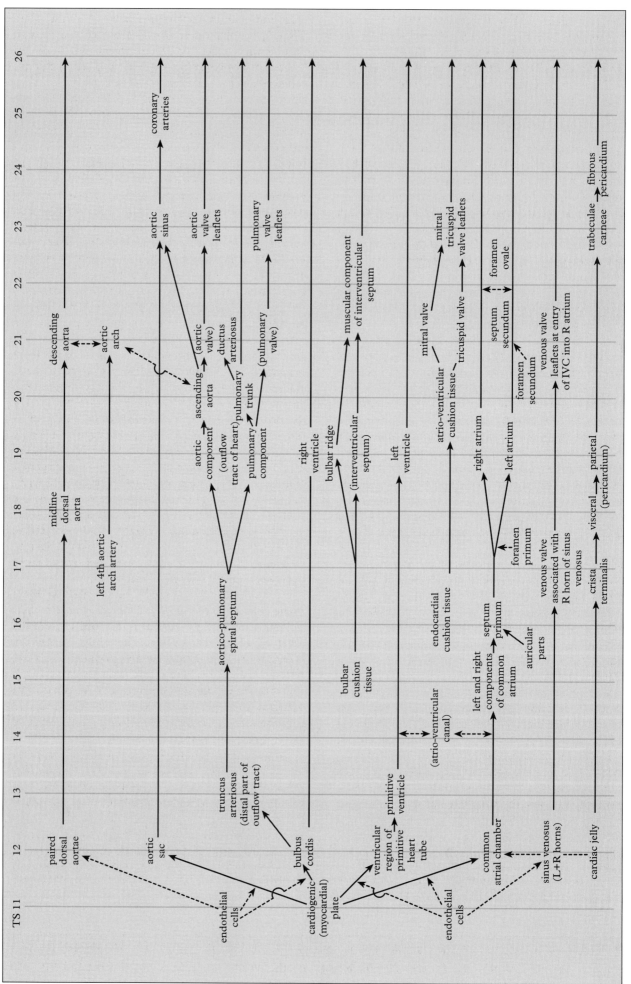

The heart.

(TS 11) that will differentiate to form the myocardium (see Figure 2.4.6) (Orts Llorca *et al.*, 1960; Van Mierop, 1969; Hay *et al.*, 1984).

In embryos with about 4–5 pairs of somites (TS 11/12), the cardiogenic plate cells and the cells which are in continuity with them and that extend caudally into the **pericardio-peritoneal canals** (the arms of the "U") have a distinct morphology: they protrude into the lumen of the intra-embryonic coelom where they can be clearly distinguished from the rest of the cells lining the coelom that retain their original squamous morphology (Kaufman and Navaratnam, 1981). It is likely that these cells in the prospective pericardio-peritoneal canals will be involved in the differentiation of the walls of the two horns of the sinus venosus, and possibly also of the caudal components of the atria of the heart. At this stage, the sinus venosus (the region where the venous drainage enters the heart) consists of only an endothelial layer.

Subjacent to the cardiogenic plate, a plexus of endothelial cells differentiates to form the right and left (lateral) endocardial heart tubes. These tubes soon fuse across the ventral midline to form the single endothelial heart tube that is seen to lie in a matrix of cardiac jelly (TS 12). As the embryo folds, the single endocardial tube is soon covered by cardiogenic plate cells so that the primitive heart consists of an inner endocardial (*i.e.* endothelial) and an outer myocardial tube with intervening jelly, the whole being suspended from the more rostrally and caudally extending foregut by a broad mesentery, the **dorsal mesocardium** (mid TS 12). The primitive heart tube is now located within the pericardial component of the intra-embryonic coelomic cavity. A similar arrangement, with the fusion of two primordia is also seen in the human embryo at a comparable stage of development (Davis, 1927).

The bilateral origin of the heart may be demonstrated in chick embryos by preventing the midline fusion of the two endocardial primordia (bisection is undertaken when the embryos possess 1–3 pairs of somites), giving rise to a condition termed *cardia bifida* in which two distinct hearts form, one on either side of the midline (De Haan, 1959). The first evidence of a heartbeat was observed in chick embryos with 10–11 pairs of somites. It was noted that, when the two heart tubes formed, the *left* heart had a higher intrinsic beat rate than the *right* up to the stage when embryos had about 18 pairs of somites. In more advanced embryos, the situation was reversed.

In mouse embryos with 5–7 pairs of somites (late TS 12), histological sections show that the median part of the cardiac rudiment (some 3–400 μm in length) passes rostrally as a wide tubular structure, subjacent to the primitive foregut, that divides anteriorly (at the rostral part of the outflow tract[1]) to form the primordia of the first pair of **branchial arch arteries**. These pass dorsally (*i.e.* posteriorly) on either side of the rostral extension of the foregut pocket (or diverticulum) to unite with the paired **dorsal aortae**. Similarly, the endothelial lining of the common atrial chamber bifurcates posteriorly to give rise to the caudal parts of the right and left endocardial tubes that pass on either side of the proximal part of the foregut pocket, with each extending to line one of the horns of the sinus venosus (Figure 4.1.1). It is generally believed that the common atrial chamber and the two horns of the sinus venosus initially develop within the mesenchyme that forms the septum transversum, but they soon become free and included within the pericardial cavity.

The two dorsal aortae develop in isolation from angiogenetic cell clusters, with one forming on each side of the notochord. When they are eventually joined by the first branchial arch arteries, the basis of the primitive circulation is established. Later, when the second and successive branchial arch arteries form, these are also directed posteriorly, and unite with the dorsal aortae to form a series of vascular channels (the **aortic arch** or **pharyngeal arch arteries**) which differentiate on either side of the primitive pharynx (see later). These vessels develop in a cranio-caudal sequence, while the dorsal aortae extend rostrally and ramify in the cephalic region, eventually differentiating into the **internal carotid arteries** (TS 16) and their branches.

By TS 12/13, the primitive heart tube has three main regions that can be distinguished by their bulging morphology: rostral is the **bulbus cordis** most of which will become the **right ventricle**, caudal to this is the **primitive ventricle**, which will become the **left ventricle**, and more caudally still is the **common atrial chamber** (or **primitive atrium**) (see Figure 4.1.1) that will later be separated into the **left** and **right atria** through the growth of the **interatrial septa**.

Irregular contractions of the primitive mouse heart occur in embryos with about 4–6 pairs of somites (Goss, 1938), well before the heart receives any form of innervation (Navaratnam, 1965). Contractility is

---

[1] Until relatively recently, the distal part of the outflow tract was subdivided into the **truncus arteriosus** and **aortic sac**, but it is now recognized that there are no morphological criteria that allow these regions to be separately distinguished. Equally, the most distal part of the bulbus cordis region (see below) is continuous with, but probably not readily distinguished from, the proximal part of the outflow tract. In some textbooks (*e.g.* Larsen, 1993) the outflow tract, or conotruncus, initially becomes subdivided into the conus cordis and truncus arteriosus, but we do not follow this terminology.

TS	1	2	3	4	5	6	7	8	9	10	11	12	13	14	15	16	17	18	19	20	21	22	23	24	25	26
E	0	1	2	3	4	4.5	5	5.5	6	6.5	7	7.5	8	8.5	9	9.5	10	10.5	11	11.5	12	12.5	13.5	14.5	15.5	16.5	17.5

Figure 4.1.1. Frontal view of the symmetrical primitive cardiac rudiment at around TS 12, a wide tubular structure which is subdivided into a number of compartments. The inflow channels, consisting of the right and left horns of the sinus venosus, pass into a common atrial chamber, believed to develop within the mesenchyme that forms the septum transversum. The more rostral part of the primitive heart, the primitive ventricle and bulbus cordis are believed to develop within the pericardial component of the intra-embryonic coelom. Distally, the outflow tract gives origin to the first aortic (branchial) arch arteries.

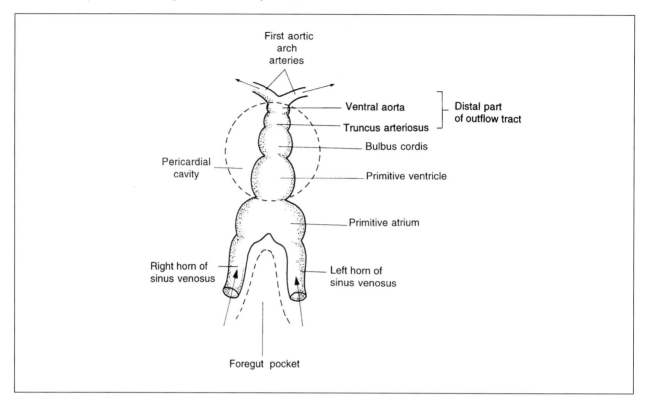

an intrinsic property of cardiac cells, even when they are isolated *in vitro* (Harary and Farley, 1963). (See below for details of the conduction system.) The heart starts pumping in embryos with about 8–10 pairs of somites (around the end of TS 12, blood cells not being seen in the circulation until TS 13). Initially, the contractions are irregular and it is not until the embryo has about 15–20 pairs of somites (TS 14) that the heart beats regularly and powerfully. During the relatively short period before the onset of regular contractions, contractile elements within the cardiac myocytes become organized into fully differentiated sarcomeres (Kaufman, 1981b; Penefsky, 1984; Rodriguez and Ferrans, 1985; Navaratnam *et al.*, 1986). The sinus venosus acts as the initial pacemaker of the primitive heart, and the wave of muscle contraction is then propagated rostrally along the tubular heart.

"Folding" or "looping" of the primitive heart (TS 13–14)

At the end of TS 12, the heart is still a largely midline structure, though its axis is principally to the left of the midline. At around the start of TS 13, the heart tube enters a "looping phase" and becomes S-shaped. At the same time, the primitive heart and its adjacent vasculature loses its original symmetry (Layton, 1985). The configurational changes associated with the early stages of looping are greatly facilitated by the loss of an extensive region of the dorsal mesocardium in the region between the inflow and outflow regions of the heart and this space or discontinuity in the dorsal mesocardium is known as the **transverse pericardial sinus** (Figure 4.1.2). In the human, the dorsal mesocardium forms in embryos with about 7–9 pairs of somites, and this ruptures in embryos with about 11–16 pairs of somites (TS 14) to give rise to the transverse pericardial sinus (Davis, 1927). In the adult mouse (and human), this passage is maintained and allows communication between the two sides of the pericardial cavity dorsal to the heart, and is located between the aorta and pulmonary trunk (the outflow vessels), in front, and the venous inflow into the heart, behind.

Looping appears to be an intrinsic property of the primitive heart tube, and is something that, despite the absence of beating, the isolated rudiment will do in culture medium containing a high level of potassium (Manasek and Monroe, 1972). It is clear that myocardial cell-shape changes play a critical role in this process (Manasek *et al.*, 1972, 1978).

During looping, the heart's initially more caudally located atrial region "ascends" and is soon seen to be located dorsal to the primitive ventricle and bulbus cordis regions of the heart and its outflow tract. At around this time, the principal regions of the primitive heart tube become demarcated from each other by the two deep **atrio-** and **bulbo-ventricular grooves**. The net effect of this is that, by TS 14, the single, straight primitive heart tube has become S-shaped and demarcated into a series of reasonably

Figure 4.1.2. This figure shows the folding of the cardiac rudiment, and the location of the dorsal mesocardium and the formation within this of the transverse pericardial sinus. a–c illustrate the increasing length of the primitive cardiac rudiment and the folding or "looping" that occurs. This is facilitated by the breaking down of a substantial part of the dorsal mesentery of the heart (the dorsal mesocardium) to give a space called the transverse pericardial sinus (TS 13). This is located between the inflow and outflow regions of the primitive heart, a communication which is readily recognized in the adult. (c) This side view of the primitive heart indicates the location of its various subdivisions at about TS 13/14. (d) This shows a frontal view of the primitive heart at TS 14 together with the direction of blood flow from the two inflow channels which enter the two atria caudally and pass out rostrally through the outflow tract. Despite the fact that externally there appears to be a right and left atrium, these represent the right and left components of the common artial chamber, as inter-atrial septation has yet to occur.

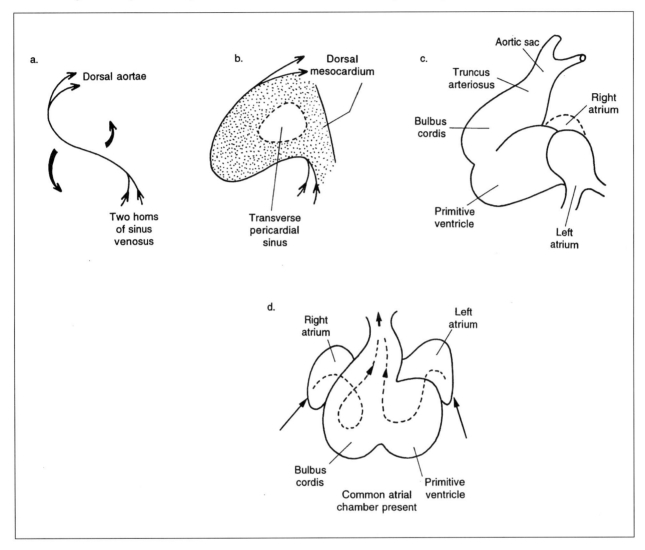

well-defined regions, each of which is destined to form a component of the four-chambered heart. The progressive changes that occur in the configuration of the primitive heart tube are illustrated in Figure 4.1.2.

Remodelling (TS 15–22), inter-atrial and inter-ventricular septation

The S-shaped heart now goes through a complex remodelling process at the end of which it has four chambers and separate inflow and outflow channels on its left and right sides. During the remodelling process, inter-atrial and inter-ventricular septation occur and the initially single outflow tract is subdivided into two channels through the formation of the **aortico-pulmonary spiral septum**. Following the completion of this septation, one channel, the **ascending aorta**, becomes the outflow tract of the left ventricle, while the second channel, the **pulmonary trunk**, becomes that of the right ventricle (Rychter, 1978). Cell death is also believed to play a critical role during these early stages of cardiogenesis (Pexieder, 1975). We now describe these processes in more detail, leaving an analysis of the various vessels that feed the heart until the next section.

Following the initial folding of the heart tube and the formation of the atrio-ventricular and bulbo-ventricular grooves, the next key events are the

TS	1	2	3	4	5	6	7	8	9	10	11	12	13	14	15	16	17	18	19	20	21	22	23	24	25	26
E	0	1	2	3	4	4.5	5	5.5	6	6.5	7	7.5	8	8.5	9	9.5	10	10.5	11	11.5	12	12.5	13.5	14.5	15.5	16.5	17.5

changes to the organization of the internal arrangement of the primitive atrium. At about TS 15, the two horns of the **sinus venosus**, which receive deoxygenated blood in the early embryo from the left and right **anterior** and **posterior cardinal veins** via the two **common cardinal veins** (or **ducts of Cuvier**) (see below and Figure 4.1.3), become incorporated into the posterior wall of the right component of the common atrial chamber of the heart (TS 15). The region of the atrium derived from the **sinus venosus** is known as the **sinus venarum** and has a smooth surface in contrast to that of the original atrium which is trabeculated (see below), while the boundary between them is marked by a well-defined ridge termed the **crista terminalis**.

At late TS 17, the first of the two components of the **atrial septum** starts to form and eventually (TS 22, when the septum secundum develops) they will together divide the common atrial chamber into its left and right components (the definitive left and right atria). Septation starts with the formation of the **septum primum**, a relatively thin structure which grows downwards from the middle part of the posterior wall of the common atrial chamber towards the **atrio-ventricular bulbar cushion tissue** which is itself growing from the region of the atrio-ventricular groove at the junction between the atrial and ventricular parts of the heart and will soon separate them (TS 21) (see below) (Fitzharris, 1981).

At about this time, the middle region (formerly termed the **truncus arteriosus** region) of the outflow tract starts to become divided into two vessels as two spiral ridges start to grow out from its wall. These eventually meet to form the **aortico-pulmonary spiral septum** (TS 16/17) (Pexieder, 1978; Fananapazir and Kaufman, 1988), a process that is mediated, in part at least, by neural crest cells (Fukiishi and Morriss-Kay, 1992; see also Kirby *et al.*, 1983; Kirby and Waldo, 1990; Takamura *et al.*, 1990). Kirby *et al.* have shown that bilateral excision of the neural crest cells at the level of occipital somites 1–3 prior to their migration causes malformations of the aortico-pulmonary spiral septum, resulting in *either* a common outflow channel *or* transposition of the great vessels. Kirby (1987) subsequently demonstrated that cranial neural crest cells form part of the wall of all of the aortic arch arteries, so that damage to these cells invariably resulted in flow abnormalities within these vessels. Cardiac neural crest cells seed the heart with parasympathetic neurons, as well as with mesenchyme. A decrease in the volume of the latter also leads to

Figure 4.1.3. An overview of the vascular arrangement in the TS 13 mouse embryo soon after the primitive circulation has been established showing all the major inflow and outflow vessels. Note that the ventrally located outflow tract gives origin to the paired, and at this stage symmetrical, branchial (pharyngeal) arch arteries, which pass rostrally and around the primitive pharynx to amalgamate with the paired dorsal aortae. In the mid-abdominal region, the paired dorsal aortae fuse to form a single dorsally located midline vessel, and this gives off the vitelline artery that supplies the primitive yolk sac (this is the region that will in due course form the primitive gut). More caudally, the dorsal aorta bifurcates to give the paired dorsal aortae, and these give rise to the paired umbilical arteries which pass through the umbilical cord (carrying deoxygenated blood) to the placenta. The inflow to the primitive heart comes principally from three sources: (i) oxygenated blood from the placenta via the umbilical veins; (ii) deoxygenated blood from most of the embryo (except from the primitive gut) which pass via the common cardinal veins (the majority of the deoxygenated blood flows via the anterior cardinal veins from the rostral part of the embryo, while a smaller volume of deoxygenated blood passes via the posterior cardinal veins from the caudal part of the embryo); (iii) a relatively small volume of deoxygented blood passes from the yolk sac – the primitive gut, via the vitelline veins. (For a frontal view of the primitive symmetrical arrangement of the aortic arch arteries and how they differentiate to form the asymmetrical system seen later in gestation, see Figure 3.3.2a–c.)

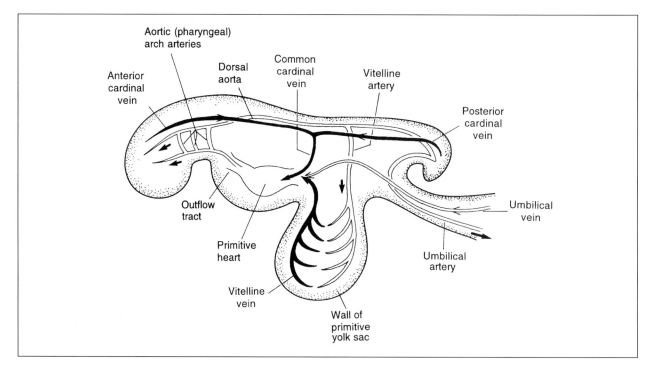

cardiac malformation, though this population of cells can undergo regeneration, so that the neural innervation of malformed hearts is usually morphologically normal.

Although the primitive (undivided) outflow tract had originally linked the bulbus cordis (future right ventricle) to the arterial system, the remodelling process taking place had left this tract straddling the region where the bulbus cordis and primitive ventricle abutted. The net result of interventicular septum formation is that the future pulmonary trunk exit of the tract is continuous with the right ventricle and the future ascending aorta with the left ventricle (see Bartelings and Gittenberger-de Groot, 1989, and below).

Before septation is complete, the two blood streams flow through the single outflow tract, but seem to be kept distinct by hydrodynamic forces. In a simple and elegant model, De Vries and Saunders (1962) demonstrated that, when two intersecting streams of water both pass in approximately the same direction, it was possible to produce a clockwise spiral. Cross sections at different levels showed the configuration of the two streams as though through the outflow tract of the heart as seen at Carnegie stage 15. With reversal of the front and back streams, reversal of the direction of the spiral could be achieved.

At about the same time as the growth of the septum primum is initiated (TS 17), the bulbar (or endocardial) cushions are formed by centres of proliferation in the mesenchymatous ring at the junction between the common atrium and primitive ventricle and give rise to a pair of elevations which become continuous with the downgrowing components of the spiral septum (see below). These also project into the single **atrio-ventricular canal**, and will soon separate the atria from the ventricles (the details about how this single canal becomes divided into the separate canals for the right and left sides of the heart remains unclear).

Towards the end of TS 18, the **epicardium** starts to become apparent. While it was formerly thought that this layer formed through the differentiation of the outer cells of the myoepicardium, this is now no longer believed to be so, and the term *myoepicardium* and its synonym *epimyocardial mantle* are no longer used. It is now clear that the cells that constitute the epicardium, or **visceral pericardium**, differentiate from mesothelial cells that derive from splanchnopleuric mesoderm and migrate onto the surface of the heart from the region of the sinus venosus or septum transversum (Ho and Shimada, 1978). These cells are continuous with the visceral pericardial cells that line the pericardial cavity.

By TS 19/20, atrial septation is well under way, with the communicating space between the two sides of the atrium being termed the **ostium primum**; this space gradually diminishes in area as the septum primum extends towards the atrio-ventricular bulbar cushion tissue. Just before the downgrowth of the septum primum completely obliterates any connection between the right and left components of the common atrial chamber, the upper part of the wall of the septum primum starts to break down (presumably by apoptosis of the component tissues in this location) to form the **ostium secundum** (TS 21). By the time that this "window" has formed, the ostium primum has completely disappeared, due to the fusion of the lower border of the septum primum with the atrio-ventricular bulbar cushion tissue. The ostium secundum is now the only communication between the right and left atria at this stage, and blood flows through it from the right to the left side (Odgers, 1935).

At about the same time, the bulbar cushions extend caudally to separate the atrial and ventricular chambers which, however, remain linked by what are now the two atrio-ventricular canals. Meanwhile, the **interventricular septum** starts to develop as the two **bulbar** (or spiral) **ridges** within the outflow tract extend caudally and they soon form the **membranous component of the interventricular septum** (TS 19), while its **muscular component** that is principally derived from the region of ventricular wall subjacent to the bulbo-ventricular groove becomes apparent a stage later (TS 20). With the completion of interventricular septation at around TS 21, the communication that previously existed between the bulbus cordis and the primitive ventricle becomes closed off (Kramer, 1942; De Vries and Saunders, 1962; Vernall, 1962; McBride *et al.*, 1981).

By about TS 21, the **spiral septum** has divided the outflow tract into two distinct channels, the proximal part of the **ascending (thoracic) aorta** which feeds the rostral part of the embryo and the **pulmonary trunk** (Figure 4.1.4). The majority of blood that flows through the trunk during the prenatal period passes directly into the **ductus arteriosus** (which forms from the left 6th branchial arch artery at TS 21) and thence into the descending aorta. Only a very small proportion of this blood is directed towards the lungs via the pulmonary arteries before birth, as there is a high peripheral resistance to blood flow in the lungs; this is initially due to their alveoli being poorly differentiated and unexpanded, and later because they are filled with amniotic fluid.

During TS 21, all the **cardiac valves** become recognizable, although they are still poorly differentiated, the ostium secundum forms, and the left fourth branchial arch artery differentiates to give the aortic arch at the distal part of the ascending aorta.

It is not until the early part of TS 22 that the

TS	1	2	3	4	5	6	7	8	9	10	11	12	13	14	15	16	17	18	19	20	21	22	23	24	25	26
E	0	1	2	3	4	4.5	5	6	6.5	7	7.5	8	8.5	9	9.5	10	10.5	11	11.5	12	12.5	13.5	14.5	15.5	16.5	17.5

Figure 4.1.4. The events occurring in the outflow region of the primitive heart. (a) It is believed that two streams of blood, one containing largely oxygenated blood and the other largely deoxygenated blood spiral around each other in this region, with only a minimal degree of intermixing. In those regions of the outflow tract where the wall is subjected to minimal pressure, bulbar ridges form which gradually grow towards each other to form the aortico-pulmonary spiral septum. This effectively completely separates the two streams of blood, so that the largely deoxygenated flow passes through the pulmonary trunk, while the largely oxygenated flow passes through the aortic outflow (*i.e.* into the ascending aorta). (b) This figure illustrates the arrangement observed at three transverse sections through the proximal, intermediate and distal parts of the outflow tract, while the small arrows indicate the location of the bulbar ridges. On the left side, the undivided (primitive) arrangement is displayed, in the middle, the intermediate arrangement is shown, while on the right side the definitive arrangement is seen, with the formation of the ascending aorta (carrying oxygenated blood) and pulmonary trunk (carrying deoxygenated blood).

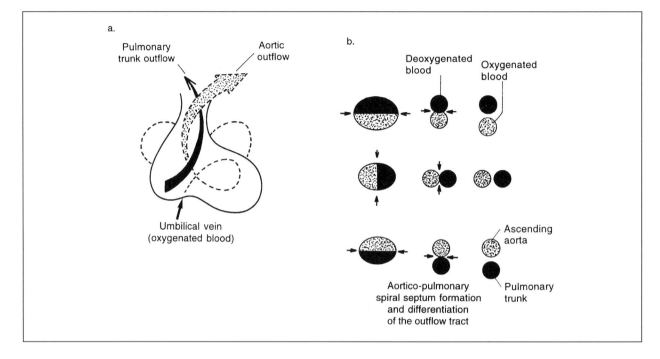

second part of the atrial septum, the **septum secundum** starts to develop when a new rudiment grows downwards from slightly to the right of the septum primum towards the atrio-ventricular bulbar cushion tissue, although its lower border never completely fuses with it (Figure 4.1.5b). The septum secundum is a thicker and less flexible membrane than the septum primum.

One of the principal functions of the lower border of the septum secundum is to deflect (i) the major volume of deoxygenated blood (derived from the right side of the upper half of the body) through the right atrio-ventricular canal into the right ventricle, and (ii) the majority of the largely oxygenated blood that flows into the right atrium from the inferior vena cava (see below) through the **foramen ovale**. This is a composite channel formed from the persistent gap between the lower border of the septum secundum and the ostium secundum. It is obliquely directed, and it is this critical feature that allows the largely *oxygenated* blood from the inferior vena cava to be directed across the midline from the right into the left atrial chamber. It is for this reason that the lower free border of the septum secundum was formerly termed the **crista dividens**, as it separates the inferior caval stream into two, the major component (~ 70%) of the oxygenated blood from the placenta flows through the foramen ovale into the left atrium, while a smaller component intermixes with the *deoxygenated* blood from the *right* superior vena cava; this flows into the right atrium and then (largely) passes directly into the right ventricle (see below).

The foramen ovale therefore allows blood with a high oxygen content to flow from the right to the left atrial chamber of the heart, bypassing the pulmonary circuit. This channel is maintained until birth when major haemodynamic changes occur that cause the two septa to press against one another, so blocking interatrial communication (see later). By about TS 24, the definitive pre-natal geometry of inter-atrial septation is finally achieved (see Figure 4.1.5). For a recent view on the factors involved in atrioventricular septation, see Webb *et al.* (1998).

It has already been mentioned that the inner wall of the common atrial chamber has "trabeculated" and "smooth" regions (Figure 4.1.6), with the trabeculated part being believed to define the original extent of the primitive (common) atrium, while the smooth part is where venous vessels have become incorporated. The smooth region on the posterior part of the right atrial wall includes drainage from the **right horn of the sinus venosus** (the inferior vena cava), the right and left superior venae cavae and the **coronary sinus**, the remnant of the left horn of the sinus venosus. On the left side, the smooth-walled region represents the site where the pulmonary veins are incorporated into the dorsal or posterior part of the wall of the left atrium (TS 23). Because of the persistence of the left superior vena cava in the mouse, the arrangement of the venous drainage to the heart is

Figure 4.1.5a–c. The process of inter-atrial septation. (a) The location of the downgrowths that develop from the upper posterior part of the middle region of the common atrial chamber towards the atrio-ventricular (a–v) endocardial cushion tissue to form the inter-atrial septa. (b) This downgrowth has two components, the septum primum which forms slightly to the left of the midline, and the septum secundum which forms just to the right of the midline. The former is a fairly flimsy membrane while the latter has a stiffer consistency. When viewed from the right side towards the left side of the common atrium, the growth of the septum primum is seen to be in three phases (i) the septum primum grows towards the a–v cushion tissue, but does not quite reach it: the communication present at this stage between the two components of the common atrial chamber is termed the ostium primum. (ii) When further downward growth obliterates the latter, continuity between the two sides is still possible because the upper part of the septum primum breaks down to form the ostium secundum. (iii) The septum secundum grows down towards the a–v cushion tissue, but fails to reach it. (c) This illustrates atrial septation after completion, and is the arrangement present up to birth. The channel between the two derivatives of the common atrium (the right and left atria) is formed in such a way that the (largely oxygenated) blood from the inferior vena cava which enters the caudal part of the right atrium is directed across the midline into the cavity of the left atrium, while the deoxygenated blood which enters the rostral part of the right atrium through the right superior vena cava (RSVC) is directed towards the right ventricle (RV), passing through the right a–v canal. The outflow of the RV is the pulmonary trunk. The blood in the left atrium passes to the left ventricle (LV) through the left a–v canal and out to the ascending aorta. The lower border of the septum secundum deflects the blood from the RSVC into the right a–v canal, and is accordingly sometimes termed the *crista dividens*. Only a limited degree of intermingling of the rostrally-directed and caudally-directed blood streams is believed to occur.

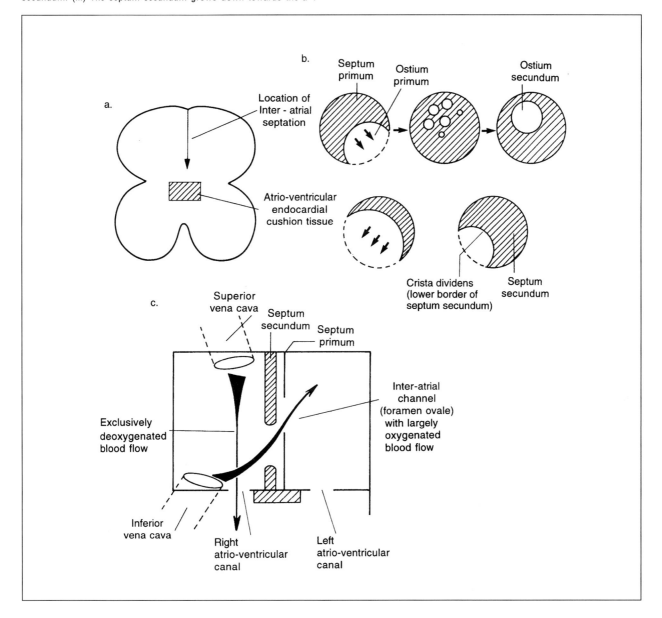

TS	1	2	3	4	5	6	7	8	9	10	11	12	13	14	15	16	17	18	19	20	21	22	23	24	25	26
E	0	1	2	3	4	4.5	5	6	6.5	7	7.5	8	8.5	9	9.5	10	10.5	11	11.5	12	12.5	13.5	14.5	15.5	16.5	17.5

Figure 4.1.6. Diagrammatic view of the posterior wall of the right and left atria indicating the trabeculated, primitive atrial-wall-derived (shown dotted), and venous-derived (non-dotted) smooth-walled components. The major venous channels that enter the cavity of the right atrium, the right superior vena cava, the inferior vena cava and the coronary sinus are also shown. On the posterior wall of the right atrium, the venous-derived part (*sinus venarum*) is bounded on its right margin by a small longitudinally-directed elevation (the *crista terminalis*). On the upper part of the posterior wall of the left atrium, the smooth-walled site of incorporation of the pulmonary veins is shown. The venous drainage into the right atrium in the mouse embryo (a) is contrasted with that seen at a comparable stage of development in the human (b). Note that in rodents, the azygos vein drains the left side of the thorax and receives the hemiazygos vein before it opens dorsally into the left superior vena cava (see Hebel and Stromberg, 1986).

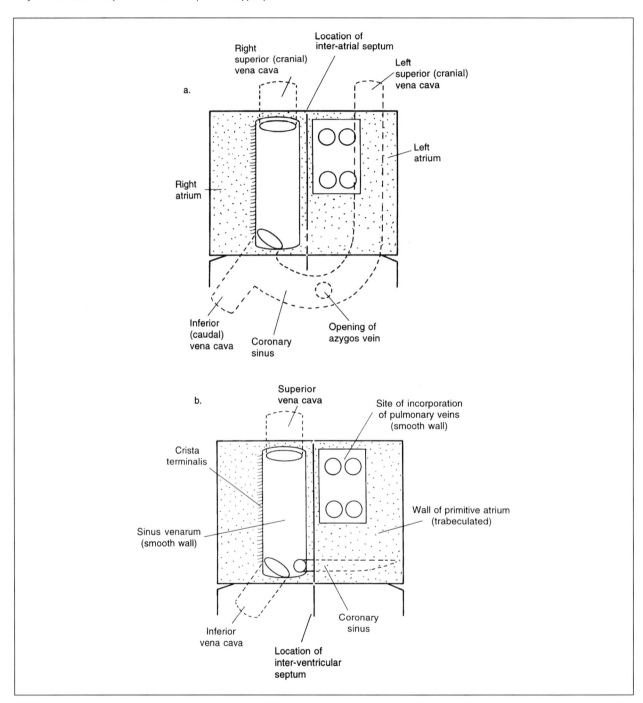

significantly different from that seen in humans (in the human, the left superior vena cava largely disappears and the venous drainage of the left side of the head and neck and left upper limb is into the brachiocephalic vein, and thence into the superior vena cava – the original right superior vena cava). The detailed differences observed in the two species are discussed below (see Figure 4.1.6).

Development of the coronary arteries

The arterial blood supply to the wall of the heart is from the coronary arteries. Studies in early chick embryos reveal that these form from vessels that sprout from the wall of the primitive aortic sinuses (Aikawa and Kawano, 1982). Similar findings are observed in relation to their development in the human (Hutchins *et al.*, 1988). The latter authors found that the connection between the proximal coronary arteries and the aorta was not observed before Carnegie stage 18 (≡ TS 21). It was their view that the coronary arteries only developed from those 2 aortic sinuses where the wall tension was increased. In the chick, initially 2–4 primitive coronary arteries appear from the right aortic sinus below the level of the cup margin, and that 1–3 form from the left one. As development proceeds, the arteries are gradually reduced in number to form a single definitive vessel on each side (Aikawa and Kawano, 1982). In the mouse, the definitive coronary arteries are first apparent at about TS 25.

Development of the conduction system

The heart is the earliest of the organs to function. While its initial contractions arise spontaneously as soon as the contractile elements within the myocardial cells become aligned to form recognizable sarcomeres, the waves of electrical activity (depolarization) spread through cell to cell contacts (Lieberman, 1985). If grown in culture, cardiac myocytes will beat in unison if they are in contact with similar cells.

A specialized part of the wall of the heart differentiates to form the so-called pacemaker region. The initial pacemaker is believed to be situated in the caudal part of the left primitive cardiac tube, but is later located on the right side of the primitive heart, once the sinus venosus region has formed. At an early stage, the primitive ventricle also has pacemaker-like activity, but this ceases to act in this way once the activity of the S-A (sino-atrial) node is established (see below). Recognition of the pacemaker region is complicated because its characteristic histological features only become apparent at about the time of birth (Walls, 1947). The Purkinje, or specialized conducting tissue, however, has histological features that allow it to be recognized somewhat earlier (De Haan, 1965).

The pacemaker region, which has a faster rate of spontaneous depolarization than the rest of the heart muscle, and initiates the cardiac impulse, becomes located in the S-A node. This node is believed to be derived from cells of the right sinus venosus, or possibly from the right common cardinal vein (Larsen, 1993).

The electrical impulse then passes from the S-A node across the wall of the atrium principally, although not exclusively, through the crista terminalis (the ridge that separates the venous-derived from the atrial-derived wall of the right atrium) to the A-V (atrio-ventricular) node. This is located in the superior part of the endocardial cushion region. From the A-V node the electrical impulse passes along specialized conduction tissue (the Bundle of His Purkinje cells), which is principally located in the muscular part of the interventricular septum, to the two ventricles. Within the septum, the Purkinje tissue bifurcates, sending one branch to the apex of the left ventricle and the other to that of the right ventricle within the septomarginal trabeculation or moderator band.

The net result is that, following the initiation of electrical activity at the S-A node, all of the chambers of the heart contract efficiently and in the correct sequence. The nodal region has a rich nerve supply, from both the vagus nerves and from the sympathetic chains (Yamauchi, 1965); while visceral afferent nerve fibres are also present (Navaratnam, 1965).

The differentiation of the conduction system has been studied in E9–12 (TS 14–20) mouse embryos from the analysis of 1 μm araldite sections (Virágh and Challice, 1977). In E9–10 mouse embryos, specialized cells are recognized in an inner layer of the dorsal wall of the atrio-ventricular canal, which connect muscle tissue in the atrial and ventricular regions. The A-V node develops from the distal end of this region of specialized tissue, and proliferates into the loose mesenchyme of the dorsal A-V cushion tissue. In E11–12 embryos, the A-V node and Bundle of His Purkinje tissue primordia interconnect with ventricular trabeculae in contact with the myocardial tissue in the region of the inter-ventricular septum (see also Walls (1947); Wenink (1976); and for ultrastructural observations on the development of the S-A and A-V nodes, see Yamauchi (1965)).

The arterial and venous systems

In the early embryo, the blood from the heart goes through the outflow tract to the branchial arches and then around the primitive pharynx to the dorsal aorta, with the majority of the oxygenated blood initially going rostrally. The venous drainages of the rostral and caudal regions of the embryo are respectively through the anterior and posterior cardinal veins that anastomose to form either the left or right common cardinal veins, each of which drains into a horn of the sinus venosus (Figure 4.1.3).

After about TS 20, when the aortico-pulmonary spiral septum divides the outflow tract, the rostral and caudal arterial supplies become separate. The majority of the oxygenated blood supply from the heart goes to the rostral part of the embryo through the ascending thoracic aorta and its branches. The smaller oxygenated supply to the caudal part of the embryo is through the pulmonary trunk via the ductus arteriosus to the post-ductal part of the dorsal aorta. The venous drainages of the rostral and caudal regions of the embryo are through the superior and

TS	1	2	3	4	5	6	7	8	9	10	11	12	13	14	15	16	17	18	19	20	21	22	23	24	25	26
E	0	1	2	3	4	4.5	5	6	6.5	7	7.5	8	8.5	9	9.5	10	10.5	11	11.5	12	12.5	13.5	14.5	15.5	16.5	17.5

inferior venae cavae which respectively derive from the right and left anterior cardinal veins and from the right posterior cardinal vein, as that of the left side regresses (Figure 4.1.7).

Because most of the oxygenated blood emerging from the heart is directed towards the developing brain, cephalic region and forelimbs, it follows that this is the principal source of the deoxygenated blood that returns to the heart. The relatively small volume of poorly oxygenated blood that initially passes to the rest of the body through the ductus arteriosus also returns to the heart via the (right) inferior vena cava. At later stages of gestation (from about TS 21/22), with the greater degree of growth of the caudal half of the embryo, the venous drainage from it becomes substantially greater. Even so, with the establishment of inter-atrial and inter-ventricular septation and the division of the outflow tract by the aortico-pulmonary spiral septum, relatively little intermixing of oxygenated and deoxygenated blood occurs.

While the venous drainage from the rostral part of the embryo via the superior venae cavae is relatively straightforward (see below), that from the caudal region is rather more complicated as there are several sources of venous blood that return to the heart. The complete inflow of blood to the early heart actually comes from three sources (Figure 4.1.3), deoxygenated blood from the body (via the left and right **common cardinal veins**) and from the yolk sac (via the **vitelline** or **omphalomesenteric veins**), together with oxygenated blood comes from the placenta (via the **umbilical veins**). Initially, both left and right umbilical veins are present as two major vascular channels, but the right vessel generally regresses (around TS 14/15), leaving the **left umbilical vein** as the only vessel that carries oxygenated blood from the placenta to the embryo/fetus. To complete the vascular circuit, deoxygenated blood passes from the embryo via the dorsal aortae and then to the right and left umbilical arteries, thence through the umbilical cord to the placenta where gaseous exchange occurs before birth.

By TS 19, the oxygenated blood from the left umbilical vein largely flows into the **ductus venosus**; this is a vessel of substantial diameter whose differentiation is discussed in *The derivatives of the septum transversum* section of Chapter 4.4, *The gut and its associated glands* and whose extra-hepatic course is soon (TS 22) located in the lower border of the **falciform ligament** (the most caudal and ventral part of the ventral mesentery of the foregut in which the liver is suspended; see Figure 4.4.1). The ductus venosus passes through the substance of the liver, to anastomose with the inferior vena cava, but its contents only intermix to a minimal degree with the blood within the hepatic venous plexuses. When the

Figure 4.1.7. Diagram illustrating the primitive venous drainage of the lower limbs, paired abdominal organs and body wall of the trunk. Initially the venous drainage of the lower half of the body, including the lower limbs, is via the left and right posterior cardinal veins. The intermediate part of their course disappears, and is replaced by the subcardinal venous system that drains the paired abdominal organs (the kidneys, adrenals and gonads) but not the gut (which initially drains into the vitelline venous system, and later to the liver via the portal system). At the same time the venous drainage of the body wall is via the supracardinal venous system. The components of these two systems on the left side tend to regress, leaving the vessels on the right side as the dominant structures. The right subcardinal vein in due course becomes a major component of the prehepatic part of the inferior vena cava.

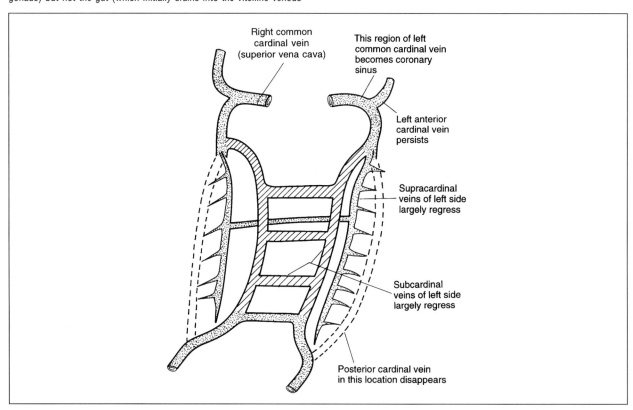

blood from this vessel emerges from the liver, it forms a major component of the blood that flows through the post-hepatic **inferior vena cava**. The latter vessel is continuous with the pre-hepatic (caudal) part of the inferior vena cava which carries the deoxygenated blood from all parts of the body caudal to the heart, except for the venous drainage of the gut which passes to the liver via the vitelline venous system which in due course differentiates into the portal vein. This arrangement consequently allows the venous blood from the entire body to return to the right atrium of the heart. For further details, see section entitled *Septum transversum*, in Chapter 2.4, on *Early organogenesis*.

From about TS 19 onwards, some of the deoxygenated blood entering the heart originates in the hepatic veins, and this represents the venous blood from the liver and alimentary tract. An additional, slightly greater volume of deoxygenated blood also drains into the post-hepatic part of the inferior vena cava from the hindlimbs and from the caudal half of the body which includes the urogenital system and body wall. Together, they represent the venous drainage of those parts of the embryo that receive their oxygen via the small volume of oxygenated blood (~30%) from the placenta that enters the right ventricle (see above).

By about TS 20, the venous drainage from the rostral part of the body is through the right and left superior venae cavae (Figure 4.1.8). In contrast to the human where only the right superior vena cava persists, rodents have both a right and left superior (or cranial) vena cava. The **right** vessel drains the right side of the head and neck, and the right forelimb into the cranial (superior) part of the right atrium as in the human. The **left** cranial (superior) vena cava drains the left side of the head and neck and the left forelimb and passes in front of both the aortic arch and the root of the left lung before it drains into the inferior vena cava just caudal to the floor of the right atrium. Before it fuses with the *inferior* vena cava, the left superior vena cava receives the venous drainage of

Figure 4.1.8. These two diagrams compare the venous drainage of the head and neck region (and the upper limbs, not shown) in the mouse (a) and in the human (b). (a) The mouse retains the primitive symmetrical arrangement, with the left and right anterior cardinal veins which subsequently differentiate respectively into the left and right superior (cranial) venae cavae. The venous blood from each side of the head and neck and the upper limb of the same side thus drains to the superior vena cava of the same side. On the right side, the superior vena cava drains into the rostral part of the right atrium. By contrast, the blood in the left superior vena cava drains into the inferior vena cava, and thence into the caudal part of the right atrium. (b) The human system (shown here at about 8 weeks of gestation) reorganizes: a large vessel, the left brachiocephalic (innominate) vein, forms which drains the venous blood from the left side of the head and neck and left upper limb into the right superior vena cava. The caudal part of the left anterior cardinal vein disappears (so that no left superior vena cava differentiates), and the left common cardinal vein diminishes in importance and drains into what was formerly the left horn of the sinus venosus. This vessel subsequently differentiates into the coronary sinus. Very rarely, if the left posterior cardinal vein persists and develops into a left inferior vena cava (in the human), this vessel drains into the coronary sinus. This allows the venous blood from the left lower half of the body to drain into the right atrium.

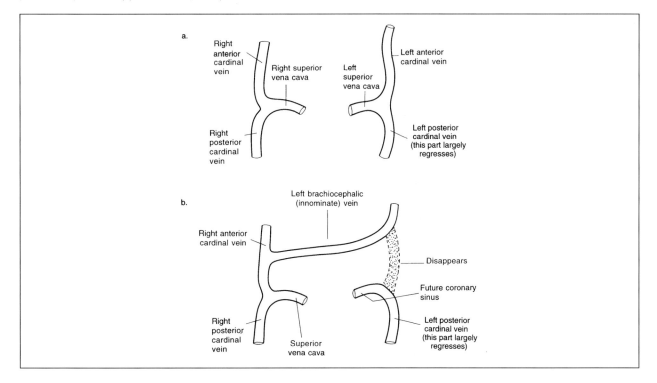

TS	1	2	3	4	5	6	7	8	9	10	11	12	13	14	15	16	17	18	19	20	21	22	23	24	25	26
E	0	1	2	3	4	4.5	5	6	6.5	7	7.5	8	8.5	9	9.5	10	10.5	11	11.5	12	12.5	13.5	14.5	15.5	16.5	17.5

the thorax (via the azygos vein), and from that point is usually referred to as the **coronary sinus** (Halpern, 1953; Hebel and Stromberg, 1986). The bronchial veins drain into the coronary sinus, as do the small cardiac veins and a dorsal vein of the heart. A simplified view of this arrangement in the mouse is shown in Figure 4.1.6a, where it is contrasted with the arrangement observed in humans (Figure 4.1.6b).

In the mouse, the deoxygenated blood from the right side of the head and neck and right forelimb goes to the right atrium via the right superior (cranial) vena cava. There, it mixes with that portion of the oxygenated blood from the inferior vena cava that has failed to pass through the foramen ovale into the left atrium. The resultant "mixed" blood (*i.e.* largely deoxygenated blood with a relatively small component of oxygenated blood), with its significantly diminished oxygen tension, passes to the right ventricle, into the pulmonary trunk and then, via the **ductus arteriosus**, into the descending aorta (*i.e.* into its "post-ductal" part) for distribution to the lower half of the body (see Figure 4.1.9). Only an insignificant volume of this blood passes, via the pulmonary arteries, to the lungs during the pre-natal period.

The presence of the left SVC inevitably means that some deoxygenated blood intermixes with the largely oxygenated blood that flows in the post-hepatic part of the inferior vena cava and through the foramen ovale into the left atrium. An additional small volume of deoxygenated blood from the vitelline veins also drains into the post-hepatic part of the inferior vena cava via the hepatic veins.

It is also worth noting that, in those rare instances (in the human, and presumably also in the mouse) where the *left* posterior cardinal vein persists, it differentiates to form the (abnormal) *left* inferior vena cava, and drains (via the coronary sinus and the oblique vein of the left atrium, the derivatives of the left horn of the sinus venosus) into the caudal part of the right atrium close to the site of entry of the *definitive* inferior vena cava (the derivative of the rostral part of the right posterior cardinal vein*)*.

Because of the persistence of the left superior vena cava in the mouse, the efficiency of the lower free border of the septum secundum in separating the oxygenated from the deoxygenated blood in rodents must be considerably less than that in humans, where almost all of the deoxygenated blood flows into the right atrium via the single superior vena cava (see

Figure 4.1.9. The principal features of the pre-term fetal circulation. The ascending aorta which contains oxygenated blood is directed towards the head and neck region (via the common carotid arteries) and to the upper limbs (via the left and right subclavian arteries). The pulmonary trunk, containing largely deoxygenated blood (principally from the head and neck and upper limbs), directs most of its flow into the ductus arteriosus, with minimal blood being directed (via the pulmonary arteries) to the lungs (which do not start to function until after birth). The ductal blood enters the thoracic aorta just distal to the origin of the left subclavian artery, and passes caudally into the descending (abdominal) aorta. As the blood in this vessel passes more caudally it supplies the tissues and organs of the abdomen and lower limbs with what little oxygen it contains, and the blood within it becomes increasingly deoxygenated. This blood then passes within the umbilical arteries (via the umbilical cord) to the placenta where it is re-oxygenated. The blood, now with a high oxygen tension, returns to the fetus (via the umbilical cord) through the left umbilical vein, and thence via the ductus venosus and post-hepatic part of the inferior vena cava to the right atrium from which it passes through the foramen ovale to the left atrium, thence to the left ventricle and into the systemic circuit.

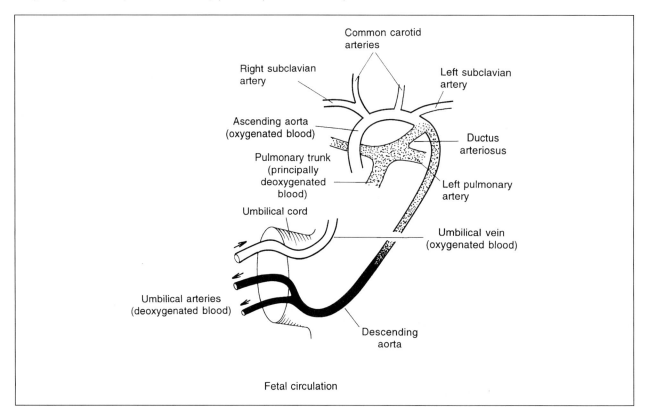

Fetal circulation

above) and then passes (largely without intermixing with the oxygenated blood that flows from the ductus venosus into the right atrium) into the right ventricle. In the human, a new vessel differentiates, the **brachiocephalic vein** which allows the venous drainage from the left side of the head and neck region and the left upper limb to pass across to the right side (Figure 4.1.8b). This vessel drains into the right superior vena cava rostral to the site where the latter drains into the right atrium. Accordingly, all of the deoxygenated blood from the upper half of the body drains into the (right) superior vena cava.

Before birth, the tissues of the lungs receive their oxygen supply and nutrition from the **bronchial arteries**, which are branches of the descending thoracic aorta or from the upper posterior (aortic) intercostal arteries. These vessels supply the bronchi as well as nourishing the connective tissue of the lungs and the visceral pleura. In the human, the bronchial veins drain via the azygous system of veins into the superior vena cava either directly (the azygos vein on the right side) or occasionally indirectly (via the accessory hemiazygos and superior intercostal vein) into the brachiocephalic vein. In the mouse, the azygos vein (from the left side) drains directly into the left superior vena cava, while the hemiazygos vein (from the right side) drains across the midline and thence into the azygos vein.

Post-birth changes

The heart

The pre-natal system described above is designed to circulate blood which has received its oxygen supply via the placenta rather than from the lungs. Because of the high degree of vascular resistance present in the primitive lungs, only a relatively insignificant volume of blood passes through the pulmonary circulation before birth. The result is that, as discussed above, blood with a high oxygen tension from the **left umbilical vein** which passes to the **right atrium** via the **ductus venosus** and then the **inferior vena cava** is deflected via the inferior border of the **septum secundum** (the *crista dividens*) through the **foramen ovale** into the **left atrium**, to the **left ventricle** and out through the **aorta** into the systemic circulation. This route ensures that those parts of the body that are differentiating rapidly, and consequently require the highest levels of oxygen (*e.g.* the developing brain, head and neck regions and the forelimbs), receive blood with the highest oxygen tension, while the rest of the body receives a mixture of oxygenated and deoxygenated blood via the pulmonary trunk, the ductus arteriosus and the dorsal aorta (Figure 4.1.9).

At birth, the blood supply from the placenta through the **left umbilical vein** ceases abruptly and the pulmonary circulation now has to function. While the physiological changes underpinning this are complex, the key event seems to be the expansion of the lungs, brought about mainly by the action of the diaphragm and by the inspiratory muscles of the thorax; this creates a strong negative pressure of 20–70 cm H_2O (Geubelle *et al.*, 1959), and its magnitude is striking as compared to the normal venous pressures of the circulatory system. Once this "opening" pressure is created, air and blood rush into the lungs, and the substitution of air for fluid in the alveoli is a factor tending to lower the pulmonary vascular resistance (Lind, 1969).

The main anatomical change that now occurs is relatively simple: as breathing starts, the **foramen ovale** closes. This closure process is entirely mechanical, and occurs when the pressure within the left atrium rises and is maintained at a higher pressure than that in the right atrium due to the need of the left ventricular output to overcome the increased peripheral resistance of the systemic circuit; this increased pressure within the left compared to the right atrium presses the septum primum onto the septum secundum, so closing and eventually sealing the foramen ovale. At the same time, the ductus arteriosus closes (see below) and the net result is that blood entering the **right atrium** now flows exclusively into the **right ventricle**, out through the **pulmonary trunk** and **pulmonary arteries** to the lungs. The oxygenated blood then returns through the pulmonary veins to the **left atrium**, is pumped into the **left ventricle** and then is directed through the **aorta** to the rest of the body (for detailed accounts of the circulatory changes that take place at birth, see Barron, 1944; Dawes, 1961; Lind, 1969; Comline *et al.*, 1973).

The redundant blood vessels

Shortly after birth, when the foramen ovale effectively closes, major changes take place within those components of the vascular system that the newborn no longer needs. Once the foramen ovale closes, blood no longer flows through the **left umbilical vein** and **ductus venosus**, and these vessels usually constrict and fibrose, forming the **ligamentum teres** and **ligamentum venosum**, respectively. Similarly, the two **umbilical arteries**, initially the two principal branches of the internal iliac arteries, also constrict and fibrose beyond the site where the two superior vesical arteries (the principal arterial supply to the upper part of the bladder) are given off, forming the **medial umbilical ligaments** (in contrast to the **median** umbilical ligament which represents the

TS	1	2	3	4	5	6	7	8	9	10	11	12	13	14	15	16	17	18	19	20	21	22	23	24	25	26
E	0	1	2	3	4	4.5	5	6	6.5	7	7.5	8	8.5	9	9.5	10	10.5	11	11.5	12	12.5	13.5	14.5	15.5	16.5	17.5

obliterated urachus, a narrow canal which, during the pre-natal period, passes from the apex of the bladder to the umbilicus).

Perhaps the most important of the major vessels that no longer serves any useful role after birth is the **ductus arteriosus**. Prostaglandins play an important role in maintaining its patency during pregnancy, and it is the reduction in the level of circulating prostaglandins as well as the change in the oxygen tension of the blood flowing through this vessel after birth that leads to its closure (Lind, 1969). Before birth, the blood that flows through this vessel from the pulmonary trunk is largely deoxygenated; from very shortly after birth, after breathing is established, virtually all of the deoxygenated blood from the pulmonary trunk is directed into the lungs. For the minutes, or possibly hours that the ductus remains open after birth, the flow through the ductus is reversed as compared to the situation before birth, with oxygenated blood entering it from the direction of the thoracic aorta.

The exact mechanism and timing of the closure of the ductus appears to vary considerably among species and an analysis of these events in the mouse (M.H.K., unpublished observations) indicates that the closure process has both immediate and long-term components. Within a matter of minutes after birth, the lumen of the ductus is seen to be almost completely obliterated, and this mainly seems to be due to a sudden and dramatic degree of hypertrophy of the layers of the tunica media of the muscle wall that occurs at this time. A decrease in the overall diameter of almost the entire length of this vessel is also seen but, though an important component of the obliteration process, this appears to be of lesser importance than the hypertrophic changes indicated above in this species. The ductus arteriosus then starts to fibrose and eventually (possibly weeks later in the mouse, and usually up to 6 months after birth in the human) becomes replaced by a fibrous band, the **ligamentum arteriosum**. Occasionally, this vessel may remain open, either as a single vascular anomaly (termed **patent ductus arteriosus**), or as one component of a spectrum of congenital abnormalities of the heart and cardiovascular system. While the latter is occasionally observed clinically in humans, similar anomalies may occur spontaneously or be induced experimentally in the mouse.

4.2 The limbs

Introduction

The limb buds form from a central core of flank mesoderm and have a covering of surface ectoderm from which the apical part of its lateral border will form the **apical ectodermal ridge** that plays a key role in the growth and differentiation of the limb. Most of the mesoderm derives from the **somatopleuric component** of the **lateral plate mesoderm** and this will form the skeletal, tendon and mesenchymal components of the limb; the remaining mesenchyme that will form the musculature comes from the **myotome component** of the **somites** which migrates into the limb buds at an early stage of their development. Here, we will first describe the general development of the forelimb mentioning how the hindlimb differs and then briefly examine some specific aspects of limb development that will include how it obtains its nerve supply and blood vessels. We also consider aspects of pectoral and pelvic girdle formation and ossification, and their articulation with the fore- and hindlimbs.

Normal development

The first evidence of forelimb bud development is seen when a ridge starts to form in the future lower cervical/upper thoracic region, opposite somites 8–12, when the embryo possesses about 15–20 pairs of somites (late TS 14), while the hindlimb bud, which develops in the lower lumbar/upper sacral region, opposite somites 23–28, is first clearly evident in the caudal part of the lateral ridge, when the embryo possesses about 30–35 pairs of somites (TS 16). For much of its development, the hindlimb is slightly delayed in its differentiation compared to that of the forelimb, although the developmental difference between them diminishes as pregnancy proceeds.

Apical ectodermal ridge (AER) differentiation of the surface ectoderm occurs along the peripheral cranial–caudal margin of the forelimb and hindlimb buds, and is first clearly seen a day or so after the limb buds first appear (TS 16 and 17/18, respectively). The primary function of the AER is to induce the underlying mesenchyme within the limb bud to grow and differentiate, and this occurs in a proximal-to-distal sequence as the bud extends. As the cells that were originally subjacent to the AER gradually get further away from the influence of the AER, they withdraw from the cell cycle, their mitotic index falls, and differentiation follows. Anterior–posterior pattern formation (*e.g.* the digits) is controlled by a region of mesenchyme subjacent to the caudal end of the AER known as the zone of polarizing activity (ZPA). The AER and the ZPA thus control proximal–distal and anterior–posterior pattern formation in the limb, while dorsal–ventral patterning is regulated by ectoderm-derived signals, and these functions have been confirmed by the results of numerous transplantation experiments. Although most of these experimental studies have been carried out with avian embryos as their limbs are accessibile for surgical manipulation *in vivo*, such evidence as is to hand suggests that the findings are equally applicable to mammalian embryos[1].

By about E10.5–11 (TS 17/18), the limb buds become increasingly prominent, and have begun their rostral "ascent", and this gradual cranial migration that now occurs in their level with respect to the embryonic axis requires some explanation. The first indication of the forelimb bud is as an enlargement at the rostral extremity of the prominent ridge that forms along the lateral part of the body wall opposite somites 8–12. The forelimb buds thus form at the level of what will become the lower cervical/upper thoracic component of the vertebral axis, but which is located initially at about the level of the mid-abdominal region because of the primitive configuration of the embryo at this stage of development.

With the growth and differentiation of the embryo, the position of the forelimb bud gradually changes, so

TS	1	2	3	4	5	6	7	8	9	10	11	12	13	14	15	16	17	18	19	20	21	22	23	24	25	26
E	0	1	2	3	4	4.5	5	6	6.5	7	7.5	8	8.5	9	9.5	10	10.5	11	11.5	12	13	14	15	16	17	18

THE ANATOMICAL BASIS OF MOUSE DEVELOPMENT

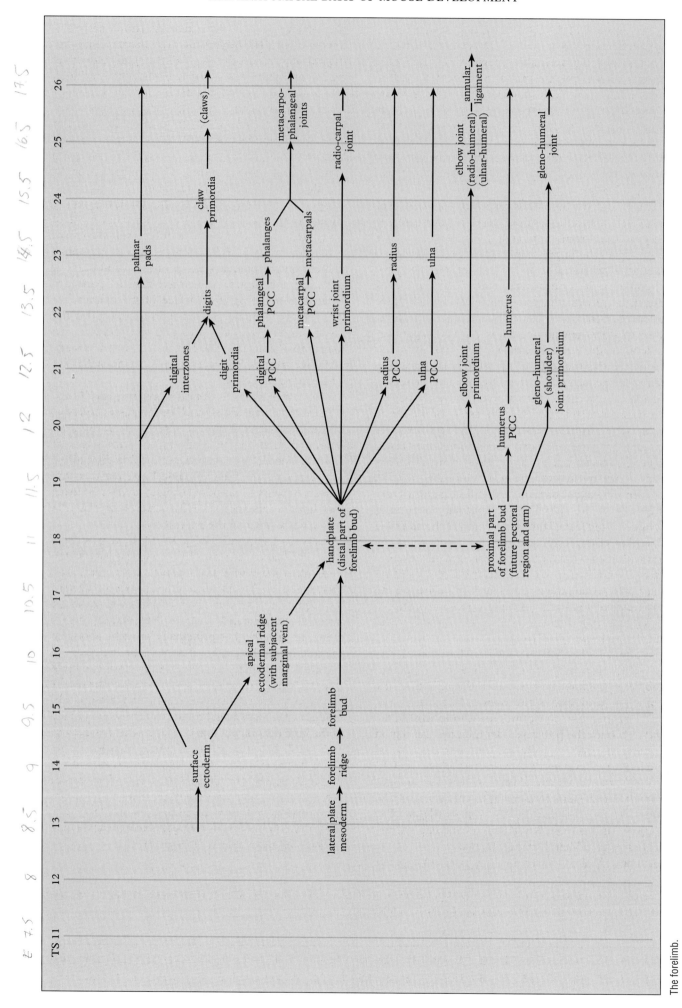

The forelimb.

that by about E11.5 (TS 19/20), it is seen to have apparently migrated cranially (or "ascended") due to elongation of the neck and "descent" of the heart opposite whose level the limb is now located. At this time, the hindlimb bud, which was first evident as an enlargement at the caudal extremity of the lateral body wall ridge opposite somites 23–28, in the mid- to distal part of the tail, is also seen to have migrated cranially, or "ascended", to its definitive position opposite the level of the pelvic region, and just proximal to the *origin* of the tail.

Thus, while the limb buds remain at the same level vis-à-vis their specific somites, the differential growth and gradual relocation of the thoracic and abdominal cavities gives the impression that the limb buds have "ascended" with respect to the thoracic and abdominal contents. It is only at the end of this period of differential growth that the limb buds achieve their definitive location – the forelimbs being located approximately opposite the upper thoracic region (at the level of the pectoral girdle, where the humerus articulates with the scapula), and the hindlimb at the level of the pelvic girdle (where the head of the femur articulates with the acetabular fossa of the hip bone).

It is also worth noting that the cavity of the hindbrain (the fourth ventricle) also ascends as development proceeds. Soon after it forms, the caudal part of this cavity extends as far down as the mid-thoracic region, but it later "ascends" with the junction between the caudal part of the hindbrain and rostral extremity of the spinal cord eventually becoming located at the level of the caudal part of the skull (in the region of the foramen magnum) where it articulates with the rostral part of the first cervical vertebral body. Differential growth of the spinal cord during embryonic development also accounts for the change in the level of its caudal (terminal) part at the time of birth (for further discussion on this topic, see Chapter 4.6 *The brain and spinal cord*).

By about E11.5 (TS 19), the limb buds are clearly divided into proximal and distal elements. In the forelimb, the proximal part includes the region of the future pectoral girdle and arm, while the distal, paddle-shaped part constitutes the **handplate**. In the hindlimb bud, the proximal part includes the region of the future pelvic girdle and leg, while the distal paddle-shaped part constitutes the **footplate**. By E12 (TS 20), the handplate shows evidence of angular contours at its peripheral margin, and these correspond to the location of the future digits. The footplate, at this time, is still paddle-shaped, and has yet to display evidence of digit formation.

At about E12.5–13 (TS 21), the **carpal region** is first delineated, and the handplate shows early evidence of **digital rays** that are separated by the **digital interzones**. By E13.5 (TS 22), the individual digits are clearly seen as the interzones have thinned due to progressive programmed cell death (or apoptosis) of the mesenchyme cells within the interzones. Up to about E14.5 (TS 22/23), all of the digits of both the forelimbs and hindlimbs are splayed out, but, by E16.5 (TS 25), digits 2–5 in the forelimb are almost parallel to each other. This change occurs in the digits of the hindlimb a day later.

By about E14.5 (TS 22/23), the tips of the palmar surface of the forelimb digits are somewhat swollen; a day later, these regions have differentiated to form prominent **tactile pads**, and these are now also present on the plantar surface of the distal parts of the hindlimb digits. Additional **digital pads** are located over the palmar surface of the metacarpo-phalangeal joints of the forelimb and over the corresponding metatarso-phalangeal joints of the hindlimb. There are five digital pads in each limb, while the hindlimb also has four **metatarsal pads** and two **tarsal pads**.

The first suggestions of **nail** (claw) **primordia** are seen on the dorsal surfaces of the terminal phalanges of the medial four digits (*i.e.* digits 2–5) of the forelimb, and on all the hindlimb digits by about E15.5. A nail primordium is not seen on the first digit of the forelimb until after E17.5. The latter digit, in the mouse, appears to be quite vestigial, and its eventual claw has the appearance of a flat plate, rather than the distinct claw seen on the other digits.

From as early as E12 (TS 19/20), a moderate degree of lateral rotation of the *proximal* part of the forelimb is seen, and this is much more evident by about E15.5 (TS 24). In order to accommodate tetrapedal gait, the *distal* part of the forearm region needs to rotate medially through 180°, and this movement (or *pronation*) is principally brought about by the rotation of the head and neck of the radius at the proximal radio-ulnar joint within the **annular ligament**. During pronation, the distal part of the radius moves in an arc relative to the ulna. In the hindlimb, medial rather than lateral rotation of the *proximal* part of the limb occurs, but this is only evident from about E17.5 (TS 26). The extent of medial rotation of the distal, but more particularly the plantar region of the hindlimb needed to achieve tetrapedal gait is also in the region of 180°. This complex change in the orientation of the proximal and distal regions of the fore- and hindlimbs to achieve tetrapedal gait is most easily understood by reference to Figure 4.2.1.

---

[1] For recent reviews on the molecular control of growth and pattern formation in the vertebrate limb, see Niswander (1996) and Gilbert (1997); for earlier work, see Tickle *et al.* (1975) and Saunders (1977).

TS	1	2	3	4	5	6	7	8	9	10	11	12	13	14	15	16	17	18	19	20	21	22	23	24	25	26
E	0	1	2	3	4	4.5	5	6	6.5	7	7.5	8	8.5	9	9.5	10	10.5	11	11.5	12	12.5	13.5	14.5	15.5	16.5	17.5

Figure 4.2.1. Simplified diagram of a ventral view of a mouse illustrating how the proximal and distal regions of the forelimb and hindlimb rotate from the primitive (or "anatomical") position seen in the embryo in order to facilitate tetrapedal gait. In the forelimb, the *upper arm* rotates laterally, while the *forearm* rotates medially, so that the thumb becomes the most medial of the digits. The latter is achieved by the rotation of the *distal* part of the radius medially, while the ulna retains its original relationship vis-à-vis the humerus. In the hindlimb, by contrast, both the proximal and distal components rotate medially, so that the large toe becomes the most medial of the digits. Because the proximal part of the hindlimb has already rotated medially, the fibula remains as a lateral relation of the tibia throughout its course. The arrows indicate the direction of rotation of the limbs from their original position in the early embryo. The limbs in the diagram are shown in their definitive position.

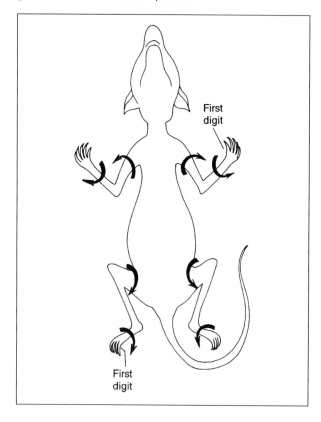

Bone formation

Ossification within the cartilage primordia of the bones of the limbs and their girdles

In the central core of the limb buds, mesenchyme condenses to form the cartilage models of the future skeletal elements, with the primary centres of ossification differentiating within them, this process being termed *endochondral ossification*. The primary centre of ossification for each bone is initially seen in the central region of the cartilaginous model, but ossification gradually extends proximally and distally until the complete shaft (or **diaphysis**) of the long bone is ossified. Separate, so-called "secondary" centres of ossification form in discrete cartilage models that are located at the extremities of the long bones (and occasionally elsewhere, at the sites of insertion of muscles) at sites that are usually distal to the cartilaginous growth plates (or **epiphyses**). Eventually the primary and secondary centres fuse following the cessation of growth of the long bones. Precartilage condensations of mesenchyme within the limbs form at E12 (TS 19/20), and are largely replaced by cartilage by about E13.5–14 (TS 22).

In the forelimb, the **humerus** has a significant primary centre of ossification by E15, and this extends rapidly along the shaft of the bone to include the site of insertion of the deltoid muscle, termed the **deltoid tuberosity** (a prominent bony protruberance to which the insertion of the deltoid muscle attaches) by E16. Both the radius and ulna have primary centres present by E15, but the cartilage primordia of the **carpal bones**, in contrast, show no evidence of ossification until about 7 days after birth. The **metacarpals ossify** in the order 3, 4, 2, 5, 1 a process that starts on about E17, with all but the first metacarpal containing ossification centres by E19; here, a centre only appears towards the end of the first postnatal week. By birth, small centres of ossification are seen in the phalanges of all of the digits, except the proximal phalanx of the first digit. The pattern of ossification of the skeletal elements of the first digit indicates that the first metacarpal/tarsal is in fact equivalent to the proximal phalanx of the other digits. Under normal circumstances therefore, the first digit only possesses two phalanges, while the others possess three; the metacarpal/tarsal element of the first digit was lost at an early stage of vertebrate evolution.

The pattern of ossification in the long bones of the hindlimb is similar to that of the forelimb, with some regions of chondrogenesis again ossifying after birth. Small ossification centres are present in both the **talus** and **calcaneum** at the time of birth, but these do not appear in the cartilage primordia of the other tarsal bones until about one week later. The first evidence of ossification in the patella is between 7 and 14 days after birth, and that in the **fabella**[2] some time later.

Timing of appearance of the primary ossification centres

The best way of following the development and extent of ossification in cartilage models is to fix the limbs in 80% ethanol, clear them and then bulk stain them using alizarin red S and Alcian blue to identify respectively the ossification centres and the cartilage models, and this process allows comprehensive measurements to be made (Patton and Kaufman, 1995). Details of the times of first appearance of the ossification centres of the long bones of the limbs and the limb girdles covering the period from E15 to 2 weeks after birth

---

[2] This pair of sesamoid bones (*i.e.* each has a centre of ossification that develops within a *tendon* that moves over a bony surface) is located within the lateral and medial heads (*i.e.* tendons of insertion) of the gastrocnemius muscle of the calf. The sesamoid surface related to another bone is covered by articular cartilage, and slides over it (for observations on the characteristic features of sesamoids, see Williams (1995)).

are given in Table 4.2.1. Diagrams showing the extent of ossification within their cartilage primordia at specific times during the prenatal and early postnatal period are given in Figures 4.2.2–4.2.6 (for additional studies in other strains of mice, see Johnson (1933), Wirtschafter (1960) and Hoshino (1967), and for comparable studies in the rat, see Strong (1925), Spark and Dawson (1928), Noback et al. (1949) and Hoshino (1967); for details of comparable studies in the human, see Patton and Kaufman (1995)).

These tables and figures provide baseline information, and allow comparisons of the degree of ossification achieved at specific stages of gestation to be made between normal material and, for example, mutant strains of mice (where there is often developmental retardation during the late prenatal period and at, or shortly after, birth). Mutant embryos as well as morphologically abnormal, and even some apparently morphologically normal embryos isolated from females exposed to a teratogenic stimulus during pregnancy can show a developmental delay of at least 24 h at E18 or E19 as compared to within-litter control embryos.

Table 4.2.1 *Time of first appearance of primary ossification centres within the forelimb and hindlimb skeleton of the mouse (from Patton and Kaufman, 1995, amended)*

	Primary centres			Primary centres	
	First seen	Present in all specimens studied		First seen	Present in all specimens studied
Forelimb			**Hindlimb**		
Scapula	NA[1]	14.5[2]	Ilium	15.5	15.5
Humerus	NA	14.5	Ischium	16.5	16.5
Ulna	NA	14.5	Pubis	16.5	16.5
Radius	NA	14.5	Femur	14.5	15.5
Carpus	26.5	26.5	Tibia	15.5	15.5
Metacarpals			Fibula	15.5	15.5
1	26.5	26.5	Calcaneus	18.5	20.5
2	16.5	16.5	Talus	18.5	20.5
3	16.5	16.5	Tarsus	26.5	26.5
4	16.5	16.5	Metatarsals		
5	17.5	17.5	1	17.5	18.5
Proximal phalanges			2	16.5	16.5
1	26.5	26.5	3	16.5	16.5
2	17.5	18.5	4	16.5	16.5
3	17.5	18.5	5	17.5	17.5
4	17.5	18.5	Proximal phalanges		
5	18.5	20.5	1	18.5	18.5
Middle phalanges			2	17.5	18.5
2	18.5	20.5	3	17.5	18.5
3	18.5	20.5	4	18.5	18.5
4	18.5	20.5	5	18.5	20.5
5	20.5	20.5	Middle phalanges		
Distal phalanges			2	18.5	20.5
1	18.5	18.5	3	18.5	20.5
2	17.5	18.5	4	18.5	20.5
3	17.5	18.5	5	20.5	20.5
4	17.5	18.5	Distal phalanges		
5	18.5	18.5	1	18.5	18.5
			2	17.5	18.5
			3	17.5	18.5
			4	17.5	18.5
			5	18.5	18.5
			Patella	33.5	33.5
			Fabellae	Not seen	—

[1] NA, information not available, since centres were already present in all samples of the earliest specimens studied.
[2] Days of pregnancy extended to include d1, 7 and 14 after birth as d20.5, 26.5 and 33.5 respectively, assuming birth occurs at E19.5.

TS	1	2	3	4	5	6	7	8	9	10	11	12	13	14	15	16	17	18	19	20	21	22	23	24	25	26
E	0	1	2	3	4	4.5	5	6	6.5	7	7.5	8	8.5	9	9.5	10	10.5	11	11.5	12	12.5	13.5	14.5	15.5	16.5	17.5

Figure 4.2.2. Key to the following 4 figures (4.2.3–4.2.6), in which the various components of the cartilaginous models of the skeleton of each of the limb bones are named: (a) the scapula and bones of the right upper arm and forearm, but excluding those of the wrist and hand, (b) bones of the right wrist, hand and its digits, (c) bones of the right half of the pelvis, excluding the sacrum, including the thigh and leg but excluding those of the ankle and foot, (d) bones of the right ankle, foot and its digits, (e) bones of the wrist, and (f) bones of the ankle.

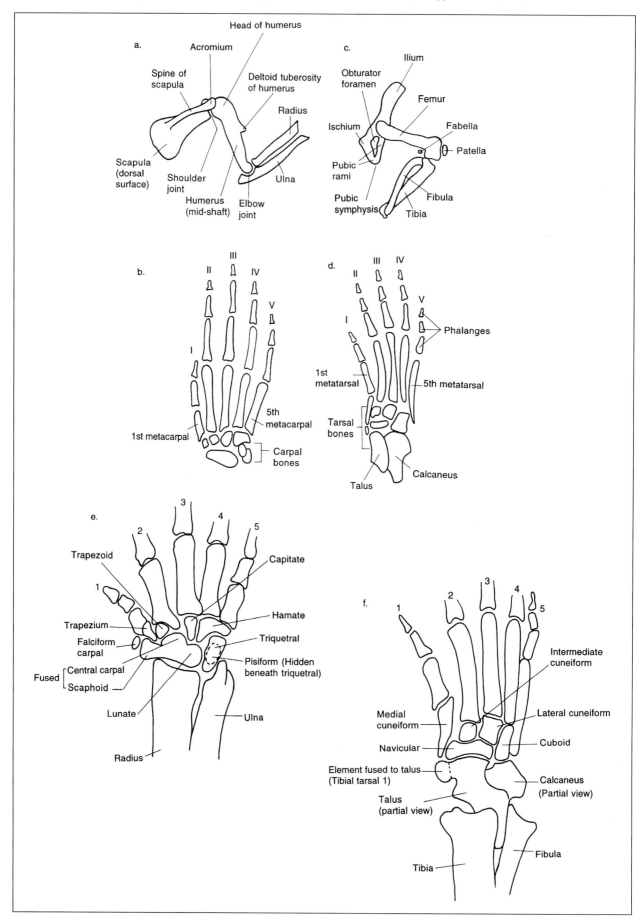

Figure 4.2.3. Location of the primary centres of ossification in the cartilage models of the skeleton of the scapula, bones of the upper arm and forearm, but excluding those of the wrist and hand. This sequence covers the period between day 15 of pregnancy (E14.5) to two weeks after birth. Note that the first of the *secondary centres* of ossification that will appear in this region is seen at the proximal end of the ulna, and that by two weeks after birth various other secondary centres have appeared. Even by this time there are still regions where the cartilage model has yet to show evidence of ossification. For key to the bones and their features, see Figure 4.2.2, and for summary of this information, see appropriate section of Table 4.2.1 (from Patton and Kaufman, 1995).

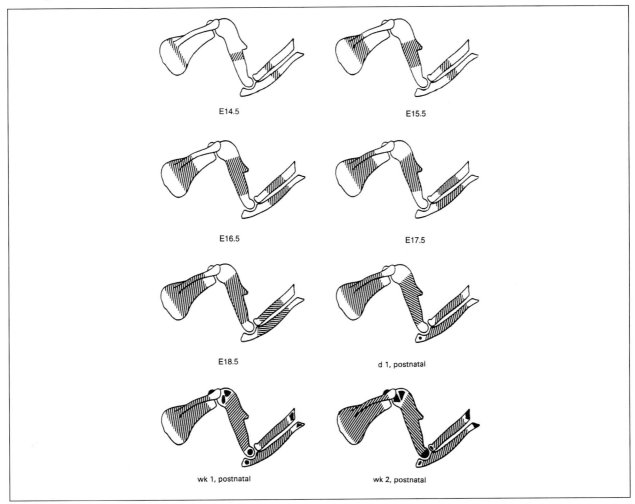

Joint differentiation

Differentiation of the joints and skeletal components of the limbs, together with the innervation of the limb musculature (see below), also occurs in a cranio-caudal and proximal-to-distal sequence. The primordium of the **shoulder (gleno-humeral) joint** is first clearly recognized during TS 21, there being only a diffuse condensation in this region during TS 20. The primordium of the **elbow joint** is also first seen at about this time, with that of the **wrist joint** a stage later (TS 22).

The first evidence of a synovial cavity within the shoulder and elbow joints is apparent at about TS 22/23. In the carpal region, the cavity of the wrist joint is first clearly seen by TS 24/25, as are the **metacarpo-phalangeal** and proximal row of **interphalangeal joints**, while the intermediate and distal interphalangeal joints appear during the following stage (TS 26).

In the hindlimb, the primordia of the **hip** and **knee joints** form during TS 21, and the various joint cavities differentiate several stages later, following a pattern similar to that in the forelimb. The synovial cavity of the hip joint is first clearly seen during TS 24, although no cavity is yet apparent in the knee or ankle joints, even though most of the components that will take part in these joints are present. By TS 25, the cavities within the knee joint and the proximal part of the ankle joint have formed, while those between the tarsal bones are first seen during TS 26, and will soon appear in the more distal joints.

The pectoral and pelvic girdles

The primordia of the various skeletal elements that

TS	1	2	3	4	5	6	7	8	9	10	11	12	13	14	15	16	17	18	19	20	21	22	23	24	25	26
E	0	1	2	3	4	4.5	5	5	6	6.5	7	7.5	8	8.5	9	9.5	10	10.5	11	11.5	12	12.5	13.5	14.5	15.5	16.5	17.5

Figure 4.2.4. Location of the primary centres of ossification in the cartilage models of the skeleton of the wrist, hand and its digits, covering the period between day 15 of pregnancy (E14.5) and two weeks after birth. By two weeks after birth, all of the carpal bones are fully ossified. For key to the bones, see Figure 4.2.2 and for summary of this information, see appropriate section of Table 4.2.1 (from Patton and Kaufman, 1995).

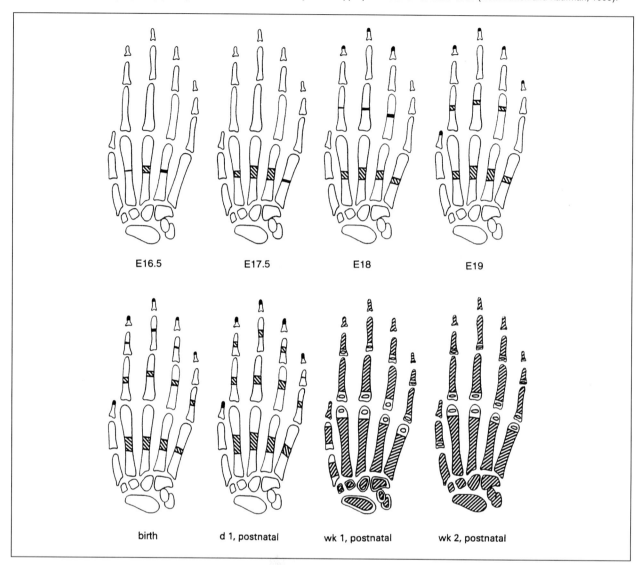

comprise the **pectoral girdle** (the **scapula, clavicle** and **sternum**) and **pelvic girdle** (the **innominate (hip) bone** and **sacrum**) are first seen during TS 21–23. The mesenchymal condensations that form the cartilage primordia of the clavicle and of the majority of the scapula can be recognized at TS 21, while that of the manubrium of the sternum is seen a stage later (TS 22), soon to be followed by that of the individual sternal elements or **sternebrae** (TS 22/23). Ossification within all elements of the sternum is first seen during TS 26.

The condensation of the pelvic skeleton occurs during TS 21, although the various cartilaginous components can only be recognized as separate entities during the following stage (TS 22). Three primary centres of ossification are involved in the formation of the hip bone: the first forms in the middle of the cartilage primordium of the iliac bone at about E14.5–15 (TS 23), while the other two appear a stage later (E16–16.5, TS 24) as small ossification centres in the central regions of the cartilage primordia of the **ischial** and **pubic bones**. Even at birth, however, extensive parts of the hip bone have yet to ossify, and these include regions of the **iliac crest**, the **acetabular cup** (or **fossa**), the superior and inferior rami of the pubic bone, and the region of the ischial tuberosity. A variable number of secondary centres may also form in the pelvic bones, and these are mainly located in the iliac crest and within the acetabular region (and form a large part of the acetabular articular surface of the adult bone). Occasionally, the anterior inferior iliac spine may form as a separate "centre", as may the pubic tubercle and crest and the symphysial surfaces of the pubic bone.

The elements that comprise the sacral and iliac components of the future **sacro-iliac joints** are initially (TS 22/23) widely separated rostrally, but more closely apposed caudally. Progressive differentiation of the sacro-iliac joint occurs during TS 24/25, with a gradual narrowing of the joint space, so that by about TS 26 no obvious gap is present between the auricular (ear-shaped) surfaces of the iliac bones and the sacrum. The sacro-iliac articulation is synovial, and located between the auricular surfaces of the sacrum and ilium which are respectively covered by fibrocartilage and by hyaline cartilage. The nature of the articulation seen here allows a small amount of anterior-posterior and rotatory movement at this joint, and occurs during flexion and extension of the trunk.

The primary centre of ossification in the **scapula** is first seen during TS 22, and soon occupies the middle third of the cartilage primordium. This ossification centre then expands, so that by birth it includes most of the blade and all but the most distal part of the spine. Several secondary centres appear postnatally, the most prominent being those associated with the **coracoid process** (the antero-lateral and upward projection of the scapula to which is attached the tendons of insertion of the short head of the biceps, the pectoralis minor and coracobrachialis muscles), in the distal part of the **acromion** (the most lateral part of the spine to which the lateral part of the clavicle articulates), along the lateral and inferior borders, and around the borders of the **glenoid fossa** (the site of articulation with the head of the humerus on the lateral extremity of the scapula). The first of these to appear is that associated with the coracoid process, and this is seen during the early postnatal period.

The majority of the shaft of the **clavicle** develops within a mesenchymal condensation that forms at about TS 21 within which one or more primary centres of ossification then develop (TS 22; this bone is thus, by definition, a "membranous" bone). Precartilaginous condensations appear in the medial and lateral parts of the clavicular primordium during TS 21 and these gradually extend around the membranous centre until the remainder of the primordium becomes chondrified (during TS 22/23). While ossification within the clavicle is initially confined to its middle-third (within the membranous region), it extends rapidly medially towards its sternal and laterally towards its scapular articulations, so that by the time of birth, all but the region close to the sterno-clavicular joint is ossified. The clavicle is absent in mammals in which the forelimbs are used principally or entirely for progression, *e.g.* the ungulates and carnivores, but it is present and well developed in animals which use the limb for prehension (for seizing and grasping), *e.g.* many rodents, the primates and man (Williams, 1995).

Figure 4.2.5. Location of the primary centres of ossification in the cartilage models of the skeleton of the right half of the pelvis (the innominate bone), but excluding the sacrum, including the thigh and leg but excluding the bones of the ankle and foot, covering the period between day 15 of pregnancy (E14.5) and two weeks after birth. The secondary centres of ossification in this part of the lower limb start to appear by about one week after birth in the distal region of the femur, proximal and distal regions of the tibia, and distal region of the fibula. By the second week after birth, the only substantial region in which ossification has yet to occur is in the inferior pubic ramus. For key to the bones, see Figure 4.2.2 and for summary of this information, see appropriate section of Table 4.2.1 (from Patton and Kaufman, 1995, amended).

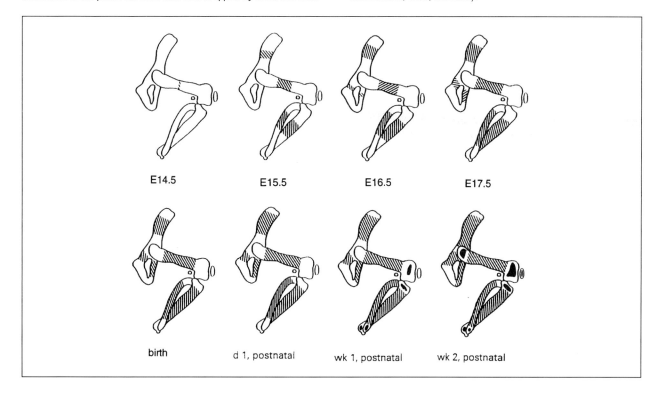

TS	1	2	3	4	5	6	7	8	9	10	11	12	13	14	15	16	17	18	19	20	21	22	23	24	25	26
E	0	1	2	3	4	4.5	5	6	6.5	7	7.5	8	8.5	9	9.5	10	10.5	11	11.5	12	12.5	13.5	14.5	15.5	16.5	17.5

Figure 4.2.6. Location of the primary centres of ossification in the cartilage models of the skeleton of the right ankle, foot and its digits, covering the period between day 15 of pregnancy (E14.5) and two weeks after birth. For key to the bones, see Figure 4.2.2 and for summary of this information, see appropriate section of Table 4.2.1 (from Patton and Kaufman, 1995).

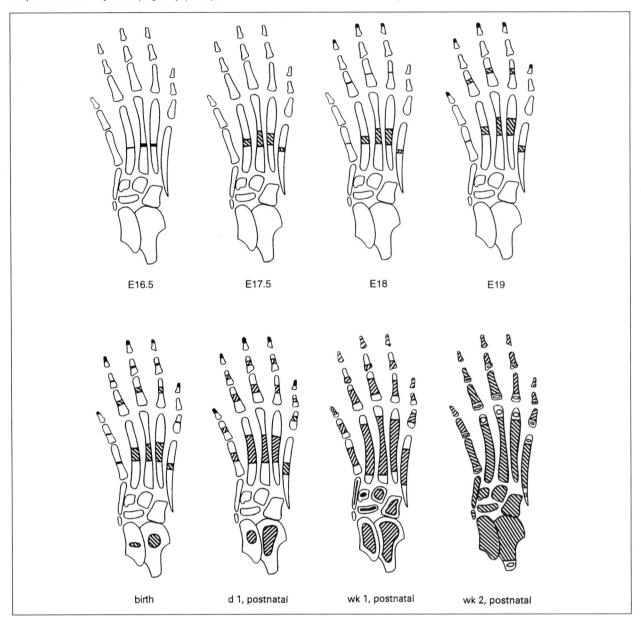

Muscle formation

The timings given in this section are for the forelimb bud, the corresponding stages in the hindlimb bud occur about one Theiler stage later.

A detailed analysis of the development and specification of the individual muscles of the limbs has yet to be undertaken, but the essential processes of their development are well known (Platzer, 1978). The critical questions in myogenesis research relate to when and how the muscle precursor cells begin to acquire their identity during gestation, at the somite stage, and how they recognize and respond to the signals that lead to their migration and differentiation. Approaches for investigating the molecular mechanisms which operate as muscle matures to produce the fibre type diversity seen in adult muscle have only recently become available (Buckingham, 1996).

Within the limbs, the musculature is derived from somitic mesoderm that migrates into the early limb bud (TS 19/20). In both the fore- and hindlimbs, the muscle masses within the limb initially develop as dorsal and ventral mesenchymal condensations (TS 20), in close relation to the pre-cartilage skeletal precursors, and early evidence of muscle cell alignment is also seen in the proximal muscle masses of the limb at this time. Innervation is initially to the mesenchymal contents in the proximal part of the limb (TS 20/21), and it is believed that innervation by the spinal nerve stimulates muscle differentiation. Even as early as E11–11.5 (TS 18/19), the nerve roots that will form the brachial plexus are readily recognized, and have already sent substantial branches into the proximal part of the forelimb bud. By E12.5 (TS 20/21), large nerve trunks are seen to enter the forelimb bud, and smaller branches pass more distally into the limb bud. The fact that early evidence of myotube formation is seen at this time provides indirect evidence that axonal contact has already occurred by this stage. Undifferentiated mesenchyme cells differentiate to form myoblasts, which then fuse together to form myotubes. It is at this stage that the latter cells withdraw from the cell cycle, and complete their development to form differentiated muscle fibres. Evidence of muscle cell alignment is seen at about TS 20; early myofibril-like structures are seen at about E13.5 (TS 22), and the first clear evidence of "striation" in the proximal limb bud muscles is seen during E14.5 (TS 22/23). A simplified diagram representing this sequence of events is illustrated in Figure 4.2.7. For a diagram which illustrates the general principles underlying the innervation of the limb muscles, see the section entitled *Nerves*, which appears later in this chapter.

The blood supply of the limbs

The essentials of the blood supply to the limbs are well known. The vascular supply to the limbs is from a series of intersegmental branches of the dorsal aorta at the levels at which the limb buds are initially situated. At about TS 16, one intersegmental vessel to the forelimb bud and another to the hindlimb bud become much enlarged, and these form the **axial arteries** to the limbs. That to the forelimb bud forms from the lateral branch of the **seventh cervical intersegmental artery**, while that to the hindlimb bud is from the **fifth lumbar intersegmental artery**. It is only from about E12.5–13 (TS 20/21) that both forelimbs receive exclusively oxygenated blood directly from the arch of the aorta. Before that time, when the origin of the left subclavian artery is still caudal to where the ductus arteriosus enters the descending aorta (previously the left dorsal aorta), and the right subclavian artery takes origin from the right dorsal aorta, they both receive a mixture of oxygenated and deoxygenated blood.[3] In both limbs, the axial artery terminates in a capillary plexus from which, at a later stage, digital arteries arise. The initial drainage of the blood supply from the fore- and hindlimbs is by the marginal veins, but these are subsequently replaced by the subclavian veins and external iliac veins, respectively.

As the development of the blood supply to the limbs is actually rather more complicated than this brief summary indicates, it is now considered in some detail.

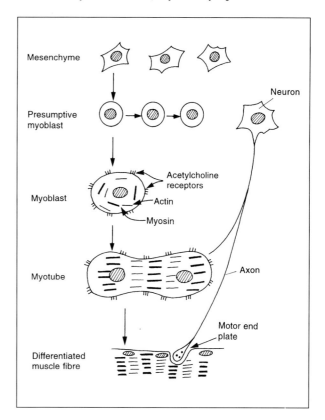

Figure 4.2.7. Diagrammatic sequence illustrating how the somite-derived mesenchyme precursors of the limb muscles undergo myofibrillogenesis to form differentiated (striated or skeletal) muscle fibres. Mesenchyme cells that are destined to form the striated muscle within the limbs initially round-up, but continue to divide: these are the *presumptive* myoblast cells. As the latter differentiate to form *definitive* myoblast cells, it is now possible to recognize both actin and myosin (cytoskeletal) elements randomly scattered within their cytoplasm, and also demonstrate the presence of acetylcholine receptors on their cell surface. Once contact is made with axonal elements, a further stage in the differentiation sequence occurs, with the fusion of adjacent myoblast cells to form *myotubes*. Within these cells, the actin and myosin elements become aligned to form recognizable, albeit primitive, sarcomeres. Such cells are therefore polynuclear, the nuclei being typically centrally located within their cytoplasm. The acetylcholine receptors at this stage are still randomly scattered over the cell surface. The characteristic features of *differentiated muscle fibres* are achieved slightly later in the sequence: these cells typically have numerous peripherally located nuclei, the cytoskeletal elements are aligned into characteristic repeated elements, termed sarcomeres, and the acetylcholine receptors become localized to a single site on the cell surface, at the motor end plate. If the myotubes are not contacted by axonal elements, they invariably degenerate.

---

[3] It is worth noting that the definitive location of the axial arteries is consistent with the basic principle which underlies the differentiation of the early embryonic and fetal circulation, where the oxygenated blood is initially selectively directed towards the head and neck (and brain), and somewhat later (during TS 20/21) is also directed to the forelimbs. If the axial vessels to the forelimbs were located any further caudally, they would receive some deoxygenated blood; in the case of the left forelimb this would be from the ductus arteriosus. The arterial blood supply to the right forelimb (once the right axial artery has "ascended" to its definitive location, and the right dorsal aorta caudal to the fourth arch artery has disappeared) only becomes uncontaminated with deoxygenated blood once its blood supply reaches it via the brachiocephalic trunk which receives its oxygenated blood supply directly from the ascending part of the thoracic aorta. The hindlimbs of course receive a mixture of oxygenated and deoxygenated blood.

TS	1	2	3	4	5	6	7	8	9	10	11	12	13	14	15	16	17	18	19	20	21	22	23	24	25	26
E	0	1	2	3	4	4.5	5	6	6.5	7	7.5	8	8.5	9	9.5	10	10.5	11	11.5	12	12.5	13.5	14.5	15.5	16.5	17.5

The arterial supply

The origin of each seventh cervical intersegmental artery that gives rise to the axial artery that will supply each *forelimb* is initially (from about TS 16, when first recognized, until TS 20/21) substantially caudal to the site of insertion of the sixth aortic arch artery into the dorsal aorta. However, this origin migrates cranially, or "ascends", until it is located approximately opposite the dorsal end of the fourth branchial arch artery (TS 21/22), at its site of attachment to the dorsal aorta. The proximal part of each forelimb axial artery differentiates to form the **brachial artery**, while its distal part forms the **interosseous artery**. Later the ulnar artery branches from the interosseous artery, while the radial artery branches, slightly more proximally, from the brachial artery (see Carlson, 1996). On the left side, the seventh cervical intersegmental artery corresponds to the stem of origin from the aortic arch of the **subclavian artery**, and this vessel represents the proximal part of the axial artery to the left forelimb. On the right side, the right fourth arch artery discharges exclusively into the subclavian artery, and this becomes continuous with the proximal part of the right axial artery.[4]

The situation is different in the *hindlimbs*: here, the axial arteries arise from the **external iliac arteries**. The original axial artery (lateral branches of the fifth lumbar intersegmental artery) supplies what becomes the dorsal aspect of the thigh region, and accompanies and supplies the sciatic nerve (via the so-called **ischiadic artery**). Later in development, a new vessel, which becomes the **femoral artery**, appears in the ventral aspect of the thigh and links distally with the original axial artery in the region just behind the knee. Proximally, the femoral artery is the continuation of the distal part of the external iliac artery, so that this vessel (rather than the internal iliac artery) subsequently becomes the principal arterial supply to the hindlimb. The proximal part of the original axial artery to the hindlimb is represented in the adult by the **inferior gluteal artery**, the **artery to the sciatic nerve**, and possibly by the distal part of the **peroneal artery**. In the region of the knee, the **popliteal artery** divides to give the anterior and posterior tibial arteries, and more distally the latter gives rise to the medial and lateral plantar arteries that supply the foot (based on human studies by Senior, 1919, 1920).

The venous system

At about E10.5 (TS 17), a prominent **marginal vein** is present just subjacent to the AER in both the forelimb and hindlimb, but most of the veins within the limbs are initially present as capillary plexuses. These soon increase in complexity and subsequently fuse and enlarge, giving rise to the definitive venous system of the limbs, with the axial venous channels that drain the limb buds being evident by about E11 (TS 17/18). The situation is complicated by the fact that the axial veins of the *forelimbs* originally drain into the cephalic portion of the **posterior cardinal veins**. Subsequently, due to the cranial migration of the forelimbs (discussed previously), these eventually drain into the **anterior cardinal veins** and, following their differentiation (in the mouse), into the left and right superior venae cavae (SVC) (for differences between the the mouse and human, see below). The proximal portion of the axial vessels that subsequently differentiate (by about TS 23) form the **axillary** and **subclavian veins**.

In the *hindlimb*, the axial venous channels differentiate at about TS 17/18 to form the **external iliac veins**. These initially drain into the right and left posterior cardinal veins, but, with the gradual obliteration of the left posterior cardinal system, both subsequently drain into the **inferior vena cava** (IVC). The caudal part of the IVC forms from the right posterior cardinal vein, while its more rostral part has a more complex origin. Part, at least, of the IVC here is believed to be of subcardinal venous origin, while other parts may be formed from the supracardinal venous system. The comparable vessels of the left side are usually completely obliterated so that all the left-sided structures caudal to the heart consequently drain across the midline into the IVC, while those that are rostral to the heart will all drain into the right atrium via the right and left SVC.[5]

---

[4] Normally, the right seventh cervical intersegmental artery migrates cranially and unites with the right subclavian trunk, but occasionally (*e.g.* when the seventh cervical intersegmental artery does not migrate sufficiently cranially, an event often associated with incomplete regression of the right dorsal aorta) the right subclavian artery can arise as a branch of the arch of the aorta just distal to the origin of the left subclavian artery. If this occurs, the right subclavian artery invariably passes to the right forelimb dorsal to the oesophagus (*i.e.* this vessel has a retro-oesophageal and retro-tracheal course). Persistence of either the entire right dorsal aorta, or parts of this vessel that normally regress may result in vascular "rings" that pass behind both the oesophagus and trachea, and may cause difficulty in swallowing (in the human, at least) because of compression (due to the presence of this aberrant retro-oesophageal vessel) on the lower part of the oesophagus.

[5] In the human, this drainage is via the right SVC alone and, since the pattern of venous drainage in the *left* forelimb is slightly different from that in the mouse, it is worth explaining why this should be the case. In both species, the venous blood from the *left* forelimb initially drains into the left axial vein, which then drains into the left anterior cardinal vein. This vessel then drains into the left common cardinal vein (the left duct of Cuvier, which was previously the left horn of the sinus venosus) which in turn drains across the midline into the right atrium. While the left SVC is retained in the mouse, the caudal part of the left SVC in the *human* regresses and is replaced by a new venous channel, the **left brachiocephalic vein**, which subsequently drains the left side of the head and neck, as well as the left upper limb, and then drains across the midline into the SVC. This latter vessel, which was previously the right anterior cardinal vein, is retained in both species. In the mouse, the left SVC passes ventral to the aortic arch and close to the root of the left lung to reach the *caudal* part of the right atrium, which it enters in close proximity to the IVC. While the caudal part of the left SVC disappears in the human, it is retained in the mouse. Before it enters the right atrium, however, it receives thoracic veins and the **azygos vein**, and from that point is equivalent to the **coronary sinus** of the human. In both species, however, the right half of the head and neck and right upper limb drain into the (right) SVC, and thence directly into the *rostral* part of the right atrium. The arrangement of the vessels that drain into the right atrium are discussed in more detail in Chapter 4.1 *The heart and its inflow and outflow tracts*.

The nerves

The innervation of the limbs is simple in principle, but the details are complex (for review, see Wolpert, 1998; see also Tosney *et al.*, 1995; Tessier-Lavigne and Goodman, 1996). The primitive muscle masses are initially innervated by **pathfinder nerves** (the pathway taken being under the control of the **growth cone** at the tip of the individual axons) that, in the forelimb, will eventually form the elements of the **brachial plexus** (TS 17). In the hindlimb, the innervating nerves will form the elements of the **lumbo-sacral plexus** (TS 20). The pathfinder axons that enter the limbs divide into anterior and posterior divisions, which respectively supply the flexor and extensor groups of muscles. The most proximal muscles in the limb are supplied by the nerves of highest segmental origin from the spinal cord, while the most distal muscles are supplied by segmental nerves of more caudal origin. It is of critical importance that movements are segmentally innervated, as this enables spinal nerves to innervate muscle groups that act together synergistically (McMinn, 1990). Indeed, it is only after the innervation of the muscle groups has taken place that joint formation can occur, as experimental studies have clearly demonstrated that limb movement plays an essential part in joint formation (Drachman and Sokoloff, 1966). In this important study, the muscles in chick limbs that acted across the sites of future joints were either denervated pharmacologically (*e.g.* with botulinum toxin), or the nerves to them were sectioned – in both situations, normal joint formation failed to occur.

Given this summary, we can now consider the general pattern of innervation of the components of the mature limb in more detail. In the forelimb, the detailed anatomy of the **brachial plexus** in the mouse is fairly similar to that in the human. In the mouse, the plexus is formed from the ventral (or anterior) primary rami of cervical (C) nerves C4–8 and thoracic (T) nerves T1 and T2, with the largest roots of the plexus coming from C6–8. In humans, however, C4 is only rarely involved, while T 2 also only occasionally contributes to the brachial plexus. The spinal nerves that enter the limb come from the cervical enlargement of the spinal cord, while the **lumbo-sacral plexus** that is involved in the innervation of the hindlimb comes from the lumbar enlargement of the cord.

The general principles which underlie the pattern of innervation in the human of the proximal muscle masses (including those to the upper arm and the rotator cuff muscles of the pectoral girdle), the arm and forearm muscles (including those whose long tendons pass distally into the hand) and most distally the intrinsic (or short) muscles of the hand, are illustrated diagrammatically in Figure 4.2.8.

The nerves that supply the individual muscles in rodents (see Hebel and Stromberg, 1986), while in many cases similar to those in humans, do not necessarily have the same root values; equally other nerves, such as the **axillary nerve**, while supplying the deltoid and teres minor muscles (as in humans), also supply the teres major muscle (which it does not in humans). The **phrenic nerve**, which supplies all but the most peripheral margin of the diaphragm, though of cervical spinal origin, but not a derivative of the brachial plexus, is usually derived from C4 and 5 in rodents (though it may have a similar origin to that in humans, where it is derived from C3–5; see Wilson, 1968).

The major nerves of the forelimb of the mouse, such as the **ulnar**, **radial** and **median nerves**, have similar though not identical functions to those in the human; in both species, the radial nerve supplies the extensor muscles of the arm/forearm (*e.g.* the triceps group, anconeus and coracobrachialis), but, in the mouse, it also supplies the short head of the biceps muscle (in the human, this is supplied exclusively by the musculocutaneous nerve). It is of interest that the flexor compartment of the forelimb invariably has a richer nerve supply than the extensor compartment so that flexor skin is thus more sensitive than extensor skin, and has a richer sensory innervation, particularly in the distal part of the limb. Similarly, flexor muscles tend to be under more precise control than extensor muscles, and this is consistent with their richer sensory and motor innervation. The most caudal nerve in a limb plexus is distributed entirely to the flexor compartment of the limb (McMinn, 1990).

The nerves supplying the muscles that act across the joints of the forelimb (such as the shoulder, elbow and carpal joints), also give sensory branches to these joints (this relationship was first described by John Hilton (1805–78), and it is accordingly known as Hilton's Law); such an arrangement provides feedback from the nerve endings (stretch receptors) in the tendons and joint capsule and allows the position in space of all components of the limb to be computed by the brain at any time. The pattern of innervation of these joints is derived directly from the information supplied in Figure 4.2.8, where the most proximal joint receives its innervation from the most *rostral* elements of the brachial plexus, while the most distal joints receive their innervation from the most *caudal* elements of this plexus. The shoulder joint, for example, thus receives its innervation from the most rostral elements which make up the brachial plexus, while the carpal, metacarpo-phalangeal and interphalangeal joints receive their innervation from the most caudal elements of the plexus. Similarly,

TS	1	2	3	4	5	6	7	8	9	10	11	12	13	14	15	16	17	18	19	20	21	22	23	24	25	26
E	0	1	2	3	4	4.5	5	6	6.5	7	7.5	8	8.5	9	9.5	10	10.5	11	11.5	12	12.5	13.5	14.5	15.5	16.5	17.5

Figure 4.2.8. The principles underlying the arrangement of the nerve supply to the muscles of the forelimb. Since the proximal group of limb muscles are the first to differentiate within the forelimb bud, they receive their nerve supply from "pathfinder" nerves which enter the limb from the most rostral elements of the brachial plexus, from the ventral primary rami (or roots) of cervical nerves 5, 6 and 7. When these axons approach the pre-muscle masses in the forelimb bud, they divide into anterior and posterior divisions which respectively supply the flexor and extensor pre-muscle masses. These subsequently differentiate to form the "rotator cuff" group of muscles that supply the shoulder region. Slightly later, as the muscle masses within the upper arm begin to differentiate, they are supplied by nerve elements from C5–C8. The upper arm muscles develop into dorsal and ventral masses, the former giving rise to the extensor muscles, the latter to the flexors of this region. The nerve supply to the forearm and hand muscles follows a similar pattern. Because similar (but overlapping) groups of nerves supply both the dorsal (extensor) and ventral (flexor) compartments, this arrangement enables similar groups of spinal nerves to innervate muscle groups that act synergistically (*i.e.* flexion is associated with the controlled relaxation of the complementary extensor group of muscles). The critical feature of this pattern of innervation is that *movements* are segmentally innervated. With this information, it is possible to determine the nerve supply to the joints of the limb since (according to Hilton's Law) "the nerve supply to a joint is the same as that which supplies the muscles that act across the joint". The root values given in this diagram are for the human, but, as the underlying *principles* are equally applicable in the mouse, they are likely to be very similar in the mouse.

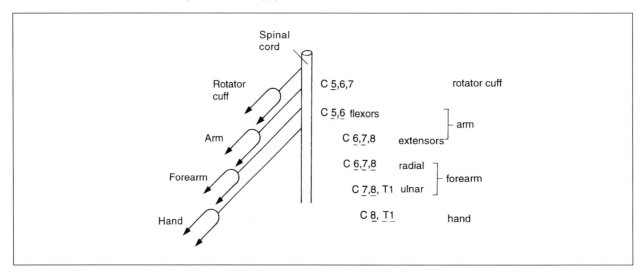

since muscles are innervated in groups according to their function, the flexors and their opposers, the extensors of, for example, the elbow joint, receive their respective innervations from the anterior and posterior divisions of plexus-derived nerves whose root values are approximately the same.

Most muscles that share a common primary action on a joint are, irrespective of their anatomical situation, all supplied by the same nerve roots from a single or occasionally from two spinal segments. Their opponents, sharing the opposite action, are likewise also supplied by the same nerve roots from a single or occasionally from two spinal segments, and the latter usually run in numerical sequence with the former (McMinn, 1990). Thus while the flexors of the arm are supplied by C5 and 6, the extensors of the arm are principally supplied by C6 and 7. Likewise, the flexors of the hip are supplied by L2 and 3, while the extensors are supplied by L4. A summary of the innervation of the muscles of the lower limb (*i.e.* their actions) and the joints that they act across is provided in Table 4.2.2

The cutaneous distribution of nerves (or **dermatome pattern**) of the forelimb is similar to that in the human, so that the limb has both a pre- and a post-axial border. The cutaneous distribution to the surface of the forelimb is from the brachial plexus, and the principles underlying the origin of the definitive arrangement are illustrated in Figures 4.2.9 and 4.2.10. The human rather than the mouse dermatome pattern is shown because this more clearly illustrates the principles involved.

The **cephalic vein** lies in a groove along the pre-axial and the **basilic vein** in a comparable groove along the post-axial border of the forelimb. The region of the shoulder is supplied by the most rostral of the cervical nerves of the brachial plexus, so that C4 supplies the region overlying the deltoid muscle, while the axillary region is supplied by the most caudal components of the brachial plexus (T1 and T2). In the most distal part of the forelimb, as in the human, the **ulnar nerve** supplies the skin on the palmar aspect of

Table 4.2.2 *Innervation of lower limb muscles and joints*[1]

Hip joint[2]	Flexion/adduction/medial rotation	L2, L3
	Extension/abduction/lateral rotation	L4
Knee joint	Extension	L3, L4
	Flexion	L5, S1
Ankle joint	Dorsiflexion	L4, L5
	Plantarflexion	S1, S2
Midtarsal/ Subtalar	Inversion	L4
	Eversion	L5, S1

[1] Data are from humans, although the same underlying principles apply to the innervation of the hindlimb muscles and joints in the mouse,
[2] Note that both the muscles that act across the hip joint, and therefore the innervation of the hip joint, is from the most rostral elements of the lumbo-sacral plexus, while the most distal muscle masses and joints of the hindlimb are innervated by more caudal elements from the lumbo-sacral plexus.

Figure 4.2.9. A diagrammatic sequence illustrating the principles underlying the establishment of the nerve supply to the skin (or dermatome pattern) of the upper limb. The nerve supply in the human is given, but the underlying *principles* involved apply equally to the mouse. In (a) the trunk region is initially envisaged as a cylinder with segments of innervation from the lower cervical and upper thoracic nerve roots from the spinal cord. As the limb bud forms (b), the basic pattern is modified to accommodate the outgrowths on either side of the body (the forelimb buds here, but the drawing is equally applicable to the hindlimb buds). A little later (c), the growth of the limb distorts the basic segmental arrangement. On the *ventral* aspect of the limb (c), the region of maximum displacement is represented by the *anterior axial line*. Across the anterior (and posterior) axial lines, the nerve supply to the pre-axial and post-axial borders of the limb are no longer seen to be in numerical sequence (compare with a,b). On the dorsal aspect of the limb bud, a similar displacement occurs, and the pre- and post-axial innervation is separated by the corresponding *posterior* axial line. Note that it is a convention that the limb is shown in the so-called 'anatomical' position, with the pre-axial border corresponding to its rostral border, and the post-axial border corresponding to its caudal border.

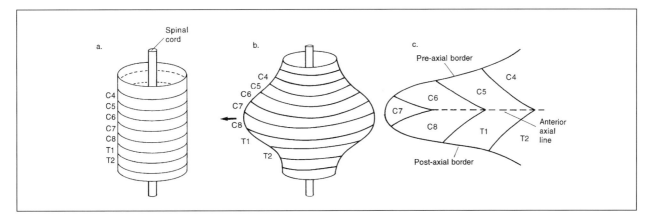

Figure 4.2.10. Diagrams illustrating the dermatome pattern of the anterior (ventral) (a) and posterior (dorsal) (b) surfaces of the upper limb (shown in the "anatomical" position) in the human adult, though, as above, the same underlying principles apply to the mouse. Note the extent and location of the *anterior axial line*, and that it is considerably longer than the *posterior axial line*. Much variation exists between individuals in the boundaries and extent of the individual dermatomes, and a considerable degree of overlap between the territories supplied by adjacent individual nerve roots is also commonly encountered. What is not observed, however, is the overlapping of dermatome territories *across* the anterior and posterior axial lines.

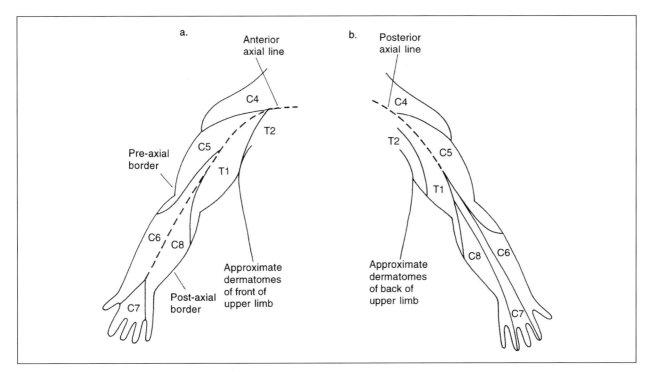

TS	1	2	3	4	5	6	7	8	9	10	11	12	13	14	15	16	17	18	19	20	21	22	23	24	25	26
E	0	1	2	3	4	4.5	5	6	6.5	7	7.5	8	8.5	9	9.5	10	10.5	11	11.5	12	12.5	13.5	14.5	15.5	16.5	17.5

the first one and a half digits, while the **median nerve** supplies the palmar aspect of the skin of the remaining digits, the dorsum of the digits and carpal region being supplied by the radial nerve and its branches. The central dermatome of the forelimb plexus is associated with the most peripheral part of the limb.

The same principles that underlie the innervation of the muscles and joints of the forelimb, as well as the dermatome pattern, apply equally to the hindlimb. The lumbo-sacral plexus is formed from the ventral primary rami of lumbar nerves L3–6, and the first sacral nerve, while the ventral primary rami of **sacral nerves** S2–4 and **coccygeal nerves** give rise to the caudal nerves that supply the muscles and skin of the tail. In the human, because of the reduced number of lumbar vertebrae compared to the situation observed in the mouse (six in the mouse compared to five in the human), the lumbo-sacral plexus in the human usually has the root values L2–5, S1.

4.3 The urogenital system

It is customary to consider the renal and genital systems together because they are the two descendants of the intermediate mesoderm and their development is, to some extent, integrated. Indeed, in the early stages of both sexes, the two systems represent a common developmental entity, although the renal and reproductive systems gradually diverge to take on their separate and largely independent roles, with only a few components being shared between them. The most important of these are the derivatives of the ducting system of the mesonephros: as it regresses, the mesonephric (or Wolffian) and paramesonephric (or Müllerian) ducts are respectively taken over by the developing male and female genital systems. It should, however, be emphasized that, apart from this utilization of one or other duct system, the two systems, though neighbours, function essentially as distinct entities.

We first consider the development of the renal system and its ducts and then describe the genital system, starting with the origin and migration of the primordial germ cells and continuing with the indifferent phase of gonadal development when the histological features of the male and female gonadal tissues and the arrangement of the ducts cannot be readily distinguished. The chapter ends with an analysis of the formation of the distinct features of male and female sexual development, both internal and external.

The renal system

Introduction

The definitive kidney present in the adult, the **metanephros**, is the third that forms in a craniocaudal sequence and is preceded by the rudimentary **pronephros** and the temporary **mesonephros** with all components differentiating from the **intermediate plate mesoderm** (TS 14). Even though the pronephros does not function in mammals, it is usually present in the form of a few transient solid or vesicular cell groups and is associated with the rostral end of the nephric duct[1] (*e.g.* Puschel *et al.*, 1992). It is still unclear whether the mesonephros is a functional entity; although it has an excretory and a drainage component, from the **nephrogenic cord** and the **nephric duct** respectively, with nephric vesicles being seen in the former, structures analogous to glomeruli are not recognized. A diagrammatic sequence illustrating the differentiation of the intermediate plate mesoderm into the nephrogenic cord, which occurs at about the same time as the lateral folding of the embryo, is shown in Figures 4.3.1a and b. The subsequent differentiation of the nephrogenic cord into, successively, the pronephros, mesonephros and metanephros, and the development of their associated duct systems is shown in Figure 4.3.2 (Fraser, 1950; Witschi, 1951; Hamilton and Mossman, 1972).

The mature urinary system is composed of four serially linked components: (i) the **metanephric (or definitive) kidneys** that represent the excretory apparatus producing urine (but have various other roles such as the production of renin, with its complex relationship to the maintenance of the blood pressure), (ii) the **ureters** which extend off the collecting system of the kidney, providing a channel for the passage of urine to (iii) the **bladder**, where it is stored, and (iv) the **urethra**, the channel through which urine is voided from the body. The urethra is, of course, also the channel through which the male ejaculates semen, and is a functional remnant of the shared development of the urinary and reproductive systems (see later).

While the ontogeny of the pro- and mesonephroi are unknown, experimentation *in vitro* has shown that the metanephros develops as the result of an only

---

[1] In the human, at least, the nephric duct largely develops caudal to the pronephroi and seems not to communicate with its vasculature.

TS	1	2	3	4	5	6	7	8	9	10	11	12	13	14	15	16	17	18	19	20	21	22	23	24	25	26
E	0	1	2	3	4	4.5	5	6	6.5	7	7.5	8	8.5	9	9.5	10	10.5	11	11.5	12	12.5	13.5	14.5	15.5	16.5	17.5

THE ANATOMICAL BASIS OF MOUSE DEVELOPMENT

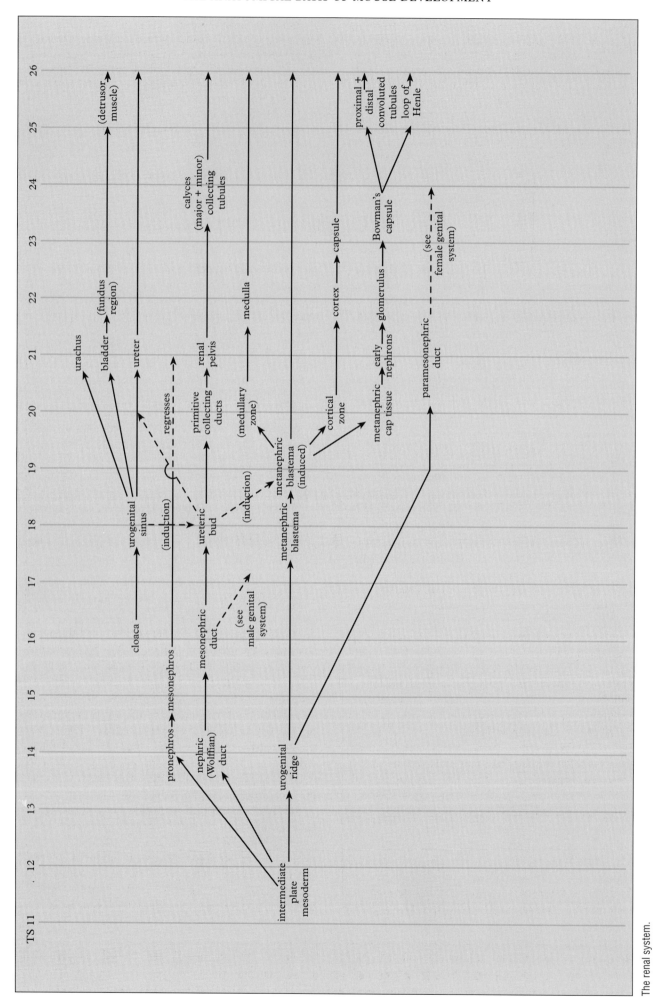

The renal system.

THE UROGENITAL SYSTEM

Figure 4.3.1. Sections through the middle of the neural axis showing the early formation of the mesoderm. (a) A pre- to early somite stage embryo (~TS 12) before the closure of the neural folds. A notochord is present, and the intra-embryonic mesoderm has formed three parallel columns of unsegmented tissue on either side of the midline. The most medial of the three columns (the paraxial mesoderm) will form the somites, the more lateral intermediate plate mesoderm will form the nephrogenic cord, the precursor of much of the urogenital system, while most peripherally is the lateral plate mesoderm that will contribute to the somatopleure and the splanchnopleure. (b) An embryo with about 10–15 pairs of somites (TS 14), after lateral folding of the body wall in the mid-trunk region has occurred. The neural tube has separated from the surface ectoderm and the three columns of mesodermally-derived tissue are now distinct entities. The paraxial mesoderm here has formed somites, the intermediate plate tissue has formed the nephrogenic cord, while the outer of the two components of the lateral plate mesoderm is closely associated with the body wall (together forming the somatopleure); while the other derivative is associated with the endoderm of the gut (together forming the splanchnopleure).

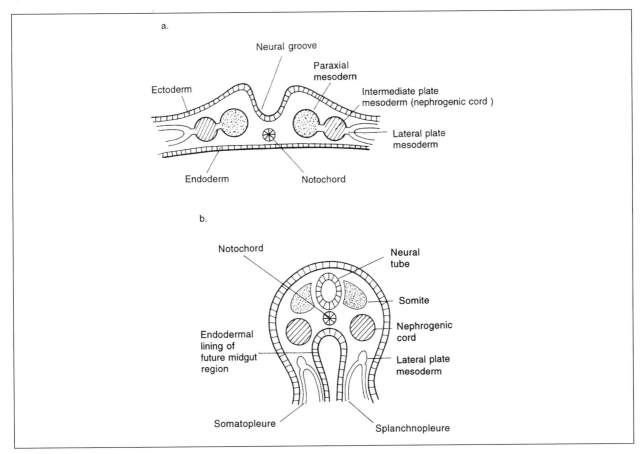

partially-understood series of reciprocal inductive interactions between the **ureteric bud**, a cranially directed outgrowth off the mesonephric duct that will form the collecting ducts, and the **metanephric blastema**, a small domain of cells derived from the intermediate mesoderm that will differentiate into nephrons and the stromal cells of the kidney (for review, see Davies and Bard, 1998). Normal renal function requires that these two anlagen first interact and then amalgamate so that the excretory products produced by the kidney can pass into the system of ducts that constitute the drainage system of the definitive kidney.

The early stages

The intermediate plate mesoderm first appears at about the time that somitogenesis begins (TS 12) and then differentiates to form the nephric cord that soon lies within the urogenital ridge (TS 14/15). Its medial component gives rise in turn to the mesodermal components of the pro-, meso- and metanephroi, while the lateral part of the cord forms the nephric duct (the drainage component) and this is first apparent in the pronephric area (as the pronephric duct). The duct extends caudally to the level of the mesonephros (where it is known as the mesonephric duct) and further caudally towards the **urogenital sinus** component (which is destined to form the bladder) of the cloaca (see later). A rostrally-directed diverticulum from the caudal end of the duct gives rise to the ureteric bud (E11) which, in due course, forms the ureter and the drainage system of the metanephros. In the male, much of the mesonephric duct rostral to the origin of the ureteric bud will be modified and in due course becomes the "drainage" system of the

TS	1	2	3	4	5	6	7	8	9	10	11	12	13	14	15	16	17	18	19	20	21	22	23	24	25	26
E	0	1	2	3	4	4.5	5	6	6.5	7	7.5	8	8.5	9	9.5	10	10.5	11	11.5	12	12.5	13.5	14.5	15.5	16.5	17.5

Figure 4.3.2. Very diagrammatic representation of the three derivatives of the nephrogenic cord. Most rostral is the pronephros that differentiates first (TS 14). Slightly later (TS 15), as the most rostral part of the mesonephros (or Wolffian body) shows early evidence of differentiation, the pronephros regresses, though the mesonephros utilizes the pronephric duct, which extends caudally towards the urogenital sinus, and becomes the mesonephric (or Wolffian) duct. The mesonephroi also regress in a rostral to caudal direction, and are replaced by the definitive kidneys (the metanephroi). These form when diverticula (the ureteric buds) extend off the duct, contact the metanephric blastema and undergo reciprocal inductions with them. By the time that the metanephroi first form, the mesonephric ducts have made contact with and drained into the urogenital sinus.

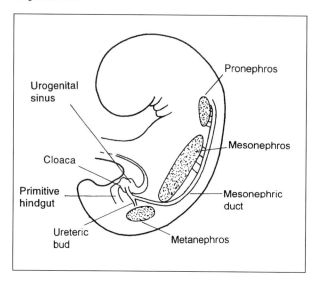

during the latter part of TS 14, when the pronephric vesicles (primordium) and their associated nephric duct are just evident in the most rostral part of the nephrogenic cord (see Figure 4.3.3), said to differentiate initially at the level of the cervical somites. At this stage of development, the cervical somites are located at approximately the same level as the forelimb buds and, like them, will gradually *ascend* until they reach their definitive position (TS 22/23). By TS 15, the first evidence of canalization is seen over much of the length of the nephric ducts as they extend caudally towards the urogenital sinus. The **urogenital ridges** are also first clearly seen at this stage, being located on either side of the **dorsal mesentery of the hindgut**. They are most pronounced in the mid-trunk region, but extend caudally, and protrude along their length into the dorsal part of the peritoneal cavity (see Figure 4.3.4). It is also at this stage that the mesonephroi are first seen within the nephric cord, although the extent of their differentiation is limited to the appearance of vesicles.

By TS 16, the pronephric tissue has regressed and the mesonephros (the "Wolffian" body) which forms in the lateral part of the urogenital ridge, has started to develop. The gonad begins to differentiate on the medial aspect of the mesonephros, the two being suspended from the posterior abdominal wall by a common broad **urogenital mesentery**. This mesentery, like those elsewhere, consists of a central core of mesenchyme tissue usually drawn from the posterior body wall from which the organs are suspended. As indicated above, such mesenteries are invariably covered by a layer of mesothelium, as are the structures that are suspended from them, and this layer differentiates to form their peritoneal covering. During this stage, the mesonephroi enlarge and the vesicles extend so that large numbers of segmentally arranged tubules are seen, with several (usually two on each

testis forming the rete testis (probably), the efferent ducts, epididymis, ductus deferens, ejaculatory ducts and the seminal vesicles; in the female, the mesonephric duct system almost completely regresses, playing no role in the reproductive process (if present, it is only in the form of aberrant tissue – see below).

The earliest stage of nephric development is seen

Figure 4.3.3. Differentiation and subsequent regression of the mammalian pronephros. The sequence illustrated here represents a hypothetical model for the differentiation of this structure as there is little evidence that the pronephros differentiates even to the extent shown here. In principle, the segmental units, or nephrotomes, form vesicles which then amalgamate and drain into a pronephric duct which grows caudally in the direction of the urogenital sinus. Shortly after its first appearance, the pronephros regresses, and its duct system extends to become the mesonephric duct. In the case of the mesonephros, two mesonephric vesicles usually form at each segmental level.

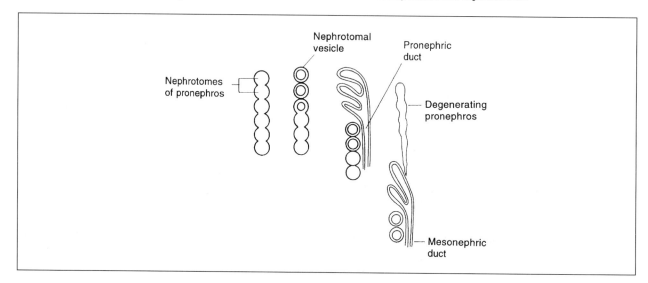

Figure 4.3.4. A representative transverse section through the mid-abdominal region in the human showing the relationship between the fully differentiated mesonephros and the gonadal ridge which develops on its medial side. In the midline, the hindgut is seen to be suspended by its dorsal mesentery. At this stage, mesonephric glomeruli form and the mesonephroi act as excretory organs that produce urine that drains via the mesonephric ducts into the urogenital sinus (the future bladder). It is believed that the paramesonephric duct is induced by the subjacent mesonephric duct to develop from coelomic mesothelium, although the underlying mechanism is unclear. Mesonephric development in the mouse is similar, but glomeruli do not form to the same extent and only the most anterior vesicles drain into the nephric duct (Sainio et al., 1997).

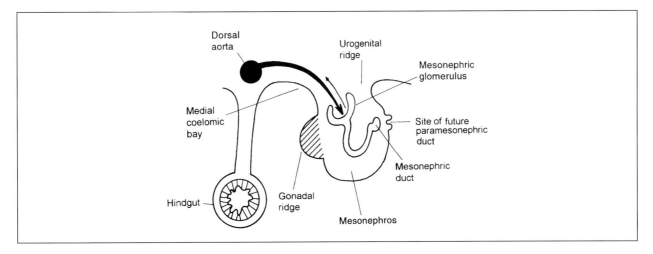

side) forming at each segmental level. Of these, the most rostral 4–6 pairs of ducts drain into the mesonephric duct, but the rest, although they differentiate into primitive nephrons, do not (Sainio et al., 1997). The pronephric duct is incorporated into the mesonephros to form the mesonephric duct, and becomes canalized (i.e. develops a lumen) along its entire length during this stage, although it has yet to reach the urogenital sinus.

By TS 17, the most rostral 4–6 pairs of mesonephric vesicles drain into the mesonephric ducts (in the male, these tubules will form the epididymal ducts, see below; Sainio et al., 1997) which now contact the wall of the cloaca in the region of the urogenital sinus, although they are not believed to fuse with it until about TS 22 when the mesonephros has largely regressed. The mesonephros achieves its maximum volume by about TS 18/19, by which time its rostral pole already shows considerable evidence of regression through apoptosis (see below). For this reason and because no well-differentiated mesonephric glomeruli are seen in *mice*, it has been suggested that the mesonephros is unlikely to function as an effective excretory organ in this species (Theiler, 1989). The mesonephroi are, however, very substantial, segmentally organized structures located on either side of the dorsal mesentery of the gut and, when fully formed, extend almost the full length of the peritoneal cavity. Because of their enormous size, it is difficult to believe that they neither function as a kidney nor serve any other useful purpose beyond acting as a base for the gonad and a duct system which is utilized by structures which differentiate at a later stage of development (e.g. as a source of the ureteric bud in both sexes and for drainage of the testis in the male – see Sainio et al., 1997).

It should be noted that in the mouse, as in the human, the mesonephric part of the nephrogenic cord is never seen to its full extent, for, while it is still growing at its caudal extremity, it is undergoing regression at its rostral pole. The differentiation of the mesonephric vesicles continues in a cranio-caudal direction[2] and, by

---

[2] In the *human*, unlike the situation observed in the mouse, the medial part of the mesonephric vesicles enlarge and then becomes invaginated by blood capillaries to form a glomerulus. The vesicles form into tubular structures which then become S-shaped. While the part adjacent to the mesonephric duct develops into the collecting tubule, the more distal element forms into the glomerular capsule, while the intermediate part forms the equivalent of the proximal and distal convoluted tubules of the adult kidney, though no loop of Henle develops. Afferent and efferent vessels are seen to differentiate in relation to the glomerular capillaries. There seems little doubt, from an analysis of its morphological features, that the human mesonephros acts as a functional unit that eliminates urine. It attains its full development by about the second month, and becomes totally non-functional by the fourth month (Potter, 1965; De Martino and Zamboni, 1966) so that there is a period when its function overlaps that of the early metanephros (Gersh, 1937). The size of the mesonephros varies considerably between species, and is believed to be related to the type of placentation present, and the degree of development of the allantois.

In the *mouse*, two distinct sets of tubules are observed during mesonephric development: four to six pairs of cranial mesonephric tubules become linked to the mesonephric duct, while the majority form more caudally but do not fuse with the mesonephric duct. All except for the most rostral tubules, which become the epididymal ducts, degenerate by apoptosis. These two systems of mesonephric ducts appear to have different regulatory mechanisms (Sainio et al., 1997).

TS	1	2	3	4	5	6	7	8	9	10	11	12	13	14	15	16	17	18	19	20	21	22	23	24	25	26
E	0	1	2	3	4	4.5	5	6	6.5	7	7.5	8	8.5	9	9.5	10	10.5	11	11.5	12	12.5	13.5	14.5	15.5	16.5	17.5

the time that the caudal pole has fully differentiated, much of the upper two-thirds of this organ has regressed.

The metanephros

The period starting at TS 18/19 is marked by the first signs of regression of the mesonephros (a process largely completed by TS 22/23), the appearance of the gonad on its medial surface, and the appearance of the ureteric bud with its extension into the metanephric blastema which differentiates from the most caudal part of the nephrogenic cord (Figure 4.3.5a–c). The ureteric buds are relatively short at this stage, but their proximal (i.e. caudal) region will lengthen as the ureter forms. As a result of the inductive interaction between the distal tip of the ureteric bud and the surrounding metanephric mesenchyme (TS 20), the former starts to bifurcate while the latter becomes partitioned into nephrogenic and stromal regions (about TS 21) (Erickson, 1968; Davies and Bard, 1998).

Over the rest of development, small domains of nephrogenic mesenchyme (blast or stem cells), now located at the periphery of the cortex, form small nephrogenic **condensations**, each of which is associated with a discrete component of the ureteric bud and will epithelialize to give rise to a **nephron**, while other mesenchymal cells form the stroma of the medulla.[3] Over the few days following its formation, each epithelialized unit reorganizes to form an S-shaped body, with the cleft distal to the nearest collecting duct attracting capillaries and differentiating into a **glomerulus** (much in the same way as the human mesonephric glomerulus – see above). This leads into a tube that extends and folds to comprise in turn a **proximal convoluted tubule**, a **loop of Henle** and a **distal convoluted tubule**, which fuses to a **collecting duct** (see Figure 4.3.6). By now, the base of the arborized ducting system is forming the **primitive renal pelvis** (TS 21), possibly by the incorporation of some of the earliest generations of its branches some of which may represent the primordia of the **major** and **minor renal calyces** (TS 23).

By TS 21, the metanephros contains an *inner* **medullary region** of loose stromal mesenchyme with collecting ducts, an *outer* primitive **cortical region**, largely consisting of nephrogenic aggregates in the early stages of differentiation, and a periph-

Figure 4.3.5a–c. Sequence illustrating the ascent of the definitive kidney from its origin in the pelvis (a) to its usual final location, at about the level of the first lumbar vertebral body (c). Note that in (a, ~TS 18), the gonad is starting to differentiate. A little later (b, TS 20), as the metanephros ascends, the mesonephros displays clear evidence of regression, though the gonad is still closely related to the mesonephros and its duct system. Somewhat later (c, TS 22), as the definitive kidney ascends even further, the mesonephros has almost completely regressed, and only in the male is the mesonephric duct utilized for the drainage of the testis, giving rise to the rete testis, efferent ducts, epididymis and ductus (or vas) deferens. This figure also shows the reorganization that occurs where the duct system meets the bladder. In (a), the mesonephric duct drains into the urogenital sinus, the ventral component of the cloaca, with the ureter still being a diverticulum off the duct. As the distal part of the mesonephric duct is resorbed, both it and the ureteric bud enter the base of the bladder in separate locations The entry of the ureters is at the upper/outer regions of the trigone (the triangular area lying between the internal urethral orifice (distally) and the orifices of the ureters), while the ejaculatory ducts (the most distal derivatives of the mesonephric ducts) enter the prostatic region of the urethra, on either side of the urethral crest (formed by the median lobe of the prostate, see Figure 4.3.11), close to the entrance of the prostatic utricle. The rostral extremity of the bladder, the urachus, connects the apex of the urogenital sinus to the umbilicus; this eventually regresses and fibroses, giving rise to the median umbilical ligament. In due course, the testes descend into the pelvis, and come to lie on either side of the bladder, their course of descent being indicated by the arrow in (c).

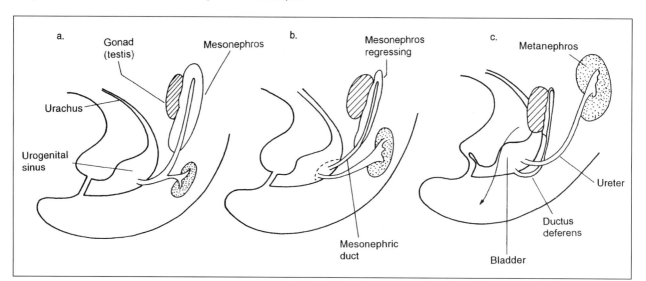

[3] Occasionally, small units of metanephric tissue may produce vestigial tubules which produce urine, but fail to join the collecting tubules. Such units form isolated renal cysts. If many of these units are present this gives rise to the condition termed congenital polycystic kidney. If the condition is extensive and bilateral it may not be compatible with survival beyond birth (Osathanondh and Potter, 1964a–d).

Figure 4.3.6a–e. Stages in the differentiation of the metanephric tissue cap, which gives rise to nephrons after inductive interactions between a terminal branch of the ureteric bud, and a discrete unit of metanephric cap tissue. The latter initially forms a nephric vesicle that then elongates and becomes S-shaped while at the same time making direct contact with, and subsequently fusing with, the distal component of the ureteric bud. The metanephric derivative gives rise to the Bowman's capsule, proximally, and more distally, to the primitive glomerulus. At about this time contact is made with a distal capillary branch of the renal artery, and drains back via a capillary knot, associated with the glomerulus, to a branch of the renal vein. The glomerulus is the functional unit of excretion of the kidney. Between Bowman's capsule and the distal part of the collecting duct, the duct system increases enormously in length to form (proximally to distally) the distal convoluted tubule, the loop of Henle and the proximal convoluted tubule (proximal because it is closest to the glomerulus). (Diagram based on Larsen, 1993.)

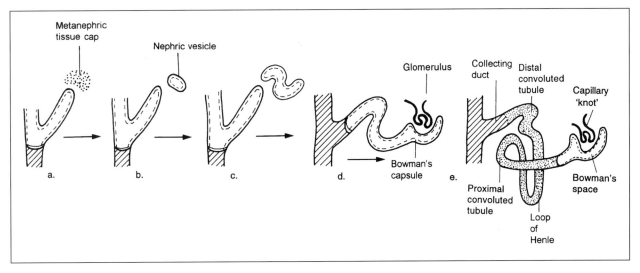

eral rind of undifferentiated **stem cells** that continue to add to the nephron population almost until birth when there are about 1000–2000 of them. As the metanephros differentiates and grows, it gradually "ascends" from its original pelvic position, so that, by the end of TS 21, its upper pole is slightly below the level of the 13th rib. As this happens, the ureters lengthen, but do not themselves yet open into the vesical part of the urogenital sinus (the future bladder, first readily recognized at TS 22). At this stage, a *common* excretory duct is present where the proximal parts of both the mesonephric duct and ureteric bud enter the urogenital sinus at a common site. At about this time, the two paramesonephric (or Müllerian) ducts form from the infolding of coelomic epithelium lateral to the mesonephric (or Wolffian) ducts on either side of the coelomic cavity.

While the metanephros is still a pelvic structure, it receives its arterial blood supply from lateral sacral branches of the aorta, or from the common iliac artery. As the kidneys ascend, they receive their blood supply sequentially from lateral stem arteries which arise from the aorta at increasingly higher levels; at their normal position they are supplied by the definitive renal arteries, which are located at the level of about the second lumbar vertebral bodies (Hamilton and Mossman, 1972). Persistence of blood vessels from lower levels than normal gives rise to *aberrant renal arteries*, and multiple renal arteries, which may be unilateral or bilateral, and are present in about 30% of humans (Merklin and Michels, 1958). These may enter at the hilum or at the upper or lower pole and represent persistent lateral splanchnic vessels. The definitive renal artery arises from the most caudal of the three suprarenal arteries, all of which represent persistent mesonephric or lateral splanchnic arteries (Williams, 1995). In cases where one, or both, kidneys fail to ascend, the arterial supply is usually consistent with its vertebral level. Thus, for example, when fusion of the lower poles of the two kidneys occurs (termed *horseshoe kidneys*), their ascent is impeded by the inferior mesenteric vessels, and their arterial supply and ureteric drainage are often aberrant.

Later development of the renal system

The first occasion that the ureters open *directly* into the urogenital sinus and becomes separate from the opening of the mesonephric duct is during TS 22; this separation is due to the gradual incorporation of the proximal part of the mesonephric duct into the posterior wall of the urogenital sinus (Frazer, 1935). While the ureters will eventually enter the bladder in the region of the supero-lateral part of the trigone (see below), the mesonephric ducts (in the male) "descend" caudally until they enter the prostatic region of the urethra as the ejaculatory ducts on the **urethral crest** (or **verumontanum**), one on either

TS	1	2	3	4	5	6	7	8	9	10	11	12	13	14	15	16	17	18	19	20	21	22	23	24	25	26
E	0	1	2	3	4	4.5	5	6	6.5	7	7.5	8	8.5	9	9.5	10	10.5	11	11.5	12	12.5	13.5	14.5	15.5	16.5	17.5

side of a minute depression, the opening of the **prostatic utricle** (the embryological remnant of the caudal end of the Müllerian or paramesonephric duct).

The part of the urogenital sinus that extends from the apex of the fundal region of the future bladder towards the umbilicus forms a tubular structure called the **urachus** (TS 21) (see Figure 2.3.5 and Figure 4.3.5). Initially, this structure connects the lumen of the urogenital sinus with that of the allantois, but the connection from the distal part, beyond the umbilicus into the umbilical cord, regresses as the allantois becomes incorporated into the placenta. The urachus eventually closes and forms a fibrous band which is recognizable in the adult as the **median umbilical ligament** which extends along the lower part of the inner aspect of the anterior abdominal wall from the apex of the bladder to the umbilicus.[4]

At TS 22, the bladder starts to differentiate from the urogenital sinus at about the time that the ureters open into the lumen of the bladder at a region that will be recognizable as the supero-lateral angles of the **trigone**, a triangular area at the base of the bladder (its inferior point is at the urethral orifice) which is apparent at TS 24. By TS 26, the bladder is particularly prominent and its wall is thick and muscular (the **detrusor muscle**), with this smooth muscle deriving from the splanchnopleuric mesoderm associated with the wall of the hindgut. The mucous membrane lining the bladder is of the transitional variety, and subjacent to this is the submucosa. In the male, the ejaculatory ducts enter the prostatic region of the urethra (see above) just caudal to the most distal part of the trigone. This *mesodermally* derived region of the bladder is later overgrown by *endodermal* tissue extending over its surface from the surrounding cells which line the rest of the wall of the bladder.

The derivation of some of these tissues is worth emphasizing. The cloaca and all of its derivatives, the distal part of the hindgut and the urogenital sinus, are of endodermal origin, while the mesonephric duct and its derivatives are of mesodermal origin. As the mesonephric duct makes contact with the urogenital sinus, it becomes progressively incorporated into the wall of the sinus, so that, in due course, the entire trigone region of the bladder is accordingly believed to be of mesodermal origin and it is this region that is, as has just been mentioned, subsequently overgrown by endodermally-derived epithelium. The embryological origin of the tissue in the region of the trigone is disputed by embryologists, though in the adult the tissues in this region differ from those elsewhere in the bladder. The origin of these tissues is discussed by Bulmer (1957).

By TS 22/23, primitive glomeruli can first be recognized within the metanephros, and large numbers of collecting tubules are also seen. It is only at about TS 24/25, however, that the kidney has well demarcated cortical and medullary components, with the glomeruli located in its outer one-third, the region of the presumptive **cortex**. By about TS 25/26, the cortical region contains well-differentiated glomeruli and substantial numbers of proximal and distal convoluted tubules, while the medullary region contains the ascending and descending components of the loops of Henle and the majority of the collecting duct sytem. Once this degree of compartmentalization is complete, the stromal cells and their matrix start to be lost from the medulla, and the **hilar region** of the kidney is now principally occupied by the **renal pelvis** and **major** and **minor calyces** into which drains the collecting duct system. Stem cells are still present around the renal periphery at birth, however, and kidney differentiation is not complete until the mouse is several weeks old.

The kidneys probably first start to function at about TS 22/23 when urine is produced and increasingly forms a major component of the amniotic fluid.[5] This is swallowed and the excretory products are absorbed via the gut, pass into the embryonic vascular system, and are eventually removed through the placenta. *In utero*, the kidneys thus provide a homeostatic mechanism for maintaining fluid balance, as well as regulating amniotic fluid volume. Amniotic fluid is initially produced by the transport of fluid across the amniotic membrane, but, at more advanced stages of pregnancy, however, the amniotic fluid is mainly derived from the urine. The urine is hypotonic with respect to the plasma, having a low concentration of electrolytes (Alexander and Nixon, 1961; Smith *et al.*, 1966). Once urine has been swallowed, the excess fluid present as well as excretory products contained in it are absorbed through the wall of the stomach and pass into the blood stream. They are then directed to the placenta before eventually passing into the maternal circulation.[6]

---

[4] In humans, abnormalities of the closure of the intra-embryonic portion of the urachus may rarely result in the formation of a urachal (or vesical) fistula (or sinus); in addition, urachal cysts may occasionally be found here.

[5] In humans, marked morphological differences have been observed between the fine structure of the mesonephros and metanephros, with, in particular, the former lacking both the juxta-glomerular apparatus and differentiated nephric tubules. Due to the absence of (i) the Henle loops, and (ii) the basal infoldings of the cell plasma membranes in the distal tubule, the mesonephric urine is extremely hypotonic (De Martino and Zamboni, 1966).

[6] In cases of urethral stricture, or in bilateral renal agenesis in which there is a complete absence of renal function, *no* urine enters the amniotic compartment and the volume of amniotic fluid present is necessarily significantly reduced (a condition termed **oligohydramnios**). By contrast, when the swallowing reflex is disturbed (*e.g.* in anencephaly or exencephaly) or in the presence of oesophageal atresia with no tracheo-oesophageal fistula (see Figure 4.3.7; Lloyd and Clatworthy, 1958), when the amniotic fluid cannot pass into the stomach, an excessive volume of this fluid may build up around the embryo/fetus (a condition termed **hydramnios** or **polyhydramnios**). For observations on the factors that control the volume of amniotic fluid, see *Postimplantation extra-embryonic development*.

Figure 4.3.7a,b. The anatomical arrangement seen when there is a segment of oesophageal atresia either in isolation (a) or associated with a tracheo-oesophageal fistula (b). In (a), a localized region of the oesophagus distal to the anterior (or ventral) bifurcation that will form the trachea and lower respiratory tract, fails to canalize. This situation is compatible with survival to birth, but all substances then taken by mouth will invariably pass into the lower respiratory tract, resulting in pneumonia and death (unless, in the human, appropriate surgical intervention is undertaken). In (b), a tubular connection forms between the trachea and the distal part of the oesophagus, termed a tracheo-oesophageal fistula; this is also a potentially lethal condition (its incidence is about 1:2500 in human infants, and mostly occurs in males) (see Footnote 6).

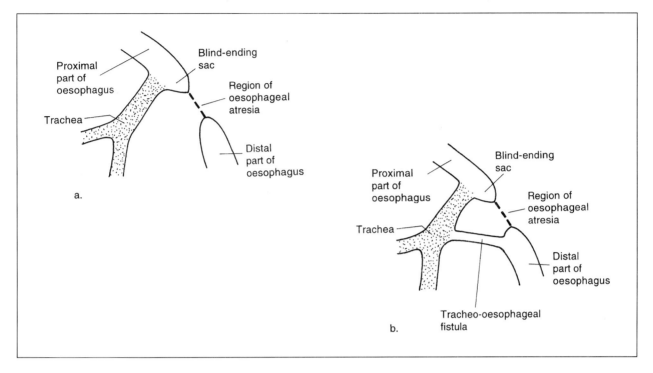

The autonomic nerve supply to the kidney

The exact time when the nerve supply reaches the various regions of the kidney is still unknown, although analysis of silver-stained sections and immunostaining (Sariola *et al.*, 1988) clearly demonstrates that extensive innervation is seen extending from the hilar region towards the cortico-medullary junction by TS 22/23. This innervation is neural crest derived, principally from the coeliac plexus, the largest of the autonomic plexuses, but also from other plexuses and ganglia in this region. Ganglia are present around the origin of the renal artery, where it branches from the aorta, and autonomic nerves progress into the kidney around the wall of its various branches. They supply the walls of these vessels, the renal glomeruli and the tubular system of the kidney, and are mostly vasomotor, in that they control the blood flow through these vessels. Their role in regulating the blood supply to the cortex and medulla is believed to be particularly important, although this is said to be achieved without affecting the glomerular circulation (Williams, 1995).

Branches from the coeliac plexus and ganglia in this region also supply the ureteric and gonadal plexuses. The supply to the ureter is particularly complex, and depends on the level; the rostral nerve supply is from the renal and aortic plexuses, the intermediate part is supplied by branches of the superior hypogastric plexus and hypogastric nerve, while the caudal nerve supply is from the hypogastric nerve and the inferior hypogastric plexus. The nerve supply as a whole influences the motility of the ureter, and fine nerve endings are first clearly seen within the wall of the ureter at about TS 23/24.

The genital system

Introduction

The genetic sex of the individual is established at the time of conception with the critical determining factor being, of course, the sex chromosome constitution as this effectively controls the subsequent differentiation of the gonads, the internal genital duct system and external genitalia (Byskov, 1986; for a brief discussion of the molecular basis of sex determination, see Gilbert, 1997). In most classes of vertebrates (but not in mammals), for example, in amphibians and fish, the phenotypic sex of an individual may change from male

TS	1	2	3	4	5	6	7	8	9	10	11	12	13	14	15	16	17	18	19	20	21	22	23	24	25	26
E	0	1	2	3	4	4.5	5	6	6.5	7	7.5	8	8.5	9	9.5	10	10.5	11	11.5	12	12.5	13.5	14.5	15.5	16.5	17.5

to female or through intersexuality to functional hermaphroditism, and involves various factors such as genetic control mechanisms, changes in social organization or temperature or changes in photoperiod (Chan, 1970).

In mammals, in an XX individual, the gonad develops as an ovary and sexual differentiation in females only takes place in the *absence* of male-determining hormonal stimulation. It is unclear at the present time whether ovarian or, more likely, placental oestrogens play any role in female gonadal differentiation, although a genetic basis is beginning to be discovered (*e.g.* Swain *et al.*, 1998). In an XY individual, the gonad develops as a testis due to testis-determining genes located on the Y-chromosome, and, once the testis has differentiated, various androgenic hormones, such as testosterone (derived from the **interstitial cells of Leydig**), and hormone-like substances, such as anti-Müllerian hormone (AMH, believed to be produced by the **Sertoli cells** that line the seminiferous tubules), control the differentiation of first the internal genital duct system and then the external genitalia (Forest, 1983). In the absence of testicular hormones, differentiation continues in the basic female direction (Jost, 1965).

There is interesting experimental evidence to sustain this conclusion: the ducts of castrated fetal male rabbits differentiate in a female direction (Jost, 1947), while there is additional evidence of male sex domination in heterosexual twins in cattle: the male twin differentiates apparently normally, whereas in the female partner (the so-called "freemartin") the external and internal sex organs and ducts are masculinized (Lillie, 1917), possibly due to the transfer of male hormones across the placenta through vascular anastomoses, of male hormones (Short, 1970). Other studies indicate that substances from the female might interfere with the development of the testis (Shore *et al.*, 1984). Another model system that has been used to investigate the factors that influence gonadal differentiation uses mouse chimaeras produced by the fusion of 4- and 8-cell stage embryos of different sexes (Tarkowski, 1961). Somewhat surprisingly, a preponderance of male embryos, and only a few hermaphrodites, are formed (Mullen and Whitten, 1971). Mammals that possess a Y chromosome (*e.g.* XY, XXY or XXXY) have testes, whereas those possessing only X chromosomes (*e.g.* XO, XX or XXX), develop ovaries (Polani, 1962).

Primordial germ cells

The first evidence of sexual development is the appearance of primordial germ cells (PGCs) in the early postimplantation stages of mouse development (Chiquoine, 1954; Ozdzeński, 1967; Snow and Monk, 1983; Ginsburg *et al.*, 1990; see also Everett, 1943, for an earlier view on the origin of the germ cells). These cells can be identified on the basis of their stage-dependent location,[7] morphology, and appearance after histochemical staining for intracellular alkaline phosphatase enzyme activity. Some 50 of these cells, presumed to be primordial germ cell (PGC) precursors, are first evident at the early primitive-streak stage (TS 10/11) in the mesodermal component of the wall of the yolk sac. It is now generally believed that they and their precursors are epiblastic in origin, being the *only* cells in the embryo at this and subsequent stages of development that are totipotential (Rohwedel *et al.*, 1996; Yeom *et al.*, 1996). In slightly more advanced primitive-streak stage embryos (TS 11/12), there may be up to 150 alkaline phosphatase-positive cells (now assumed to be PGCs) that are located in the mesodermal component of the yolk sac, at the base of the allantois, and at the caudal end of the primitive streak (McKay *et al.*, 1953; Chiquoine, 1954). The cytoplasm of these cells has a characteristic basophilic appearance, particularly when stained with toluidine blue (Zamboni and Merchant, 1973). Female germ cells can usually be recognized by the presence of the Barr body, representing the inactivated X chromosome (Ohno, 1963; Gartler *et al.*, 1980).

At about TS 12, these cells start to migrate by what is thought to be amoeboid-like movement (Spiegelman and Bennett, 1973; Clark and Eddy, 1975) from the wall of the yolk sac towards the base of the allantois and then, possibly as a result of morphogenetic movement of the tissues (Jeon and Kennedy, 1973; Snow and Monk, 1983), to the hindgut endoderm, with the fastest cells reaching it at the early somite stage (up to about 10 pairs of somites, TS 12/13). In embryos with 11–20 pairs of somites (TS 13/14), most of the PGCs are associated with the hindgut endoderm, though a few may still be found in the mesodermal component of the wall of the visceral yolk sac. In embryos with about 15–20 pairs of somites (TS 14), a few PGCs have reached the hindgut mesentery, while, at a stage later (TS 15), about two-thirds of the PGCs present

---

[7] The extra-gonadal origin of the primordial germ cells was first demonstrated by Everett (1943) who transplanted genital ridges from different stages of mouse embryogenesis under the kidney capsule of adult mice. When transplants were isolated from E9.5–10 embryos no germ cells were observed, whereas transplants from E11–14 embryos contained germ cells. As development proceeds, increasing numbers of PGCs are found in the urogenital ridges, so that, by the early limb-bud stage (TS 16/17), they are (usually) exclusively located in its genital component (on the medial side of the mesonephros where the gonad will subsequently differentiate – see Figure 4.3.8), while the mesonephros itself and its duct are located on the lateral aspect of the urogenital ridge (for further details, see Tam and Snow, 1981; Snow and Monk, 1983; Ginsburg *et al.*, 1990). It should be noted that, during this period of PGC migration, there is no clear relationship between the genetic sex of the embryo and either the location or the total number of PGCs present. Even at slightly later stages of gonadal differentiation, up to about E12–12.5 (TS 20), there is no obvious difference in either the histological or gross morphological appearance between the developing ovary/testis.

are associated with the hindgut endoderm, about one-third are in the hindgut mesentery and a few have now reached the medial part of the urogenital ridge where the gonad differentiates (see Figure 4.3.8).

Germ cells that do not succeed in reaching the gonads by the time of sex differentiation usually disappear rapidly, though they may occasionally survive in extragonadal sites, giving rise to teratomas (Stevens, 1975; McIlhinney, 1983). While migrating, and within the gonad, germ cells increase in number mitotically, so that by E13.5, there may be about 25,000 primordial germ cells present (Byskov, 1986).

The early "indifferent" gonad stages

The first evidence of a **gonadal primordium** is seen at about TS 18 on the medial aspect of the urogenital ridge, although the gonads of the two sexes are still indistinguishable, both histologically and morphologically (the so-called "indifferent" stage). By about TS 19, when the mesonephros is at its largest, with its tubules and ducts being particularly prominent, and while the metanephros is just starting to form, the gonad has grown and now contains large numbers of primordial germ cells (Brambell, 1927).

During TS 19, the gonads are covered by a layer of coelomic mesothelium that proliferates and invades the underlying mesenchyme so forming the **cortical cords** or **primary sex cords** (see Figure 4.3.9) (this mesothelium was formerly termed the *germinal epithelium* because it was erroneously believed that it was here that the primordial germ cells originated (Allen, 1904; Kingery, 1917; Franchi *et al.*, 1962)). In the male, the primary sex cords proliferate within the future medullary region of the testis, eventually forming the seminiferous tubules (see below), and the cortical region largely regresses. In the female, in contrast, the germ cells concentrate in the cortical region of the developing gonad and the medulla regresses.

By TS 20, the mesonephros begins to regress, while the gonads, whose sex is still indistinguishable, continue to increase in volume, becoming elongated and sausage-shaped. It is only during TS 21 with the mesonephros having largely regressed that the gonads show the first evidence of sexual

Figure 4.3.8. The migration of primordial germ cells as they pass from the wall of the hindgut, along its dorsal mesentery towards the gonad which differentiates on the medial aspect of the mesonephros. Just lateral to the mesonephric duct may be seen the paramesonephric (or Müllerian) duct which subsequently gives rise, in the female, to the oviduct, uterus and its cervix and the upper part of the vagina. In the male, the paramesonephric duct largely regresses, though it is believed to give rise to several named structures (for details, see Figure 4.3.11). The mesonephros and gonad, as well as the mesonephric and paramesonephric ducts, are suspended from the posterior abdominal wall by a wide urogenital mesentery.

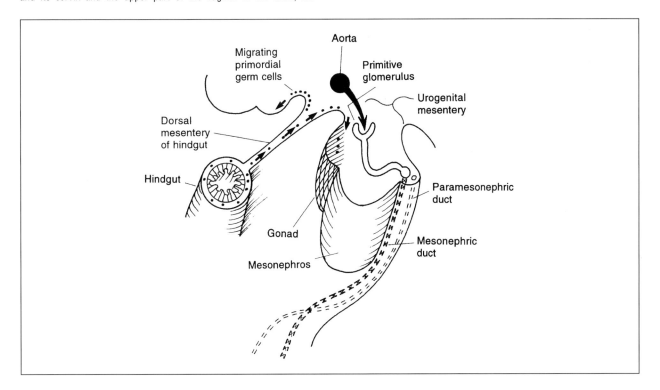

TS	1	2	3	4	5	6	7	8	9	10	11	12	13	14	15	16	17	18	19	20	21	22	23	24	25	26
E	0	1	2	3	4	4.5	5	6	6.5	7	7.5	8	8.5	9	9.5	10	10.5	11	11.5	12	12.5	13.5	14.5	15.5	16.5	17.5

THE ANATOMICAL BASIS OF MOUSE DEVELOPMENT

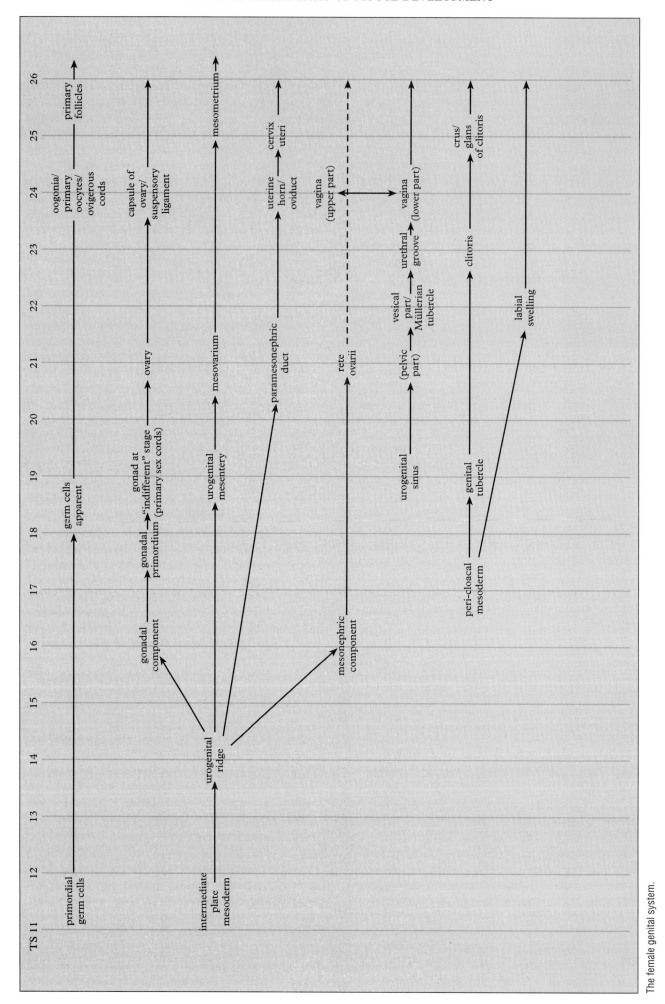

The female genital system.

Figure 4.3.9. As the mesonephros regresses, the first evidence of gonadal differentiation is seen, with the ingrowth from the surface of the primitive gonad of primary sex (or cortical) cords into the subjacent mesonephric tissue. Depending on the sex chromosome constitution of the primordial germ cells, the gonad will become either an ovary or a testis. In both sexes, the first evidence of gonadal differentiation is the presence of *primary* sex cords. In the male, these differentiate further to form the seminiferous tubules, while in the female they regress, and at a later stage are replaced by *secondary* sex cords. Up to about E12.5 in the mouse, it is not possible to determine the sex of an embryo from a histological analysis of the gonads or their internal genital duct system. This early stage is often known as the indifferent gonad stage.

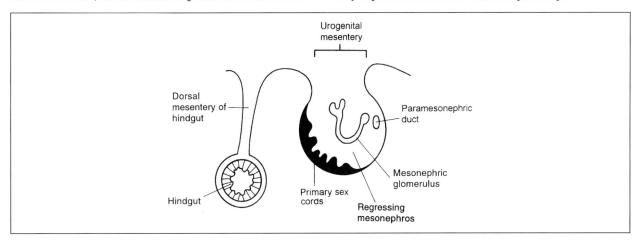

dimorphism, although both the ovaries and the testes at this stage have a similar size and shape. In the female, the cortical cords are just recognizable in the ovaries which have a granular or "spotty" appearance (early oogonia). In males, however, the testes become "striped", because the **testicular cords** (which will form the **seminiferous tubules**) become substantially more pronounced (for further details, see later).

In females during TS 21 (E12.5), germ cells gather in clusters that are uniformly distributed throughout the ovarian tissue,[8] and a thin rim resembling tunica albuginea tissue is present just beneath the surface epithelium of the ovary. In males, the extending testicular cords incorporate the germ cells (future spermatogonia) and mesenchymal cells (future Sertoli cells).

By the end of TS 21, the mesonephros has all but regressed in both sexes (*e.g.* Figure 4.3.10), leaving only a few seemingly disorganized tubules, although the **mesonephric** and more laterally located **paramesonephric (Müllerian) ducts** are clearly seen. These run inferomedially towards the urogenital sinus, and at this stage both sets of ducts are canalized throughout their entire length. The factors that induce the paramesonephric duct to develop are unclear, but it has long been believed that it may be induced by factors emanating from the mesonephric duct (Didier, 1973). While the mesonephric ducts are more medially located than the paramesonephric ducts in the rostral part of their course, they are more laterally located at their site of insertion into the urogenital sinus (see Figure 4.3.10). The cranial extremity of the paramesonephric duct opens into the peritoneal cavity in many species to form the abdominal os of the oviduct that will later differentiate to give rise to the fimbriated os.

By about TS 22/23, the ovaries tend to be more elongated than ovoid in shape, and a little smaller than the testes at this stage of development. One characteristic feature of ovaries is the presence within them of germ cells entering meiotic prophase (Peters, 1970); comparable events do not occur in the testis until puberty (Steinberger and Steinberger, 1975).

The female gonads and reproductive duct system

A progressive change in the relationship between the ovaries and the kidneys now occurs with the "ascent" of the kidneys towards their normal position in the

---

[8] For observations on the ultrastructural features of the ovaries in fetal and early postnatal mice, see Odor and Blandau (1969). In this study, the differentiation of oogonia, oocytes and follicle cells was studied in mouse embryos between E12 and mice on the 3rd postnatal day. During this period, an increase in the number of mitochondria was observed within the oocytes, and this was associated with an increase in the size of the Golgi complex. This period also coincided with the first appearance of the zona pellucida. In previous rat studies, Franchi and Mandl (1962) had noted that only about one-third of follicles analysed in the day 3 rats had a continuous but narrow zona pellucida; most possessed a discontinuous zona, while in others at the same time, no evidence of a zona was observed (see also Baker and Franchi (1967) for analysis of oogonia and oocytes in human fetal ovaries). Chiquoine (1960), in an ultrastructural analysis of the ovaries of newborn mice suggested that the zona was formed from follicle cells, a view no longer held to be the case, as others have demonstrated that the glycoproteins of the zona are oocyte-derived (Wassarman, 1988).

TS	1	2	3	4	5	6	7	8	9	10	11	12	13	14	15	16	17	18	19	20	21	22	23	24	25	26
E	0	1	2	3	4	4.5	5	6	6.5	7	7.5	8	8.5	9	9.5	10	10.5	11	11.5	12	12.5	13.5	14.5	15.5	16.5	17.5

Figure 4.3.10. The early female gonad (shown in sagittal section) and its associated internal genital duct system. The ovary contains cortical cords at its periphery (or cortex) where all of the primordial germ cells are located. In the medullary region, close to the hilus of the ovary, a primitive tubular system (the rete ovarii) may be present which is thought to derive from mesonephric duct tissue. These ducts, if present, are seen in the mediastinum of the ovary, but are, at most, only poorly represented (the equivalent structure in the male (the rete testis) is usually more prominent). On the lateral aspect of the ovary, the mesonephros has almost completely regressed, though its duct is still present and may give rise to a variety of aberrant structures, although it usually completely regresses. In the female, the paramesonephric duct gives rise to the oviduct (with its opening into the peritoneal cavity – the abdominal os), the *two* uterine horns and cervix (in the mouse and in all non-primate mammalian species) and the upper part of the vagina. At this stage, both the mesonephric and paramesonephric ducts enter the urogenital sinus. The diverticulum of the mesonephric duct, the ureteric bud, is not (for simplicity) shown here.

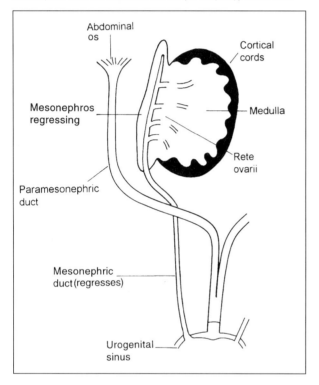

rostral part of the abdominal cavity, and "descent" of the ovaries towards their normal position (TS 23/24). This is in the caudal part of the abdominal cavity, often just behind and slightly lateral to the lower pole of the kidney (in the human, by contrast, the ovaries "descend" more caudally, in the direction of the pelvis). This ascent of the kidneys may be due to the disproportionate growth of the caudal as compared to the rostral half of the embryo that occurs over this period, while the ovaries seem to be relatively "fixed" in their position, due to the anchoring effect of the **ovarian ligament** and possibly the **round ligament of the uterus** (or ligamentum teres), both of which are believed to be the homologues of the **gubernaculum** which serves a similar function in the male with regard to the descent of the testis (see below). Any "descent" of the ovaries is, however, very limited as compared to that of the testes, as the definitive position of the ovaries (which is achieved at this stage) is close to the inferolateral surface of the kidneys.

By TS 24, the ovaries are partially enveloped on their posterior aspect by Müllerian duct tissue which will form the capsule of the ovary and, slightly more distally, the **oviducts** (or **Fallopian tubes**). The more caudal (inferior) parts of the two Müllerian ducts, which will form the two **uterine horns**, descend towards the midline and meet just behind and below the neck of the bladder, at the site of the future **cervix uteri**. The upper one-third or so of the **vagina** is also believed to be of Müllerian duct origin, while its lower two-thirds probably derives from the **urogenital sinus** (Cunha, 1975; O'Rahilly, 1977; for review, see Kaufman, 1988), although there is no single view on this topic. The origin of the cervix in the human is equally contentious, although most authorities believe that it is of paramesonephric duct origin (Koff, 1933; Forsberg, 1965), while the endodermal lining of the vagina and cervix is thought to be derived from the endodermal epithelium of the definitive urogenital sinus. The embryological origin in the mouse is in principle similar to that in the human, taking into account the obvious species differences (*i.e.* the lack of an ovarian capsule and single (rather than double) body of the uterus present in the human female).

The most caudal part of the Müllerian (or paramesonephric) ducts makes contact (around TS 21) with a small thickening of the posterior wall of the urogenital sinus called the **sinusal tubercle** which subsequently elongates and canalizes to form the **sinuvaginal** (or **sino-vaginal**) **bulb**. Shortly afterwards, the Müllerian ducts (which unite in this location to form a common canal) fuse with the rostral part of the sinuvaginal bulb to form a **common genital canal** (or **uterovaginal canal**). More caudally, the uterovaginal canal becomes occluded by the **vaginal plate** (whose origin is unclear). This also elongates and subsequently canalizes and is believed to form the inferior part of the vagina (McKelvey and Baxter, 1935). The site of union of the vaginal plate with the urogenital sinus migrates caudally, and an endodermal membrane at this site persists as the **hymen** of the vagina (Koff, 1933). Although this barrier usually breaks down before puberty, it may persist as an imperforate hymen, resulting in an accumulation of the contents of the uterus/vagina which would normally be desquamated at the completion of the oestrus cycle. This condition would clearly also act as a barrier to pregnancy.[9]

Initially (TS 21), the paramesonephric ducts are identical in appearance and position in the two sexes (Glenister, 1962), only later showing evidence of sexual dimorphism, due to the influence of gonadally-derived sex hormones. In females, the mesonephric ducts largely degenerate due to the absence of a positive influence of sufficiently high levels of circulating androgens, while the paramesonephric system persists

---

[9] In the human, if this condition (termed haematocolpos or haematocolpometra) is present, it results in the accumulation of menstrual blood in the vagina and uterus.

because there are no factors present to destroy it (Jost et al., 1973; Edwards, 1980a).

Although the mesonephric ducts normally regress completely in the female, aberrant tubular or cystic tissue believed to be of mesonephric duct origin is, however, occasionally encountered along the original line of descent of the mesonephric duct within the **mesovarium** (the derivative of the urogenital mesentery in the female) and along the **mesometrium** (or **broad ligament**, from which the uterine horns are suspended) to the urogenital sinus.[10] For an overview of the development of the female genital tract, see Kaufman (1988), and for observations on the embryological remnants often found in the broad ligament, see Duthie (1925). The **rete ovarii** (which consists of aberrant tubular tissue located in the hilar region of the ovary) is also believed to be of mesonephric duct origin. The rete is subdivided into three parts: extraovarian, connecting and intraovarian rete (Byskov, 1978; Byskov and Lintern-Moore, 1973).

The male gonads and reproductive duct system

By TS 21 (E12.5), the male gonad displays an increased degree of vascularity compared to the ovary, particularly at its peripheral margin, and this can be seen slightly before the testes develop their "striped" appearance which is initially due to the ingrowth of the coelomic epithelium. These form septa that subdivide the gonad perpendicular to its long axis and give rise to the primitive testicular cords (Clermont and Huckins, 1961) which will become the **seminiferous tubules** (Pelliniemi, 1975; Mackay et al., 1993). These cords enclose the germ cells (the future spermatogonia or prespermatogonia [Gondos, 1980]) and somatic mesenchymal cells (the future Sertoli cells). Both the testicular cords and the rete testis (which consists of anastomosing tubular tissue located in the hilar region of the testis that "drains" sperm from the straight tubules of the testis into the epididymis) are believed to be mesonephric in origin (Zamboni et al., 1981), but indirect evidence suggests that the germ cells do not appear to be necessary for testicular cord formation (Byskov, 1986).

Sertoli cells play a critical role in facilitating sperm maturation (spermiogenesis phase), and are the primary target for follicle stimulating hormone (FSH),[11] as well as having a wide range of hormone-producing and other roles (for reviews, see Fawcett, 1975; Setchell, 1978). Between these cords is **interstitial tissue**, some of which differentiates to form the **Leydig cells**. The androgenic hormones, particularly testosterone, that they produce are extremely important in inducing the internal genital duct system and external genitalia to differentiate in the direction of maleness (Eik-Nes, 1969), and their differentiation appears to be dependent on the formation of the testicular cords. Similarly, anti-Müllerian hormone (AMH, also termed Müllerian inhibiting substance or MIS) is now produced in the precursors to the Sertoli cells in response to testis-determining factor (TDF) encoded on the sex-determining region of the Y chromosome (SRY). Such hormones plays a dual role in facilitating the differentiation of the derivatives of the mesonephric (Wolffian) ducts, while at the same time causing the regression of the paramesonephric (Müllerian) duct derivatives (Jost and Magre, 1988; for general review, see Gilbert, 1997). By this stage, the mesonephros has all but regressed, leaving only a network of fine tubules (the **rete testis**) in the region of the **mediastinum** (or **hilar region**) of the **testis** which is continuous with the **efferent ductules** which are, in turn, continuous with the other components of the drainage system of the testis (see below).

By TS 22, primitive, uncanalized seminiferous tubules can first be recognized in the testis and, in sectioned material, large numbers of **primordial germ cells** (now at the **spermatogonial** stage), some of which are in division, are embedded within them. Between the tubules are modest amounts of interstitial tissue, although it is not possible at this stage to distinguish between the various cell types present. The mesonephric and Müllerian ducts, both canalized, now run in parallel inferomedially towards the urogenital sinus, the Müllerian ducts being the more lateral of the two throughout the upper part, and the more medial in the caudal part of their course. In the region of the rete testis, the derivatives of the mesonephric tubules are forming the **efferent ductules** that will later connect the seminiferous tubules with the upper part of the **epididymis** which is itself derived from the most rostral mesonephric tubules (Sainio et al., 1997). The urogenital mesentery is relatively wide at this stage, and once the embryo is recognized as a

---

[10] Such aberrant tissue of mesonephric duct origin may be present in the female as the epoophoron or paroophoron; the male homologues of these structures are the testicular ends of the efferent ductules and the paradidymis, respectively, and in both sexes the inferior ductules (ductuli aberrantes inferiores). Some of the caudal part of the mesonephric duct may persist in the female into adult life as Gartner's duct cysts (for reviews, see Meyer, 1909; Duthie, 1925; Koff, 1933; Gruenwald, 1941).

[11] FSH stimulates growth and secretion of oestrogens in the ovarian follicles and spermatogenesis in the testis (see Williams, 1995), although its action in the male is indirect: FSH stimulates the sustentacular (or Sertoli) cells to release androgen-binding protein (ABP) which prompts the spermatogenic cells to bind and concentrate testosterone which, in turn, stimulates spermatogenesis (see Marieb, 1995).

TS	1	2	3	4	5	6	7	8	9	10	11	12	13	14	15	16	17	18	19	20	21	22	23	24	25	26
E	0	1	2	3	4	4.5	5	6	6.5	7	7.5	8	8.5	9	9.5	10	10.5	11	11.5	12	12.5	13.5	14.5	15.5	16.5	17.5

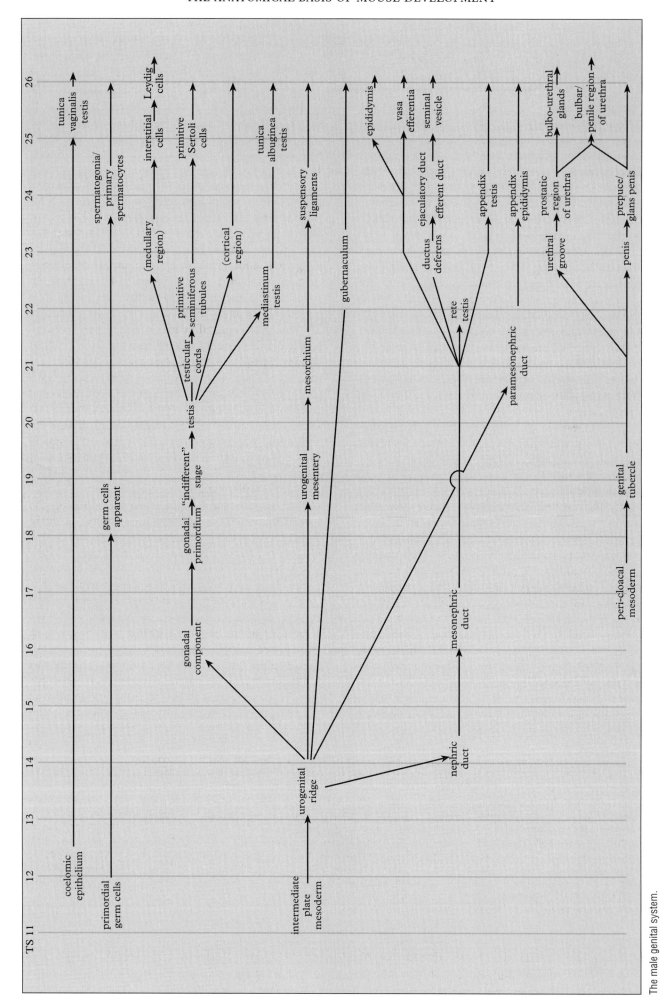

The male genital system.

male, is termed the **mesorchium**. The testes, now ovoid, are close to the antero-lateral surface of the adrenals and to the upper halves of the kidneys.

By TS 22/23, the relationship between the testes and the kidneys is beginning to change. This is partly due to the "ascent" of the kidneys, probably due to the disproportionate growth of the rostral compared to the caudal half of the embryo, and the anchoring effect of the **gubernaculum testis**. This fibro-muscular band of tissue connects the lower pole of the testis to the inner aspect of the scrotal sac (originally, the labio-scrotal sac); it is believed to play a critical role in guiding the testis in its descent from its site of origin in the upper lumbar region initially into the pelvis (see later), and subsequently (after puberty in the mouse, and only following sexual arousal) through the inguinal canal into the scrotum.[12] However, the precise role of the gubernaculum in the descent of the testis is controversial (Backhouse and Butler, 1960).

The testes are now located close to the antero-lateral surface of the lower half of the kidneys. The first histological evidence of differentiation of the outer region of the testis to form its fibrous capsule (the **tunica albuginea**) is now seen, and the mesenchymatous elements involved are just subjacent to its outer coelomic epithelial covering (the **tunica vaginalis**). It is at this stage that the external genitalia first show evidence of differentiation (see below).

By TS 24, the testes have "descended" towards the inlet of the pelvis, and are now located on either side of the lower part of the abdominal cavity at the level of the upper half of the bladder. Furthermore, the various components of the **epididymis** can now clearly be recognized: its duct is disposed in numerous coils, while its tail drains into the newly formed **ductus deferens**, possibly the most important derivative of the mesonephric duct in the male. Towards the upper pole of the epididymis, a minute oval sessile body, the **appendix testis,** is seen. This is generally considered to be one of the derivatives of the Müllerian duct, the other being the **prostatic utricle** (or **uterus maculinus**), although it has been suggested that the latter may arise from the urogenital sinus, and may be the male homologue of the sinuvaginal bulb (very rarely this structure may form a diminutive uterus and may even possess bilateral uterine tubes. Apart from the occasional presence of aberrant tissues, see above, the Müllerian duct completely regresses in the male). The function of these two structures is unclear as is that of the small stalked appendage termed the **appendix epididymis**, located at the blind cranial end of the mesonephric duct (Jones and Scott, 1958).

At around TS 24, the **urethra** starts to form and, in the male, its various components have a more complex origin than that in the female. The **prostatic region** proximal to the site of entry of the **ejaculatory ducts** forms from the vesical part of the urogenital sinus, while the epithelium of the spongy urethra forms from the phallic part of the urogenital sinus, and is continuous with the urethral groove which differentiates on the ventral part of the phallus. The lining of the urethral groove forms the **urethral plate**, which subsequently separates from the surface to form the blind-ending **penile urethra**, while the most distal part of the urethra (in the region of the glans penis) forms from an ectodermal ingrowth from the tip of the phallus. These two subsequently meet, fuse and become canalized. In the male, therefore, the region of the urethra distal to where the ejaculatory ducts enter it has a shared urinary and reproductive function.

By TS 25, the **seminal vesicles** are present, forming as diverticula at the caudal ends of the mesonephric ducts and located just proximal to where the ducts enter the urogenital sinus in the region now represented by the **prostatic region of the urethra**. The seminal vesicles will secrete a fructose-rich fluid which adds bulk to the ejaculate (seminal plasma) as well as providing an important source of nutrients for the sperm, while additional fluid is provided by the secretions of the **prostate gland** (which forms at TS 26, see below). The most caudal region of the mesonephric ducts, where they pass through the wall of the prostatic region of the urethra, differentiate into the **ejaculatory ducts** (for the gross arrangement of the seminiferous tubules and genital duct system in

---

[12] In the mouse, the testes only pass into the scrotum following sexual arousal and then, following mating, return to the safety of the pelvic cavity. This differs from the situation in the human male where the testes usually descend through the inguinal canal to the level of the superficial inguinal "ring" at or very shortly after birth, descending into the scrotum shortly afterwards where they are retained. In the human it is said that the optimal temperature for successful spermatogenesis to occur is slightly lower than core body temperature; if the testes (bilaterally) remain as intra-abdominal structures, spermatogenesis fails to occur and the germ cells may undergo malignant transformation to form teratocarcinoma cells. Clearly, this must be species specific as in most mammals the testes only descend into the scrotum during sexual arousal, and the germ cells do not normally (except in a few specific strains) undergo malignant transformation. In the human, if a testis fails to descend into the scrotum, it may be located at any point along its normal line of descent, or may be located ectopically, for example, in the so-called inguinal pouch. If located within the inguinal canal, or in the inguinal pouch, the testes are vulnerable to external trauma, and surgical intervention may be necessary to facilitate their descent into the scrotum; if the latter cannot be achieved, it may be appropriate to remove the testis. If the problem is bilateral then exogenous hormone therapy will have to be instituted after the time puberty would normally have been initiated.

TS	1	2	3	4	5	6	7	8	9	10	11	12	13	14	15	16	17	18	19	20	21	22	23	24	25	26
E	0	1	2	3	4	4.5	5	6	6.5	7	7.5	8	8.5	9	9.5	10	10.5	11	11.5	12	12.5	13.5	14.5	15.5	16.5	17.5

the male, see Figure 4.3.11). By now, the testes have descended further, and are located well below the "brim" of the pelvis, being guided in their descent by the **gubernaculum testis** (see above) which passes through the **inguinal canal**, and is anchored to the inner aspect of the scrotum. The arrangement of the duct system in the male and its relationship to the ureter are shown in Figures 4.3.5 and 4.3.11. As development proceeds, the ureteric openings enter the upper dorso-lateral aspects of the urogenital sinus whereas, possibly due to differential growth of the intervening tissues (Frazer, 1935), the mesonephric ducts open into the urethra in a much more caudal location. An alternative suggestion is that the mesonephric ducts may become absorbed into the dorsal wall of the inferior part of the trigone, in the region of the prostatic urethra.

The cellular components of the seminiferous tubules show increased evidence of differentiation with primitive **Sertoli cells**, that are attached to their basement membrane, being interspersed with numerous **spermatogonial cells**. It is not yet possible to distinguish between the mesenchyme cells and the precursors of the interstitial cells (of Leydig) in the interstitial region between the seminiferous tubules.

By TS 25/26, the testes have descended well into the pelvis and lie on either side of the bladder with the various components of the epididymis now seen to be their postero-lateral relations. The seminiferous tubules have become canalized and their lumens contain large numbers of spermatogenic cells. The epididymis and ductus deferens are now readily recognized and the wall of the latter is substantially thicker than previously. By TS 26, the disposition of the testes and internal genital duct system is similar to that seen in the immediate post-natal period.

Figure 4.3.11. Representative sagittal section through the testis, showing its relationship to its internal genital duct system. At its periphery, the testis is covered by a fibrous capsule (tunica albuginea), and immediately subjacent to this the tissue of the testis is divided up into a large number of radially arranged compartments, each of which contains a seminiferous tubule. In the testis, the cortical region largely regresses while the medulla forms the functional component of the gonad. This contrasts with the situation in the female, where it is the cortex that is of critical importance, being the site where the primordial follicles are located, while the medullary region largely regresses. During ejaculation, the seminiferous tubules drain their contents (the mature spermatozoa and the fluid in which they are bathed) via the efferent ducts into the epididymis. From the latter the contents pass into the ductus (or vas) deferens, and thence via the ejaculatory duct into the prostatic region of the urethra where additional fluid is added to the ejaculate from the prostatic glands. Shortly before the ductus deferens terminates, it gives off a diverticulum that differentiates to form the seminal vesicle. The latter is of critical importance because it adds fluid (and bulk) to the ejaculate as well as energy sources including fructose. Two structures that are believed to be derived from paramesonephric duct tissue are also shown in the diagram, the appendix testis and the prostatic utricle (or uterus masculinus). Their function has yet to be established. An appendix *epididymis* may also be present as a distal extension of the mesonephric duct, being located close to the appendix testis, and this is believed to be of mesonephric duct origin. The various components of this diagram are not drawn to scale.

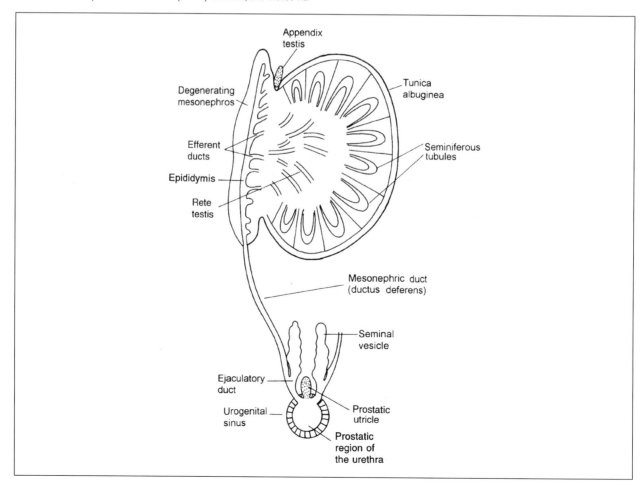

The **prostate** develops at about TS 26 from three paired urethral primordia, a dorso-cranial part, termed the **coagulating gland** which is located on the concavity of the seminal vesicle, to which it is attached by a common connective tissue sheath, and ventral and dorso-lateral primordia. The duct of the coagulating gland opens into the prostatic region of the urethra close to that of the ejaculatory duct. The ducts from the other parts of the prostate open into the region of the prostatic urethra into a shallow depression on either side of the **seminal colliculus**, close to the narrow opening of the prostatic utricle. The homologous structures to the prostate in the female are believed to be the urethral and paraurethral glands. The coagulating gland and prostate show only minimal evidence of differentiation during the early post-natal period. In adult rodents, the secretions from the coagulating gland coagulate the ejaculate in the proximal part of the female tract, forming the vaginal plug; a rodent feature that has no equivalent structure in humans.

The external genitalia

The terminal part of the hindgut which forms at about TS 15 becomes slightly dilated to form the **cloaca** (TS 17) which accordingly possesses an endodermal lining. As the tail fold develops, the region where the cloaca makes contact with the subjacent overlying ectoderm forms the **cloacal membrane**. Here, there is no intervening intra-embryonic mesoderm; a similar situation is observed at the rostral extremity of the gut tube in the buccopharyngeal membrane. The cloacal membrane gradually comes to be situated on the ventral surface of the embryo, and its caudal margin becomes continuous with the caudal attachment of the connecting stalk. Mesodermal cells from the caudal part of the primitive streak pass around the sides of the cloacal membrane, raising the overlying ectoderm to form the two **genital folds**. The cloacal membrane comes to lie in a shallow depression termed the **proctodaeum** (TS 19).

The externally located genital tubercle is also first apparent at TS 19, although the details of the external components of the urogenital system only start to become apparent at around TS 21–23 (E13–14). The cloacal membrane partitions to give the ventral urogenital sinus and a dorsal hindgut region at about TS 18, although these are not physically separated until a stage later when they are divided by the downgrowth of the **urorectal septum** and become the dorsally located **anal membrane** and the ventrally located **urogenital membrane** (see also Chapter 4.4 *The gut and its associated tissues*), with the fibrous **perineal body** forming in the septal midline (TS 21).

The anal membrane defines the boundary between the ectodermal and endodermal components of the anal region, with the endodermal region giving rise to the upper two-thirds of the anal canal while the lower one-third is of ectodermal origin (and is derived from the anal pit which is first recognized at about TS 19). The peripheral part of the cloacal membrane together with its associated mesenchyme gives rise to the **cloacal sphincter**, which in turn forms the **external anal sphincter** (TS 22) and the **urogenital diaphragm** with its associated muscles, all of these being supplied by the **pudendal nerve**. The urogenital membrane has an internal and an external component; the internal part of this endodermally-derived epithelium also forms part of the lining of the **urinary bladder** (TS 21/22), with the **urachus** (TS 21) at its apex, and the proximal part of the urethra (which also has a distal surface ectoderm component – see above), while the external part forms most of the external genitalia.

The early stages in the differentiation of the external genitalia are similar in the two sexes which both have a **genital tubercle** (apparent at TS 19), a **urethral groove** (first evident in the vesical part of the urogenital sinus at about TS 23) and the **labial** (or **labio-scrotal**) **folds** (TS 22) located on either side of the urogenital membrane (the derivatives of the genital folds). In the female, the **clitoris** and **glans clitoridis** form from the tubercle at about TS 23. while the **urethra** differentiates from the urethral groove a stage later (TS 24). The urethral folds do not fuse, but form the **labia minora**, while the two **labio-scrotal folds** (TS 22), initially located on either side of the cloacal membrane, do not fuse across the ventral midline but form the **labia majora**. In the male, by contrast, the **urethral groove** forms the penile urethra, while the two **labio-scrotal folds** (TS 22) fuse across the midline in the region of the cloacal membrane to form the scrotum. The **glans penis** differentiates from the distal part of the genital tubercle at about TS 23.

Because of the relative immaturity of the embryo at the time of birth, it is quite difficult to distinguish between the two sexes from a cursory inspection of the external genitalia in newborn mice, although the distance between the anus and **genital papilla** (the genital tubercle or immature phallus) is somewhat less in the male than the female. For an overview of the development of the external genitalia in the human embryo, see Spaulding (1921), and for observations on the development of the penis in the mouse, see Murakami (1987). One significant difference exists between the penis and the clitoris in rodents and these structures in the human. The penis in rodents possesses skeletal tissue (absent in the human) and erectile tissues chiefly comprising

TS	1	2	3	4	5	6	7	8	9	10	11	12	13	14	15	16	17	18	19	20	21	22	23	24	25	26
E	0	1	2	3	4	4.5	5	6	6.5	7	7.5	8	8.5	9	9.5	10	10.5	11	11.5	12	12.5	13.5	14.5	15.5	16.5	17.5

the **os penis** and corpus cavernosum penis, respectively (Murakami and Mizuno, 1984, see also Hummel *et al.*, 1966). The os penis in rodents consists of a proximal and distal segment. In the clitoris of mice, a membranous bone homologous to the distal half of the proximal segment of the os penis is formed (Glucksmann *et al.*, 1976).

The autonomic nerve supply to the gonads

The autonomic nerve supply to the gonads accompanies their respective arteries. The innervation to the testis is from branches of the renal and aortic plexuses, and the innervation of their distal parts is supplemented by branches from the hypogastric plexuses, with these nerves then going on to innervate the epididymis and ductus deferens. Large nerve trunks are associated with the efferent ducts of the testis by TS 22/23, although few nerves are seen coursing within the interstitial tissue of the testis at this time. The ovary and oviduct are innervated by branches from the renal and aortic plexuses, with the innervation to their inferior parts being reinforced by branches from the hypogastric plexuses.

The innervation to the gonads and their associated structures consists of afferent and efferent sympathetic fibres; the efferents are said to be vasomotor and from the lower thoracic spinal segments and the parasympathetic innervation from the hypogastric plexuses probably has a vasodilator function (Williams, 1995).

Implications for genetic analysis

With increasing interest being paid to pre- and postnatal mice with induced genetic deficiencies (*e.g.* knockout mice) that are sterile, it is worth noting that this phenotype can be caused by interfering with a wide range of reproductive processes (*e.g.* spermatogenesis/oogenesis, seminiferous tubule formation and folliculogenesis). While histological/immunohistochemical analysis of the gonads in these individuals clearly provides key data, it should also be borne in mind that interference with gonadal differentiation, at almost any level, is likely to have *secondary* consequences on the development of both the internal genital duct system and on the external genitalia. Equally, although it would depend to a considerable degree on the *primary* effect produced by the abnormal genome, there may be *additional* consequences with respect to the development of the renal system.

Although analysis of components of the genital system before about E12.5 will not be particularly informative, particularly on seminiferous tubule or follicular differentiation, it may well shed light on interference with germ cell differentiation. Here, analysing embryos from as early as the primitive streak stage for the presence (or absence) of intracellular alkaline phosphatase-positive staining cells in regions where germ cells might reasonably be expected to be present (see above) might well prove to be extremely informative.

Where germ-cell differentiation is normal and animals survive to puberty or beyond, interference with seminiferous tubule formation or folliculogenesis may well lead to sterility, irrespective of the underlying lesion. Moreover, deficiencies in seminiferous tubule development will almost certainly have secondary consequences: a failure of secretion of androgenic hormones (produced by the Leydig cells in the interstitial tissue), and hormones and hormone-like substances (such as AMH normally produced by Sertoli cells lining the seminiferous tubules) will inevitably have downstream effects on the development of the internal genital duct system and external genitalia. In addition, the selective absence of Sertoli cells from otherwise normal seminiferous tubules will also interfere with spermiogenesis (if it has not been interfered with at an earlier stage of germ cell differentiation).

Where a reasonable degree of postnatal development occurs in sterile males, it is particularly instructive to investigate not only the gross morphology of the internal genital duct system, but also that of the external genitalia. In some cases, the latter may display such characteristic features of pseudohermaphroditism as, for example, testicular feminization where, despite the normal production of androgenic hormones and AMH by the male, the target organs fail to respond to their presence, and the external genitalia consequently closely resemble the appearance seen in the female.

Because of the inter-relationship that exists at early stages in the development of the reproductive and urinary/renal systems, it is also instructive to follow the fate of the derivatives of the intermediate mesoderm, as components which might normally be expected to regress, such as the mesonephroi, may occasionally fail to do so, despite the presence of morphologically normal and apparently normally functioning metanephroi (*i.e.* definitive kidneys). Such a situation has occasionally been observed in knockout mice where the deficient gene was not previously known to play any part in urogenital development (MHK, unpublished observations).

4.4 The gut and its associated tissues

Introduction

The gut develops from the definitive endoderm, initially appearing as an infolding on the ventral surface of the *unturned* embryo. By the time that the turning process is completed (TS 14), the foregut and hindgut pockets have extended cranially to the buccopharyngeal membrane and caudally towards what will become the cloacal membrane. As the foregut, midgut and hindgut regions develop, further interactions take place between the endodermally-derived outgrowths from the primitive gut tube and adjacent mesenchyme, and these lead to the formation of such tissues as the salivary glands (outgrowths of the rostral part of the oropharynx – see Chapter 4.5 *The mouth and nose region*), the various derivatives of the branchial pouches (see Chapter 3.3 *The branchial arch system*), the lower respiratory tract, the extra- and intra-hepatic components of the biliary tree, and the duct system of the pancreas (see below).

Furthermore, parts of the primitive gut are held in place by dorsal and ventral mesenteries and the changes that occur in the conformation of different regions of the gut as they elongate or expand in volume to accommodate their future role after birth allows these mesenteries to define specific domains in the peritoneal cavity. The most important of these is associated with the rotation of the stomach; this leads to the isolation of a defined region of the peritoneal cavity to form the **lesser sac**, with the rest of the peritoneal cavity then being termed the **greater sac**. The aggregation of a number of mesodermal precursors within the dorsal mesentery of the stomach gives rise to the **spleen**, while other complex interactions in the **septum transversum** (which forms within the ventral mesentery of that region of the gut which will become the stomach) lead to the formation of the **liver** (see Chapter 2.4 *Early organogenesis*).

This chapter is therefore a convenient place in which to consider the gut and those tissues and organs closely associated with it in the mouse embryo.[1] We first therefore consider the early development of the gut as a whole, then examine the separate fates of the foregut, midgut, and hindgut (where there is some overlap with Chapter 4.3 on *The urogenital system*) and finally consider the development of the liver, biliary apparatus, spleen and pancreas. In all cases, the analysis of tissue development is accompanied by an outline of their vascular and (where appropriate) their nerve supply.

Early gut development

All of the definitive endoderm cells that line the primitive gut are believed to be derived from the anterior end of the primitive streak. During TS 11, a slight indentation just caudal to the future pericardial region appears in the ventral midline of this endoderm region, and this is the first indication of the **foregut pocket**. By TS 12 (early-somite-stage embryos), the rostral part of the primitive gut is represented by the **foregut diverticulum**. This extends rostrally from a wide entrance pocket or portal at the site of the future foregut–midgut junction, dorsal to the future pericardial region, into the cephalic region of the embryo, being separated from the overlying neural tissue by mesenchyme in which is located the notochord. The future **midgut region**, in turn, extends caudally as far as the entrance to the **hindgut portal** (only clearly defined towards the end of TS 12), and this is continuous caudally with the future hindgut region.

---

[1] The sequential stages in the development of the various components of this system in the *human*, are reviewed by O'Rahilly (1978); for lists of appropriate references to studies on which the findings are based, see also O'Rahilly and Müller, 1987.

TS	1	2	3	4	5	6	7	8	9	10	11	12	13	14	15	16	17	18	19	20	21	22	23	24	25	26
E	0	1	2	3	4	4.5	5	6	6.5	7	7.5	8	8.5	9	9.5	10	10.5	11	11.5	12	12.5	13.5	14.5	15.5	16.5	17.5

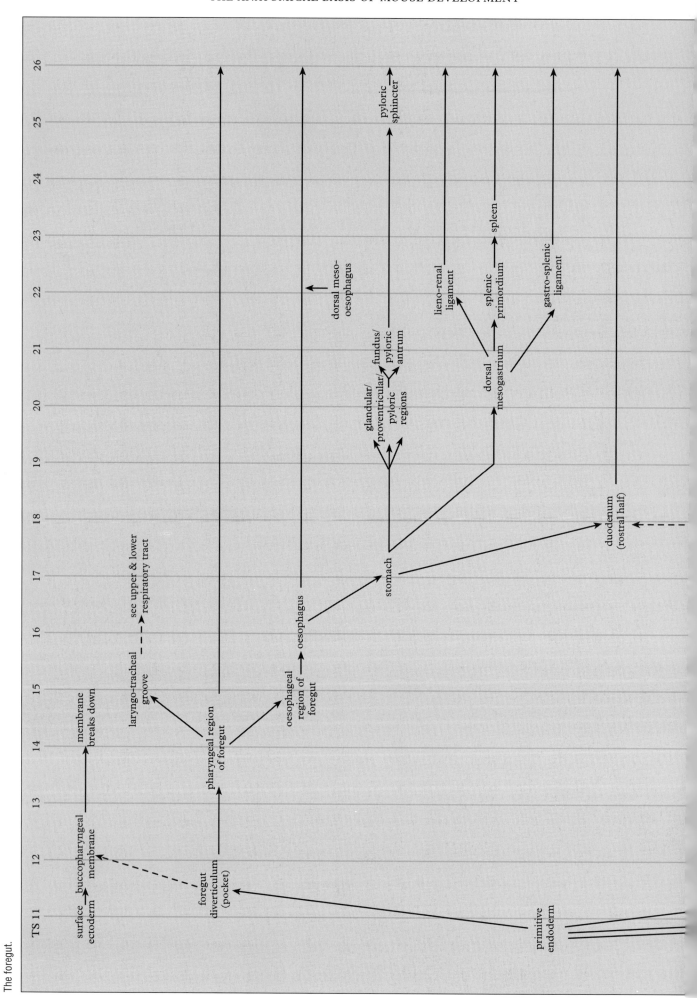

The foregut.

THE GUT AND ITS ASSOCIATED TISSUES

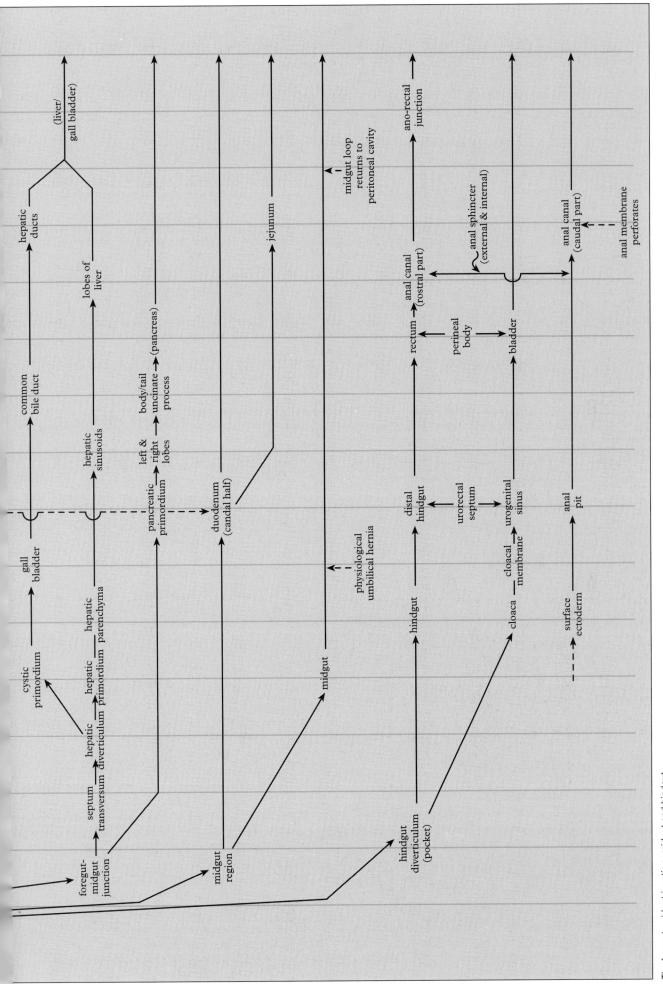

The foregut–midgut junction, midgut and hindgut.

The foregut

The foregut diverticulum will differentiate to form the majority of the oral cavity, the pharynx, the oesophagus, the lower respiratory tract, the stomach and the rostral part of the duodenum. The nasal cavity (or nasopharynx) also forms from the rostral part of the pharyngeal region, although it is later separated from its oropharyngeal part with the elevation and subsequent fusion of the palatal shelves (see Chapter 4.5 *The mouth and nose region*). Initially, however, the foregut extends as a blind diverticulum into the headfold region, and its endodermal lining is only separated from the ectodermally-derived neuro-epithelial cells of the neural groove by the cells of the **notochordal plate** and a relatively small number of cephalic mesenchyme cells. During the latter part of TS 12, the rostral part of the foregut diverticulum widens laterally to form the **first branchial pouches**. It is worth noting that, at this stage, the side-to-side width of the rostral part of the foregut is always considerably greater than its dorsal–ventral diameter except in the region just rostral to the entrance to the **foregut portal** (in a region which is destined to form the majority of the oesophagus, the stomach and rostral half of the duodenum) where its anterior–posterior diameter is only marginally narrower than its transverse diameter.

During TS 13/14, the ectodermally-lined **stomatodaeum** or **mouth pit** approaches the most rostral part of the endodermally-lined foregut diverticulum, the location where these are apposed being termed the **buccopharyngeal** (or oropharyngeal) **membrane**. It is important to note that no mesenchyme intervenes between the surface ectoderm of the mouth pit and the subjacent endodermal lining of the oropharynx. A similar arrangement will later be seen at the caudal end of the gut tube, in the region of the cloacal membrane (TS 17), where no mesenchyme intervenes between the surface-ectodermal-derived anal pit (proctodaeum) and the endodermal lining of the caudal extremity of the hindgut.

At about the same time, the ectodermally-derived roof of the oral cavity, just anterior to the buccopharyngeal membrane, thickens and indents to form **Rathke's pouch**. This indentation gradually deepens and elongates (TS 18/19), and eventually loses its connection with the roof of the oral cavity to form the various components of the anterior pituitary (TS 19/20; see Chapter 4.6 *The pituitary*). The buccopharyngeal membrane breaks down during TS 14, so removing the barrier between the rostral part of the foregut and the oropharynx. By TS 14/15, the first three branchial pouches with their dorsal and ventral components become more clearly defined (see Chapter 3.3 *The branchial arch system*).

During the early part of TS 14, a thickened and indented region (termed the **foramen caecum**) is seen in the midline in the floor of the rostral part of the pharyngeal region of the foregut, located between the second and third branchial arches. This is the first indication of the **thyroid diverticulum** (or primordium). (For the subsequent development of the thyroid gland, see Chapter 4.5 *The mouth and nose region*.)

During TS 14/15, the caudal part of the foregut gradually narrows to form the future **oesophageal region**, and at about this time a well-defined groove (the **laryngo-tracheal groove**) appears on the ventral aspect of the floor of the oropharynx just caudal to the fourth branchial arch component of the **hypobranchial eminence** that will form the epiglottis and pharyngeal part of the tongue (see Chapter 4.5 *The mouth and nose region*). This groove subsequently deepens to form a diverticulum whose proximal part differentiates to form the **trachea**, while its distal part bifurcates (TS 15) to give rise to the right and left **main bronchi**. During TS 15/16, the distal part of these bronchi become surrounded by splanchnic mesenchyme and together they form the **lung buds** (see the *Lower respiratory tract*, later this chapter). A little later (TS 17), the lumen of the oesophagus caudal to the tracheal diverticulum narrows to form a tubular structure, in contrast to the lumen in the region of the future **stomach** which shows early evidence of dilatation.

By about TS 14/15, the primitive foregut distal to the pharyngeal region has become a straight tube along much of its length, while the region caudal to the diaphragm where the stomach forms is the only part of the primitive gut which retains remnants of both a *dorsal* mesentery (the dorsal mesogastrium) and a *ventral* mesentery, the latter in the form of the **falciform ligament** and **lesser omentum**. These are located ventral and dorsal to the liver, although the boundaries of the liver are only poorly defined before TS 17/18. The arrangement of the subdivisions of the dorsal and ventral mesenteries of the primitive gut within the peritoneal cavity, together with the principal arterial blood supply to the fore-, mid- and hindgut in this region are shown in Figure 4.4.1. (For observations on the differentiation of the dorsal mesentery during human embryonic and fetal development, see Yokoh, 1970).

From TS 17/18 onwards, the **stomach** ceases to be a midline structure and becomes an almost exclusively left-sided organ, except for its proximal and distal junctions with the oesophagus and duodenum, respectively, which retain their midline position. This change is mainly due to the considerable expansion in the volume of the hepatic primordium (the future liver) that occurs at this time. From TS 18 to about TS 22/23, the stomach not only elongates and expands, with its long axis eventually becoming orientated transversely, rather than longitudinally as it was initially, but also rotates through about 90° clockwise on its longitudinal axis when viewed from above (see Figure 4.4.2a–d). Since the dorsal wall expands to a greater degree than the ventral wall, the stomach develops a greater (dorsal) and a lesser (ventral) curvature. As a consequence of this rotation, the original *left* side of the primitive stomach forms its *anterior* wall, while its former *right* side forms its *pos-*

Figure 4.4.1. An approximately median sagittal section through the early (~TS 17/18) abdomen and pelvis showing the primitive gut, its arterial blood supply from the aorta and the dorsal and ventral mesenteries within which are the liver, spleen and pancreas. The distal part of the foregut (the oesophagus) pierces the diaphragm and its lumen expands to form the stomach which is suspended from the anterior (ventral) upper abdominal wall by a ventral mesentery. This is subdivided into two parts by the liver, into the lesser omentum, close to the lesser curvature of the stomach, and into the falciform ligament, between the ventral surface of the liver and the anterior abdominal wall. The spleen develops within the dorsal mesentery of the stomach (dorsal mesogastrium), while the pancreatic duct (unlabelled) and common bile duct enter the gut at the foregut–midgut junction. At this stage of development, the hindgut is in continuity with the urogenital sinus (the dorsal and ventral components of the cloaca).

The arterial blood supply to the sub-diaphragmatic part of the foregut is from branches of the coeliac trunk. The midgut loop has a dorsal mesentery, termed "the" mesentery, but no ventral mesentery, and this region of the gut is supplied by branches from the superior mesenteric artery. The distal (anti-mesenteric) extension of the midgut, the vitelline duct, is supplied by the vitelline artery, an additional branch of the superior mesenteric artery. The rostral (or proximal) half of the duodenum is of foregut origin, and supplied by a branch of the coeliac trunk, while its caudal (or distal) half is supplied by a branch of the superior mesenteric artery. The hindgut, all but its most distal part being supported by a dorsal mesentery, is supplied by arterial branches from the inferior mesenteric artery; the umbilical arteries are seen to be branches of the latter that pass via the umbilical cord to the placenta.

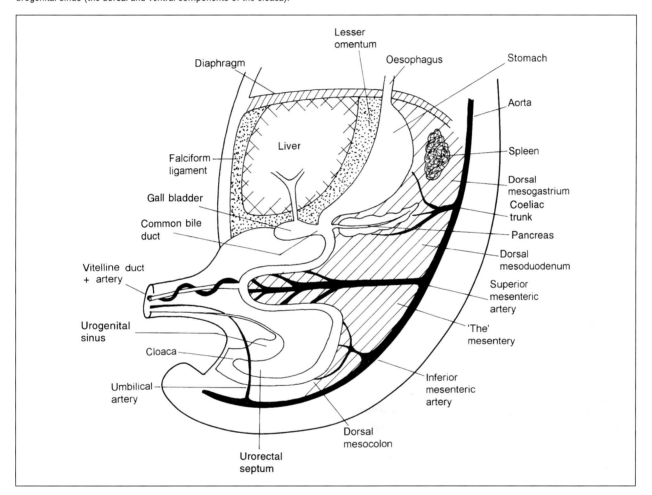

terior wall. This equally applies to the trunks of the vagus nerve, where the left vagal trunk which descends on the left side of the oesophagus becomes the anterior vagal trunk once it passes through the diaphragm, and the right vagal trunk becomes the posterior vagal trunk. In the human embryo, rotation of the stomach occurs at the 12 mm stage. The pyloric region may be recognized by about 3 months of gestation, though full development of this region does not occur until much later (Bremner, 1968).

As the stomach rotates, a small volume of the peritoneal cavity becomes localized (TS 17) behind or dorsally to it, and this space is known as the **lesser sac** (or omental bursa) to contrast it with the **greater sac**, the remaining (greater) part of the peritoneal cavity. The lesser sac is bounded superiorly by the inferior surface (or left dome) of the diaphragm (see Chapter 2.4 *Intraembryonic cavities*), anteriorly by the posterior wall of the stomach, posteriorly by the peritoneal covering of the posterior abdominal wall subjacent to which is the body and tail of the pancreas and the left kidney, and inferiorly by the inner part of the greater omentum (Kanagasuntheram, 1957).

The **epiploic foramen** (or entrance to the lesser

TS	1	2	3	4	5	6	7	8	9	10	11	12	13	14	15	16	17	18	19	20	21	22	23	24	25	26
E	0	1	2	3	4	4.5	5	6	6.5	7	7.5	8	8.5	9	9.5	10	10.5	11	11.5	12	12.5	13.5	14.5	15.5	16.5	17.5

sac) links the lesser sac with the main part of the peritoneal cavity. This change in the configuration of the stomach with the formation of the lesser sac is illustrated in Figure 4.4.2d. Two sagittal sections through the lesser sac indicate the embryonic relationship between the transverse colon and the posterior wall of the greater omentum in this location (Figure 4.4.3a), and the definitive arrangement (Figure 4.4.3b). A diagram illustrating a transverse section through the entrance to the lesser sac in a mature embryo is shown in Figure 4.4.3c.

The stomach in the mouse starts to regionalize at about TS 21 and eventually subdivides into the smaller **cutaneous**, or **proventricular** (non-glandular) **region** at the left of the site of entrance of the oesophagus (at the gastro-oesophageal junction), and the larger **fundic** (glandular) **region** to the right of this entrance which leads into the **pyloric region** whose **sphincter** is first seen at about TS 26. A clearly delineated line of transition is seen between the cutaneous and fundic regions of mucous membrane which is clearly visible on histological sections through the stomach at this and subsequent stages of development.

While the stomach is rotating, its dorsal mesentery (or **dorsal mesogastrium**) extends to form the **greater omentum**. Since the **spleen** forms by the aggregation of mesodermal tissue within the dorsal mesogastrium (TS 21/22, see below), this organ is therefore enclosed within two layers of peritoneum. That part of the dorsal mesogastrium which is located between the stomach and the spleen is thus called the **gastro-splenic ligament**, while the part of the mesentery located between the spleen and its attachment to the posterior abdominal wall (close to the ventral/anterior surface of the left kidney) is termed the **lieno-renal ligament**.

The body and most of the tail region of the **pancreas** are retroperitoneal (*i.e.* behind the peritoneum), but the most distal part of the tail almost reaches to the hilar region of the spleen where the vessels and nerves enter. This organ is enclosed within two layers of peritoneum and, in the mouse, is a narrow *ribbon-like* organ which is closely associated with the dorso-lateral surface of the stomach (in humans, the spleen is roughly oval in shape and slightly concave in its hilar region). The distal part of the splenic artery, which is embedded for much of the proximal part of its course in the dorsal part of the pancreas, passes via the lieno-renal ligament towards the spleen before breaking up into a number of separate vessels just before entering its hilar region. These are termed *end-arteries* because each supplies a discrete region of this organ, there being no inter-communication between these vessels. If one becomes occluded, then the region of the spleen supplied inevitably dies (termed splenic infarction).

The midgut

It is not possible to be certain when the midgut first appears as a distinct entity because, when the fore- and hindgut pockets can first be seen (TS 11/12), there is no clear line of demarcation between them and the future midgut region. The rostral and caudal extents of the primitive midgut region, which is located on the ventral surface of the unturned embryo and fully exposed to the fluid within the secondary yolk sac cavity, are only defined a few hours later at about TS 12 when the entrances to the foregut and hindgut pockets can be recognized. As the embryo progresses through the turning sequence, the margins of the initially flattened layer of midgut endoderm become elevated and then apposed across the ventral midline to form a simple tube, though its middle part initially remains widely open at the so-called **umbilical ring** where its margins are continuous with those of the proximal part of the wall of the visceral yolk sac. The primitive midgut extends between the future foregut–midgut and midgut–hindgut junctions, and differentiates to form the caudal half of the duodenum, the jejunum, the ileum, the caecum, and the ascending and majority of the transverse colon almost to the splenic flexure.

From about TS 12 to about TS 14/15, the diameter of the umbilical ring progressively diminishes (this also happens to be the region where the intra- and extra-embryonic coelomic cavities meet, and it is bounded at this stage by the peripheral margin of the body wall where it is in continuity with the amnion – this relationship is clearly seen in Figure 2.3.1). At about the same time (TS 14), the neck of the **yolk sac** progressively narrows and eventually forms the **vitello-intestinal** (or **vitelline**) **duct**, which extends (during the latter part of TS 14) from the mid-point of the midgut, distally into the **umbilical cord**. By TS 15, the lumen of this duct is obliterated, although its proximal remnant may still be recognized at TS 17 because of its attachment to the apex of the midgut loop. This, the first stage of development, is ended when the most distal part of the midgut loop is seen to be located within the **physiological umbilical hernia**.

From about TS 17 (E10.5), the midgut lengthens enormously and so allows regional differentiation to take place. Because there is insufficient space within the peritoneal cavity for this to occur (the liver, in particular, and stomach occupy most of the upper half of the peritoneal cavity, while the **mesonephroi**, though retroperitoneal structures, expand forwards to occupy much of the rest of the available space), the greater part of the midgut loop herniates into the proximal part of the umbilical cord forming the so-called physiological umbilical hernia. Within this, the distal part of the midgut loop initially appears to rotate[2] through

---

[2] It should be said that differences of opinion exist as to whether it is appropriate to discuss the changes that occur in the configuration of the midgut loop in terms of degrees of rotation since the proximal and distal parts of the root of the mesentery of the midgut, at the duodenal–jejunal and ileo-caecal junctions, are fixed in position throughout these events. While clearly an over-simplification of the events observed, the use of the concept of degrees of rotation when considering the relocation of the various parts of the midgut loop with respect to the fixed abdominal contents at least provides an indication of the dynamic nature of the events occurring at this time.

Figure 4.4.2a–d. The rotation of the stomach. (a, TS 17). Initially, a slight swelling appears in the sub-diaphragmatic part of the foregut that will form the stomach, and its long axis is along the length of the gut tube (the aorta indicates the location of the dorsal midline). A transverse view shows that it is suspended by both a ventral and a dorsal mesentery, and that the left and right vagal trunks pass distally on either side of it. (b, ~TS 19). During the next stage of stomach differentiation the dorsal surface of the stomach expands to a greater extent than its ventral surface, so forming its greater and lesser curvatures. (c, ~TS 21). When the stomach rotates through 90° (viewed from above), its original left side becomes its anterior surface, while its original right side now becomes its posterior one. The dorsal mesentery (the dorsal mesogastrium) also increases in length and is now termed the greater omentum. (d, ~TS 23). In due course, the site of origin of the dorsal mesentery migrates to the left (eventually being located directly in front of the left kidney). At the same time, the posterior surface of the liver also becomes extensively adherent to the posterior abdominal wall. As a consequence of these two events, a small volume of the peritoneal cavity becomes enclosed behind the stomach, and extends caudally into the cavity created by the greater omentum. This small, isolated space is now termed the omental bursa (or lesser sac), to contrast it with the rest of the peritoneal cavity (or greater sac), while its entrance is called the epiploic foramen. As the stomach rotates and its long axis becomes transverse, what was originally the left vagal trunk now becomes the anterior vagal trunk, while the former right vagal trunk now becomes the posterior vagal trunk.

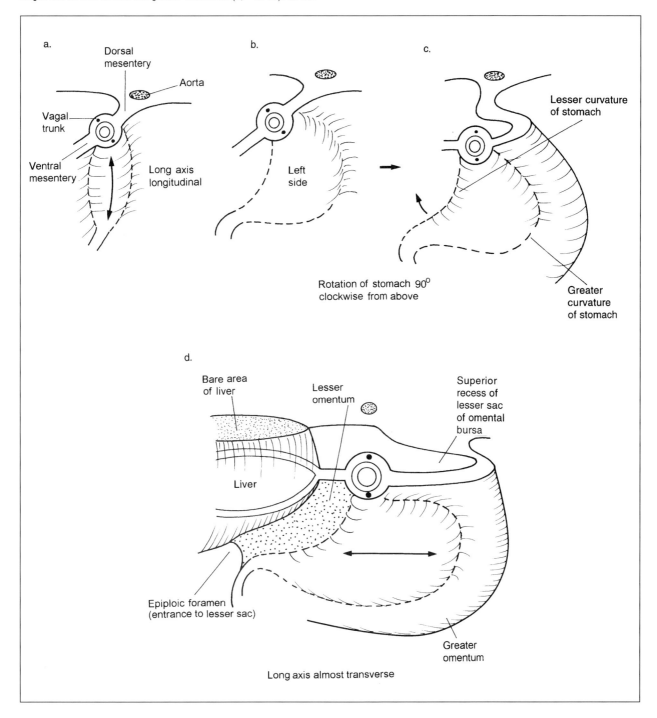

TS	1	2	3	4	5	6	7	8	9	10	11	12	13	14	15	16	17	18	19	20	21	22	23	24	25	26
E	0	1	2	3	4	4.5	5	6	6.5	7	7.5	8	8.5	9	9.5	10	10.5	11	11.5	12	12.5	13.5	14.5	15.5	16.5	17.5

Figures 4.4.3 a–c. The relationship between the lesser sac of the omental bursa and the rest of the peritoneal cavity (the greater sac) is shown in approximately median sagittal (a, b) or transverse (c) views through the peritoneal cavity. (a) During embryogenesis, the lesser sac extends rostrally up to the level of the diaphragm (termed its superior recess), while its inferior recess extends caudally to include the entire space delineated by the peritoneal lining of the greater omentum. In the adult (b), the two peritoneal layers lining the inferior recess fuse together, and effectively obliterate this region of the lesser sac so that all that remains of the lesser sac is its superior recess. In the adult, the transverse colon is incorporated into the posterior free border of the greater omentum. (c) A transverse section through the peritoneal cavity at the level of the entrance to the lesser sac (via the epiploic foramen) (arrow). This view shows the relationship between the structures near the superior recess. Note that the free border of the lesser omentum contains the portal vein and hepatic artery ascending towards the liver, and the common bile duct descending towards the greater duodenal papilla (usually) at the foregut–midgut junction, though in the mouse the bile duct commonly enters the duodenum at numerous sites rather than at a single site as in the human. The lesser omentum, part of the ventral mesentery of the stomach, forms part of the anterior wall of the lesser sac. The dorsal mesentery of the stomach (not shown here) is subdivided by the presence of the spleen into the gastro-splenic and lieno-renal ligaments. Posterior relations of the superior recess of the lesser sac are seen to be the inferior vena cava, the aorta and the left kidney.

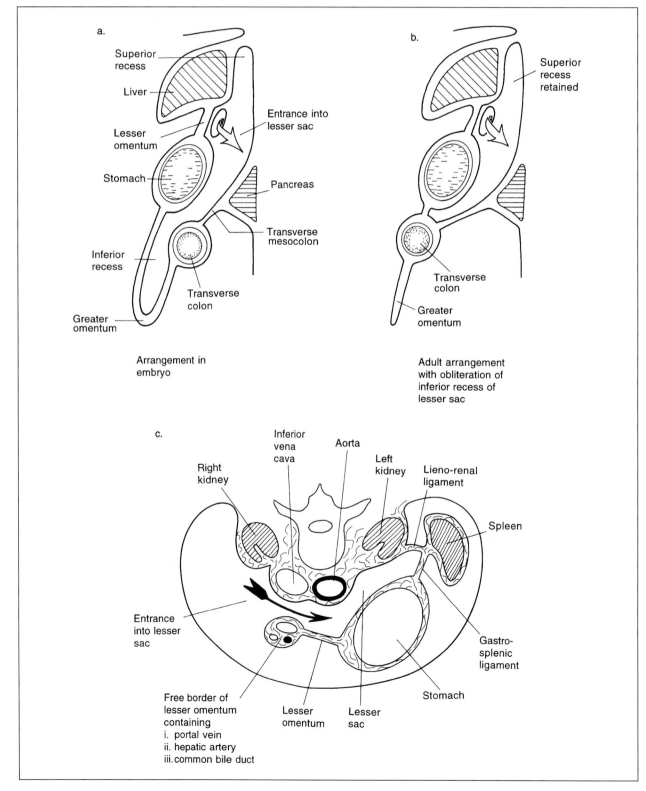

Figure 4.4.4a–d. A simplified view of the differentiation of the midgut from TS 17 to birth. During the early stages of midgut differentiation, there is insufficient space within the peritoneal cavity for the elongation and differentiation of the midgut. This is mainly due to the liver, which occupies much of the upper part of the peritoneal cavity, the rest being taken up by the stomach, while much of the lower half of the peritoneal cavity is taken up by the two (retroperitoneal) mesonephroi. The growing midgut therefore herniates into the proximal part of the umbilical cord, the ventral extension of the intra-embryonic coelomic cavity, which opens up to accommodate the resulting loop of gut. While it is within this region of the cord, now termed the *physiological* umbilical hernia, the distal part of the midgut loop rotates through 90° in an anti-clockwise direction when viewed from the front of the embryo, using the vitelline duct as the axis of rotation. Note that the proximal and distal extents of the midgut loop (at the foregut–midgut and midgut–hindgut junctions, respectively) retain their position *vis-à-vis* the posterior abdominal wall, as these are two fixed points to which *the* mesentery is attached.

Some days later, when the space within the peritoneal cavity increases, mainly because the mesonephroi regress and are replaced by the smaller metanephroi and because the lower part of the embryo enlarges as compared to the upper half (the liver does not grow much over this period), the midgut loop gradually returns to the peritoneal cavity. As it does so, its distal part rotates through a further 180° in an anti-clockwise direction when viewed from the front of the embryo. During this phase, the proximal half of the loop elongates far more than the distal half and forms the distal half of the duodenum, the jejunum, the ileum (i.e. the small bowel components) and the ascending and proximal part of the future transverse colon (the large bowel component). It is at this stage (TS 23/24) that the appendix and dilated caecal region are first clearly recognized, and eventually rotate until they descend into their definitive location in the right iliac fossa (before birth).

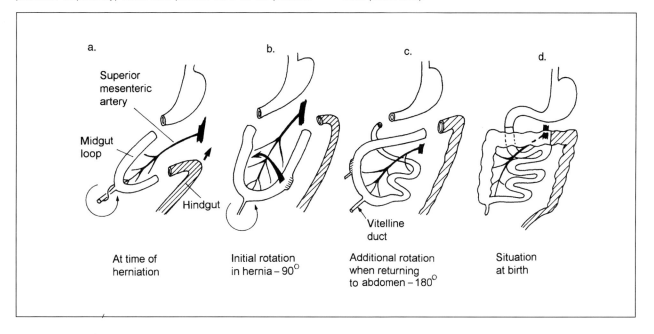

about 90° in an anticlockwise direction, when viewed from the front, and then, as it returns into the peritoneal cavity (TS 24), it continues to rotate through a further 180° in an anticlockwise direction. This sequence of events is illustrated in Figure 4.4.4a–d.

The midgut loop is only able to return into the peritoneal cavity (during TS 24, some 5 days later) because the available space within it has enlarged, mainly because the rate of increase in the volume of the liver is less than that of the peritoneal cavity, but also because the mesonephroi have almost completely regressed and been replaced by the considerably smaller **metanephroi** (or **definitive kidneys**).

Once the distal part of the midgut loop has fully returned into the peritoneal cavity, its various parts assume their definitive locations. The ascending and descending regions of the colon become fixed to a considerable degree to the posterior abdominal wall, losing their dorsal mesentery in the process, while the rest of the midgut remains suspended by its dorsal mesentery (now termed *the* mesentery) (Frazer and Robbins, 1916). It is worth pointing out that the luminal diameter of the caecum and proximal part of the ascending colon are proportionately considerably greater in the mouse than are the comparable regions in the human.[3]

All of the primitive midgut and the majority of the primitive hindgut (see below) retain their dorsal mesentery, though, following rotation of the midgut, several parts of this region of the gut lose their dorsal mesentery when they become adherent to the

---

[3] The histological development of the human small intestine was studied by Berry (1900). He noted that there was no evidence of intestinal villi by about 7 weeks of gestation (17 mm crown–rump length embryo); villi were, however, observed to be exclusively present in the duodenum by about 8–9 weeks (26 mm embryo), but along the length of the entire small intestine by about 12 weeks (80 mm fetus).

TS	1	2	3	4	5	6	7	8	9	10	11	12	13	14	15	16	17	18	19	20	21	22	23	24	25	26
E	0	1	2	3	4	4.5	5	6	6.5	7	7.5	8	8.5	9	9.5	10	10.5	11	11.5	12	12.5	13.5	14.5	15.5	16.5	17.5

posterior abdominal wall. This is particularly so for most of the duodenum (from just distal to the gastro-duodenal junction) and the ascending colon. In relation to the hindgut, the distal part of the rectum also gradually loses its dorsal mesentery as it passes caudally towards the upper part of the anal canal. In the peritoneal cavity, those regions of the bowel that are suspended from the posterior abdominal wall by a mesentery are covered on all sides (except where they are attached to the dorsal mesentery) by a layer of peritoneum. In those regions where the bowel has effectively lost its dorsal mesentery, the gut is only covered on its anterior surface by peritoneum (these regions of the gut are therefore fixed to the posterior abdominal wall and said to be *retroperitoneal*).

During TS 22/23, the duodenum and the most proximal part of the midgut including the duodenal–jejunal junction lose their dorsal mesentery and gradually become retroperitoneal, being fixed at a site just to the right of the midline. At the same time, the dorsal component of the anterior/proximal part of the greater omentum overlaps the ventral/anterior surface of the transverse colon and its mesentery and fuses with them (see Figure 4.4.3b). The luminal diameter of the distal part of the foregut, the future proximal part of the duodenum, does not increase in size over this period, and its dorsal mesentery shortens and eventually disappears during TS 22/23 (see above). It has been suggested (Kanagasuntheram, 1960) that the final degree of rotation of the midgut loop is probably dependent upon the position of the terminal part of the duodenum in relation to the root of the midgut mesentery. It is believed that withdrawal of the midgut loop into the abdomen is due to contraction of the longitudinal musculature of the duodenum and proximal jejunum. For observations on the morphogenesis of the duodenal villi in the fetal rat, see Mathan *et al.* (1976).

The hindgut

While the *proximal* part of the hindgut differentiates into the distal part of the **transverse colon** (in the region of the splenic flexure) and **descending colon**,[4] the *distal* part of the hindgut gives rise to the **sigmoid colon**, the **rectum** and upper two-thirds of the **anal canal** (dorsally), and the **urogenital sinus** (ventrally), with these two latter regions being subsequently partitioned by the downgrowth towards the **perineum** (this is the region of the pelvic floor bounded anteriorly by the pubic symphysis, laterally by the ischial tuberosities, and posteriorly by the coccyx) of the **uro-rectal septum** during TS 18/19. The primitive **cloacal membrane**, which is first seen during TS 17, accordingly becomes subdivided respectively into the **anal membrane** (dorsally) and **urogenital membrane** (ventrally) by the downgrowth of the urorectal septum into the urogenital sinus (Figure 4.4.5). Within the septum, an aggregation of fibrous tissue forms that is termed the **perineal body** (first seen at about TS 21); this is also called the **central tendon of the perineum** because of its pivotal role in supporting the pelvic contents, acting in much the same way as the central tendon of the respiratory diaphragm does in supporting the heart and thoracic contents. The separation of the urogenital sinus from the primitive hindgut by the downgrowth of the urorectal septum is shown in Figure 4.4.5. As the rectum is principally formed in association with the downgrowth of the urorectal septum, which occurs in the human during the 8th week, failure of separation of the anal canal occasionally occurs, for example, in cases of an imperforate anus. This may be associated with a rectovesical or rectovaginal fistula.

Soon after the hindgut diverticulum has first been delineated (about TS 12), it extends caudally into the tail region (the so-called post-anal component) and gradually enlarges. At about TS 14/15, the hindgut divides into a rostral (or definitive) component relative to the cloacal/anal region, and the relatively narrow and elongated post-anal component which initially possesses a dilated caudal extremity (TS 14/15). The post-anal region of the hindgut is particularly clearly seen once the cloaca is well delineated (from about TS 16/17) when it extends as a narrow diverticulum almost reaching the tip of the tail. By about TS 19/20, however, the post-anal component has usually completely regressed.

The anal membrane (first seen at TS 19) provides a junctional zone between the primitive hindgut and the ectodermal invagination termed the **anal pit** or **proctodaeum.** Unlike the buccopharyngeal membrane between the **oral pit** (or **stomatodaeum**) and the foregut, the site of the anal membrane remains delineated after its breakdown (TS 25/26) as the **pectinate line**, which, because of its relatively poor vascularity, was formerly also known as Hilton's *white line* (Figure 4.4.6, and below for observations on the arterial supply and venous drainage of this junctional zone). The **external anal sphincter** forms at TS 22 from a partition of the cloacal sphincter which itself

---

[4] During early human embryonic development (9 weeks), the colon is a simple tube lined with pseudostratified columnar epithelium. By 10 weeks, up to eight longitudinal ridges start to form a series of longitudinal mucosal folds that are evident by about 11–12 weeks, and form the broad primary villi. These become subdivided by about 17 weeks to form the smaller secondary villi, and are covered by simple columnar epithelium. Their mechanism of formation is discussed by Bell and Williams (1982). The glandular cells of Brunner's glands are seen at about 14–15 weeks, and resemble those seen in the adult (Moxey and Trier, 1978). For observations on the histological development of the intestinal villi in the chick embryo, see Coulombre and Coulombre (1958a) and Burgess (1975); for additional studies on the differentiation of the human large intestine, see Johnson (1913).

Figure 4.4.5. The subdivision of the primitive cloaca into the urogenital sinus (ventrally) and distal part of the primitive hindgut (dorsally). Initially, the urogenital sinus and distal hindgut are two components of the cloaca, but are separated by the downgrowth (arrow) of the urorectal septum (a, TS 17). After the septum fuses with the cloacal membrane (b, TS 18/19), these become separated into two distinct entities: the urogenital sinus becomes the definitive bladder, while the hindgut component gives rise to the distal part of the hindgut and proximal (i.e. upper) part of the anal canal. The point where the urorectal septum makes contact with the cloacal membrane differentiates to form the perineal body (or central tendon of the perineum, TS 21), and serves to subdivide the cloacal membrane into the urogenital membrane anteriorly, and the anal membrane posteriorly.

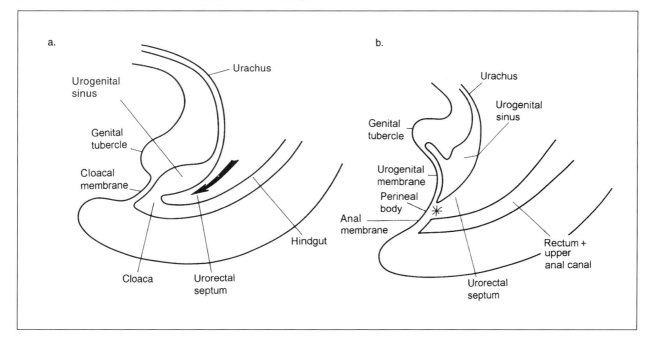

forms from the peripheral part of the cloacal membrane[5] (see Chapter 4.3 *The urogenital system*).

Gut function

Numerous attempts have been made to establish the timing of onset of functioning of the various parts of the gastrointestinal tract and its associated organs, and the modern era of research in this area began with the observations by Hess (1913), with his analysis of gastric secretion in human infants at birth, and those of Johnson (1910) who studied the mucous membrane of the oesophagus, stomach and small intestine in human embryos. Keene and Hewer (1929) also studied the timing of first appearance of a range of digestive enzymes.

The contents of the intestine, or **meconium**, was first so-named by Aristotle, who thought, because of its likeness to material obtained from the poppy (mekonion), that it kept the fetus asleep in the uterus (Grand *et al.*, 1976). Meconium consists of intestinal secretions, desquamated cellular products, as well as swallowed amniotic fluid. Small quantities of serum proteins as well as many enzymes are also present (Green *et al.*, 1958; Eggermont, 1966). As a result, one of the early tests for the diagnosis of cystic fibrosis was based on the fact that in affected infants there was an increase in the serum albumin content of their meconium (Green *et al.*, 1958).

The length of the intestine in the human is said to increase by 1000-fold from the 5th to the 40th week of gestation, with the small intestine being about four times the length of the colon. At full term, the small intestine is approximately three times the crown–heel length of the fetus (Scammon and Kittelson, 1926).

The human caecum is first identified at about 5 weeks, and is followed by a phase of rapid growth and elongation and its eventual fixation in the right lower quadrant of the abdomen. Once the gut has re-entered the abdominal cavity, it is fairly constant in position following its fixation (Fitzgerald *et al.*, 1971a,b).

---

[5] According to Johnson (1914), the human cloaca is never open to the outside. With the downgrowth of the urorectal septum, the cloaca is subdivided to form the urogenital sinus ventrally and the hindgut dorsally at about the 16 mm stage (CS 18, equivalent to TS 21). This event also subdivides the cloacal membrane into the urogenital membrane anteriorly and the anal membrane posteriorly. The latter breaks down at the 23 mm stage (CS 20/21, equivalent to TS 22), though there is some variability as to when this event occurs. The external sphincter appears quite early, and is seen at about the 23 mm stage, though the anlagen of the muscle fibres are present as early as the 12.5 mm stage (CS 17, equivalent to TS 20).

TS	1	2	3	4	5	6	7	8	9	10	11	12	13	14	15	16	17	18	19	20	21	22	23	24	25	26
E	0	1	2	3	4	4.5	5	6	6.5	7	7.5	8	8.5	9	9.5	10	10.5	11	11.5	12	12.5	13.5	14.5	15.5	16.5	17.5

Figure 4.4.6. The two components of the anal canal and their blood supply. The anal canal contains a rostral and a caudal component. The former derives from the primitive hindgut, and has typical hindgut or *endodermal* features (*e.g.* columnar epithelium associated with goblet cells), while the latter derives from an invagination of surface *ectoderm* (the *proctodaeum* or *anal pit*), whose epithelium is of the stratified squamous type. The arterial blood supply to the hindgut-derived territory is from the inferior mesenteric artery (the main arterial branch of the aorta that supplies the embryonic hindgut) via branches of the *superior rectal artery*. By contrast, the arterial blood supply to the anal pit-derived territory is from the internal pudendal branch of the internal iliac artery via the *inferior rectal artery*. The junctional zone between these two territories is *relatively avascular*, and is termed the pectinate line. The venous drainage of these two territories is also related to their origin: that of the hindgut-derived region is initially through the *superior rectal vein* to the inferior mesenteric vein and then to the portal vein, while that of the anal pit-derived region is initially through the *inferior rectal veins* to the internal pudendal veins and thence to the internal iliac veins, and finally via the common iliac veins to the inferior vena cava.

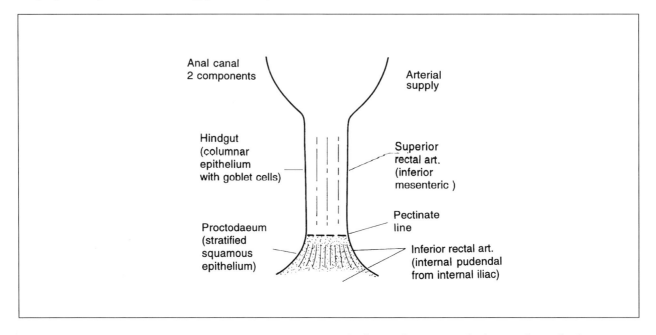

The blood supply to the gut

It is difficult to state with any degree of certainty when the gut first establishes its definitive arterial blood supply and venous drainage. From the earliest stages of gut differentiation, however, it is possible to recognize evidence of a fine capillary network within its walls, while the **vitelline venous plexus** is clearly associated with the midgut wall by about TS 15/16, draining into the **portal vein** by TS 19. Although the vitelline arteries are first seen as early as TS 12, their principal derivatives supplying the midgut and hindgut, the superior and inferior mesenteric arteries, are difficult to distinguish before about TS 22/23.

The principal subdivisions of the primitive gut tube each have a discrete blood supply, with branches of the **coeliac trunk** supplying the region of the primitive foregut distal to the lower one-third of the oesophagus, including the stomach and proximal half of the duodenum. Although in developmental terms the primitive foregut region also includes the derivatives of the pharynx (see Chapter 3.3 *The branchial arch system*), the lungs and lower respiratory tract receive their arterial blood supply directly from the aorta at levels proximal to the origin of the coeliac trunk (see *The lower respiratory tract and lungs*). The latter has three branches, the left gastric, the hepatic and the splenic arteries which supply the pancreas, the liver and the spleen as well as the region of the alimentary tract indicated above.

The arterial supply to the oesophagus is essentially similar to its venous drainage where the lower one-third of the oesophagus is supplied by an **oesophageal branch of the left gastric artery**, while its venous drainage is via oesophageal branches of the left gastric vein which drain into the portal venous system. The venous drainage of the proximal two-thirds of the oesophagus, by contrast, is into the azygos veins which in turn drain into the systemic venous system, while its arterial supply is from oesophogeal branches of the descending thoracic aorta, supplemented by branches of the inferior thyroid artery. The junctional zone between the upper two-thirds and lower one-third of the oesophagus is one of several sites where the portal and systemic venous systems anastomose; the vessels at these sites can become enormously dilated should the portal venous pressure significantly rise.

The **superior mesenteric artery** supplies the derivatives of the primitive midgut, which extends from the middle of the second part of the duodenum, and therefore includes the distal half of the duodenum, the jejunum and ileum, the caecum and the ascending and proximal two-thirds of the transverse colon. Branches of the **inferior mesenteric artery** supply the region of the hindgut, which extends distally from the junction between the proximal two-thirds and the distal one-third of the transverse colon, and includes the descending and sigmoid colon, the rectum and upper two-thirds of the anal canal.

The primitive hindgut is initially continuous with

the **urogenital sinus** (see Chapter 4.3 *The urogenital system*), but becomes separated from it with the downgrowth towards the cloacal membrane of the urorectal septum (see above). As has also been noted previously, the pectinate line is located at the junction between the upper two-thirds of the anal canal (a hindgut derivative) and the lower one-third of the anal canal (a derivative of the proctodaeum or anal pit). Both the arterial blood supply and venous drainage of the anal canal may be predicted from the different embryological origins of the caudal region of the hindgut *above* the pectinate line, and the ectodermally-derived anal pit *below* it (Figure 4.4.6).

While the **superior rectal arteries** (which are branches of the inferior mesenteric artery) supply the caudal part of the hindgut, the region derived from the anal pit receives its arterial blood supply from the **inferior rectal arteries**, branches of the internal iliac arteries. Accordingly, the venous drainage of these two regions is via the superior rectal veins (which drain via the inferior mesenteric vein to the portal vein), and via the inferior rectal veins which drain into the internal iliac veins and thence into the systemic circulation; this like the lower region of the oesophagus is thus an important site of porto-systemic anastomoses.[6]

The nerve supply to the gut

While the exact timing of the innervation of the gastrointestinal tract and its derivatives has yet to be established in detail in the mouse, the general pattern of innervation is now clear, based on the timing of appearance of neural structures in the human embryo (see below). Analysis of silver-stained histological preparations reveals that an extensive nerve plexus is already present in the region of the oesophagus by E11 (TS 17/18), and that similar plexuses are also present in the lung buds at this time. Less extensive evidence of neural networks is also observed more distally in association with the wall of the gut at all levels, this being particularly clearly seen in transverse sections of the midgut and hindgut regions at this time.

All components of the autonomic nervous system are derived from neural crest cells, the postganglionic parasympathetic neurons coming from the **occipito-cervical** (or cranial outflow) and **sacral neural crest** outflow, while the sympathetic neurons are derived from the **thoraco-lumbar neural crest**.

Chick–quail studies indicate that the neural crest cells giving rise to the cervical ganglia originate from the cervical region, while the thoracic and lumbar ganglia form from crest cells that originate in the corresponding levels of the neural tube. Pomeranz *et al.* (1991) have shown using a retroviral marker that crest cells originating in the occipito-cervical (vagal) and sacral regions migrate into the wall of the gut to innervate the entire length of the gastrointestinal tract, forming the parasympathetic ganglia (TS 22) and supplying parasympathetic motor innervation to the viscera (TS 22). More recently, Durbec *et al.* (1996) have used genetic markers to show that a pool of crest cells from the post-otic region of the hindbrain give rise to most of the enteric nervous system (and the superior cervical ganglion), while a second lineage from trunk neural crest forms the more caudal sympathetic ganglia and also contributes to the foregut enteric nervous system.

Sensory fibres in the vagus (X) cranial nerve are distributed to the gastrointestinal tract as far as the splenic flexure, with the preganglionic vagal fibres passing with the oesophagus through the diaphragm into the abdomen through the anterior and posterior vagal trunks. As noted previously, due to the rotation of the stomach, the *left* becomes the *anterior* while the *right* becomes the *posterior* vagal trunks. The fibres that supply the gastrointestinal tract terminate on postganglionic neurons in the **myenteric** (Auerbach's) and **submucosal** (Meissner's) **plexuses**, while the short postganglionic fibres innervate smooth muscle and gland cells. The nerve terminals are stimulated by a variety of factors, that include distension of the stomach and intestine, contraction of the smooth musculature of the wall of the gut tube, and irritation of its mucosa. Both gut motility and glandular secretion are modified by reflex action involving vagal afferent and efferent neurons. The descending and sigmoid colon, the rectum and urinary bladder are supplied by pelvic splanchnic branches of the second, third and fourth sacral nerves. Reflexes involving the latter (*i.e.* the sacral portion of the parasympathetic system) and the corresponding segments of the spinal cord are involved in the emptying of the bladder and the large bowel, and these functions are subject to voluntary control.

The smooth muscle and secretory cells of the viscera come under the dual influence of the sympathetic and parasympathetic divisions of the

---

[6] While these sites are not normally patent, this may arise when, for example, the drainage through the portal vein is severely reduced, as may occur in portal hypertension secondary to liver disease, and the portal venous blood of necessity returns to the inferior vena cava through the systemic venous system. In the human, the anastomoses in the anal canal (a condition termed haemorrhoids or piles) commonly become patent and dilated when the individual is constipated, in this instance secondary to raised intra-abdominal pressure during defaecation, and not infrequently in advanced pregnancy when the normal venous drainage of the gut is compressed by the presence of the enlarged uterus.

TS	1	2	3	4	5	6	7	8	9	10	11	12	13	14	15	16	17	18	19	20	21	22	23	24	25	26
E	0	1	2	3	4	4.5	5	6	6.5	7	7.5	8	8.5	9	9.5	10	10.5	11	11.5	12	12.5	13.5	14.5	15.5	16.5	17.5

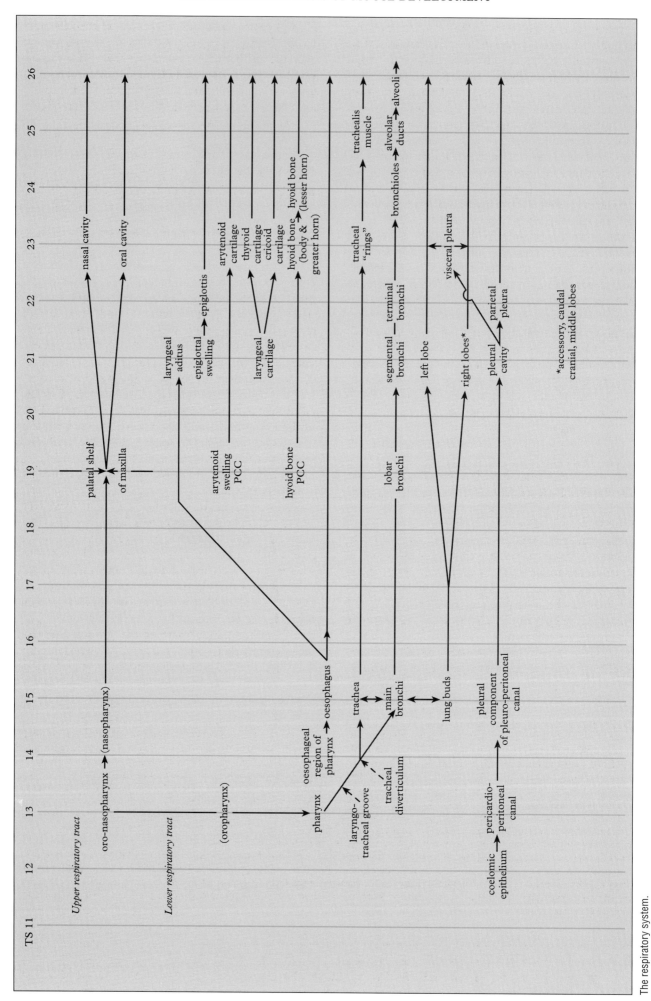

The respiratory system.

autonomic nervous systems.[7] They are functionally antagonistic to one another, and the balance that exists between them maintains normal visceral activity. Preganglionic fibres directed towards the abdominal and pelvic viscera pass through the sympathetic trunk and its splanchnic branches, and these fibres terminate on postganglionic neurons located in plexuses surrounding the main branches of the abdominal aorta that include the **coeliac plexus** and the **superior** and **inferior mesenteric plexuses**. The bronchial tree receives its sympathetic innervation from the thoracic outflow, while the stomach and small intestine are innervated via postganglionic fibres whose nuclei are located in the **coeliac ganglion**. The sympathetic branches to the stomach are distributed with the coeliac artery, and those to the bowel almost as far distally as the splenic flexure are distributed with branches of the superior mesenteric artery.

The sympathetic innervation to the small intestine is also via postganglionic fibres whose nuclei are located in the **superior mesenteric ganglia**, while postganglionic neurons from this source also innervate the proximal part of the large bowel. The innervation of the caudal part of the hindgut and bladder are via postganglionic neurons whose nuclei are located in the **inferior mesenteric ganglia**.

The parasympathetic supply to the descending colon, rectum, bladder and other pelvic organs is via **pelvic splanchnic nerves** of S2 to S4, which terminate on postganglionic neurons in and near the organs to be innervated.

Neural crest-derived **chromaffin cells** are widely scattered in the embryonic sympathetic nervous system, being found in the sympathetic ganglia and the visceral sympathetic rami, and are particularly abundant on the anterior surface of the abdominal aorta, where they form the **abdominal paraganglia** (of Zuckerkandl). The principal group of neural crest-derived chromaffin cells that persist into adult life are, however, those that form the **adrenal medulla**.

These become surrounded by coelomic epithelium derived from the posterior abdominal wall (principally from the medial coelomic bay) which differentiates into the **adrenal cortex**. In the early postnatal period, much of the inner layer of the adrenal cortex regresses, so that only the outer layer remains and differentiates into the post-natal adrenal cortex.

Systems and organs associated with the gut

The lower respiratory tract and lungs

The **lower respiratory tract** initially develops from an anteriorly directed diverticulum in the floor of the **pharyngeal region** of the foregut and is thus endodermal in origin (Figure 4.4.7a–d). The earliest evidence of this system is the appearance of a groove (the **laryngo-tracheal** or **tracheo-bronchial groove**) located immediately caudal to the **fourth pharyngeal arch** component of the hypobranchial eminence (which gives origin to the root of the tongue and the epiglottis, see Figure 4.4.7a) (TS 15). Shortly after its appearance, the foregut caudal to it, which forms the definitive oesophagus, lengthens and its lumen narrows.

The region around the laryngo-tracheal groove will in due course form a T-shaped cleft which is destined to become the entrance to the tracheal diverticulum (the so-called **laryngeal aditus**, see Figure 4.4.7b). This is bounded rostrally by the epiglottis, and on either side by a pair of ridges. As the groove deepens, the caudal portions of these ridges fuse and, as this happens, the oesophagus becomes increasingly separate from the caudal part of the groove and the diverticulum that develops from it. Rostro-laterally the aditus becomes bounded on either side by the

---

[7] Differentiation of the human oesophagus has been studied by Smith and Taylor (1972). By about 5 weeks (8 mm embryo), the epithelium contains a double layer of cells while occasional neuroblasts were observed lining the inner surface of its circular muscle. Longitudinal muscle is first present by the 24 mm stage. At no time was total occlusion of this region of the gut observed. Myenteric plexuses were seen by about 9 weeks (35 mm fetus), and ganglion cells observed by about 13 weeks (65 mm fetus). The oesophageal epithelial cells were ciliated by about 10 weeks (Arey, 1974), and evidence of peristalsis observed by the end of the first trimester of pregnancy.

In the region of the stomach, circular muscle appears during the 8th–9th weeks, and neural plexuses also form in the body and fundus regions at this time. Gastric acidity was not seen much before birth in the human (Ebers et al., 1956).

In the small intestine, Auerbach's nerve plexus appears at about 9 weeks (35 mm fetus) and Meissner's plexus at about 13 weeks (65 mm fetus) (Read and Burnstock, 1970). Lymphopoiesis is first observed at about 15 weeks (110 mm fetus), and Peyer's patches are usually well developed by about 20 weeks (160 mm fetus) (Cornes, 1965). These patches are characteristic aggregations of lymphoid tissue seen on the mucosa in the small intestine, and are composed of many lymphoid nodules closely packed together. Meissner's and Auerbach's plexuses are present in the colon by the 8th and 12th weeks, respectively, and it is now possible to make the diagnosis of aganglionosis in a premature infant by the absence of ganglia in this region, for example, by the histochemical staining of tissue for acetylcholinesterase activity which would be noticeably absent in areas containing no ganglion cells (Aldridge and Campbell, 1968).

TS	1	2	3	4	5	6	7	8	9	10	11	12	13	14	15	16	17	18	19	20	21	22	23	24	25	26
E	0	1	2	3	4	4.5	5	6	6.5	7	7.5	8	8.5	9	9.5	10	10.5	11	11.5	12	12.5	13.5	14.5	15.5	16.5	17.5

Figure 4.4.7. The development of the lower respiratory tract: (a,b) the opening of the trachea; (c,d) the development of the lungs. (a, TS 15) A view of the floor of the primitive pharynx shows the location of the laryngo-tracheal groove in relation to where the hypobranchial eminence will appear (a branchial arch derivative which will form the posterior part of the tongue). (b, TS 19). The tissue around the laryngo-tracheal groove increases in volume to form (anteriorly) the future epiglottis, and (laterally) the two arytenoid swellings. The laryngeal aditus represents the opening from the pharynx into the trachea. (c, TS 15) The ventrally-directed groove gives rise to the tracheal diverticulum. (d, TS 18/19) This differentiates into the trachea and subsequently gives rise to the two main bronchi whose distal parts undergo successive bifurcations to form the bronchial components of the two lung buds.

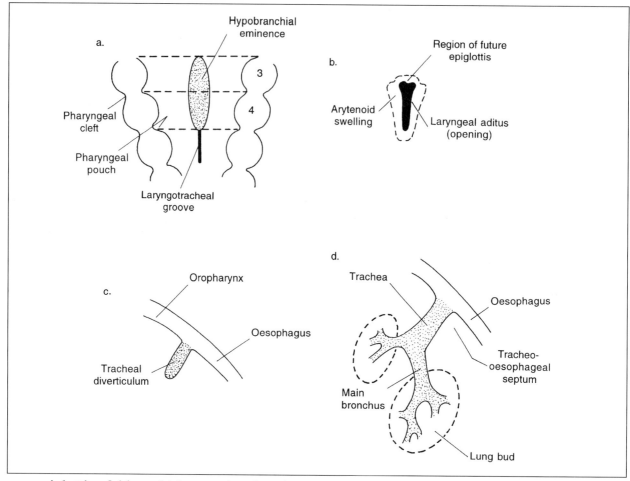

ary-epiglottic folds, which contain the tiny **cuneiform cartilages**, and more caudally the ridges thicken to form the **arytenoid swellings**, within which differentiate the **arytenoid** and **corniculate cartilages** (see O'Rahilly and Tucker, 1973). The detailed timing of these events is discussed below.

The laryngo-tracheal groove rapidly deepens to give the endodermally-lined **tracheal diverticulum**, whose proximal part will form the **trachea**, while the distal part soon bifurcates to give the two main **bronchi** (Figure 4.4.8c, d). Slightly later (TS 15/16), mesenchyme tissue is seen to surround the distal parts of the two main bronchi, and the two laterally-directed "bulges" that form are the first indication of the **lung buds**. The mesenchyme that surrounds the developing trachea and bronchi gives origin to the bronchial musculature, the cartilaginous rings and the pulmonary connective tissue (for a review of these early stages, see Emery, 1969). Occasionally, incomplete separation of the oesophageal region of the foregut from the developing respiratory system occurs, and accounts for the rare finding of aberrant pieces of lung tissue and cartilage at the gastro-oesophageal junction and, more rarely, in other parts of the foregut, and even extending to the first part of the jejunum (Emery, 1969).

The initially symmetrical lung buds bulge laterally into the **pericardio-peritoneal canals** which are cranio-caudally directed. They are covered by coelomic mesothelium which also lines the future pleural cavities to form the visceral and parietal layers of the pleura, respectively. When first apparent, the lung buds are inferior to the common atrial chamber of the heart, but they gradually "ascend", and by about TS 25/26 are at about the same thoracic level as the heart, some lung tissue being posterior and others antero-lateral to it.

By TS 16, the **pulmonary arteries** develop as branches from the paired **sixth pharyngeal (branchial) arch arteries**. They then descend towards the lung buds (TS 17/18), and eventually enter them through their hilar region (by TS 19). It should be noted that the arterial supply to the tissues of the lungs is from the **bronchial arteries**, which are initially derived from segmental branches from the descending thoracic aorta. In the mouse, the bronchial veins drain into the coronary sinus; in the human, the right bronchial veins join the end of the azygos vein whereas the left bronchial veins may join either the left superior intercostal or hemiazygos vein (Williams, 1995). By TS 17/18, the organization of the left and right lung buds and their bronchi has now become asymmetric, mainly because a single lobe forms on the left side, whereas four distinct lobes develop on the right. The first evidence of this change is the rapid division of the right main bronchus to form four lobar (secondary) bronchi, although they do not become distinct until about TS 21/22. These secondary lobes are termed **cranial, middle, caudal** and **accessory**, although from as early as TS 22 much of the accessory lobe is located across the median plane, largely to the left of the midline and below all but the most caudal part of the left and right ventricles. The situation observed in the mouse is different from that observed in the human, where the left lung usually has two lobes (an upper and a lower lobe) separated by the oblique fissure, while the right lung becomes subdivided into 3 lobes (an upper, middle and a lower lobe) by an oblique and a transverse (or horizontal) fissure. The difference in the arrangement of the principal subdivisions of the lungs in the mouse and in the human is illustrated in Figure 4.4.8a, b. The medial surface of the left lung is excavated by the presence of the heart, and its anterior border is relatively sharp and tongue-shaped (and accordingly termed the **lingula**).

Although the bronchial tree has branched considerably and formed many **segmental bronchi** by TS 21, the parenchyma of the lung buds still consists of homogeneous cellular tissue with no obvious

Figure 4.4.8. (a) Posterior and anterior views of adult mouse lungs displaying their lobar arrangement. Note that the mouse has a single (undivided) left lung, whereas the right lung is subdivided into four lobes (cranial, middle, caudal and accessory), each with its own main bronchus. A considerable volume of the accessory lobe lies across the midline and is directly posterior to the caudal part of the right atrium and much of the right ventricle of the heart. (b) Lateral and anterior views of human adult lungs displaying their lobar arrangement. The left human lung has two lobes, with the upper and lower being separated by the oblique fissure. The right lung has three lobes, with the upper and a middle ones being separated by the transverse fissure, while the oblique fissure separates the lower lobe from the middle one. The medial surface of the upper left lobe is deeply indented to accommodate the heart, most of which is located on the left side of the thoracic cavity. The left lung also has a relatively sharp anterior border.

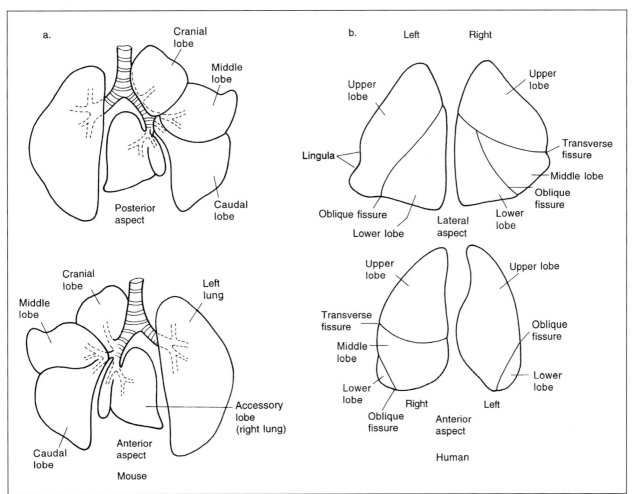

TS	1	2	3	4	5	6	7	8	9	10	11	12	13	14	15	16	17	18	19	20	21	22	23	24	25	26
E	0	1	2	3	4	4.5	5	6	6.5	7	7.5	8	8.5	9	9.5	10	10.5	11	11.5	12	12.5	13.5	14.5	15.5	16.5	17.5

characteristic features. The segmental bronchi supply the so-called **broncho-pulmonary segments** of the lungs. These are roughly pyramidal in shape, their apices towards the hilum and their bases lying on the surface of the lung. Each has its own segmental blood supply and venous drainage, and they act as distinct functional entities. The lymphatic drainage is also largely segmentally arranged, and drains to the hilar lymph glands. By now, the volume of the lungs has expanded considerably, particularly posteriorly, and encroaches on the upper half of the posterior part of the liver, being separated from it by the left and right domes of the **diaphragm**.

The normal histological architecture of the lungs is first seen at TS 22/23 as the terminal bronchi and bronchioles differentiate; further branching then occurs over the next few stages with luminal diameters getting progressively smaller.[8] By TS 26, the histological appearance of the tissues of the lungs is similar to that seen just before birth as the **alveoli** are now present. Although the cells lining them are initially low columnar, they become progressively more squamous so as to facilitate gaseous exchange between the air within them and the blood in the capillary network of the pulmonary vasculature immediately after birth. The histological appearance of the lungs in TS 26 embryos usually shows the alveoli partially or in some cases fully expanded with air. This is an artefactual appearance, and is due to the fact that the embryos breathe during the interval between their isolation and fixation.

The intrauterine growth of the lung is usually divided into three stages:

(a) *The glandular period*: this occupies the period up to about TS 16 (in the human, it is up to the end of the fourth month). During this period, the lung consists of a loose mass of connective tissue with an actively proliferating central lobular mass lined by columnar cells.

(b) *The canalicular period*: this occupies the period up to about TS 24 during which time the bronchi are actively dividing, sometimes dichotomously (*i.e.* into two equal branches), but also into unequal branches (in humans, this is the middle one-third of gestation). The relative amount of connective tissue diminishes and the lungs become more vascular. The duct epithelium tends to become more cuboidal.

(c) *The alveolar period*: This is the last 2 or 3 days of mouse development (TS 25 onwards) when the alveoli form (in humans, this starts at around 6 months,[9] but the time of onset is variable – between 20 and 26 weeks). The epithelium of the alveoli becomes increasingly squamous (or flattened). Extensive capillary plexuses are observed in the full-term unexpanded lung. Aeration of the collapsed lung at birth consists of the rapid replacement of intra-alveolar fluid with air (Geubelle *et al.*, 1959).

By about TS 26, the volume of the **pleural cavity** is seen to be considerably larger than that occupied by the lungs; this allows the former to accommodate the rapid expansion of the alveoli within the lungs that follows the ingress of air from the first breath. When this occurs, the lungs fill the pleural cavities, although the intervening space after birth normally contains sufficient pleural ("lubricating") fluid to allow the two layers to glide over each other smoothly during breathing, its contents before birth are unclear. It should be recalled that, up to the time of the first breath, the fetus is bathed in amniotic fluid that fills both the upper as well as the lower respiratory tract. With the first breath this fluid is removed from within the bronchial tree and alveoli of the lungs, and as a consequence, the peripheral resistance within the lungs to blood flow through the pulmonary vasculature rapidly diminishes. This plays a critical role in facilitating the changes that occur in the dynamics of the circulation of the newborn, but particularly the closure of the foramen ovale by the septum primum. (The changes that occur in the circulation at the time of birth are discussed in detail in Chapter 4.1 *The heart and its inflow and outflow tracts*.)

The histological appearance of the trachea and walls of the main bronchi is initially similar to that of the oesophagus, but, by about TS 19, the difference between them becomes increasingly marked, with the differentiation of the so-called *respiratory* (ciliated) epithelium associated with large numbers of mucus-secreting goblet cells that line all parts of the respiratory tract. The mesenchymatous condensations of pre-cartilage, the primordia of the "**rings**", are first seen in the trachea and main bronchi at about TS 22/23 and differentiate into true cartilage by about TS 24.

---

[8] The pattern of branching in the human fetal lung between 10 weeks and term was studied by Bucher and Reid (1961a). They reported that, by 16 weeks, the bronchial tree was completely lined with epithelium, and that by 24 weeks the epithelium was interrupted by the ingrowth of capillaries. They also noted that the bulk of bronchial formation occurred between 10 and 14 weeks, and that in this period about 70% of all branches had formed; later, the formation of new epithelialized bronchioli was minimal.

By 14 weeks, glandular density was also at its maximum (Bucher and Reid (1961b), but many *new* glands form between 14 and 28 weeks. Cilia are present in the trachea and main bronchi at about 10 weeks, and in most of the peripheral airways by 13 weeks. For additional information, see Pattle (1969) and O'Rahilly and Boyden (1973), and for observations on the control of branching in the mouse lung, see Hilfer *et al.* (1985).

[9] In the human embryo, alveoli are usually first encountered from about 24–28 weeks of gestation. In a morphometric analysis by Hislop *et al.* (1986), the alveoli were first counted and measured at 29 weeks, and gradually increased in diameter with increasing gestational age. Lung volume increased four-fold between 29 weeks and term, and doubled again between term and four months after birth. At birth the lungs had an average of 150 million alveoli, about 50% of the expected adult number. The surface area of the alveoli is 3–5 m^2 at birth, about one-twentieth of the adult value.

The entrance to the lower respiratory tract begins to differentiate at about TS 19 when the **arytenoid swellings** first appear. These are located on either side of the laryngeal entrance (or aditus), while the swelling associated with the future **epiglottis** is located more anteriorly, and differentiates from the caudal part of the **fourth pharyngeal arch component of the hypobranchial eminence**. By TS 23/24, these pre-cartilage components of the fourth arch differentiate into the **thyroid** and **cricoid cartilages** of the **larynx**.

The nerve supply to the lungs and the control of respiration

It is believed that groups of neurons in the reticular formation in the pons and medulla regulate visceral functions through connections with the nuclei of the autonomic outflow; in the case of respiration, this is through motor neurons in the phrenic nucleus and thoracic region of the spinal cord. Respiratory "centres" have been identified by electrical stimulation within the brain stem of experimental animals. Two "centres", one for *inspiration* and the other for *expiration*, are located in the reticular formation of the medulla, while normal respiratory rhythm is controlled through a "centre" in the pontine reticular formation. Inspiration is initiated by stimulation of neurons in the inspiratory "centre" through a rise in carbon dioxide in the circulating blood, via the stimulation of the diaphragm and intercostal muscles. Chemoreceptors in the carotid bodies that are located close to the origin of the internal and external carotid arteries, and other chemoreceptors adjacent to the aortic arch, respond to a decrease in oxygen tension in the circulating blood. Impulses from these receptors stimulate the appropriate regions of the respiratory "centres" in the brain stem to bring about an increase in the rate and depth of respiratory movements (Barr, 1979; Williams, 1995).

The autonomic innervation of the lungs is from sympathetic and parasympathetic branches of the anterior and posterior pulmonary plexuses. Subnuclei within the **tractus solitarius**[10] receive tracheal, laryngeal and pulmonary afferents, and have an important role in the control of respiration (Kalia & Richter, 1985). Preganglionic parasympathetic fibres arise in the dorsal nucleus of the vagus and travel to the pulmonary plexuses and thence pass distally within the lungs via the pulmonary branches that accompany the bronchial tree.

Efferent fibres pass to the bronchial muscles, and afferent fibres pass back from the bronchial mucous membranes and alveoli. The parasympathetic supply is generally considered to be bronchoconstrictor, while the sympathetic supply is bronchodilatory, relaxing the bronchial smooth muscle. While individual nerve fibres may be seen as early as TS 19/20 within the tissues of the lung buds, passing between the main bronchi, substantial nerve trunks are not seen before TS 21/22.

The nerve supply to the muscular tissue of the lower respiratory tract is from the **vagus** (X) cranial nerve, via its superior and recurrent laryngeal branches which respectively supply the derivatives of the fourth and sixth pharyngeal arches (see Chapter 3.3 *The branchial arch system*).

The derivatives of the septum transversum

The **septum transversum**, which develops within the ventral mesentery of the stomach, is first recognized at about TS 13/14; it will in due course give rise to the **parenchyma of the liver**, elements of which are recognizable as early as TS 15/16, as well as the central tendon of the diaphragm (TS 20/21), though the precursors of the septum-transversum-derived components of the central tendon, as well as the crura and domes of the diaphragm, are first evident from about TS 17/18. The septum transversum probably also gives rise to the fibrous pericardium, though its precursors are not readily recognized before about TS 24. The **cystic (gall bladder) primordium** is first seen at about TS 15 and is derived from the **biliary bud**, which will also form the **intra-hepatic component of the biliary system** (see also *The intraembryonic cavities* in Chapter 2.4 *Early organogenesis*).

The septum transversum has a well-developed blood supply, and, initially, both the vitelline and umbilical veins pass into it, subsequently draining via the hepatic veins into the two horns of the sinus venosus. In the dorsal margins at each side of the septum transversum are the two common cardinal veins, and these also communicate with their respective horns of the sinus venosus, before draining into the *common atrial chamber* of the heart, and subsequently

---

[10] The **tractus solitarius** consists of caudally directed fibres whose cell bodies are located in the inferior ganglia of the vagus and glossopharyngeal nerves, and in the geniculate ganglia of the facial nerve. These fibres terminate in the **nucleus of the tractus solitarius**. The vagal and glossopharyngeal afferents play an important part in visceral reflexes, those from the glossopharyngeal and vagus nerves transmitting impulses for taste from the posterior one-third of the tongue and from the region of the epiglottis and pharynx. The fibres from the facial nerve transmit taste from the anterior two-thirds of the tongue. Fibres from this nucleus project to the cortical area for taste located at the lower end of the general sensory region of the parietal lobe.

TS	1	2	3	4	5	6	7	8	9	10	11	12	13	14	15	16	17	18	19	20	21	22	23	24	25	26
E	0	1	2	3	4	4.5	5	6	6.5	7	7.5	8	8.5	9	9.5	10	10.5	11	11.5	12	12.5	13.5	14.5	15.5	16.5	17.5

(following atrial septation) into the caudal part of the right atrium. While the definitive blood supply and venous drainage of the liver is complex, the general principles involved may be understood if the normal sequence of events is followed. This is shown diagrammatically in Figure 4.4.9a–d.

Initially, the primitive gut is seen to be drained by the left and right vitelline veins which drain venous blood into their respective horns of the sinus venosus. Adjacent to these two vessels run the left and right umbilical veins which drain oxygenated blood from the placenta to their respective horns of the sinus venosus (see Figure 4.4.9a). The drainage of these two pairs of vessels to the primitive heart is disrupted by the differentiation of the septum transversum (the future liver) in the ventral mesentery of the stomach: the hepatic diverticulum (or biliary bud), an outgrowth from the foregut–midgut junction grows into and bifurcates within the septum transversum, giving rise to the intra-hepatic component of the biliary system, as

Figure 4.4.9. The development of the venous drainage of the gut and the venous system within the liver. (a, TS 12/13) Before the development of the septum transversum in the region of the foregut–midgut junction, the primitive gut is drained by a pair of vitelline veins. A pair of umbilical veins, carrying oxygenated blood from the placenta, runs on the lateral side of the vitelline veins, draining with them towards the two horns of the sinus venosus and thence to the heart. A little later (b, TS 15), the septum transversum develops in the region of the foregut–midgut junction, and is invaded by a diverticulum of the gut termed the *hepatic diverticulum*. The latter has both an extra-hepatic component (that gives rise to the gall bladder and common bile duct) and an intra-hepatic component (that gives rise to the biliary system that ramifies within the liver). (c, TS 18/19) As the liver differentiates, the vascular system is modified: the two vitelline veins enter and ramify within the septum transversum, forming the hepatic sinusoids, while the two umbilical veins each give rise to two branches, the medial, intra-hepatic branch is directed towards the liver, while the other has an extra-hepatic course (and is soon lost). (d, TS 20) The extra-hepatic part of the left vitelline vein regresses, while the right vitelline vein becomes the portal vein into which the entire venous drainage of the gut drains. The two extra-hepatic branches of the umbilical veins also disappear, together with the hepatic branch from the right umbilical vein. The only retained branch from the umbilical system is that from the left side which now passes through the liver where it forms a large diameter channel, the *ductus venosus*, on the way to the post-hepatic part of the inferior vena cava, and thence into the caudal part of the right atrium. The definitive relationship between the portal vein and the duodenum is also shown here.

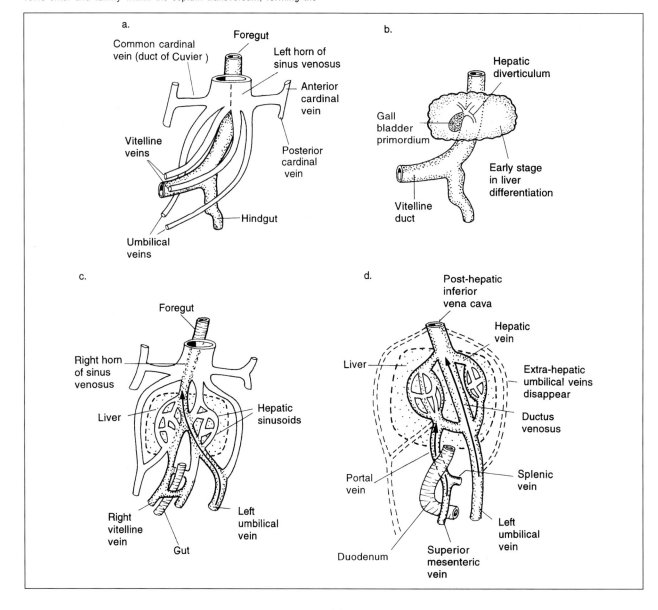

well as giving rise to a number of extra-hepatic components (the gall bladder, common bile duct etc.).

The liver

At TS 13/14, the septum transversum displays few of the features that will characterize the definitive liver. Nevertheless, and despite its relatively small volume, it provides the scaffolding into which various venous channels will subsequently differentiate to form the hepatic sinusoids, and the primitive biliary apparatus will ramify. It is located, at this stage, across the midline close to the ventral surface of the entrance and proximal part of the foregut pocket. Its peripheral boundaries are fairly difficult to discern, though it is bounded rostrally by the caudal surface of the common atrial chamber of the primitive heart.

By TS 15/16, the septum transversum has increased in volume to a limited extent, although it still possesses relatively few recognizable hepatic features. This is, however, the first stage at which the cystic (gall bladder) primordium is clearly recognized as a distinct entity separate from the hepatic biliary primordia located within the septum transversum. The cystic primordium is also seen to drain into the lumen of the duodenum through a short channel which is destined to elongate and form the **common bile duct**.

By TS 16/17, while the peripheral margins of the primitive liver are still only poorly defined, this organ is seen to occupy both the right and left sides of the upper part of the abdominal cavity on either side of, and ventral to, the median foregut dilatation which is destined to form the stomach; the cystic duct is also readily recognized at this stage. With the lateral displacement of the stomach to the upper left part of the abdominal cavity (initiated during TS 19 and consolidated during TS 20) the liver tends to become a more right-sided structure than formerly (see below). The location of the future liver and its associated vascular supply in relation to the foregut–midgut junction is illustrated in Figure 4.4.9.

During TS 18/19, the liver increases enormously in volume and occupies almost the entire upper half of the available space in the abdominal cavity: much of its central part is, however, occupied by the enormously dilated venous channel, the **ductus venosus**.[11] The peripheral boundaries of the liver are now well defined, and it is seen to be subdivided into a considerable number of lobes, with deep fissures between them. The enormous enlargement of the liver at this stage is principally accounted for by its increased haematopoietic activity, it being the main *embryonic* source of primitive nucleated red blood cells, taking over this role from the blood islands within the visceral yolk sac shortly after the amalgamation of the yolk sac and embryonic circulations. This is because the **haematopoietic stem cells** which originate in the blood islands migrate into and colonize the septum transversum very shortly after its differentiation.

During TS 21/22, the **liver** becomes suspended from the ventral wall of the stomach by the **lesser omentum** and from the anterior abdominal wall by the **falciform ligament** (see Figure 4.4.10). Initially, the stomach is supported by a dorsal and a ventral mesentery. Within the former, the spleen develops, while the liver enlarges within the latter, and both mesenteries become subdivided into two parts as a consequence of the presence of these organs: the spleen separates the dorsal mesentery of the stomach (or dorsal mesogastrium) into lieno-renal and gastro-splenic components, while the ventral mesentery becomes subdivided (due to the presence of the liver) into the lesser omentum and falciform ligament. With the rotation of the stomach, the spleen comes to occupy the left side of the upper part of the abdominal cavity, while the liver (which was originally a midline structure) becomes a predominantly right-sided structure (Figure 4.4.10).

Within the lower border of the falciform ligament runs the **left umbilical vein** (which carries oxygenated blood from the placenta). The contents of this vessel then pass directly into the ductus venosus and thence to the right side of the heart. After birth, when this venous channel no longer serves any useful role, its lumen becomes obliterated, and the fibrous band that replaces it is termed the **ligamentum**

---

[11] The ductus venosus acts as a conduit that allows most of the oxygenated blood from the left umbilical vein to pass into the right hepato-cardiac channel and thus to by-pass the hepatic sinusoids. The resultant changes that occur in the configuration of the right horn of the sinus venosus also influence the orientation of the entrance of this vessel into the right atrium. Two venous valves form here in association with the dorsal wall of the right atrium at about TS 21. These valves are located where the right hepato-cardiac channel enters this chamber of the heart. At this stage, the hepato-cardiac channel represents the principal precursor of the post-hepatic part of the inferior vena cava. Systemic (largely deoxygenated) venous blood, principally from the trunk and lower limbs, but also from the paired body organs (the kidneys, adrenals and gonads) also enters this channel through what will become the definitive inferior vena cava, but this input of deoxygenated blood only marginally decreases its oxygen tension.

After birth, once the connection with the placenta is severed, the ductus venosus regresses, and is eventually replaced by a fibrous band, the *ligamentum venosum*. It is also worth noting that the pre-hepatic part of the left umbilical vein also regresses shortly after birth, and is replaced by a fibrous band termed the *round ligament of the liver* (or ligamentum teres hepatis). In the adult, this is located in the lower border of the falciform ligament, and runs from the umbilicus to the inferior border of the liver.

TS	1	2	3	4	5	6	7	8	9	10	11	12	13	14	15	16	17	18	19	20	21	22	23	24	25	26
E	0	1	2	3	4	4.5	5	6	6.5	7	7.5	8	8.5	9	9.5	10	10.5	11	11.5	12	12.5	13.5	14.5	15.5	16.5	17.5

Figure 4.4.10. The relationship between the structures in the dorsal and ventral mesenteries of the stomach (a transverse section through the upper part of the peritoneal cavity at about TS 22). The stomach is now in the left (Figure 4.4.2) and the liver is in the right part of the cavity. The spleen develops within the dorsal mesentery of the stomach, and subdivides it into *lieno-renal* and *gastro-splenic* components, while its ventral mesentery is also subdivided into the *falciform ligament* and the *lesser omentum* by the presence of the liver. Note that, with the rotation of the stomach, the dorsal mesogastrium becomes displaced from its original midline position towards the left side, and becomes an anterior relation of the left kidney. Note also that the splenic artery, a branch of the coeliac trunk from the aorta, passes to the spleen via the lieno-renal ligament.

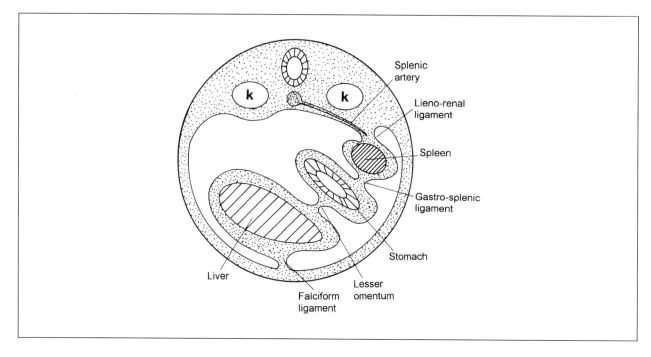

teres (hepatis). In, for example, cases of portal hypertension, this vessel may recanalize, or small veins in its immediate vicinity may open up, allowing communication between the portal venous system and branches of the epigastric veins (belonging to the systemic system) in the region around the umbilicus (in the human, if such an arrangement develops, this may occasionally have the appearance of the flowing hair associated with the head of medusa, and this site of porto-systemic anastomosis at the umbilicus is accordingly termed a *caput medusae*).

The liver is situated in the upper part of the abdominal (peritoneal) cavity, in close contact with the diaphragm, and largely protected from the trauma to which it is vulnerable by the lower part of the rib cage. The relationship between the liver and its peritoneal covering is only readily understood when it is appreciated that this organ not only develops within the ventral mesentery of the stomach, but also abuts onto the lower (caudal) surface of the central tendon of the diaphragm. The central part of the rostral surface of the liver has no peritoneal covering, and is accordingly termed its **bare area**. The peripheral boundary of the latter is termed the **coronary ligament** (because it "sits" like a *crown* on the rostral surface of the liver), and this also represents the site where the *parietal* peritoneum (which lines the peritoneal cavity) meets the *visceral* peritoneum that covers the majority of the surface of the liver. The region where the two meet (at the coronary ligament) is termed the peritoneal reflexion (see Figure 4.4.11a,b).

As noted above, the liver is initially, at least in developmental terms, a midline structure but, as it grows, it rapidly expands and, mainly because (from TS 19) the stomach increasingly occupies much of the left side of the upper part of the abdominal cavity, it swings to the right, taking the ventral mesogastrium with it. Deep fissures divide the liver principally into **left**, **middle** (or intermediate) and **right lobes**. The middle lobe extends cranially and to the right, overlapping the right lobe from which it probably derives, while the left lobe consists of a large left lateral lobe and a smaller left medial lobe. The left and right (including the middle) lobes are separated by a deep fissure that contains the insertion of the falciform ligament, and its location defines the developmental origin of the various lobes of the liver: all components of the liver located to the left of the falciform ligament are derived from the left lobe, while all located to its right come from the right lobe. The **caudate process** (or lobe) is derived from the right lobe, and projects dorsally and to the right. Two additional **papillary processes** (or lobes) extend on either side of the oesophagus, and are associated with the rostral surface of the stomach.

The overall shape of the mouse liver is different from that in the human, where the right part has a considerably greater volume than the left. The only clearly defined subdivision in the human liver is the **caudate lobe** which is located between the insertion of the lesser omentum and the inferior vena cava. The **quadrate lobe**, which is less well-defined in the human than in the mouse, is located between the bed of the gall bladder and the falciform ligament. In both

Figure 4.4.11a, b. The formation of the liver shown in transverse sections through the upper part of the peritoneal cavity. (a, TS 15) At this stage the liver, whose parenchyma largely forms from the septum transversum, is suspended in the ventral mesentery of the stomach that is subdivided into the lesser omentum, between the stomach and liver, and what will become the falciform ligament, between the liver and the anterior abdominal wall. (b, TS 21) The liver is now a right-sided structure with the stomach being located in the upper left part of the peritoneal cavity. The visceral peritoneum covering the majority of the surface of the liver does not cover its rostral part, and this region is therefore termed the *bare area*. The periphery of this area is bounded by the coronary ligament, and this is also the site where the visceral peritoneum is continuous with the parietal peritoneum lining the peritoneal cavity, and is thus the site of the peritoneal reflexion here.

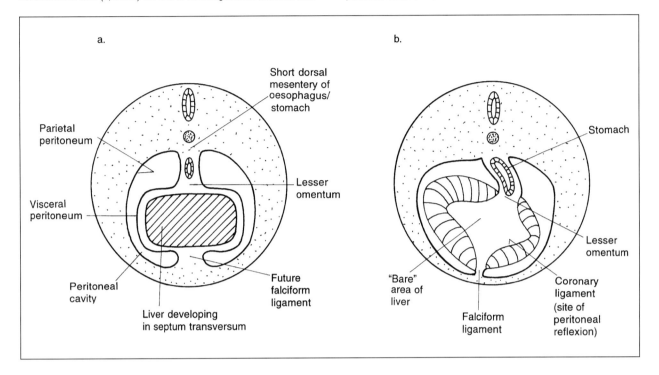

species, the **right** and **left hepatic ducts** unite to form the **common hepatic duct**, and this unites with the **cystic duct**, which drains the gall bladder, to form the **common bile duct** which, in the mouse but not the human, receives a number of pancreatic ducts along its course. It should be noted that, although both the mouse and the human possess a gall bladder, the rat does not (for accounts of the stages in the development of the human liver, see Severn, 1971, 1972).

Arterial supply and venous drainage of the liver

As will be seen, certain vessels such as the right umbilical vein and the extra-hepatic courses of the left and right umbilical veins become resorbed and disappear. As a result, the right vitelline vein forms the **portal vein**, while the left umbilical vein (which is retained) largely bypasses the substance of the liver, forming the **ductus venosus**. Rostral to the liver, the left horn of the sinus venosus decreases in importance (see below), while the right horn becomes the principal drainage of the liver, carrying a mixture of oxygenated blood (from the ductus venosus) and deoxygenated blood. The right horn thus represents the venous drainage of the gut which passes via the portal vein to the liver and thence, via the hepatic veins, to the post-hepatic part of the inferior vena cava. The deoxygenated blood carried via the pre-hepatic part of the inferior vena cava also joins the post-hepatic part of the inferior vena cava where it intermixes with the oxygenated blood from the ductus venosus before passing into the caudal part of the right atrium. During early embryogenesis, the venous drainage from the body (excluding the gut and liver) passes to the heart via the two common cardinal veins.

From as early as about TS 13/14, the vitelline veins form a substantial venous plexus within the septum transversum (the precursors of the hepatic sinusoids), and the blood from this plexus passes predominantly into the **hepato-cardiac channel** of the original right vitelline venous trunk; while the equivalent vessel on the left side regresses and eventually disappears. The left horn of the sinus venosus also becomes much reduced in size, and eventually forms the **coronary sinus** which also drains into the caudal part of the right atrium (see Figure 4.4.9c and d).

Because of its critical role in detoxification of substances absorbed through the wall of the gut, the maturing liver has a double blood supply. Its arterial blood supply is to the parenchyma of the liver, and is from the **common hepatic artery** (TS 21). This

TS	1	2	3	4	5	6	7	8	9	10	11	12	13	14	15	16	17	18	19	20	21	22	23	24	25	26
E	0	1	2	3	4	4.5	5	6	6.5	7	7.5	8	8.5	9	9.5	10	10.5	11	11.5	12	12.5	13.5	14.5	15.5	16.5	17.5

vessel usually arises from the **coeliac trunk**, but may arise from the superior mesenteric artery or from the aorta, and divides into a right and left branch in the **porta hepatis** (the hilar region of the liver). Occasionally the right hepatic artery may arise from the superior mesenteric artery, while the left hepatic artery may arise from the left gastric artery (in humans, at least). These latter vessels may replace the normal vessels, or exist in addition to them. The other vascular supply to the liver consists of the venous drainage of the gut, and is carried to it via the **portal vein**; this also divides into a right and left branch in the porta hepatis.

The right and left branches of both the hepatic artery and the portal vein are of approximately equal size, and lie together as they ramify to the right and left lobes of the liver accompanied by tributaries of the biliary system in the so-called **portal triads**. The vessels from the two sides do not communicate, nor do the larger arterial branches within each half (i.e. these are therefore "end" arteries), thus allowing the possibility of hepatic infarction should these vessels for any reason become occluded.

The portal venous blood passes to the sinusoids between the liver cells where it mixes with the arterial blood. The "mixed" blood then drains to the vein at the centre of each lobule, and the blood within them drains towards the hepatic veins, and this arrangement allows intermixing of the blood from the left and right lobes. The venous blood drains directly into the inferior vena cava via a large central vessel and left and right vessel. They help to suspend the liver to the inferior vena cava in the region of the central tendon of the diaphragm through which this vessel passes (and to which it is circumferentially attached), before it terminates in an opening in the floor of the right atrium guarded by a ridge (McMinn, 1990). For diagrams illustrating the venous drainage of the gut and the development of the venous system within the liver between TS 12/13 and TS 18/19, see Figure 4.4.9 a–c.

The autonomic nerve supply to the liver and biliary system

Relatively little appears to be known about the innervation of the liver, and even less about the timing of its innervation, beyond the fact that the hepatic nerves arise from the hepatic plexus, and that they contain both sympathetic and parasympathetic (vagal) fibres. The nerves enter the liver at the porta hepatis and, as elsewhere, largely accompany the blood vessels; additional autonomic nerves accompany the bile ducts and their branches. Other myelinated and nonmyelinated fibres pass to the liver in its peritoneal folds (for further details, see Sutherland, 1965, 1966, 1967).

Spleen

Aggregates of mesodermal tissue associate in the **dorsal mesogastrium** during TS 21/22 to form the **spleen**. In the mouse, this organ has a narrow ribbon-like form, and becomes suspended between the greater curvature of the stomach and the posterior abdominal wall by the **gastro-splenic** and **lieno-renal ligaments** (Figure 4.4.10). It is also worth noting that the arterial blood supply to the spleen from the abdominal aorta, the splenic artery (a branch of the coeliac trunk), passes to this organ via the lieno-renal ligament. As an evolutionary remnant of the fact that the spleen forms from the amalgamation of (usually) five or more aggregates of mesodermal tissue within the dorsal mesentery of the stomach, the splenic artery breaks up into a similar number of branches in the hilar region, each *end* artery supplying a discrete unit of splenic tissue. Occasionally, these units fail to amalgamate with the main body of the spleen, and are termed **splenunculi**, and each is supplied by a discrete branch of the splenic artery (for an ultrastructural analysis of the human spleen during the early fetal period, see Vellguth *et al.*, 1985).

According to Lewis (1956), who studied the vascular circulation in the spleen of rabbit embryos between 11 and 28 days of gestation, from the earliest stages studied, arterial vessels, veins, venous sinusoids and pulp spaces developed from a plexus of tissue spaces in the mesenchymal anlage. The arteries in the late embryonic stages form loops and arcades within the organ from which the branches supplying the pulp arise. This arrangement ensures that the distribution of blood is at equal pressure in different parts of the developing pulp tissue.

The spleen initially serves as an important organ of haematopoiesis, second only in rank to the liver, though its volume is clearly far smaller than that of the liver. In due course, however, it gradually becomes invaded by lymphoid tissue, many of the cells being T-lymphocyte precursors, and the spleen then increasingly takes on an immunological role.

The pancreas and gall bladder

Three diverticula bud from the foregut–midgut junction. One of these, the **biliary bud** (TS 15), elongates and repeatedly branches in association with the tissue of the septum transversum to form the **intra-** and **extra-hepatic components of the biliary system**, differentiating into the gall bladder and its associated ducts (TS 19). The other two form the **dorsal** (larger, TS 18) and **ventral** (smaller, TS 19) **pancreatic buds** (or primordia). These enlarge and amalgamate to form the **definitive pancreas** (TS 20) which is in the upper dorsal part of the abdominal cavity. (For a diagrammatic sequence that illustrates the normal development of the *human* pancreas, see Figure 4.4.12.) The lower part of the

Figure 4.4.12a–c. The formation of the pancreas, and the relationship between the pancreatic and common bile ducts. The *normal* human arrangement is shown, where a single pancreatic duct and common bile duct (usually) enter the middle of the second part of the duodenum at the greater (or major) duodenal papilla. The situation in the mouse is more complex, because there are often numerous sites where fine branches of the bile duct enter the second part of the duodenum. (a, TS, 18/19) A dorsal (larger) and a ventral (smaller) pancreatic primordium form within the dorsal and ventral mesenteries of the duodenum (*i.e.* the mesoduodenum), each with its own duct system. A transverse section through the duodenum shows the initial relationship between the ducts of the dorsal and ventral pancreatic primordia and the common bile duct. (b TS, 19/20) The ventral pancreatic primordium migrates dorsally towards the dorsal primordium, bringing the site of entrance of the common bile duct with it. (c, TS 20) In the definitive arrangement, the two primordia amalgamate, so that the body and tail derive from the dorsal primordium, while the lower part of the head and uncinate process derive from the ventral primordium. Similarly, the dorsal and ventral pancreatic ducts usually amalgamate (but see footnote 12 in the main text).

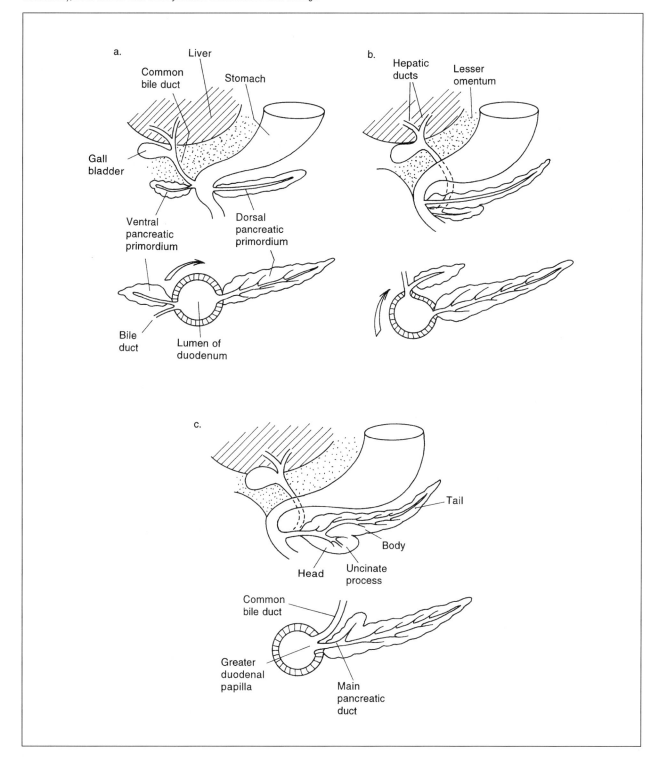

TS	1	2	3	4	5	6	7	8	9	10	11	12	13	14	15	16	17	18	19	20	21	22	23	24	25	26
E	0	1	2	3	4	4.5	5	6	6.5	7	7.5	8	8.5	9	9.5	10	10.5	11	11.5	12	12.5	13.5	14.5	15.5	16.5	17.5

head and uncinate process of the **pancreas** (derived from the ventral pancreatic rudiment, or right lobe) are largely located in the hollow of the duodenum (associated with the left side of the mesoduodenum), and encroach on the proximal part of the mesentery of the jejunum (the mesojejunum). The upper part of the head and the body of the pancreas (from the dorsal pancreatic rudiment, or left lobe) is associated with the dorsal surface of the stomach as it is embedded in the dorsal part of the greater omentum, while the **tail region** (also from the dorsal rudiment) is directed towards the hilum of the spleen. The splenic artery which supplies it is usually partially embedded in the dorsal surface of the pancreas, and then passes towards the hilar region of the spleen in the lieno-renal ligament (see above). Glandular tissue is first recognized in the pancreas during TS 20, but is particularly readily seen during TS 21.

The mouse pancreas is relatively larger than its human counterpart,[12] and its duct system is different. While the basic principles underlying the development of this organ in the mouse are similar, the arrangement observed in this species is clearly somewhat more complex than that in the human. At least 2 and occasionally up to 5–8 pancreatic ducts, may be present, and all drain into the common bile duct, which is surrounded along almost its entire length by pancreatic tissue. The first, and usually the largest of the collecting ducts invariably drains the derivatives of the dorsal pancreas, but additional small pancreatic ducts occasionally open directly into the duodenum.[12] Early evidence of **islet** (of **Langerhans**) formation is seen at about TS 24, although the characteristic **beta-cells** (involved in the secretion of insulin) are not recognized until about TS 26.

Arterial supply and venous drainage of the pancreas

Most parts of the pancreas receive their arterial blood supply from the coeliac trunk, principally via numerous small branches which pass to the neck, body and tail of the pancreas from the splenic artery as it runs along its superior border. The anterior part of the head and uncinate process receive their arterial blood supply from the **anterior** and **posterior pancreaticoduodenal arteries**. Branches of the **gastroduodenal artery** also supply the head of the pancreas, and these anastomose with the inferior pancreaticoduodenal branch of the superior mesenteric artery. The arterial supply to the posterior part of the pancreas is quite variable, and it may receive branches from the superior mesenteric, middle colic, hepatic or even the coeliac artery. A vascular supply to the dorsal and ventral pancreatic primordia is observed at an early stage of their differentiation (TS 18–20).

The venous return is via numerous small veins to the splenic vein and, in the region of the head of the pancreas, either directly to the portal vein via the superior pancreaticoduodenal vein or indirectly to the portal vein via the inferior pancreaticoduodenal vein which initially drains into the superior mesenteric vein.

The autonomic nerve supply to the pancreas

The nerve supply to the pancreas is complex and derives from the coeliac ganglion and plexus, entering with the arteries. The sympathetic nerves pass from the coeliac ganglion, while the parasympathetic nerves pass from the right vagus. The fibres are both vasomotor (sympathetic) and also pass to the parenchyma of the pancreas, the latter nerves being both sympathetic and parasympathetic. It is believed that there is a fine neural control of both endocrine and exocrine function, and many fibres are seen (in

---

[12] In the human, the two pancreatic primordia appear at the 4 mm stage, and these fuse at about the 16 mm stage (about 7 weeks of gestation). Cytological differentiation of the pancreas has been studied by Conklin (1962) and Liu and Potter (1962). These researchers have observed that, at the 30 mm stage, the pancreatic parenchyma consists of a system of epithelial tubes; these are diffusely branched and there are clumps of cells at their endings that are the precursors of the acinar and islet epithelia.

By the 56 mm stage, agyrophilic cells (the precursors of the α-cells) are seen, and by the 90 mm stage δ-cells and finally β-cells are apparent (see also Pearce, 1903, who first described the development of the islets from the primitive pancreatic tubules). At term, β-cells predominate in the islets, with fewer α-cells and only a few δ-cells at the periphery. Insulin is present in all fetuses by the 80 mm stage (Adesanya et al., 1966).

The proximal part of the main pancreatic duct in the human is usually derived from the main duct of the ventral pancreatic rudiment, while the rest of the duct forms from the duct of the dorsal pancreatic rudiment. Usually, the common bile duct and the main pancreatic duct drain into the lumen of the duodenum in a single site, the greater duodenal papilla. Often, however, the original ducts of the dorsal and ventral pancreatic primordia are retained, and when this occurs there is also a *lesser* (or minor) duodenal papilla invariably located *proximal* to the greater duodenal papilla, where the duct of the ventral rudiment (the accessory duct) drains into the lumen of the duodenum.

In addition, there is sometimes an **accessory** (or additional) duct (of **Santorini**), which drains the lower part of the head of the pancreas and crosses the main pancreatic duct before it opens directly into the duodenum at the **minor** (or lesser) **duodenal papilla**. In the adult, this is usually about 1–2 cm proximal to the site of entrance of the main pancreatic duct at the **major** (or greater) **duodenal papilla** and the two ducts frequently communicate with each other. The common bile duct also opens into the duodenum at this site, known as the **hepatopancreatic ampulla** (of **Vater**). For a diagrammatic sequence that illustrates the normal development of the *human* pancreas, see Figure 4.4.12. While the basic principles underlying the development of this organ in the mouse are similar, the arrangement observed in this species is clearly somewhat more complex than that in the human. Occasionally, if the ventral pancreatic bud bifurcates to form two units, while the one rotates clockwise around the duodenum in the normal way, the aberrant unit rotates anticlockwise around the duodenum, forming a ring of pancreatic tissue around the gut (termed an **annular pancreas**). This may cause duodenal obstruction and, *in the human*, is more commonly encountered in males than in females.

the adult) to enter the islets with the arterioles (Coupland, 1958).

Arterial supply and venous drainage of the gall bladder

The arterial blood supply to the gall bladder is via the **cystic artery**, usually a branch of the right hepatic artery, but this is supplemented by numerous small branches from its hepatic bed (*i.e.* the site where the gall bladder is closely adherent to the inferior surface of the right lobe of the liver). The cystic artery reaches the neck of the gall bladder, and branches from it ramify over its surface. Occasionally, it may arise directly from the main trunk of the hepatic artery, from the left branch of that vessel or from the gastroduodenal artery. The venous return is via numerous small veins in the bed of the gall bladder towards the substance of the liver, and thence into the hepatic veins. Occasionally, one or more cystic veins may be present and, if this is the case, these drain into the right branch of the portal vein.

TS	1	2	3	4	5	6	7	8	9	10	11	12	13	14	15	16	17	18	19	20	21	22	23	24	25	26
E	0	1	2	3	4	4.5	5	6	6.5	7	7.5	8	8.5	9	9.5	10	10.5	11	11.5	12	12.5	13.5	14.5	15.5	16.5	17.5

4.5 The mouth and nose region

Introduction and overview

The mouth and nose form from the oro-naso-pharyngeal region of the early embryo as a result of an extremely complex set of morphogenetic interactions that remodel the early mouth pit and anterior foregut region and turn it into the external and internal entrances to the alimentary and respiratory tracts. The first stage in this process is the breaking down of the buccopharyngeal membrane between the mouth (or oral) pit and the most rostral part of the pharyngeal region of the foregut (TS 14). This occurs at about the same time as Rathke's pouch, the precursor of the future anterior part of the pituitary, develops in the roof of the ectodermal part of what is still the oropharynx.

The external components of the mouth and nose start to form when the nasal placodes (TS 15) develop into pits that deepen to become the nasopharynx which is soon continuous with the oropharynx. Medial and lateral nasal processes surrounding the nasal (or olfactory) pits (TS 16) extend and meet to form the region around the nostril (TS 18); the rest of the nose is formed from the downgrowth of the medial part of the frontal process. The inferior borders of the medial and lateral nasal processes meet and fuse with the supero-medial borders of the two maxillary processes (from the first branchial arches) to form the outer region of the upper lip and region of the face below the eye, while the central region of the upper lip (the **philtrum**, TS 21) is believed to form from either the most caudal (or labial) part of the frontal process, or from the most medial (labial or pre-maxillary) parts of the maxillary processes (for an overview of the external appearance of the cephalic region of the mouse embryo between TS 15 (E9.5) and TS 19 (E11.5), see Figure 4.5.1a–h).[1] Relatively little is known of the factors involved in the initial development of the facial primordia, though some information is available from studies carried out in avian embryos (*e.g.* Wedden *et al.*, 1988).

The anterior skeletal component of the upper jaw is the pre-maxilla; this underlies the philtrum, and forms the primary palate. The pre-maxilla is believed to originate either from the caudal part of the frontal process or from the medial extensions of the maxillary processes (TS 20); the region of the premaxilla is eventually, in the mouse, associated with the upper incisor teeth. Also at about TS 20, the mandibular processes extend and meet to form the lower jaw and its bony element, the mandible, is initially formed adjacent to Meckel's cartilage. Ossification of the mandible (TS 22/23) subsequently develops around, but principally lateral to, this cartilaginous element which is also of first arch origin.

The key internal component of the region, however, is the secondary palate that forms when the two vertically directed **palatal** (or **palatine**) shelves of the maxillae that had first appeared at TS 18 or so on either side of the oropharyngeal cavity subsequently elevate and start to extend medially (TS 21); they finally meet in the midline at the centre where they fuse to one another and to the downgrowing nasal septum (TS 23/24). The palate thus forms both the roof of the oropharynx and the floor of the nasopharynx. At about the same time as the palatal shelves form, the lingual swellings develop from branchial arch material on the ventral surface of the oropharynx at about TS 18/19, and, over the next few stages, the tongue grows rapidly. As the mouth tissues develop, so do the structures that develop within the nose, with the appearance of the vomeronasal organs, one on either side of the nasal septum, the conchae and the nasal capsule (TS 20). As the structural components of the mouth and nose continue to develop, a set of associated tissues also make their appearance and these include the teeth and a series of glands (*e.g.* the thymus and thyroid glands, and the submandibular, sublingual and parotid salivary glands).

As so many events are taking place at the same time in neighbouring regions, it is not easy to integrate the development of the nose and mouth. Rather than considering the system as a whole and describing it on a stage-by-stage basis, the approach that we therefore take is to first discuss the early development of the oro-nasal-pharyngeal region as a whole in some

---

[1] The appearance of the upper lip varies considerably among mammals, and for a classification of these various forms, see Boyd (1933).

Figure 4.5.1a–h. Drawings of pairs of frontal and lateral views of the cephalic region of mouse embryos principally to display the differentiation of the fronto-nasal derivatives and the branchial/pharyngeal arches. a and b, TS 15 (E9.5), embryo with about 25–30 pairs of somites; c and d, TS 17 (E10.5–11), embryo with about 35–39 pairs of somites; e and f, TS 18 (E11–11.25), embryo with about 40–45 pairs of somites; g and h, TS 19 (E11.5), embryo with about 43–48 pairs of somites.

Key:
1. Maxillary component of first branchial arch
2. Mandibular component of first branchial arch
3. Entrance to Rathke's pouch (future anterior pituitary)
4. Region of olfactory placode (nasal part of fronto-nasal process)
5. Optic eminence (region overlying optic vesicle)
6. Second branchial cleft (groove)
7. Second branchial arch (hyoid arch)
8. First branchial cleft (groove)
9. Medial nasal process
10. Lateral nasal process
11. Lens pit
12. Nasal (olfactory) pit
13. Region overlying telencephalic vesicle
14. Groove between lateral nasal process and maxillary component of first branchial arch (naso-lacrimal groove)
15. Corneal ectoderm overlying lens vesicle

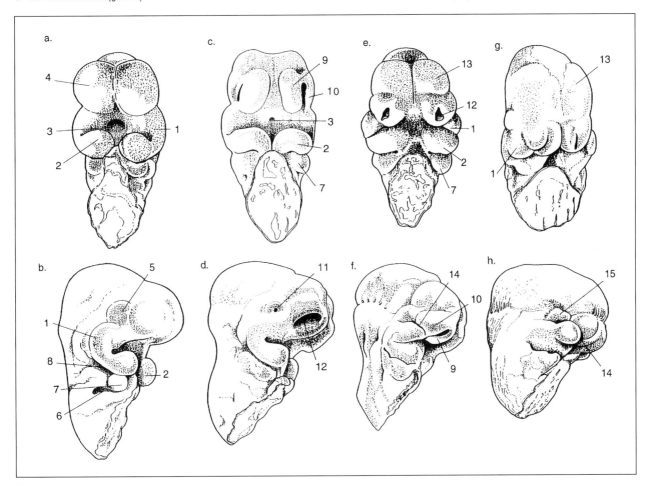

detail, and then consider separately the development of its major components which include the palate, the internal and external regions of the nose, the tongue, the thyroid, salivary and other glands, and the teeth, thus accepting that there inevitably has to be some duplication of information.

The early stages

The earliest sign of mouth development is the formation of the ectodermally-derived **stomatodaeum** (**mouth pit**; TS 13) which is directed towards the most rostral part of the endodermally-lined gut tube, the two soon being separated by the bilayered **buccopharyngeal membrane** which has no intervening mesodermally-derived tissue and which appears during TS 13, but breaks down a stage later. Although the continuity that then occurs is between two tissues of very different embryological origin, evidence of this boundary zone is soon lost. This situation contrasts markedly with that in the comparable region at the junction of the caudal end of the gut tube and the **proctodaeum** (or **anal pit**), where the relatively unvascularized **pectinate line** (which is located at the junction between the upper two-thirds

TS	1	2	3	4	5	6	7	8	9	10	11	12	13	14	15	16	17	18	19	20	21	22	23	24	25	26
E	0	1	2	3	4	4.5	5	5.5	6	6.5	7	7.5	8	8.5	9	9.5	10	10.5	11	11.5	12	12.5	13.5	14.5	15.5	16.5 17.5

and lower one-third of the anal canal) remains as the remnant of this boundary zone.

The earliest stage at which a mouth can be said to be present is thus TS 14 when the buccopharyngeal membrane breaks down and the amniotic cavity surrounding the embryo becomes continuous with the the broad pharyngeal region at the rostral part of the foregut tube, on either side of which are the paired **branchial pouches**. At about this stage, pituitary development is initiated when an ectodermally-lined diverticulum (**Rathke's pouch**) grows cranially from the roof of the oropharynx, towards the floor of the third ventricle of the brain (TS 14). This pouch eventually loses contact with the roof of the pharynx and forms the various components of the **adenohypophysis** (*i.e.* the pars anterior, intermedia and tuberalis of the pituitary gland (TS 20) – for further details, see the section on *Pituitary*, in Appendix to Chapter 4.6 *The brain and spinal cord*).

Before about E9–9.5 (TS 15), there is only a mouth region that consists exclusively of the **oropharynx**, and the first sign of nose development is the differentiation of the **olfactory placodes** (TS 15; possibly under an inductive influence from the future olfactory regions of the prospective forebrain). These gradually become indented in their central regions to form the **olfactory pits** which subsequently hollow out to form the **nasopharynx** when their lumens become continuous with that of the oropharynx. The oropharynx and nasopharynx only become separated again after the **primary palate** has formed and the two **palatal shelves** of the maxillae have fused to form the **secondary palate** (TS 22/23 – see below). During TS 16, the olfactory pits become narrow slits and are bounded on their medial and lateral borders by the prominent **medial** and **lateral nasal processes**. The inferior border of the nostril which forms from the fusion of the two **nasal processes** fuses with the immediately inferiorly located supero-medial border of the maxillary process to form the lateral part of the upper lip, and the region of the face below the eye.

The exact origin of the **intermaxillary segment** (the skeletal component of which forms the primary palate) is unclear: the skeletal part subjacent to the philtrum (see above) derives either from the caudal part of the nasal process or from the medial extensions of the maxillary processes, and is first recognized at about TS 20. This region of the upper jaw (also termed the **premaxilla**) is, in the mouse, associated exclusively with the upper incisor dentition. While the skeletal part of the intermaxillary segment mainly develops into the primary palate, its labial (*i.e.* upper-lip-derived) component gives rise to the philtrum region of the upper lip at about TS 22. In the mouse, this is represented by a deep cleft, whereas in the human, the philtrum is represented by the grooved median part of the upper lip.

During TS 19, the nasal component of the oronasal cavity which is entered through the **external** (or anterior) **nares** (or **nostrils**) expands dramatically, while its entrance portal diminishes to a narrow slit-like orifice. Within the oral cavity, the **palatal** (or **palatine**) **shelves** of the maxillae start to elevate and extend medially on their way to forming the secondary palate. At about the same time, small indentations or pits, which subsequently form the **primary choana** (or **primitive posterior naris**/nasal aperture) also start to form on either side of the anterior part of the roof of the oropharynx, and represent the openings of the olfactory pits into the oropharyngeal cavity. These will, after the formation of the palate, form the **definitive** (or **secondary**) **choanae** or posterior nasal apertures where the posterior part of the nasal cavity is in continuity, on either side of the nasal septum, with the oral cavity. It is only at TS 21, however, that the anterior and posterior nares become linked to form the nostrils.

The **lingual swellings** also appear in the floor of the oropharynx at about TS 19, and the dorsal surface of the developing tongue mass approaches the roof of the oropharynx close to the entrance to Rathke's pouch. As the tongue expands at a quite prodigious rate over the period TS 19–22, it soon takes up most of the space in the oro-/nasopharynx even though this has now deepened considerably. Further, the width of the tongue in its anterior third is almost matched by that of the primitive nasal septum which now forms from the downward growth of the central part of the roof of the oropharynx. As the lumen of the primitive primary choana enlarges (TS 21), it delineates both the postero-lateral boundary of the primary palate and the antero-lateral border of the primitive nasal septum. Because of the considerable width of the anterior part of the nasal septum, it is difficult in sectioned material to distinguish it from, and indeed it appears to blend into, the superior surface of the primary palate.

The palate

While the exact embryological origin of the primary palate (TS 20) is unclear, it does appear to develop from the **intermaxillary segment** (TS 19) which is thought to form from the amalgamation of the inferior border and the immediately surrounding tissue of the medial nasal processes, though it may form from the amalgamation of the antero-medial borders of the maxillary processes.

The palatal processes of the maxillae first appear (TS 17/18) as two narrow shelves running antero-posteriorly at the upper border of the lateral walls of the oropharynx. Up to about TS 20/21, their anterior one-third remains blunt and relatively poorly developed, while the posterior two-thirds or so have sharper borders which become increasingly more vertically directed. As the external nares link to the oropharynx, the volume of the palatal shelves increases and, by TS 22/23, the shelves elevate as they rotate upwards, becoming horizontal along their entire length. A stage later, the anterior half to two-thirds of their medial borders become apposed and fuse together both on their antero-medial surface and with the primary palate to form the **secondary** (or, more correctly, the **definitive**) **palate**. The final fusion that now occurs during TS 23/24 is that between the midline of the upper surface of the palate and the inferior border of the nasal septum, so dividing the primitive nasopharynx into two definitive nasal cavities. A series of diagrams showing representative coronal sections through the middle region of the oropharynx during this period of early palatal shelf formation in the human embryo clearly demonstrates that, not only are these shelves initially vertically directed, but that they initially form the lateral boundaries of the tongue (see Figure 4.5.2a). With their elevation to the horizontal position, but prior to their fusion, they almost completely cover the dorsal surface of the tongue (see Figure 4.5.2b) and, once they have fused across the midline, the definitive palate is seen to form the roof of the oropharynx (or oral cavity) (see Figure 4.5.2c).

Palate-shelf elevation is a complex process and seems to involve both mechanical and morphogenetic mechanisms (Ferguson, 1987, 1988). The former include the reflex opening and closing movements of the lower jaw which, because they allow the lower jaw and dorsal surface of the tongue to descend below the lower margins of the palatal shelves, seem to facilitate the initial elevation of the palatal shelves and their maintenance in a horizontal position (Humphrey, 1969). The morphogenetic mechanisms here are the so-called "internal shelf forces", which involve an increase in blood supply and tissue rigidity brought about by the hydration of such extracellular matrix components as hyaluronic acid and glycosaminoglycans (Brinkley and Morris-Wiman, 1987), and which, together with changes to the shelf proteins just before and during shelf elevation, facilitate the elevation and maintenance of the palatal shelves in the horizontal position. As the dorsal surface of the tongue becomes more caudally located, the medial (more correctly termed the *inferior* because the orientation of palatal shelves is initially vertically directed) borders of the growing palatal shelves elevate, soon become closely apposed and then fuse across the midline. This fusion initially takes place at the ventral (anterior) end of the palatal shelves, but apposition probably occurs simultaneously along much of their length, and then proceeds dorsally (*i.e.* posteriorly) until, by about TS 24, fusion is complete. The most posterior part of the palate has an inverted V-shape, there being no equivalent of an uvula which hangs down from the middle of the posterior border of the soft palate as seen in humans. Once the cartilage primordium of the hard palate has fully differentiated (TS 24), the non-cartilaginous region posterior to its dorsal border represents the equivalent of the soft palate, though this part is not mobile as in the human.

Cartilage forms within the mesenchymal core of the palatal shelves,[2] with early ossification centres initially being seen at their lateral borders (TS 22/23) and these gradually extend medially into the rest of the secondary palate. At about TS 24, precartilaginous condensations start to appear in the anterior half of what will become the hard palate, although condensations of precartilage/cartilage are already present by this time within its posterior half. The most posterior part of the palate does not contain a cartilaginous core, and is termed the soft palate (see above).

The bony (or hard) palate is covered by a thin layer of mucous membrane which is itself continuous posteriorly with the **soft palate**. The postero-lateral borders of the soft palate extend downwards and backwards between the oral and nasal parts of the pharynx. The mucous membrane covering the upper part of the posterior surface of the hard palate is columnar and ciliated (having the characteristic features of "respiratory epithelium") and is continuous with the olfactory epithelium lining the nasal cavities. Elsewhere, particularly on its oral surface and on the soft palate, the mucous membrane contains numerous mucous glands, while taste buds are present in the epithelium of its oral surface.

Another notable feature of the palate are the **rugae** associated with its oral surface. These **ridges** are first apparent on the inferior (epithelial) surface of the most lateral part of the palatal shelves at about TS 21. They can clearly be seen in scanning electron micrographs at this stage (Kaufman, 1992), but are probably too subtle to be recognized in histological sections. By TS 23, however, they have become increasingly obvious, and extend towards the medial borders of the shelves, while, by TS 24, they almost reach the midline, at the site of shelf fusion. In the definitive arrangement, nine rugae of various sizes are seen, only the most anterior three of which are continuous across the midline (Sakamoto *et al.*, 1989).

Once (secondary) palatal shelf fusion is complete at around TS 24, the oropharynx is effectively separated

---

[2] The *cartilaginous* primordium of the hard palate which precedes evidence of ossification is clearly seen in the palatal shelves of embryos at TS 24, and appears to be different from the situation observed in the human embryo at the equivalent stage of development, where ossification is described as occurring within a *mesenchyme* precursor, as would be expected if the palatal shelves of the maxillae are components of the membranous viscerocranium.

TS	1	2	3	4	5	6	7	8	9	10	11	12	13	14	15	16	17	18	19	20	21	22	23	24	25	26
E	0	1	2	3	4	4.5	5	6	6.5	7	7.5	8	8.5	9	9.5	10	10.5	11	11.5	12	12.5	13.5	14.5	15.5	16.5	17.5

Figure 4.5.2a–c. Three stages in the formation of the secondary palate in the human embryo, displayed as a series of ventral views of the palatal region together with matched *coronal* sections through the oropharyngeal region. At the earliest stage illustrated (a, equivalent to TS 18), the palatal shelves are vertically directed on either side of the tongue mass. Shortly afterwards (b, TS 20), and associated with the initiation of reflex opening and closing movements of the mouth, the lower jaw 'descends', bringing the tongue down with it. This enables the palatal shelves to be redirected into a horizontal orientation over the dorsum of the tongue. The medial edges of the palatal shelves grow towards the midline and fuse (i) with each other (c, TS 23/24), and (ii) with the lower (inferior) border of the nasal septum (TS 24), resulting in the formation of the definitive secondary palate. The latter also fuses anteriorly with the primary palate, a much smaller midline structure believed to be of fronto-nasal origin, and a small incisive foramen marks the location of their site of fusion in the midline. The most dorsal "mobile" component of the palate is termed the soft palate, in the middle of which is the uvula. With the apposition and fusion of the palatal shelves the nasopharynx (above) is separated from the oropharynx (below). With the fusion of the nasal septum to the superior surface of the secondary palate in the midline, the nasopharynx becomes effectively divided into the two discrete nasal cavities; only in the most posterior part of the oropharyngeal region are the nasopharynx and oropharynx in continuity, close to the site where the pharyngo-tympanic (Eustachian) tube opens into the nasopharynx. The bony element within the nasal septum is the vomer. Note that the dotted regions show domains of ossification.

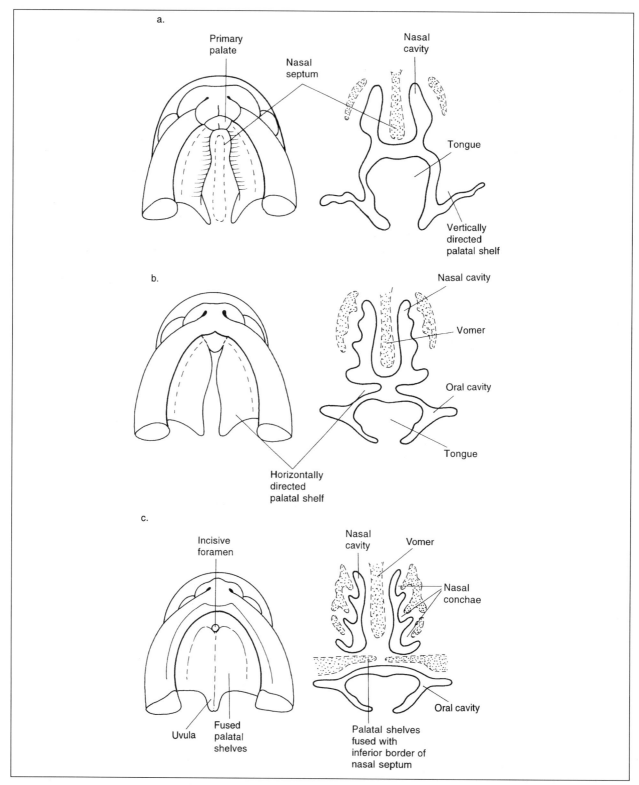

from the nasopharynx, although the two remain in communication at the postero-lateral borders of the primary palate through two slit-like orifices, the **incisive canals**, which are the remains of the primitive posterior nares, and at the posterior nasal apertures (see above). A midline funnel-shaped opening, the **incisive fossa**, is seen on the oral surface of the palate at the junction between the primary and the secondary palate (see above) during TS 23 and this leads into the two canals which transmit the greater palatine vessels and nasopalatine nerves. This median V-shaped space gradually diminishes so that, by TS 25, it is reduced to two minute slit-like orifices (the incisive canals), which are located on either side of the midline. Occasionally these are retained as minute foramens in the **hard palate** and, when present, link the oral and nasal cavities. Normally, however, the bony canal is retained, while the epithelial discontinuities become filled-in, closing off any communication between the oropharynx and nasopharynx in this location.

The other bony component of the hard palate is made up by the horizontal plates of the **palatine bones** which are united anteriorly with the palatine processes of the maxillary bones by the palatomaxillary sutures. Like the other facial bones, the palatine bones are said to be ossified in membrane: they are complex in shape, having horizontal and perpendicular plates, and pyramidal, orbital and sphenoidal processes (Williams, 1995).

The nasal cavities, nasal septum and vomeronasal organ

The nose takes up its final form when the *inferior* border of the nasal septum, which starts to form at about TS 19, fuses in the midline with the upper surface of the palate, so subdividing the primitive nasopharynx into the right and left **nasal cavities** (TS 24). The most posterior part of the inferior border of the nasal septum does not, however, fuse with the upper surface of the palate, and in this location, communication persists between the right and left halves of the nasopharynx.

An important structure that starts to develop within the nasal septum at TS 20 is the **vomeronasal (Jacobson's) organ**. This is an auxiliary olfactory organ that receives a special (vomeronasal) branch of the olfactory (I) nerve terminating in the **accessory olfactory bulb**, an outgrowth of the forebrain, and which may respond to a narrower range of chemical/olfactory stimuli than the olfactory receptors located elsewhere in the nasal cavity.[3] The vomeronasal organs first appear as a pair of invaginations in the ectodermally-derived nasal mucous membrane (which at this stage is still relatively undifferentiated olfactory epithelium) on either side of the lateral walls of the lower anterior part of the primitive nasal septum. As its olfactory epithelium soon loses contact with that present elsewhere in the nasal cavity, the vomeronasal organ is, in essence, a blind tubular pouch. The organ enlarges during TS 21/22, causing the lateral wall of the nasal septum to bulge into the nasal cavity. These bulges eventually (TS 26) occupy much of the anterior part of the nasal septum, being located on either side as well as just below the lower border of the cartilaginous primordium of the nasal septum.

As early as TS 20, **conchae** (the future **turbinate bones**) start to form as olfactory-epithelium-covered mesenchymatous shelves on the lateral walls of the most rostral part of the nasopharynx. At the time of birth, the turbinate bones are covered by olfactory epithelium and occupy much of the upper half of the nasal cavities. Their principal function is to humidify and warm the air that passes through the nasal passages (the upper respiratory tract) and thence into the lower respiratory tract. By TS 21, the most superiorly located of the conchae are seen to be directly subjacent to the **olfactory lobes** of the brain, with only the mesenchyme that will subsequently form the **cribriform plates** of the **ethmoid bones** intervening. It is through minute channels in these plates that the olfactory (I) nerves pass upwards from the olfactory epithelium to the olfactory lobes of the brain.

At around TS 21/22, the olfactory epithelium in the superior and posterior part of the nasopharynx, subjacent to the cribriform plates, starts to show clear evidence of differentiation (TS 21/22), and the conchae and their overlying olfactory epithelium continue to differentiate during TS 22/23, with the conchae extending further forward into the nasal cavity. Considerable numbers of serous glands now start to form in the tissues subjacent to the bases of the conchae, with their ducts opening into the nasal cavities.

During TS 23/24, the mesenchymal shelves of the conchae develop a cartilaginous core that will subsequently ossify to form the turbinate bones, while their overlying surface epithelium becomes particularly

---

[3] The vomeronasal organ is particularly well developed in rodents, but is either lost or vestigial in higher mammals where olfaction is less important for survival and reproduction than the other senses. In the human, this organ reaches its fullest development by about the end of the fifth month. It, with its nerve, usually completely regresses, but vestigial remnants have been observed in the adult (Hamilton and Mossman, 1972).

TS	1	2	3	4	5	6	7	8	9	10	11	12	13	14	15	16	17	18	19	20	21	22	23	24	25	26
E	0	1	2	3	4	4.5	5	6	6.5	7	7.5	8	8.5	9	9.5	10	10.5	11	11.5	12	12.5	13.5	14.5	15.5	16.5	17.5

convoluted, so substantially increasing their surface area; even so, the conchae still only occupy the dorsal half of the nasal cavity at TS 26. Note that the inferior half of the posterior part of the nasopharynx contains no olfactory epithelium, and is merely a passageway connecting the nasal cavity (anteriorly) with the oropharynx (posteriorly). The difference in the gross and histological morphology of these two regions of the posterior part of the nasopharynx can be particularly well seen in coronal sections through this region.

The external nose and nasal cartilages and bones

As the internal parts of the nasal system develop, changes also occur in its external components. After the olfactory placodes have differentiated (TS 15) and as the olfactory pits form (TS 15/16), the medial and lateral nasal processes develop and are first clearly delineated during the early part of TS 16. The combined infero-lateral borders of these processes become apposed to the medial (or anterior) border of the maxillary component of the first branchial arch, initially being separated from it by the **naso-lacrimal groove**. During TS 18, this groove deepens and eventually runs diagonally antero-inferiorly from the medial angle (or inner canthus) of the eye, between the anterior border of the maxillary process and the postero-lateral border of the lateral nasal process, to a site just medial to the future angle of the mouth. The naso-lacrimal groove later closes (TS 20/21) to form the **naso-lacrimal duct**.

From TS 17/18, the entrances to the olfactory pits become progressively narrower, so that by TS 20/21 they are reduced to a pair of narrow slits. Similarly, the considerable increase in the volume of the telencephalic hemispheres that occurs from about TS 18/19 tends to emphasize the transverse groove now delineating the upper boundary of the nasal region (below) from the region overlying the telencephalic region (above). Over the next few days, this boundary zone is gradually lost while the nasal region increases in volume and elongates. Similarly, with the increasing depth of the philtrum, there is a progressive diminution in the volume of the *external* nasal region which surrounds the external naris.

A pre-cartilaginous **nasal capsule** is first seen at about TS 20 when it surrounds most of the olfactory system (this is the evolutionary remnant of the cartilaginous capsule whose main role was to protect the olfactory sensory apparatus in primitive vertebrates, see 4.8 *Development of the skull*). At this early stage, it is incomplete on its inferior and lateral surfaces, and continues to be so even after the palate has formed, whereas, in the most posterior part of the nasopharynx, it forms an almost complete protective capsule that extends posteriorly, being located just above the site of communication between the nasopharynx (above) and the oropharynx (below). Even at TS 26, the capsule is still cartilaginous and the region around the external naris remains so, giving rise to the ventral and dorsal nasal (or alar) cartilages. As mentioned earlier, the superior part of the nasal capsule becomes modified and eventually ossified to form the **cribriform plate of the ethmoid bone** through which the olfactory (I) nerves pass *upwards* to the olfactory lobes of the brain.

The nasal bones are components of the viscerocranium and form within separate membranous precursors overlying the anterior part of the nasal capsule. Ossification in this maxillary part of the nasal region is first seen at about TS 22/23, and is mainly seen just lateral to the nasal capsule. Ossification centres in the premaxilla and in the rest of the maxilla, which also make major contributions to this region of the facial skeleton, appear at about TS 23/24, while ossification within the nasal bones is first seen, at about TS 24/25.

The tongue

The epithelial and mesenchymal components of the tongue develop on the floor of the oropharynx from primordia that develop from the first four branchial arches (mainly from the first, third and fourth which largely overgrow the derivative of the second arch). The first of these primordia forms during TS 18 as a pair of discrete swellings (the **lateral lingual swellings**) that are associated with the dorsal surface of the mandibular component of the first branchial arch. These swellings enlarge rapidly and will eventually form most (the anterior two-thirds) of the surface of the tongue and its subjacent mesenchyme tissue. Another, but much smaller, midline primordium is also associated with the dorsal surface of the first pharyngeal arch and this is the primordium of the **tuberculum impar** (sometimes termed the median lingual swelling) and also forms at TS 18. The tuberculum impar was formerly believed to be the precursor of the anterior two-thirds of the tongue, but it is now generally accepted that this is a transient structure that does not contribute to any significant part of the adult tongue, other than perhaps the intermolar eminence (see below). The lateral lingual swellings are relatively easy to recognize in SEM micrographs and in sagittal histological sections through the floor of the oropharynx, but are harder to identify in transverse histological sections, while the other tongue primordia (the derivatives of the 2nd, 3rd and 4th pharyngeal arches) are almost impossible to distinguish as individual entities – whatever approach is used.

The two lateral lingual swellings continue to enlarge and their medial borders fuse across the midline (TS 19), the site of fusion initially being marked by a deep dorsal median sulcus and later by the presence of a median, vertically-directed, fibrous septum (TS 22). An additional smaller pair of swellings now form in association with the ventro-medial ends of the second pharyngeal arch (these later fuse together to form a single midline swelling, that is sometimes termed the **copula**). The first arch derivatives soon amalgamate with these second arch tissues (TS 20), and slightly later (TS 21) with tissues of third and fourth arch origin (*e.g.* the epiglottis, TS 19, see below). The combined derivative of the third and fourth pharyngeal arches is termed the **hypobranchial eminence** (TS 19) and it, together with the first arch derivatives, completely overgrow the copula from the second arch.[4] A diagrammatic sequence illustrating two stages in the development and a third illustrating the definitive arrangement of the tongue is shown in Figure 4.5.3a–c.

From about TS 21 onwards, it is not possible to follow with any confidence the fate of the individual components of the tongue, although they may be traced, indirectly, from a knowledge of the nerve supply to the various regions of the tongue [the cranial nerve supply to the derivatives of the first, second, third and fourth pharyngeal arches is from cranial nerves V, VII, IX and X, respectively (see Figure 3.3.5b)]. This indicates that the derivatives of the first branchial arch give rise to the surface epithelium (endoderm-derived) and mesoderm of the anterior two-thirds of the tongue, while that of the posterior one-third is derived from the combined derivatives of the third and fourth branchial arches, the derivative from the fourth arch giving rise to the most dorsal or caudal part of the tongue and region of the epiglottis.

General sensation for the surface of the anterior two-thirds (the oral or pre-sulcal part) of the tongue is from the **lingual nerve**, a branch of the mandibular division of the trigeminal (V) nerve from the first arch. The nerve supply for special sensation (*i.e.* taste) to the anterior two-thirds of the tongue is via the **chorda tympani branch** of the facial (VII) nerve (which is distributed with the lingual nerve), and indicates that the tissues of the second pharyngeal arch (whose nerve supply is from the facial (VII) nerve) are incorporated into this part of the tongue. The posterior (or pharyngeal, or post-sulcal) part of the tongue is largely of third arch origin, as both general and special sensation to this region is via the glossopharyngeal (IX) nerve. The hypoglossal (XII) nerve supplies all of the extrinsic muscles of the tongue, except the palatoglossus which is supplied by the glossopharyngeal (IX) cranial nerve. The hypoglossal nerve also supplies the occipital somites (whose cells migrate into the tongue mass and differentiate there to form its intrinsic musculature). The fourth arch component of the tongue gives rise to its most caudal part, its base, as well as the region around the **epiglottis** as it is supplied by the vagus (X) nerve (via the internal laryngeal part of its superior laryngeal branch). The **median circumvallate papilla** (**MCP**) is first seen around E21/22 as a downgrowth of epithelial cells from the dorsal surface of the tongue, and even at this time is associated with numerous fine nerve endings[5] (AhPin *et al.*, 1989). The MCP becomes more prominent and eventually has a central dome-like region surrounded by a deep sulcus that is supplied bilaterally by the glossopharyngeal (IX) nerves, while on its lateral walls are soon located a large number of accessory **filiform** and **fungiform papillae**. The **filiform papillae** (also called conical papillae because of their conical epithelial cap) tend to cover the presulcal area of the dorsum of the tongue, and appear during TS 22. These papillae are minute structures arranged in rows which run parallel with the sulcus terminalis, except at the apex of the tongue where they run transversely. A stage or two later, numerous **fungiform papillae** (which appear to be miniature versions of the MCP) are clearly seen to be widely distributed on the surface of the tongue, mainly at its sides and apex. In the midline, a crescentic fold of mucous membrane, the **frenulum** of the tongue, connects the inferior surface of the anterior part of the tongue to the floor of the mouth (TS 23), and on either side of it are located the **sublingual papillae** or **caruncles**. On the surface of each of these are the orifices of the ducts of the sublingual/submandibular salivary gland complex (for additional details, see later).

By TS 22, the tongue extends forwards over the dorsal surface of the anterior part of the fused mandibular processes while a large swelling, the **intermolar eminence**, is now seen protruding from the dorsal surface of the tongue almost at the junction between its anterior two-thirds and posterior one-third. Although its exact origin is unclear, this

---

[4] It should be noted that the terminology used to describe the precursors of the various pharyngeal (branchial) arch derivatives of the tongue depends on the authority cited. While the descriptive account and terminology given here follows that of Hamilton and Mossman (1972) and more recently that of Larsen (1993), those found in Gray's Anatomy (Williams, 1995) are slightly different. Thus, according to the latter authority, "caudal to the median tongue bud, a second median elevation, termed the *hypobranchial eminence* (or *copula* of His *i.e.* Wilhelm His (1831–1904)), forms in the floor of the pharynx, and the ventral ends of the fourth, the third and, later, the second visceral arches converge into it . . . . In the process the third arch elements grow over and bury the elements of the second arch, excluding it from the tongue".

[5] The presence of the *single* large midline circumvallate papilla in the mouse contrasts with the situation in the human where they vary from eight to twelve in number, and are 1–2 mm in diameter. The latter form a V-shaped row immediately in front of and parallel with the **sulcus terminalis** (or terminal sulcus). The rat also has a single circumvallate papilla whereas the Japanese dormouse has three, and the porcupine and anteater have two (for references, see AhPin *et al.*, 1989).

TS	1	2	3	4	5	6	7	8	9	10	11	12	13	14	15	16	17	18	19	20	21	22	23	24	25	26
E	0	1	2	3	4	4.5	5	6	6.5	7	7.5	8	8.5	9	9.5	10	10.5	11	11.5	12	12.5	13.5	14.5	15.5	16.5	17.5

Figure 4.5.3a–c. Three stages in the development of the tongue. At TS 19 (a), mesodermally-derived swellings form in the floor of the oropharynx, and are covered by a layer of endoderm. From the first pharyngeal arch, two large (lateral lingual) swellings form, one on either side of the midline, and are associated with a smaller midline swelling – the median lingual swelling (or the tuberculum impar). The swelling associated with the floor of the second arch is termed the copula, while the hypobranchial eminence is formed from the fusion of third arch- and fourth arch-derived components.

By about TS 20 (b), the two lateral lingual swellings have enlarged and overgrown both the median lingual swelling and the derivative from the second arch; all of these become incorporated into a single entity which subsequently forms the non-muscular tissue of the anterior two-thirds of the tongue. The two halves of this region are separated by a median sulcus superficially, and deep to this is located a median fibrous septum (TS 22). The hypobranchial eminence subsequently forms the non-muscular tissue of the posterior one-third of the tongue, including the tissues at the base (or pharyngeal region) of the tongue; the epiglottis is also believed to be of fourth arch origin. The musculature of the tongue derives from cells of the occipital somites (see text).

The definitive arrangement is illustrated in (c, TS 22): the region of the anterior two-thirds of the tongue constitutes its *oral* part, while the component of the tongue derived from the hypobranchial eminence constitutes its *pharyngeal* part. The terminal sulcus is located at the junction between the anterior two-thirds, and posterior one-third of the tongue. In the mouse, only a single median circumvallate papilla is present. This contrasts with the situation in the human, where a row of circumvallate papillae form at the junction between the anterior two-thirds and posterior one-third of the tongue, just in front of the terminal sulcus.

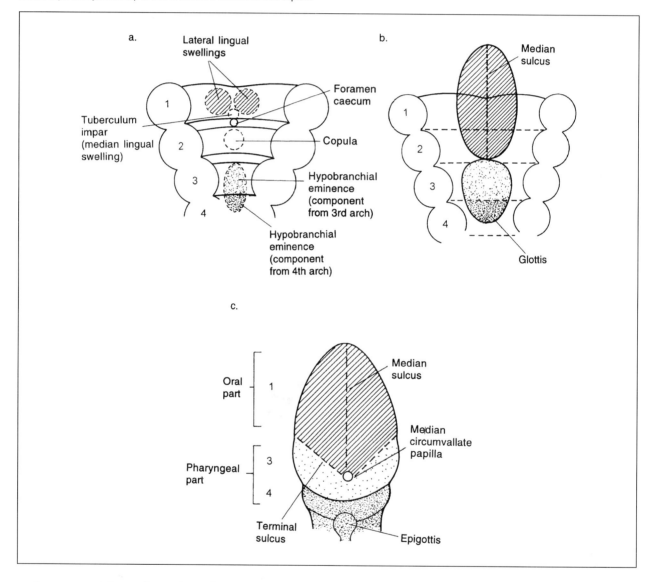

eminence may form from the tuberculum impar. At the most caudal (posterior) part of the root of the tongue, an additional median swelling appears (TS 19/20), and this is the primordium of the **epiglottis** which will guard the entrance to the larynx. As the epiglottis enlarges, it develops a cartilaginous core (TS 22/23) whose base is attached to the inner aspect of the middle part of the thyroid cartilage.

During TS 24, it is possible to recognize both the **intrinsic musculature** of the tongue (with its transverse and vertical components) and the pairs of muscles that comprise the **extrinsic musculature** of the tongue (*i.e.* genioglossus, hyoglossus, styloglossus, chondroglossus (sometimes described as part of the hyoglossus) and palatoglossus), with the genioglossus making up much of the tongue bulk. The principal sites of attachment of these muscles are the hyoid bone and the mandible. This musculature is now thought to be almost exclusively derived from the cells of the **occipital myotome** (3 or possibly 4 units of paraxial mesoderm that are located in the occipital region[6]) that invade and intermingle with the neural-crest-derived and branchial-arch-derived mesoderm. These muscles are all supplied by the hypoglossal (XII) nerve, except the palatoglossus which is more closely associated with the soft palate in both situation and function, and which is accordingly supplied by the glossopharyngeal (IX) nerve.

The thyroid gland

At about TS 13, an endodermal thickening, said to be that of the thyroid primordium, is first seen in the midline of the floor of the pharynx just caudal to the region of the first branchial arch which forms the **tuberculum impar**. Clear evidence of the solid nature of the primordium is, however, only apparent at about TS 17/18 when an endodermally-lined indentation or diverticulum (the **foramen caecum**) is seen in the midline of the dorsum of the tongue at the junction between the combined derivatives of the first and that of the second branchial arches, just caudal to the tuberculum impar. This indentation then deepens and extends caudally to form the proximal (rostral) part of the **thyroglossal duct**, which descends in the tissues in the front of the neck, at the same time usually losing its connection with the floor of the pharynx (TS 18/19).

By TS 21, the thyroid primordium has formed from the walls of the duct, and is a solid midline structure near the anterior surface of the larynx, that soon (TS 22) bifurcates to give two thyroid lobes, each containing large numbers of buds, on either side of the **primitive laryngeal cartilage**. By TS 24, these lobes have expanded considerably, extending caudally from near the posterior parts of the thyroid and cricoid cartilages to the rostral part of the trachea. A narrow **isthmus**, usually in front of the first or second tracheal "ring", connects the lower parts of the two lobes and this isthmus may later descend to the level of the third or fourth tracheal "ring" (about TS 26).[7] At this stage, **colloid-filled follicles** and C- or **parafollicular cells** are disseminated within the thyroid, and these are believed to originate from the ultimobranchial bodies that form from the ventral part of the fourth branchial pouches (see Chapter 3.3 *The branchial arch system*).

Van Heyningen (1961) determined the time of onset of thyroid function in a series of mouse embryos. He noted that the ability to accumulate iodine from the circulation, to bind this iodine organically and to form colloid appears between E15 and E16, whereas the ability to produce thyroxine is temporally associated with the first evidence of follicles and appears between E16 and E17. It thus appears that the development of thyroid function in the mouse takes place between E15 to E17.

It is of interest that two types of thyroid follicles are seen in the mouse, the *typical* variety and a less well-documented second variety. The embryological origin of the various epithelial cell types found in the second type of thyroid follicle was studied by Wollman and Hilfer (1978). In order to determine their origin, ventral pharyngeal outpocketings and the ultimobranchial outpocketings were isolated before they had fused to the thyroid gland (from E12 mouse embryos). These explants were transplanted under the kidney capsule of adult mice and allowed to grow there for several months. The ventral pharyngeal outpocketings only gave rise to typical thyroid follicles, whereas the ultimobranchial outpocketings gave rise to a small number of small-sized follicles believed to be of the second variety. No *typical* thyroid epithelium was found in association with the latter. These authors suggested that their findings indicated that the C-cells in the normal follicles are derived from the ultimobranchial contribution, and that the second type of thyroid follicle was also derived from the same source. The *typical* thyroid epithelium found in the usual follicles was largely, or possibly entirely, of ventral pharyngeal origin.

The relationship between the thyroid and the

---

[6] Although the explanation for the nerve supply to the different regions of the tongue is indirect, being from an analysis of the cranial nerve supply to these regions, considerable doubt now exists as to whether the extrinsic musculature of the tongue originates from this source (see Bates, 1948, Deuchar, 1958). It is now believed, as was once considered to be the case, that these muscles develop *in situ* (Frazer, 1926; Williams, 1995).

[7] There are a number of anatomical variants associated with thyroid development found in the human, but that are also likely to be occasionally present in the mouse. Because the thyroglossal duct "descends" at an early stage of neck development, before the mesodermal condensation that forms the body of the hyoid bone has fully differentiated, a fibrous band of tissue of thyroglossal duct origin may sometimes connect the isthmus of the thyroid gland with the body of the hyoid bone which develops around the thyroglossal duct (see Figure 4.5.4). Part of the caudal end of this duct may also persist to form an accessory or **pyramidal lobe** of the thyroid gland which is attached to the upper part of the isthmus of the thyroid and located along the line of descent of the thyroglossal duct. Occasionally, "rests" of thyroid tissue may develop in the substance of the tongue near the point of initial attachment of the thyroglossal duct at the foramen caecum (termed a "**lingual**" **thyroid**), or along the line of descent of the duct in the form of an isolated **thyroglossal cyst** or multiple cysts. Rarely, the thyroid gland may descend into the superior mediastinum and have a retrosternal location (termed a **retrosternal goitre**). These aberrant sites of thyroid tissue, located along the path of descent of the thyroglossal duct are shown in Figure 4.5.4. The developmental history of the thyroglossal duct and its caudal derivative, the pyramidal process or lobe of the thyroid, was studied in a considerable number of human embryos by Sgalitzer (1941), and the variations in its differentiation are discussed in this paper.

TS	1	2	3	4	5	6	7	8	9	10	11	12	13	14	15	16	17	18	19	20	21	22	23	24	25	26
E	0	1	2	3	4	4.5	5	6	6.5	7	7.5	8	8.5	9	9.5	10	10.5	11	11.5	12	12.5	13.5	14.5	15.5	16.5	17.5

Figure 4.5.4. Diagrammatic view of the tongue and neck region to illustrate the descent of the thyroid tissue (i.e. thyroglossal duct) from the region of the foramen caecum (an indentation on the dorsum of the tongue at the junction between its anterior two-thirds and posterior one-third), to its definitive location with its isthmus across the ventral midline (usually in front of the first tracheal ring) and its two lobes one on either side of the laryngeal cartilages. The two lobes of the gland are usually fairly bulky, and are located on either side of the cricoid cartilage and first tracheal ring; they normally extend rostrally to about the middle of the posterior border of the thyroid cartilage. The two lobes of the thyroid gland may extend caudally slightly further, and cover most of the second tracheal ring.

The line of descent of the thyroid is shown by the dotted arrow in the diagram. While the thyroid gland normally descends caudally only as far as its definitive location, it occasionally descends into the superior mediastinum (and may even be located exclusively within the upper part of the thoracic cavity). If this occurs, it may be located behind the manubrium of the sternum and is then termed a *retrosternal* goitre. Occasionally, thyroid tissue fails to descend, and may be located as a mass in the dorsum of the tongue at the junction between its anterior two-thirds and posterior one-third; it is then termed a *lingual* thyroid. More commonly "rests" of thyroid tissue occur, usually in the form of *thyroglossal cysts* or ectopic thyroid tissue along the path of descent of the thyroglossal duct. Relatively commonly, a *pyramidal lobe* is present, this is a rostral medially directed lobe of thyroid tissue attached to the isthmus of the gland.

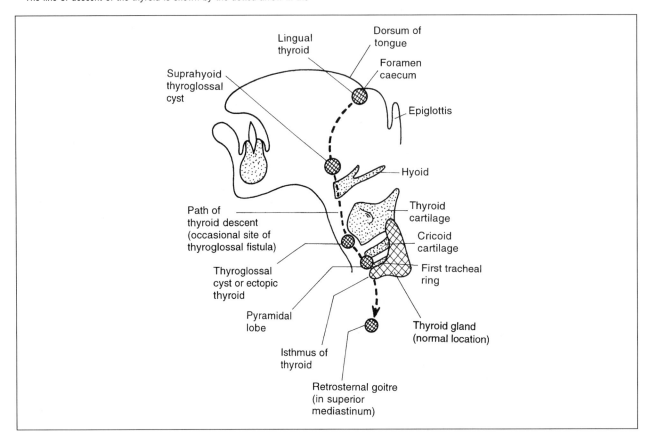

parathyroid and thymus glands and the other derivatives of the pharyngeal (branchial) arch apparatus (*e.g.* the tonsils) are discussed in more detail in Chapter 3.3 *The branchial arch system*.

The sublingual, submandibular and parotid salivary glands

The primordia of the **sublingual/submandibular salivary gland** complexes are each seen (TS 21/22) as a single mass of mesenchyme surrounding a small endodermal bud which extends ventrally from the floor of the mouth on either side of the tongue. As it invades the mesenchyme, inductive interactions between the bud and its surrounding mesenchyme (Bernfield *et al.*, 1984) cause the bud to bifurcate and the branches then start to arborize. Each glandular mass rapidly enlarges and soon (TS 22/23) contains a typical, centrally located, arborizing arrangement of lobules with their associated ducts, each of which drains into a central duct that itself drains into the ventral part of the oral cavity, the whole being surrounded by a mass of apparently undifferentiated mesenchyme (Arey, 1974). The main ducts from the medial and lateral parts of the glandular mass respectively form the definitive **submandibular** and **sublingual ducts**. The paired ducts are clearly seen on coronal histological sections through the region of the base of the tongue, and initially pass directly to the floor of the mouth, but they later elongate (TS 23) as they extend forwards towards the pair of elevated papillae (termed the sublingual caruncles) where the ducts eventually open into the floor of the mouth on either side of the frenulum of the tongue (see above). Lobules bifurcate every 8 h or so, with morphogenesis ceasing at around TS 25/26 (Bernfield *et al.*, 1984). Although the sublingual and submandibular glands are morphologically similar during embryogenesis, the latter is the more medially located, has a far greater volume, and, in the adult, has a darker colour. The **anterior lingual mucous**

glands open via several minute ducts on the inferior surface of the apex of the tongue.

The primordium of the **parotid gland** forms around TS 22/23 and, when fully developed in the mouse, is considerably smaller and flatter than the sublingual and submandibular glands. It lies below the external acoustic meatus, between the mandible and the rostral insertion of the sternocleidomastoid muscle, and lies mostly on the surface of the masseter muscle on the deep aspect of the cheek.[8] While the parotid gland contains only serous alveoli which are involved in the production of its secretions, both the sublingual and submandibular glands are of the so-called "mixed" variety and contain both serous and mucous alveoli, although most of the alveoli in the sublingual and submandibular glands are said to be of the latter variety. The rate of production of these glandular secretions is under both nervous and hormonal control.

The exact embryological origin of the parotid gland is unclear, although most authorities indicate that it arises from the (*ectodermally-derived*) epithelial lining of the mouth on the deep aspect of the cheek. It is initially present in the form of a blind-ending tube (or diverticulum) which maintains a ventral link with the epithelium of the mouth, and grows dorsally into the substance of the cheek. The proximal part of the diverticulum persists as the parotid duct, while the distal blind-ending part proliferates to form the gland. The duct eventually opens inside the cheek opposite the second upper molar tooth. The embryological origin of the other salivary glands, which initially form from solid outgrowths from the buccal epithelium in the mesodermal substance of the floor of the mouth between the tongue and the developing gums, are also of uncertain lineage, but are believed to be *endodermal* in origin.

Accessory salivary glandular structures are scattered throughout the mouth region, large numbers being found on the tongue, in the tonsillar region and particularly in the soft palate. Others, found in the posterior part of the mucous membrane overlying the inferior surface of the hard palate and in the vestibule of the mouth (in the region between the cheeks and the teeth), are mainly mucous secreting.

The lips

These are ectodermally-derived structures whose glandular mucous membrane is continuous with that of the cheeks, gums and part of the floor of the mouth and palate. It is also continuous with the glandular mucous membrane of the nasal cavities and paranasal sinuses. When the maxillary processes fuse with the outer borders of the medial and lateral nasal processes (TS 19/20) a continuous ridge is established from the superficial part of which the upper lip starts to develop at about TS 20. Shortly afterwards (TS 20/21), when the clefts between these various processes diminish and the transverse width of the aperture of the oral cavity decreases, this is associated with the formation of the cheeks.

The lower lip forms from the free edge of the mandibular process by the separation of its superficial portion from the remainder of this process by a groove termed the **labio-gingival sulcus**. Deep to this another groove develops, the **linguo-gingival sulcus** which separates the mandibular portion of the developing tongue mass from the more superficial part of the rest of the mandibular process. The primitive **gum** of the lower jaw forms between the lip externally, and the tongue internally, and within the gum the lower teeth differentiate. While the lateral part of the upper lip forms in a similar manner, but from the *lateral* part of the maxillary process, the origin of its *medial* portion is less clear, but is believed to differentiate from the lower and most medial part of the fronto-nasal process, with both upper and lower lips being formed by TS 21. It has been suggested that there may be a contribution to the deep part of this region from medial extensions of the maxillary processes (Frazer, 1931; Boyd, 1933). The origin of the philtrum is equally unclear, but may be formed from the ingrowth of maxillary mesoderm.

By TS 21/22 the facial features become increasingly recognizable as that of a rodent, with the establishment of the nose, upper and lower lips, cheeks and slightly later (TS 22/23) the eyelids, as well as the primordia of the external ear, the first evidence of which is seen during the early part of TS 21.

---

[8] In the human, the parotid gland is the largest of the salivary glands, and extends superiorly as far as the region overlying the temporomandibular joint, and here the auriculotemporal nerve is embedded either in the gland or in the capsule around it. The relationship of the gland to other structures is of particular importance: the external carotid artery enters the posteromedial surface of the gland and divides into its terminal branches within its substance, the retromandibular vein forms in the upper part of the gland by the amalgamation of the maxillary and superficial temporal veins, but, most importantly, the facial (VII) nerve traverses the gland, and the five divisions or branches of this nerve as well as the parotid duct emerge from its anterior border. The emergence of the 5 branches of the facial nerve in this location was formerly likened to the appearance of a goose's foot (and accordingly termed the *pes anserinus*, even though the goose only has three toes!).

TS	1	2	3	4	5	6	7	8	9	10	11	12	13	14	15	16	17	18	19	20	21	22	23	24	25	26
E	0	1	2	3	4	4.5	5	6	6.5	7	7.5	8	8.5	9	9.5	10	10.5	11	11.5	12	12.5	13.5	14.5	15.5	16.5	17.5

Figure 4.5.5a–e. Diagrammatic sequence representing five stages in the development of a representative incisor tooth, though the same principles apply to the development of the molar dentition (diagrams based principally on Larsen, 1993). (a, TS 21) The ectodermal epithelium of the oral cavity is seen to invaginate to form the *enamel organ* (or dental primordium), while an aggregate of subjacent mesenchyme gives rise to the *dental papilla*. The connection between the epithelium of the oral cavity and the enamel organ elongates (b, TS 22) and eventually degenerates (c) and finally disappears. During the cap stage (b), the *dental lamina* embraces the *dental papilla* like a "cap", and this will differentiate to form the various components of the definitive tooth, with the mesenchyme that surrounds the dental papilla condensing to form the *dental sac*. As the dental papilla gradually invaginates into the dental lamina, the former develops into the *pulp of the tooth*, and this constitutes the "bell" stage. At the late bell stage (c), the first evidence of vascular and neural invasion into the dental pulp is seen. The inner enamel epithelium differentiates to form a layer of *ameloblasts* that produce *enamel*, while the outermost layer of the dental papilla differentiates to form the *odontoblasts* which produce *predentine*, which later calcifies at its periphery to produce *dentine*, which will later form the bulk of the definitive tooth. Shortly after birth, the tooth erupts (d) but is still briefly covered by a layer of gum epithelium. The pulp becomes increasingly invaded by vascular tissue, while the root of the tooth becomes embedded in alveolar bone. The inner layer of cells of the dental sac differentiates into *cementoblasts* which secrete *cementum* which covers the dentine of the root. The outermost cells in this region are involved in the formation of the *periodontal ligament* which "fixes" the tooth into its bony socket. In the definitive arrangement (e) the tooth erupts through its outer covering of gum (or gingival) tissue, thus exposing the enamel layer of the tooth, subjacent to which is the dentine. The *cemento-enamel junction* is seen to be at the boundary between the anatomical crown and root of the tooth.

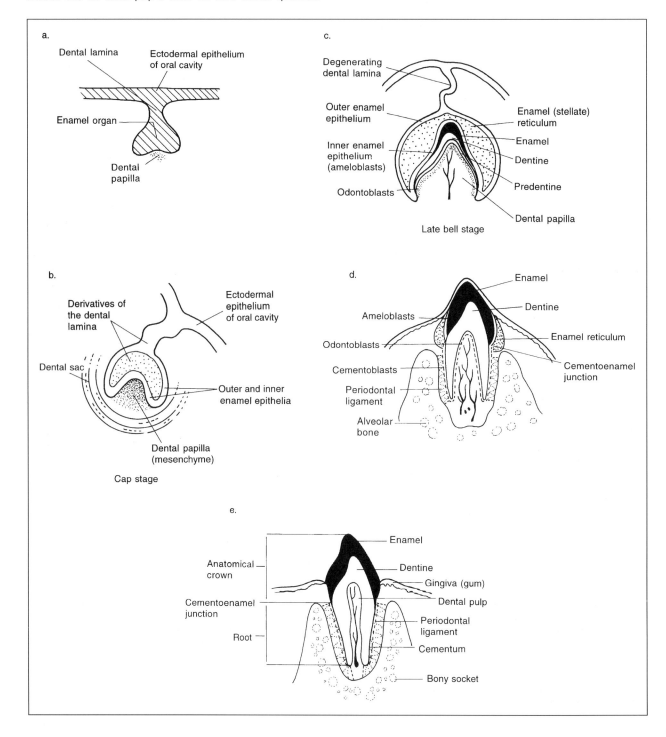

The vibrissae

The vibrissal (tactile or sinus) hairs are of a specially modified type, and are the first of the hair rudiments to appear, initially as surface elevations in the maxillary region during TS 20, while by TS 21 it is evident that they are arranged in a series of about 6 horizontal rows on each side of the future upper lip and nose [in the rat they are distributed in 8–10 horizontal rows (Hebel and Stromberg, 1986)]. Their size decreases from dorsally to ventrally, and, but restricted to the upper rows, from caudally to rostrally. Such hairs are also found in other locations on the face, such as on the lower lip and chin; one prominent pair is found above the upper eyelid, one pair is located in the infraorbital region overlying the caudal part of the masseter muscle and another pair is located just behind the lateral margin of the mouth overlying the buccinator muscle. Their principal function is in facilitating the orientation of the animal, being particularly modified for tactile reception. Those of the upper lip are innervated by branches of the infraorbital nerve which anastomoses with branches of the facial (VII) cranial nerve and innervates the skeletal muscles of the follicle. The sinus hairs in other parts of the facial region are innervated by other branches of the trigeminal (V) and facial nerve via other local nerves.

By TS 22, the primordia of the vibrissae are clearly recognized, with those in regions other than the upper lips often being particularly prominent. The first evidence of penetration of these hair follicles through the surface is observed during TS 23, though at this stage the majority are at an early stage of their eruption. A day later, however, during TS 24/25, nearly all have erupted, the base of each follicle being surrounded by a mound of cellular debris. The total number of vibrissae associated with the upper lips is in the region of 100 at the time of birth.

Hair follicles elsewhere are much smaller in size than the vibrissae, and are first recognized during TS 22, but more clearly delineated during TS 23/24. They do not, however, erupt until at or shortly after birth. This extensive crop of fine fetal hairs, or **lanugo**, covers the entire body, but the newborn mouse apart from the vibrissae lacks normal hairs, and it takes some weeks before it achieves an obvious covering of hair.

The teeth

The teeth form as composite structures from the oral epithelium and from neural-crest-derived mesenchyme associated with the **mandibular** and **maxillary bone rudiments** of the jaw (see Chapter 3.3 *The branchial arch system*). Each tooth primordium derives (early TS 21) from an infolding of the ectodermally-derived, stratified, surface epithelium of the gum region of the mouth. Although the primordia initially have a fairly homogeneous morphology, they soon (late TS 21) develop an inner, less dense core bounded by an outer epithelial layer (the whole unit being termed the **dental lamina**) which is almost completely surrounded in its basal region by a less clearly defined layer of neural-crest-derived mesenchyme (the **dental papilla**), with the dental lamina and dental papilla composite forming the **tooth germ** (or **bud**). The connection between the epithelium of the oral cavity and the enamel organ elongates, eventually degenerates (Figure 4.5.5c) and finally disappears.[9]

As the lamina starts to surround the papilla, the tooth rapidly goes through the **cap stage** which becomes the **bell stage** when the dental lamina surrounds the dental papilla that will form the central core of the tooth. In due course (TS 25/26), the outer layer of cells of the dental papilla differentiates to form the **odontoblast cells** that will initially produce predentine and subsequently dentine, while the dental lamina differentiates to form the **enamel organ** that will produce enamel. This organ has three layers, an inner enamel epithelium overlying the dental papilla, a **middle enamel epithelium** or **stellate reticulum**, and an **outer enamel epithelium** (this complex sequence of events is illustrated diagrammatically in Figures 4.5.5a–e, with further details of tooth morphogenesis being given in the legend).

By TS 22, the primordia of both the upper and lower first molar and incisor teeth are clearly delineated. The stellate reticulum which is formed by the downgrowth of the oral epithelium and which partially surrounds the dental papilla does not, however, develop until about TS 24 when centres of ossification first appear within the maxillae and mandible[10] where they surround the base of the **incisor** and **molar tooth** primordia. By TS 26, both the first and second upper molar tooth primordia are present, with second molar development

---

[9] In the human, an additional bud from this connection and represents the primordium of the permanent tooth, a feature not present in the mouse, where only a single dentition forms.

[10] The mandible, although an essential component of the mouth region, is considered in Chapter 4.8 *Development of the skull*, while the development of Meckel's cartilage, around and principally lateral to which the mandible forms, is also considered in Chapter 3.3 *The branchial arch system*.

TS	1	2	3	4	5	6	7	8	9	10	11	12	13	14	15	16	17	18	19	20	21	22	23	24	25	26
E	0	1	2	3	4	4.5	5	6	6.5	7	7.5	8	8.5	9	9.5	10	10.5	11	11.5	12	12.5	13.5	14.5	15.5	16.5	17.5

being 3–4 days behind that of the first, while the upper third and the lower second and third molars have yet to appear.

The teeth which show the greatest degree of differentiation are the very large upper and lower incisors and these have differentiated to a considerable extent by birth, and are almost at the stage illustrated in Figure 4.5.5d, although they have yet to erupt. At the time of birth, no pulp cavity is present.

In the rat, the incisors erupt on about the 10th day postnatally, the molars on the 19th (first molar), 22nd (second molar), 35th to 40th day (third molar) and by 6 weeks the entire set is in use (Schour and Massler, 1949). The postnatal timing and sequence of the eruption of the dentition in the mouse is believed to be very similar. It should be noted that in the mouse, unlike in the human, only one set of teeth is formed (equivalent to the primary dentition in the human). Equally, and again unlike the situation in the human, the incisors grow continuously throughout life (Addison and Appleton, 1915) and if, for example, the jaws become malaligned, or because the normally apposed tooth fails to erupt, teeth fail to wear down (as normally occurs), and become enormously elongated, so that the animal eventually dies of starvation because its feeding is impaired (Hammett and Justice, 1923).

4.6 The brain and spinal cord

Introduction

The rostral part of the primitive neural tube forms the brain, while the caudal part differentiates into the spinal cord. Soon after the neural tube closes, the neural lumen in the head region forms three linked vesicles, these are the **prosencephalon**, the **mesencephalon** and the **rhombencephalon**, and the tissue surrounding them will form the fore-, mid- and hindbrain (see Figure 4.6.1) from which will emerge the cranial nerves (see Table 4.6.1). The subsequent development of the brain and its associated central nervous system is complex because a great many specialized regions develop in a relatively small volume, as well as, in developmental terms, a relatively short period of time, while its description is made hard by the fact that the nomenclature is complex, with the names of specific regions changing as they differentiate.

There is a further problem with analysing brain development and this is that, as development proceeds, early features disappear and new ones form and it is not always easy to analyse the history or even to identify the individual components into which an initially fairly homogeneous structure partitions. Consider, for example, the segmentation of the central nervous system: although this aspect is later hidden, both morphological and expression data show that the very early brain and, to a lesser extent, the spinal cord are segmented structures, with direct evidence for this being seen in the transitory prosomeres of the forebrain and the rhombomeres of the hindbrain. Later this phenomenon can only be recognized through the 12 cranial nerves of the brain and their associated ganglia, parts of which originate in the neural crest associated with them, while other parts derive from ectodermal placodes.[1] It is noteworthy here that the segmented regions of the spinal cord, each of which

Table 4.6.1 *The origin and principal functions of the cranial nerves*

Brain region	Associated cranial nerves	Principal functions
TELENCEPHALON	Olfactory (I)	Olfaction/smell
DIENCEPHALON	Optic (II)	Vision
MESENCEPHALON	Oculomotor (III)	
METENCEPHALON	Trochlear (IV)	
	Trigeminal (V)	Sensory – head region including anterior two-thirds of tongue
		Motor – muscles of mastication
	Abducens (VI)	Motor supply to extrinsic ocular muscles
	Facial (VII)	Muscles of facial expression
	Vestibulocochlear (VIII)	Hearing/balance
MYELENCEPHALON	Glossopharyngeal (IX)	Sensory to posterior one-third of tongue and pharynx
	Vagus (X)	Innervation of pharynx, larynx, thoracic and abdominal viscera
	Accessory (XI)	Accessory to vagus – supplies muscles concerned with swallowing; also motor to sternocleidomastoid and trapezius muscles
	Hypoglossal (XII)	Motor supply to intrinsic and most extrinsic muscles of the tongue

[1] Evidence from studies on chick embryos suggests that the facial (VII) ganglion, the inferior ganglion of the glossopharyngeal (IX) and inferior ganglion of the vagus (X) nerve partly originate from ectodermal epibranchial placodes which develop at the dorsal ends of the first three branchial clefts.

TS	1	2	3	4	5	6	7	8	9	10	11	12	13	14	15	16	17	18	19	20	21	22	23	24	25	26
E	0	1	2	3	4	4.5	5	6	6.5	7	7.5	8	8.5	9	9.5	10	10.5	11	11.5	12	12.5	13.5	14.5	15.5	16.5	17.5

Figure 4.6.1. The early differentiation of the brain and spinal cord. (a, TS 13) The formation of the three primary brain vesicles, the *prosencephalon* (or primitive forebrain), *mesencephalon* (or primitive midbrain) and *rhombencephalon* (or primitive hindbrain). (b, TS14) The rostral part of the primitive forebrain (known as the *telencephalon*) forms the two *telencephalic* (or future cerebral) *hemispheres*, while the more caudal part of the forebrain forms the *diencephalon* (whose principal derivative is the thalamus). On either side and between these tissues the optic vesicles differentiate. The primitive mesencephalon consolidates to form the midbrain, and its luminal diameter correspondingly decreases. The rostral part of the primitive hindbrain forms the *metencephalon* (whose basal plate and alar plate regions gives rise to the *pons* and the *cerebellum*, respectively), and caudal part forms the *myelencephalon* (which becomes the *medulla oblongata*). (c, ~TS 18) The neural tube along the embryonic axis now differentiates into the spinal cord and the cavities of the forebrain, midbrain and hindbrain also evolve. Those within the telencephalic vesicles become *the lateral ventricles* and are linked to what has become the *third ventricle*, with the connections between the lateral and third ventricles being termed the *interventricular foramina*. The cavity of the midbrain becomes dramatically reduced in diameter, and is now termed the *cerebral aqueduct*, while the cavity within the hindbrain becomes the *fourth ventricle*. The neural lumen also narrows and is now termed the *spinal canal*. The entire ventricular system of the brain and the spinal canal is now filled with cerebrospinal fluid which is formed by the choroid plexus, mainly that of the lateral and fourth ventricles. This circulates within and around the brain and spinal cord providing it with nutrients and at the same time acting as a buffering system to reduce the effect of external trauma.

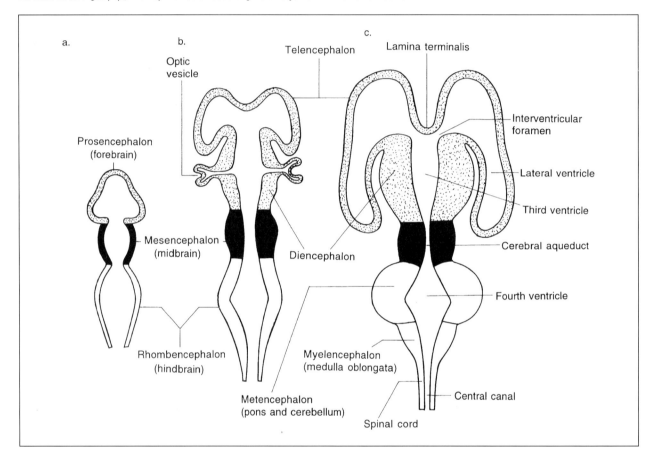

is associated with a spinal nerve and a neural-crest-derived dorsal root ganglion may, on the basis of *hox* codings and other data, reflect mesodermal rather than neural tube segmentation (for review, see Gilbert, 1997).

Another important example of this difficulty derives from the layering of the neuronal tissue: at a relatively early stage of neural differentiation (~TS 18), the primitive neural tube becomes divided into three concentric layers: an *inner* ependymal layer, an *intermediate* mantle layer and an *outer* marginal layer. This basic arrangement extends along the length of the neural axis (the primitive spinal cord) and is still recognizable even in the adult, up to the rostral extent of the hindbrain. In the midbrain and forebrain, however, neural differentiation, but particularly cell migration from one part of the brain to another rapidly disguises the basic arrangement so that the mature tissue geometry hides the original developmental processes.

A further example of the difficulty encountered here lies in recognizing many of the *functional* components of the brain. In certain very localized areas of the brain, large numbers of cell bodies with a similar function aggregate together to form specific "nuclei".[2] In other regions of the central nervous system, further aggregations of axonal processes may be observed, many of whose individual components become covered (during the postnatal period) by a myelin sheath, to form peripheral nerves or nerve trunks. Within the brain, in particular, this regionalization is very marked, with "grey" areas forming

---

[2] When similar aggregates of cells form outside the central nervous system, they are usually termed "ganglia", and tend to be arranged segmentally, as in the cranial part of the neural axis (to form *cranial* ganglia) or along the spinal part (to form *spinal* ganglia).

where the cell bodies are localized, while the "white matter" mainly consists of myelinated cell processes. This compartmentalization between grey and white areas is less easily seen if sections of the brain and spinal cord and components of the peripheral nervous system are stained with haematoxylin and eosin, but is more obvious when these tissues are stained either with silver (which specifically stains the cell bodies and axons) or other stains which specifically stain the myelin sheath material. Indeed, when haematoxylin and eosin stain is used, it is often difficult to see peripheral nerve trunks, and usually imposible to recognize small peripheral nerves and nerve fibres.

To provide the reader with some background to appreciate the analysis of brain development, this introduction is followed by a brief summary describing how the neural tube is partitioned along its rostral–caudal axis and into concentric layers to give separate components, and their distinct functions are briefly mentioned. The next and main part of this chapter contains fairly detailed sections[3] describing the early development of the neural tube as a whole and, following partition, of each of the three main regions of the brain, again pointing out the functions of the various regions as they become apparent. Because the development of the spinal cord goes hand in hand with that of the brain, we mainly integrate the description of the cord development with that of the brain, particularly the hindbrain, to avoid duplication of text. There is, however, a brief section on some specific aspects of the development of the spinal cord towards the latter part of the chapter which ends with a brief appendix describing the development of the pituitary gland.

The basic subdivisions of the central nervous system

The earliest appearance of what can be considered brain tissue is the formation of the two **neural folds** in the cephalic region (the headfolds) of the presomite, so-called headfold-stage embryos (TS 11), but it is only during the early part of TS 14 that they close over the cephalic region to demarcate the most rostral part of the neural tube that is destined to form the brain. The first evidence of the three sub-divisions of the primitive brain is the partitioning at TS 15 of the neural lumen of this region into three linked vesicles (those of the prosencephalon, mesencephalon and rhombencephalon). At TS 15/16, the forebrain vesicle forms two new **telencephalic vesicles** from its antero-lateral regions whose (lateral) ventricles are linked through the **interventricular foramina** to the lumen of the remainder of the prosencephalic vesicle, now called the **third ventricle**. At around this stage, the lumen of the mesencephalic vesicle becomes the **cerebral aqueduct**, while that of the rhombencephalic vesicle is renamed the **fourth ventricle** of the hindbrain, there being continuity between all these ventricles and the **lumen** (or **central canal**) of the **spinal cord**.

It should also be mentioned that there is head paraxial mesoderm lateral to the developing cranial neural tube and it is now generally accepted that this forms a series of 7 segmental units termed somitomeres. These are associated with the caudal part of the forebrain, the midbrain and rostral part of the hindbrain. More caudally, four additional mesodermal units are associated with the caudal half of the hindbrain, and the four segmental spinal nerves from the myelencephalon associated with these **occipital somites** form the rootlets of origin of the hypoglossal (XII) cranial nerve (Table 4.6.1).

Forebrain

By about TS 15, the essential geometry just described is in place, but the caudal region of the prosencephalon has already become the **diencephalon** (TS 13), part of which associates with the optic (II) nerve, while the remainder differentiates at TS 20 into the **thalamus, epithalamus** and **hypothalamus** (which, in developmental terms, is directly connected with the posterior part of the pituitary, through the pituitary stalk (or infundibulum), while, as will be seen later, the anterior pituitary is an ectodermal derivative of the roof of the oral cavity, termed Rathke's pouch).

The anterior lateral parts of the prosencephalon become telencephalic tissue which will later form the cerebral cortex, while its most rostral/inferior part has a major olfactory function as it is associated with the olfactory (I) cranial nerves. Since olfaction is much more important than vision in the mouse, a correspondingly large part of the rostral region of the forebrain is subsumed for this key function. In the human brain, the corresponding region is only poorly developed, being represented by the two olfactory bulbs and their central connections. In both species, the olfactory nerves pass centrally from the olfactory epithelium that lines the superior part of the nasal cavity, through the cribriform plate of the ethmoid bone ("cribriform", because it is like a "pepper-pot",

---

[3] The authors regret the duplication of information here, but have tried to make successive sections complete, albeit at increasing levels of detail.

TS	1	2	3	4	5	6	7	8	9	10	11	12	13	14	15	16	17	18	19	20	21	22	23	24	25	26
E	0	1	2	3	4	4.5	5	6	6.5	7	7.5	8	8.5	9	9.5	10	10.5	11	11.5	12	12.5	13.5	14.5	15.5	16.5	17.5

THE ANATOMICAL BASIS OF MOUSE DEVELOPMENT

The forebrain and pituitary.

THE BRAIN AND SPINAL CORD

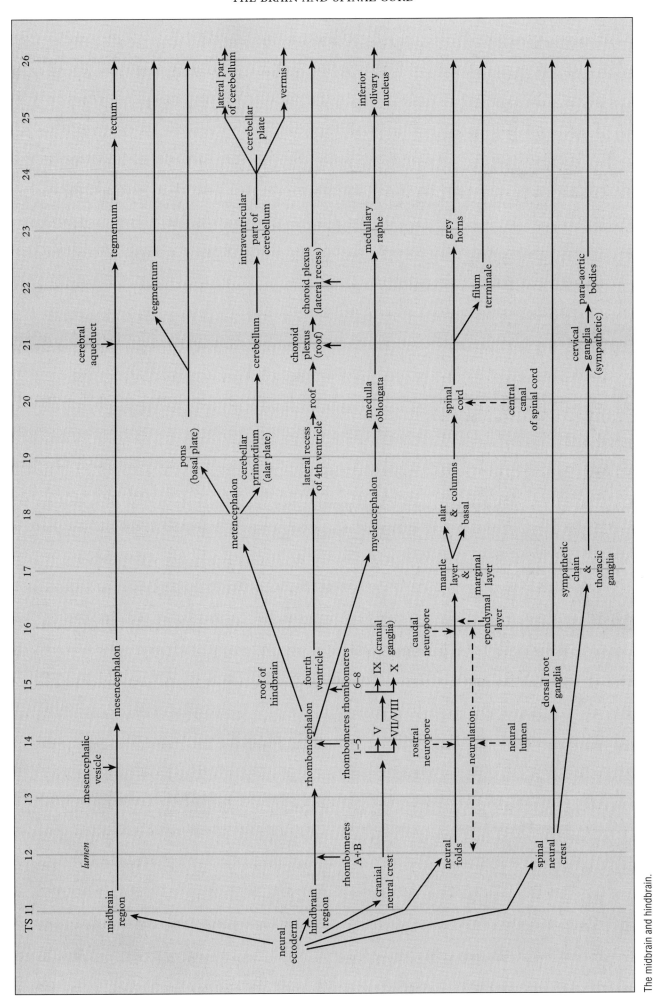

The midbrain and hindbrain.

with numerous small holes through which the olfactory nerves pass upwards) to the olfactory cortex (in the human, the olfactory bulbs), see Chapter 4.5 *The mouth and nose region*.

Midbrain

The midbrain (**mesencephalon**), the smallest region of the primitive brain, links the diencephalon of the forebrain with the **pons** of the hindbrain (its most rostral part) and its role seems limited to processing signals from the **oculomotor** (III), **trochlear** (IV) and **trigeminal** (V) nerves (see Figure 4.6.1). The mesencephalic vesicle gradually reduces in size (TS 24–26) and eventually narrows to become the **cerebral aqueduct** which links the third ventricle of the forebrain (the derivative of the neural canal in the diencephalic region of the prosencephalon) with the fourth ventricle of the hindbrain (the derivative of the neural canal in the region of the rhombencephalon).

Hindbrain

The rostral parts of the hindbrain (**metencephalon**) dorsal and ventral to the fourth ventricle respectively differentiate into the **cerebellum** (TS 21; this enlarges and eventually overgrows the roof and both sides of the rostral half of the rhombencephalon) and the **pons** (which forms the floor of the rostral half of the rhombencephalon) while its caudal region (or **myelencephalon**) differentiates to form the **medulla oblongata** (TS 21). These tissues, together with that of the midbrain, are collectively known as the brainstem and they control such essential functions as the rate of the heartbeat, breathing and involuntary gastrointestinal properties. The mature pons contains the nuclei of the **abducens** (VI), **facial** (VII) and **vestibulocochlear** (VIII) nerves (Table 4.6.1). It also attaches to and links with the cerebellar hemispheres through the **middle cerebellar peduncles**, and these links are important in maximising the efficiency of motor activities.

The **cerebellum** is connected to the midbrain and pons by the **superior** and **inferior cerebellar peduncles**, as well as the middle ones (see above). Although the cerebellum has a substantial sensory input, it is principally a motor part of the brain, involved in the maintenance of equilibrium and co-ordination of muscle action. Its three divisions, named according to their evolutionary/phylogenetic origin, are the **archicerebellum** (or **vestibulocerebellum**) which is concerned with the maintenance of equilibrium, the **paleocerebellum** (or **spinocerebellum**) which helps control muscle movements (*e.g.* in postural changes and locomotion), and the **neocerebellum** (or **pontocerebellum**) which is involved in the control of fine movements, particularly those based on learning. These divisions cannot, however, be resolved into morphologically distinct components before birth.

The medulla oblongata links the pons with the spinal cord and is associated with cranial nerves IX–XII. Its other functions are as an important additional source of cerebrospinal fluid (CSF); this is made within the choroid plexus of the roof and lateral recesses of the fourth ventricle and from the third ventricle as well as the medial surface of the lateral ventricles, although the choroid plexus in the lateral ventricles is the largest and most important source of CSF (see later). Its other functions are as a tract for long dorsal column (ascending) fibres carrying sensory information principally from the surface of the body on their way to the cerebral cortex, the thalamus or cerebellum, and for descending fibres originating in the midbrain and particularly the cerebral cortex to peripheral structures (*e.g.* muscles) for the fine control of movement.

The spinal cord

The rostral part of the cord is of course continuous with the caudal part of the hindbrain (and its lumen, the spinal canal, is in continuity with the fourth ventricle) so that there are clear structural similarities between the two. Segmented spinal nerves start to form at TS 16/17, while the trunks of the peripheral nervous system are also first observable at TS 17. The concentric layering of the neural tube into the ependymal, mantle and marginal layers and the appearance of the alar and basal columns occurs at around TS 17/18, while the the floor- and roof-plate appear at TS 19. After these developments have taken place (TS 19/20), the neural tube caudal to the cervical flexure is then, by convention, termed the spinal cord.

The layering of the neural tube

At about TS 16, the whole length of the neuroepithelium (or neural ectoderm) of the neural tube that surrounds the central lumen starts to differentiate into three distinct layers: an outer **marginal** (TS 17), an intermediate **mantle** (TS 17) and an inner **ependymal** layer (TS 16), the last of these being the first to differentiate, and forms the lining of the lumen of the spinal cord and ventricles of the brain. In the spinal cord and hind brain, the marginal and mantle layers, respectively, form the anuclear white (outer) and nucleated grey (intermediate) matter. The definitive arrangement is slightly different in the fore- and midbrain: while the tissues originate in the same way, the migration of mantle layer neuroblasts toward the periphery of the brain (from about TS 22) means that the "grey" matter comprises the *outer* and the "white" matter the *intermediate* layers of the brain.

The primitive brainstem and spinal cord both initially contain readily recognizable basal and alar columns whose nuclei are located in the grey (*i.e.* mantle) matter, these two columns being separated by the **sulcus limitans**. The basal columns are involved with *motor* functions, and hence are associated with the ventrally located outgoing (or efferent) nerve fibres, while the alar columns receive incoming (or afferent) dorsally-located *sensory* inputs via the dorsal root ganglia. A diagrammatic sequence illustrating how the basic (*i.e.* primitive) arrangement seen in the neural tube develops to form the definitive arrangement seen in the spinal cord is shown in Figure 4.6.2.

The basic organization seen in the spinal cord is also found in the hindbrain where its primitive and definitive subdivisions (into the alar and basal derivatives) although much modified are nevertheless still readily recognizable, and are illustrated diagrammatically in Figures 4.6.3a,b, respectively.

At about TS 21 (E13), the brain and spinal cord become covered by the three protective membranous layers known as the meninges (see below for detailed description). The innermost layer is the **pia mater** and this is intimately attached to the surface of the brain and spinal cord to such an extent that it is not readily separated from it. It forms the inner boundary of the **subarachnoid space**, while the intermediate **arachnoid meningeal layer** (arachnoid mater) forms its outer boundary and this is, in turn, surrounded by the thick, externally located **dural layer** (dura mater).

Figure 4.6.2. The differentiation of the neural tube to form the definitive spinal cord. (a, TS15) Soon after neurulation, neural crest cells migrate laterally (arrows). The neural tube soon subdivides into a dorsal and a ventral component, the former (or *alar plate*) will become the sensory half of the grey matter, while the latter (or *basal plate*) will become the motor component of the grey matter, the two being separated by the *sulcus limitans*. (b, TS 17) A day later, the neural tube has developed three layers: an inner *ependymal,* a central *mantle* and an outer *marginal* layer and its lumen is then termed the *central* (or *spinal*) *canal*. (c, TS 19) The outline of what is now the early spinal cord changes, with the basal half tending to be wider than the alar half, while the ependymal layer is particularly thin in the regions of the *roof plate* and *floor plate*. (d, TS 25) The definitive arrangement has now nearly been achieved: the relative cross-sectional area of the spinal cord compared to that of the embryo tends to diminish proportionately at each level of the cord, as does the diameter of the lumen of the spinal canal. (e, TS 26) The *definitive* arrangement shown in a transverse section through the thoracic region. The morphology is much as in (d), but additional features are shown: note the presence of *motor neuron cell bodies* in the ventral horn of the grey matter of the spinal cord, and it is from these cells that axons emerge, passing into the ventral root towards the mixed (segmental) spinal nerve. The *dorsal root ganglion* is associated with the dorsal root which carries sensory fibres (axons) from the periphery towards the dorsal horn of grey matter of the spinal cord where they contact the processes of *internuncial neurons* (for example, in the simple reflex arc), though they are particularly prevalent in the dorsal horn. The peripheral margin of the spinal cord consists of white matter due to the presence of nerve fibre bundles within it. The small *intermediate* (or *lateral*) *horn*, containing sympathetic efferent neurons, is only present in the thoracic and upper lumbar regions of the spinal cord. The white matter in the dorsal region of the spinal cord is subdivided into two columns by a *dorsal median septum*, while the ventral part of the spinal cord is indented by a *ventral fissure*. This basic arrangement of the spinal cord into grey and white matter, with dorsal and ventral horns each of which serve quite different functions, and which are separated by a sulcus limitans, is extended rostrally into the hindbrain region (For details of these structures in the hindbrain, see Figure 4.6.3a,b.)

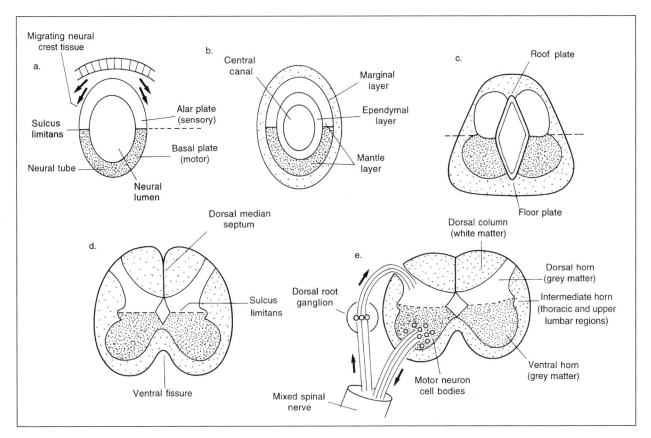

Figure 4.6.3. The organization of the hindbrain in the region of the medulla oblongata. This diagram shows that the basic arrangement seen in the spinal cord can be recognized in the brainstem. (a) Note that the spinal canal of the hindbrain (the *fourth ventricle*) opens out, giving this region of the brain its characteristically thin roof. The dorsal horn (or alar plate) and the ventral horn (or basal plate) are respectively associated with the sensory tracts and motor tracts of the cranial nerves. A small portion of the alar plate becomes dissociated from the rest, and migrates ventrally to form the *olivary nucleus*, but still retains its sensory function. The white matter in the spinal cord periphery becomes exclusively laterally and ventrally located in the brainstem. (b) The location of the sensory and motor tracts of all of the cranial nerves (except the olfactory (I) nerve). Those with *sensory* components are located in one of the three tracts formed by the subdivision of the alar plate. Some are exclusively associated with the perception of general stimuli from the body surface (II, V, VIII) including those of vision, touch (in the facial region) and hearing, while others are associated with taste perception (VII, IX, X) and sensory stimuli from the gastro-intestinal tract (X). The basal plate is also subdivided into three tracts, each of which has exclusively *motor* functions. The most dorsal is associated with cranial nerves that are motor to the pharyngeal (or branchial) arches (V, VII, X). The intermediate tract is associated with autonomic functions of the respiratory and intestinal tracts and the heart (IX, X, XI), while the most ventral of the tracts contains those cranial nerves associated with the control of the somite-derived muscles, such as the extrinsic ocular muscles (III, IV, VI) and the tongue (XII).

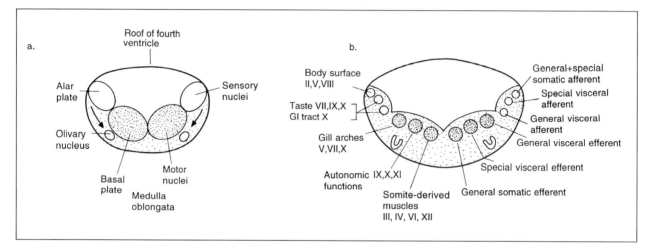

Early neural tube development (TS 11–16)

By TS 12, the two **neural folds** in the cephalic region (the headfolds) that had appeared in presomite embryos (TS 11) have enlarged and elevated, and they soon become closely apposed across the midline in the cephalic region, with the region overlying the forebrain–midbrain junction being the first to fuse (TS 12/13). This is followed very shortly afterwards by a restricted site of fusion in the most inferior/ventral part of the prospective forebrain. Slightly later, the neural folds overlying the anterior part of the prospective forebrain close in a zipper-like process until the forebrain vesicle (as well as the optic vesicles) has completely formed.[4] At about this time, the neural folds overlying the midbrain start to become apposed across the midline and fuse; the closure process from this site extends caudally, and meets the cervical neural folds that are closing rostrally (the first region along the neural axis where apposition and then fusion of the neural folds occurs is in the future mid-cervical region of the cord). When these two domains of fusion meet, the hindbrain vesicle forms (for further details, see Kaufman, 1979; Jacobson and Tam, 1982). Fusion then occurs caudally, again in a zipper-like fashion, with the **caudal neuropore** closing at about TS 16.

There may be some variability in this closure process; Geelan and Langman (1977) report that the first evidence of closure occurs in the cervical region, and that this proceeds rostrally towards the rhombencephalon until it reaches a site just caudal to the otic pits. Shortly afterwards, the neural folds in the prosencephalic region fuse, and that from this site closure occurs caudally in the direction of the mesencephalon until it reaches the rostral part of the rhombencephalon. Fusion then proceeds between these two independent sites of closure.

Although the fore-, mid- and hindbrain regions form at about the same time (TS 14), the first clear evidence of boundaries between them is the separation of the lumen of the cranial region of the neural tube into the three linked vesicles (those of the prosencephalon, mesencephalon and rhombencephalon) and, although there are considerable changes in morphology, this essential topology then remains unaltered. The forebrain vesicle thus becomes the third ventricle when, at TS 15/16, its antero-lateral regions form the two **telencephalic vesicles** (whose lumens are later called the **lateral ventricles** of the cerebral hemispheres) that are linked through the **interventricular foramina** to the **third ventricle** which is itself connected through the **cerebral aqueduct** (the diminished vesicle of the midbrain) to the **fourth ventricle** of the hindbrain and so to the

---

[4] No site in the mouse exactly corresponds to the rostral neuropore of the human embryo (which almost exactly overlies the most ventral part of the prospective forebrain) since the pattern of closure of the cephalic neural folds is quite different (the exact pattern of neural fold closure in the cephalic region varies quite considerably even between closely related species).

lumen (or **central canal**) of the **spinal cord** (see Figure 4.6.1a–c).

The primitive brain region, or rostral extremity of the neural axis, becomes demarcated from the future spinal cord, the derivative of the caudal part of the neural axis at the **cervical flexure** (TS 18), and subsequently the forebrain and midbrain regions of the primitive brain become demarcated from the primitive hindbrain by the **midbrain flexure** (TS 14). A little later, the hindbrain region is indented by the additional **pontine flexure** (TS 15). The changes in the configuration of the primitive brain are illustrated diagrammatically in Figure 4.6.4a–c where it can be seen that the particularly prominent trigeminal (V) and facio-acoustic (VII/VIII) ganglionic complexes, are associated with the primitive hindbrain, and that the largest components of the primitive brain are the derivatives of the diencephalon and telencephalon.

While the early brain is undergoing its profound morphological changes, the neural tube caudal to the cervical flexure is developing less dramatically and, during these early stages, the major events taking place are the emigration of neural crest cells and the initial emigration of the segmental spinal nerves (TS 16).

The forebrain

External morphology

Fusion of the cephalic neural folds at around TS 13/14 leads to the formation of the forebrain (prosencephalic) vesicle with its **optic vesicles** and their **stalks** which are symmetrical postero-laterally-directed evaginations of this vesicle (see Chapter 4.7 *Sensory organs – the eye*). Prior to the fusion of the forebrain neural folds, the **optic sulci** (TS 12), a pair of grooves directed postero-medially towards the diencephalic part of the prosencephalon (the future third ventricle), meet at the site of the future **optic chiasma** (see later). Initially, the rostral and caudal parts of the primitive forebrain are relatively flattened, with the central part being wider than the rest due to the presence of the optic evaginations. This initial shape may well reflect the topography of the closure of the neural tube (the process of neurulation, and the cellular shape changes involved are probably mediated via alterations in the configuration of their intracellular cytoskeletal elements). Later shape changes are mainly brought about by differential growth in the cephalic part of the neural axis, although the increased pressure that builds up in the closed rostral part of the neural axis (as a consequence of neural luminal occlusion, see Kaufman, 1983c, 1986) may also play a role here (see Coulombre, 1956; Coulombre and Coulombre, 1958b; Desmond and Jacobson, 1977).

The telencephalic vesicles start to form during TS 15 and, by TS 16, have enlarged and become separated in the ventral midline by the **lamina terminalis**. During TS 16–19/20, the diameter of the interventricular foramina becomes much smaller than those of the telencephalic vesicles, so that, by TS 20, they are relatively narrow and slit-like. The two telencephalic vesicles, which were originally inferior and lateral to the third ventricle, expand upwards and enlarge in volume until they cover most of the midbrain region, while the space occupied by the third ventricle correspondingly decreases. This relative increase in the volume of the telencephalic vesicles compared to the diminution in the volume of the third ventricle and mesencephalic vesicle has a dramatic effect on the external appearance of the head region: by TS 21, these derivatives of the forebrain occupy almost half of the volume of the entire cephalic region.

Two recent sets of information on forebrain morphogenesis that are based on molecular studies should be mentioned here as they complement standard histology. The first concerns segmentation, and gene-expression data have recently demonstrated that the early forebrain is segmented into prosomeres that are analagous to the rhombomeres of the hindbrain (Bulfone *et al.*, 1993). These segments are already present and recognizable in appropriately stained sections at E12 (TS 20), but it is as yet unclear as to when they first form and when their boundaries become too diffuse to be recognized and they are lost. The second observation concerns rostral–caudal organization in the forebrain and, again, patterns of gene expression demonstrate that, as early as E7.5, there are distinct longitudinal regions here that are then separated by a series of transverse borders (Shimamura *et al.*, 1995) that are not apparent from the analysis of standard histological sections and whose role is still unclear.

The olfactory and optical input

The **olfactory placodes** also appear at about TS 15/16, differentiating in the surface ectoderm overlying the ventro-lateral surface of the telencephalic vesicles, with a considerable volume of cephalic mesenchyme initially separating the placode and the subjacent neural ectoderm. The situation is thus different from the events associated with lens formation, where the optic vesicle seems to contact and then induce the overlying ectoderm; it is still unclear whether olfactory differentiation requires an inductive interaction (for observations on the olfactory system, see Chapter 4.5 *The mouth and nose region*).

TS	1	2	3	4	5	6	7	8	9	10	11	12	13	14	15	16	17	18	19	20	21	22	23	24	25	26
E	0	1	2	3	4	4.5	5	6	6.5	7	7.5	8	8.5	9	9.5	10	10.5	11	11.5	12	12.5	13.5	14.5	15.5	16.5	17.5

Figure 4.6.4a–c. The "folding" and development of the cephalic part of the primitive neural tube. (a, TS 14) Soon after the closure of the rostral part of the neural tube, the lumen in the region of the future brain starts to dilate, particularly in the future forebrain region. Two pronounced flexures are soon evident, the *midbrain flexure*, at the junction between the primitive midbrain and hindbrain, and the *cervical flexure* which forms at the junction between the distal part of the future hindbrain and the most rostral part of the future spinal cord. The principal features of note are the optic vesicles, a pair of outgrowths from the diencephalic region of the primitive forebrain, and the two large cranial neural crest aggregations associated with the future hindbrain region, the trigeminal (V) pre-ganglion and combined facio/acoustic (VII/VIII) pre-ganglion complex. (b, TS 15) The dorsal part of the primitive hindbrain region is now indented by the *pontine flexure*, while the telencephalic vesicles have developed as lateral outgrowths of the primitive forebrain, and the lateral part of the optic vesicles have indented, becoming "secondary" optic cups. (c, TS 18) The primitive brain is now partitioned into the forebrain, itself subdivided into the diencephalon from which emerge the two telencephalic vesicles (the future cerebral hemispheres) and the optic cups. The *mesencephalon* differentiates into the midbrain, while the primitive hindbrain becomes subdivided rostrally into the *metencephalon* (which will form the cerebellum (dorsally) and pons (ventrally)), and the *myelencephalon* which differentiates into the medulla oblongata.

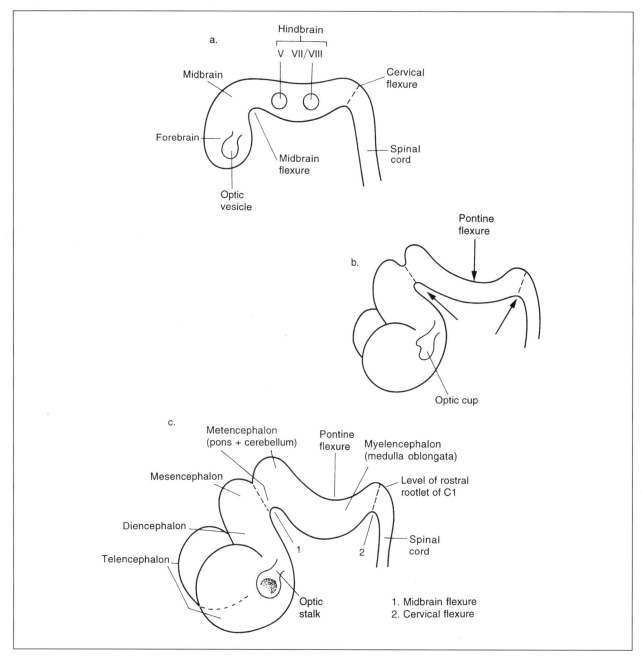

At about TS 20, the primordium of the **vomeronasal organ** (of Jacobson) forms as a paired invagination of nasal mucosa located in the lower anterior portion of the **nasal septum**; this specialized organ is particularly well-developed in the mouse, and contains pheromone receptors essential for such social and reproductive processes as searching for a mate (for further details, see Chapter 4.5 *The mouth and nose region*). It has been suggested that vomeronasal receptors respond to odours in much the same way as olfactory receptors, although possibly over a narrower range of chemical types. The nerve fibres and bundles from this organ converge to form the **vomeronasal nerve** (c. TS 24) which terminates in an *accessory* olfactory bulb situated on the dorsomedial aspect of the main bulb (see Williams, 1975).

During TS 21, the most rostral and ventral part of the telencephalic vesicle that will become the three-layered olfactory cortex (and which is derived from the **paleocortex**) expands forward, coming to overlie the olfactory epithelium in the roof of the primitive nasal cavity. This region of the forebrain represents the future **olfactory lobe** of the brain, and numerous bundles of non-myelinated neurons which constitute the **olfactory** (I) **nerves** pierce the nasal capsule (which in this region will form the **cribriform plate of the ethmoid bone**) as they pass upwards from the **olfactory epithelium** and terminate in the **olfactory cortex**. It is not until about TS 23, however, that the olfactory region has formed a pair of reasonably circumscribed **olfactory bulbs** that are located directly above the roof of the **nasal cavity** and can be distinguished from the rest of the neopallial cortex by the deep transverse groove that separates them (the olfactory cortex is particularly prominent in the mouse – see above). By the end of TS 21, the lumen of the lateral ventricle is linked to that of the future olfactory lobe, although the cavity within it diminishes in size (TS 26) and eventually disappears due to the growth of the olfactory bulb.

The development of the visual cortex (which is located on the medial aspect of the posterior part of the occipital pole of the **neopallial cortex** (TS 20), associated with the **calcarine fissure**, also called the **geniculo-calcarine** or **striate cortex**, see below) and the optic (II) nerves occur as rather late events in the mouse. The volume of the eyeball increases between TS 22 and TS 23, as does the diameter of the optic nerve as a result of the differentiation of large numbers of ganglion cells within the neural retina which send axonal fibres through the optic nerve to the brain (for further details, see Chapter 4.7 *Sensory organs – the eye*).

Diencephalon derivatives

The posterior, dorsal region of the third ventricle is called the **diencephalon** and its major derivatives (TS 20) are the **thalamus, hypothalamus** and **epithalamus** (whose most important derivative is the **pineal gland**, first visible at about TS 23, see below). The derivatives of the diencephalon regulate a wide variety of behaviours. Parts of the thalamus process all forms of sensory inputs (except olfactory), and transmit them (after some modification) to the sensory cortex where localization occurs, while other parts are concerned with emotion, pain, memory and instinctive behaviour. The hypothalamus is concerned with the autonomic control of cardiovascular activity, respiratory and alimentary functions, and with regulating hormone levels, as well as playing a role in eating behaviour and autonomic emotional responses (through the limbic system). Gonadotrophins from the pituitary gland help to regulate reproductive physiology, while the pineal gland influences circadian and diurnal rhythms. (For further details, the interested reader should refer to a standard text-book of neurophysiology.)

An important structure associated with this region of the brain is the **pituitary gland**. At about TS 16, the **infundibular recess** starts to form from the posterior part of the floor of the third ventricle and, by about TS 20/21, gives rise to the **neurohypophysis** and its associated **infundibulum** or **pituitary stalk**. The neurohypophysis will amalgamate (about TS 23) with the **adenohypophysis** (derived from **Rathke's pouch**, an ectodermally-derived diverticulum which grows upwards from the roof of the oral cavity) to form the anterior part of the pituitary gland (see *Appendix* for details of its morphogenesis).

During TS 20, the left and right components of the thalamus, the largest of the diencephalic derivatives, start to develop within the wall of the dorso-lateral part of the third ventricle, and, by TS 21, almost meet in the midline (this is associated with the collapse of the lumen of the third ventricle in this location). They will later (TS 25/26) contact each other there at the **inter-thalamic adhesion (massa intermedia)**, just below the **anterior commissure** which consists of a large bundle of association fibres (those connecting the paired olfactory bulbs and the symmetrical zones of the frontal and temporal cortex) that cross the midline in the **lamina terminalis**.

The **hypothalamus**, though enlarging, shows little evidence of differentiation at TS 21, and is separated from the thalamus (above it) by a deep groove (the **hypothalamic sulcus**). Equally, the bulging thalamus is separated from the epithalamus (above it) by the **epithalamic sulcus**. The **striatum**,[5] the thicker basal portion of the developing cerebral hemisphere, also undergoes rapid enlargement at this time and will subsequently develop into the **corpus striatum** (TS 20), while the thinner-walled, supra-striatal portion (or **pallium**) forms the primordium of the cerebral cortex. The striatal

---

[5] The term "striate" is used to describe this area as connecting bands of grey matter (*i.e.* myelinated fibre bundles) can be seen on myelin-sheath-stained sections that pass from one nucleus to the other through the cellular masses and the white matter of the **internal capsule**. The function of the corpus striatum has been determined indirectly from the clinical manifestations of people with degenerative disorders involving this region of the brain. They show various forms of involuntary movements which may be either brisk, jerky and purposeless (**choreiform movements**) or slow, sinuous and aimless and especially involve the distal muscles of the limbs (**athetoid movements**).

TS	1	2	3	4	5	6	7	8	9	10	11	12	13	14	15	16	17	18	19	20	21	22	23	24	25	26
E	0	1	2	3	4	4.5	5	6	6.5	7	7.5	8	8.5	9	9.5	10	10.5	11	11.5	12	12.5	13.5	14.5	15.5	16.5	17.5

Figure 4.6.5a, b. The ventricular system of the adult human brain. (a) A lateral view of the left lateral ventricle. Each telencephalic vesicle (Figure 4.6.1c) has expanded laterally (both rostrally and caudally) from the diencephalic part of the primitive prosencephalon to form a future cerebral hemisphere. Its (lateral) ventricle is continuous with the third ventricle (in the diencephalon) through an interventricular foramen (of Monro), and the third ventricle is linked to the fourth ventricle of the hindbrain through the cerebral aqueduct (the derivative of the mesencephalic vesicle). The entire system is lined with ependyma and contains cerebrospinal fluid. The lateral ventricle is roughly C-shaped and subdivided into a central part (the body) and three horns – anterior, posterior and inferior. The body is located in the parietal lobe, from which the anterior horn extends into the frontal lobe, the posterior horn into the occipital lobe and the inferior horn into the temporal lobe of the brain. The lateral view of the third ventricle shown here clearly delineates a number of recesses – optic, infundibular, pineal and suprapineal. Of these, only the optic recess is bilateral, and extends into the lumen of the optic stalk, the rest being midline evaginations from the third ventricle into the pituitary stalk, into the stalk of the pineal body and a diverticulum of the posterior part of the roof of the ventricle. The postero-inferior part of the floor of the third ventricle communicates with the rostral part of the fourth ventricle via the cerebral aqueduct. In this view, only the median aperture of the latter is illustrated, but there are also two lateral apertures situated at the ends of the two lateral recesses (see Figure 4.6.5b). (b) A simplified dorsal (superior) view. The symmetrical arrangement of the lateral ventricles is only readily appreciated from this view, although the location of the interventricular foramina and the cerebral aqueduct is hidden by the body/anterior horn of the lateral ventricle and by the suprapineal recess, respectively.

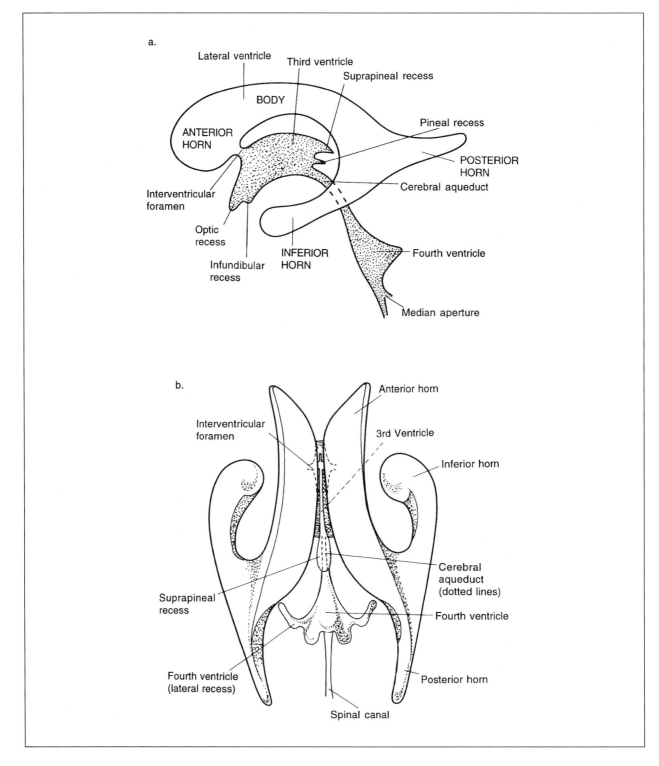

elevations bulge into the floor of the cavity of the lateral ventricles and eventually form discrete masses of grey matter termed the **caudate** (TS 22) and **lentiform** (TS 25) **nuclei** (the latter subdivides into the **putamen** and the **globus pallidus** or **pallidum**). The term striate body or corpus striatum is often used almost synonymously with the basal ganglia, and covers the claustrum, caudate, putamen and globus pallidus.

At about this time, the choroid invagination (or fissure) forms in relation to the roof of the third ventricle (TS 20) and along the length of each lateral ventricle, and is soon (TS 21) associated with the formation of the **choroid plexus**[6] of blood vessels at these sites, although at this stage the choroid plexus in these regions is less florid than that associated with the roof of the fourth ventricle.

The region of brain in the floor of the third ventricle that surrounds the infundibular recess (that was originally the entrance to the lumen of the posterior pituitary) becomes the pituitary stalk (or infundibulum), and the entire pituitary gland at this stage (TS 22/23) is located both posterior and superior to the **optic chiasma** (which lies in the midline close to the inferior surface of the hypothalamus). The optic chiasma is the region where the two optic (II) nerves meet in the midline, and where about 50% of their fibres cross from one side to the other, the partial crossing of these fibres being essential for binocular vision. Distal to the optic chiasma, the *optic tracts* pass backwards, initially to the **lateral geniculate bodies** (or nuclei) of the thalamus before passing in the *optic radiation* to the **visual** (or **geniculo-calcarine**) or **striate cortex**.

During TS 22/23, as both the thalamus and hypothalamus continue to expand, progressively encroaching on the third ventricle, the epithalamus first becomes apparent. The expansion of the three regions of the diencephalon clearly influences the definitive shape of the lateral ventricles which now have a superiorly located "body" with anteriorly and posteriorly directed "horns", although the caudal (inferior) part of the posterior horn is only seen at about TS 25/26. The general configuration of the ventricular system and its various components is illustrated in Figures 4.6.1a–c, and the definitive arrangement shown in Figure 4.6.5a, b.

The **pineal primordium** (or **epiphysis**) forms around a hollow outgrowth in the epithalamic region of the diencephalon (TS 22/23) roof, and initially possesses a small lumen which is continuous with the dorsal part (or roof) of the third ventricle. By TS 24/25, when the wall of the outgrowth has thickened, choroid plexus material associated with the pineal recess extends into the proximal part of its lumen. By TS 26, the pineal gland is directed posteriorly and rostrally, and its walls are relatively thin. Over the next few days, these walls differentiate and eventually obliterate the lumen.

The cerebral cortex

As time proceeds, the walls of the telencephalic vesicles (the future cerebral hemispheres) differentiate. As mentioned earlier, the walls of the primitive brain and spinal cord divide into an inner ependymal, an intermediate mantle and an outer marginal layer at TS 15/16 and, by about TS 17/18, the ependymal layer appears to constitute the majority of the thickness of the walls of the brain and primitive spinal cord. The roof of the telencephalic vesicles starts to stratify as cells from the mantle layer migrate into the overlying marginal zone to form the **neopallial cortex** (TS 20) which will become the outer grey layer of the cerebral hemispheres. Its superficial layer will, in due course, contain axons, dendrites and glia overlying two or more cellular layers and be bounded below by the white matter. The cortex now consists of a superficial and relatively narrow cortical (nuclear) layer separated from the subjacent mantle layer by a narrow anuclear intermediate zone that becomes progressively wider over the next few days. More superficially, beyond the nuclear layer of the neopallial cortex, a further relatively anuclear layer represents the marginal zone. Other neuroblasts and glioblasts will migrate into the intermediate zone which also stratifies.

The two cerebral hemispheres, which were originally laterally-directed outgrowths of the primitive forebrain, differentiate to form the region of the brain where the higher cortical centres are localized. The cerebrospinal fluid-filled lateral ventricles located within the cerebral hemispheres now contain moderate amounts of choroid plexus which arise from their medial walls (~TS 21). It is worth noting that the external appearance of the full-term mouse brain is completely smooth: there are no sulci or gyri (i.e. grooves and bulges) on the surface, and so even the

---

[6] These rudiments of the choroid plexus consist of a layer of ependymal cells covered (in the roof of the fourth ventricle) by an external layer of pia to form the **tela choroidea** whose folds herniate through the ependymal lining of the fourth ventricle. Numerous capillaries, some of which have wide pores (termed "fenestrated"), lie beneath the pia of the choroid, and it is from the blood that passes through these capillaries that cerebrospinal fluid is *actively secreted* rather than produced as a blood filtrate. In the third and lateral ventricles, the choroid plexus herniates through the choroid fissure. The definitive choroid plexus has a convoluted, cauliflower-like appearance. Its surface is covered by a specialized cuboidal monolayer epithelium whose cells have a distinct basement membrane and apical microvilli. Tight junctions between these cells form the *blood–brain barrier* (for further details, see below).

TS	1	2	3	4	5	6	7	8	9	10	11	12	13	14	15	16	17	18	19	20	21	22	23	24	25	26
E	0	1	2	3	4	4.5	5	6	6.5	7	7.5	8	8.5	9	9.5	10	10.5	11	11.5	12	12.5	13.5	14.5	15.5	16.5	17.5

mature mouse brain looks different from the cortical region of the brain of late human fetuses where these surface features are first seen between the 6th and 7th months of gestation.

While the superficial appearance of the mouse brain is much less complex than its human equivalent, at all stages of its differentiation, it nevertheless shares certain common features. The most important of these is the configurational changes that occur during the late pre- and early post-natal period in the mouse (equivalent to mid-gestation in the human) where the cerebral hemispheres grow enormously in size, compared to the rest of the brain, whose volume only increases to a relatively limited extent. Much of this growth is essentially dorsal and caudal, the net result being that the diencephalon and midbrain, as well as the rostral part of the hindbrain, are gradually overgrown, and nearly completely covered. More caudally, what brainstem is not overgrown by the cerebral hemispheres becomes overgrown by the cerebellum which eventually covers the entire roof of the hindbrain.

The meninges[7] and choroid plexus

The meninges are three layers that form from the mesenchyme surrounding the brain and spinal cord. These condense (early TS 21) to form a covering (the primitive meninx), the outermost part of which thickens (TS 21/22) to form the **dura** (or more correctly the **dura mater**[8]) while the innermost layer remains extremely thin, becoming the **pia-arachnoid** (or **leptomeninges**), the two being separated by the narrow **subdural** space. Further spaces then appear within the leptomeninges which are filled by cerebrospinal fluid (CSF) made by the choroid plexus (see above) and these spaces soon coalesce (TS 22/23) to form the **subarachnoid space** which is bounded by the other two layers, the delicate, avascular **arachnoid** (*i.e.* "arachnoid" mater *i.e.* like a spider's web) and the thin, vascular **pia mater**. In the region of the future calvarium (cranium or skull cap), the mesenchyme condenses as two distinct layers which are separated by a narrow gap (early TS 21); while the inner layer forms the **primitive meninx**, the outer layer represents the condensation in which ossification will later occur (TS 22/23), giving rise to the so-called **membrane bones** of the skull.

The main role of the meninges is to supplement the physical protection given to the brain and spinal cord by the skull and vertebral column. The protection provided by the meninges mainly comes from the strength of the dura, which is a dense, firm layer consisting of collagenous connective tissue, and the cushioning effect of the CSF in the subarachnoid space. While the dura is believed to be mesodermal in origin, the pia-arachnoid is thought to be neural crest-derived (see above).

The pia mater contains a fine meshwork of blood vessels that closely adheres to the surface of the brain and spinal cord and is readily seen as early as TS 21. This vascular pia mater becomes invaginated, particularly where it associates with the supero-medial wall of the lateral ventricles and with the roof of the third and fourth ventricles, to form the choroid plexus which produces cerebrospinal fluid. This core of vascularized connective tissue contains many capillaries and is covered by a surface layer of simple cuboidal or low columnar epithelium which derives from the ependymal lining of the ventricles. As the core of connective tissue contains many wide-bore capillaries, the total surface area of the choroid plexus is considerable. The capillaries in the choroid plexus are unusual in that the junctions between the epithelial cells have wide pores and are semi-permeable so that some components of the plasma readily pass into the cerebrospinal fluid by diffusion, while other components need to be actively transported through the epithelial cells. The detailed mechanism of production of cerebrospinal fluid has yet to be fully determined.

The CSF made within the choroid plexus flows through the ventricular system (which is continuous with the spinal lumen), leaving it through the **median** and **lateral apertures** (the so-called foramina of Magendie and Luschka, respectively) of the fourth ventricle. From there, it enters the spinal subarachnoid space, and also passes upwards over the surface of the brain. The CSF then circulates throughout the subarachnoid space, whose width is quite variable, being broadest at the base of the brain

---

[7] The meninges are discussed here in the context of the forebrain, but the text is intended to cover their development throughout the brain and spinal cord. Chick–quail transplantation studies indicate that the meninges of the hindbrain and spinal cord are entirely mesodermally-derived (Le Lièvre, 1976), while earlier experimental studies in *Amblystoma* suggest that the pia-arachnoid is neural crest derived (Harvey and Burr, 1926): this finding was later confirmed in *Rana* and in the chick (Harvey *et al.*, 1933). More recent chick–quail studies (Le Lièvre, unpublished, cited by Le Douarin, 1982) suggest that at the level of the prosencephalon both the dura and pia-arachnoid are crest-derived.

[8] The terminology used here is particularly descriptive, in that "mater" is Latin for "mother", so that "*dura* mater", means "hard mother", while "*pia* mater" means "tender mother" and "leptomeninges" (the pia-arachnoid) means "slender, thin or delicate meninges"; "meninges", is the plural of "meninx", or membrane; while *arachnoid* is from the Greek (arachne) word for a spider, because it resembles a spider's web.

The **subarachnoid space**, which contains the cerebrospinal fluid, is located between the pia and the arachnoid, and varies considerably in width, being least where the pia adheres to the irregular contours of the brain (over the summits of the gyri) and greatest in the region of the major sulci and at the base of the brain and in the lumbo-sacral region of the spinal canal (in the region of the **cauda equina**, formed by the lumbo-sacral nerve roots). The regions where the subarachnoid space contains particularly substantial amounts of cerebrospinal fluid are called **subarachnoid cisterns**. The meningeal layers of the subarachnoid space extend for a short distance around the cranial nerves and the spinal nerve roots.

(the largest of these spaces is termed the **cerebellomedullary cistern** which lies between and behind the cerebellum and medulla oblongata, and this is in continuity with the **pontine** and **interpeduncular cisterns**) and in the lumbo-sacral region of the spinal canal. CSF is principally resorbed by the **arachnoid** (or pacchionian) **granulations** which are located in association with the **superior sagittal dural venous sinus**.

During TS 21, it is also possible to recognize the condensation of dural mesenchyme that will form the **tentorium cerebelli** which intervenes and provides both a support and a baffle between the future occipital lobes of the cerebral cortex and the cerebellum. These mesenchyme condensations, together with the **falx cerebri**, and **falx cerebelli** which also develop at about this time (but in the midline as a vertically-directed dural condensation located between the two cerebral and cerebellar hemispheres) also help prevent excessive movement of the brain that might occur in association with violent and potentially catastrophic movements of the head. By TS 21, the *inner* layer of the dura is well delineated, although most of this layer remains as undifferentiated mesenchyme. Within the *outer* broad layer of mesenchyme, the precursors of the dural venous sinuses (see below) are first seen during TS 23/24.

The dura closely adheres to the inner layer of the **periosteum**, the tough fibrous membrane, that encloses the cranial cavity. The periosteum contains the **meningeal arteries**, the blood vessels which mainly supply the underlying bone. The largest of these vessels, the **middle meningeal artery**, is a branch of the maxillary artery, while smaller meningeal arteries are formed from branches of the ophthalmic, occipital and vertebral arteries. These arteries are accompanied by **meningeal veins**, which arise from the diploic veins located in the marrow cavity (or diploë) of the flat bones of the skull, most of which leave the cranial cavity by the **foramen spinosum** (with the middle meningeal artery) or through the **foramen ovale** and drain into the **pterygoid venous plexus**; others drain directly into cranial venous sinuses.

The veins draining the brain (rather than the periosteum) empty into the **dural venous sinuses** (that form at TS 22) from which blood flows into the **internal jugular veins**. The walls of the venous sinuses consist of dura mater and periosteum, and are lined by endothelium. **Emissary veins** connect the dural venous sinuses with veins outside the cranial cavity, the parietal and mastoid emissary veins being the largest of these channels.

The midbrain

The midbrain or mesencephalon, the smallest of the major subdivisions of the brain, links the diencephalon of the forebrain with the pons of the hindbrain and its role seems limited to processing signals from the oculomotor (III), trochlear (IV) and trigeminal (V) nerves. The roof of the midbrain is known as the **tectum** (and consists of the paired inferior and superior colliculi), while each side is known as a **cerebral peduncle** – these should not be confused with the *cerebellar* peduncles which are described in the section on the *hindbrain*). The major subdivisions of the midbrain are (i) the **tectum**, (ii) the **basis pedunculi** consisting of a dense mass of descending fibres, (iii) the **substantia nigra**, a prominent mass of grey matter located immediately dorsal to the basis pedunculi, and (iv) the remainder of the midbrain, termed the **tegmentum**, which contains fibre tracts, the particularly prominent **red nuclei**[9] and subthalamic nuclei, and a less well organized region of grey matter which surrounds the cerebral aqueduct. (For further details, the interested reader should refer to an appropriate text-book of neuroanatomy.)

The mesencephalic vesicle gradually reduces in size to become the cerebral aqueduct which links the third ventricle of the forebrain with the fourth ventricle of the hindbrain. By about TS 24/25, the most rostral part of this vesicle is relatively narrow in all dimensions (and may justifiably be called the cerebral aqueduct), while the caudal part is still dilated, particularly where it is in continuity with the rostral part of the fourth ventricle. By TS 26, however, the volume of the fourth ventricle and the diameter of the caudal part of the mesencephalic vesicle have both decreased in relative size, with the diminution in the latter mainly being due to the considerable enlargement of the dorsal and lateral parts of the cerebellum. At this stage, the dorsal–ventral diameter of the rostral part of the cerebral aqueduct is greater than its transverse width, with the reverse being the case at the caudal end.

---

[9] This name derives from the nucleus being readily identifiable in the fresh brain by its slightly reddish colour. It has a high content of iron and is covered by a sort of capsule consisting of afferent and efferent fibres. The nucleus has a caudal magnocellular and a rostral parvicellular part, and includes a number of afferent and efferent connections as well as input from the cerebellum and cerebral cortex, while fibres also project from it to the spinal cord. Its function does, however, remain largely unclear. For further details, see Massion (1967).

TS	1	2	3	4	5	6	7	8	9	10	11	12	13	14	15	16	17	18	19	20	21	22	23	24	25	26
E	0	1	2	3	4	4.5	5	6	6.5	7	7.5	8	8.5	9	9.5	10	10.5	11	11.5	12	12.5	13.5	14.5	15.5	16.5	17.5

The hindbrain

The rostral part of the hindbrain (rhombencephalic) vesicle forms at about TS 13/14, soon after the cephalic neural folds have fused down to the level of the otic pits, although most of the hindbrain neural folds do not fuse until TS 14 when its vesicle is then termed the fourth ventricle. This is voluminous and characteristically diamond-shaped, and eventually extends caudally from the narrow channel of the cerebral aqueduct to the equally narrow central canal of the spinal cord. Up to TS 17, however, analysis of conventionally-stained, serially-sectioned material does not readily demonstrate the extent of the fourth ventricle: there is no clear line of demarcation between it and the rostral part of the neural lumen as both have a considerable luminal diameter. Equally, there is no clear line of demarcation between the caudal extent of the hindbrain and the rostral extent of the spinal cord at this stage.

Somewhat later (TS 17/18), the rostral part of the hindbrain starts to differentiate into the **metencephalon**, the ventral (or basal) half of which will form the pons (TS 19), while the dorsal (or alar) half gives rise to the cerebellum. The part of the primitive hindbrain caudal to a transverse line across the widest part of the fourth ventricle, between its two lateral extremities, forms the **myelencephalon** (TS 18) and this soon differentiates into the **medulla oblongata** (TS 20). The ependymal, mantle and marginal zones, visible in sections of hindbrain from about TS 16, are well defined in this region by about TS 19/20, and are clearly delineated from about TS 21 onwards. From about TS 17/18 until TS 19, a large part of the floor of the hindbrain is formed from the mantle layer. The marginal layer, though relatively thin at this stage, is also more noticeable in the hindbrain than elsewhere in the cephalic region.

The first clearly defined landmarks to form in the hindbrain are **rhombomeres A and B**. They first appear at about TS 12, and then segment, so that by about TS 14, the first six (and most obvious) rhombomeres have formed as reasonably well-defined segmental units in the central (i.e. medial) half of the walls of the rostral part of the metencephalon, almost reaching to the midbrain, and extending caudally to about the level of the otic vesicles. More caudally still, there are two additional rhombomeres that are less well-defined and more extensive than their more rostral counterparts; their boundaries are particularly difficult to discern at this stage, as the caudal part of the hindbrain merges into the upper cervical region of the spinal cord. Rhombomere morphology is lost by TS 20.

At TS 19/20, it is possible to identify the roots and ganglia of the **trigeminal** (V), **facial** (VII), **auditory** (**vestibulocochlear**) (VIII), **glossopharyngeal** (IX), **vagus** (X) and the cranial roots of the **accessory** (XI) cranial nerves (though the **trochlear** (IV) cranial nerve, which, like the trigeminal (V) nerve, initially evolves as a midbrain derivative, is not easily recognized at this stage), all of which derive from the rhombomeres and are associated with the pons and medullary region of the hindbrain. This is also the first stage when the fourth ventricle (or rhombencephalic vesicle) has an obvious **lateral recess** (from which it derives its shape and hence its Latin name). The **abducens** (VI) and **hypoglossal** (XII) cranial nerves develop from the metencephalon and myelencephalon, respectively, but are not readily recognized at this stage.

Genetic analysis has now shown that the neural crest (NC) cells migrating from some of the rhombomeres (R) colonize the branchial arches. NC cells from R2, 4 and 6 migrate into the first, second and third branchial (pharyngeal) arches, while other rhombomeres (e.g. R3 and 5) appear to yield either no NC cells, or cells of very limited migration capability, mainly because most of their cells undergo apoptosis (see also Chapter 3.3 *The branchial arch system*). The information presently available suggests that the trigeminal (V) ganglion is associated with R2, although there may also be a small contribution from R3. The facio-acoustic (VII–VIII) ganglion complex is associated with R4, though R5, which is located at about the level of the rostral half of the otic pit/vesicle, may also contribute to the acoustic (VIII) nerve. The glossopharyngeal (IX) ganglion is associated with R6, though there may be a small contribution from R7 which is principally associated with the vagus (X) nerve, although it may also contribute to the cranial part of the accessory (XI) nerve. The spinal part of the accessory (XI) nerve is associated with R8. The role of R1 is still obscure. This arrangement is shown diagrammatically in Figure 4.6.6.

By TS 20 and as rhombomere identity is lost, the pons and the medulla oblongata start to differentiate, while, during TS 20/21 the **rhombic lip** thickens (this is the dorsal part of the **alar plate** or lamina of the metencephalon and will form the **interventricular portion of the cerebellum**). During TS 20/21, the close relationship between the trigeminal cranial nerve ganglion (more rostrally), the facial and acoustic ganglia (which are close to the pre-cartilage primordium of the **petrous part of the temporal bone**), and (more caudally) the glossopharyngeal, vagal and accessory cranial nerve ganglia and the pons become easier to recognize in histological sections. (For a review of the development of the cranial nerve ganglia in the rat, see Altman and Bayer (1982).)

At TS 21, the junction between the fourth ventricle and the central canal of the spinal cord can now be readily recognized, as can the choroid plexus in the roof of the fourth ventricle which forms from an invagination of the vascular **pia mater** (the **tela choroidea**), see above, and, together with the choroid plexus that develops in association with the medial walls of the lateral ventricles, plays a critical role in the production of cerebrospinal (CSF) fluid (see *The meninges and choroid plexus* above). By TS 22,

Figure 4.6.6. The brainstem showing the relationships between the pharyngeal arches, the rhombomeres (or neuromeres), the somitomeres and the cranial nerves with their motor nuclei (diagram based on Larsen, 1993). The location of the paraxial mesodermal segments in the cephalic region are shown to the right of the diagram. The other landmark in this region, the otocyst, is also shown. Note that the roots of cranial nerves V and VII/VIII emerge from the middle parts of rhombomeres 2 and 4, respectively, while the roots of cranial nerves IX and X emerge from the boundaries between rhombomeres 6 and 7, and 7 and 8, respectively. The vagus nerve has an extensive origin from the 7th and 8th rhombomeres, and the cranial root of the accessory (XI) nerve arises from the caudal part of rhombomere 8, with its spinal roots originating in the upper cervical region of the spinal cord. The origin of the neural crest that migrates into pharyngeal arches 1–3 is also shown: that from the mesencephalon and rostral metencephalon migrates into pharyngeal arch 1, while that from the rostral part of the myelencephalon migrates into pharyngeal arch 2, and that from the caudal myelencephalon migrates into pharyngeal arch 3. No neural crest tissue is believed to originate from rhombomeres 3 and 5.

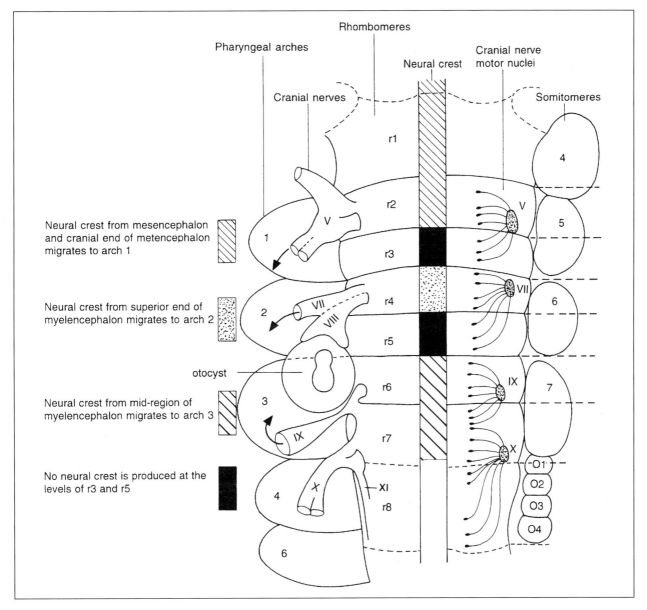

the choroid plexus here is particularly florid, while at TS 22/23 further choroid plexus forms in the lateral recesses of the fourth ventricle and it is likely that it is choroid plexus in this location rather than that in the lateral and third ventricles which subsequently becomes the main site of CSF production in the mouse (in the human, it is the choroid plexus in association with the lateral ventricles that are the main source of CSF).

During TS 22/23, the **cerebellar primordium** markedly enlarges, with its interventricular part, in particular, dramatically increasing in size. By the end of TS 23, this region has enlarged to such an extent that it encroaches on the lumen of the lateral recesses of the fourth ventricle where it will form the **cerebellar plate** (TS 24/25), while the thin roof of the fourth ventricle becomes increasingly covered by the **cerebellum** as it expands caudally.

TS	1	2	3	4	5	6	7	8	9	10	11	12	13	14	15	16	17	18	19	20	21	22	23	24	25	26
E	0	1	2	3	4	4.5	5	6	6.5	7	7.5	8	8.5	9	9.5	10	10.5	11	11.5	12	12.5	13.5	14.5	15.5	16.5	17.5

The transition from the caudal part of the medulla to the upper cervical region of the spinal cord is also more easily seen at TS 25 than previously, as are the rootlets of the cranial and spinal parts of the accessory (XI) cranial nerve. While it is just possible to see the lumen of the caudal extremity of the fourth ventricle in transverse sections through this region, the central canal of the cervical region of the spinal cord has narrowed to such an extent that it is almost completely obliterated in histological sections.

By TS 26, the cerebellum has become subdivided into a median dorsal part (the **vermis**) and two lateral parts (its so-called hemispheres), but few other gross morphological features are recognizable at this time. The pons and the medulla have also substantially enlarged, while, as a consequence of the growth and differentiation of the pons, cerebellum and medulla, the volume of the fourth ventricle has dramatically decreased.

The cerebellar peduncles

The cerebellum is connected to the midbrain and pons by the superior, middle and inferior peduncles through which pass major nerve tracts. The **superior cerebellar peduncles** contain mainly efferent fibres originating in the **globose, emboliform** and **dentate nuclei**, which are discrete aggregates of nerve cells within the cerebellum, and enter the brain immediately caudal to the inferior colliculi of the midbrain. The latter largely consist of two nuclei located in its tectum, or roof, and are part of the auditory pathway to the cerebral cortex. The majority of these fibres cross the midline and mostly continue forward to the ventral lateral (or VL) nucleus of the thalamus, from which fibres mainly project to the motor areas of the cortex (a few fibres end in the **red nucleus** – an oval mass of grey matter (see above) in the anterior part of the tegmentum and extending into the posterior part of the subthalamic region, while others end in the reticular formation).

The **reticular formation** of the brainstem is the part that is not well organized into nuclei and tracts. It is the region which receives data from all of the sensory systems, and has efferent connections with all levels of the central nervous system, and makes a significant contribution to various aspects of brain function including to regions that regulate visceral functions, and plays a role in the sleep–arousal cycle. Visceral functions are regulated through connections with nuclei of the autonomic outflow; that for respiration is through motor neurons in the phrenic nucleus and thoracic region of the spinal cord, while cardiac "centres" have been recognized in the brainstem of experimental animals. A minority of these fibres are, however, afferent to the cerebellum: most are in the spino-cerebellar tract, although a few run from the red nucleus or from the mesencephalic nucleus of the trigeminal (V) nerve.

The **middle cerebellar peduncle** (or **brachium pontis**) consists entirely of pontocerebellar fibres which are distributed to the cortex of the cerebellar hemispheres. Fibres from each of the **inferior cerebellar peduncles** enter the cerebellum from the caudal part of the pons and its main part is the **restiform body** in which the olivocerebellar fibres projecting from the **inferior olivary nucleus** in the lateral part of the medulla oblongata are the most numerous. The part of the peduncle forming the lateral part of the fourth ventricle is known as the **juxtarestiform body** and mainly contains afferent fibres from the vestibular nerve and nuclei, although a minority are efferent fibres concerned with the maintenance of equilibrium.

Histological atlases of the developing rodent brain are available, and should be consulted if further information is required (mouse: Schambra (1992), Franklin (1996); rat: Paxinos *et al.* (1991), Altman and Bayer (1995); Alvarez-Bolado and Swanson (1996); for an atlas of the developing human brain, see O'Rahilly and Müller (1994)).

Blood supply of the brain

The blood supply to the brain is from two principal sets of "feeder" vessels, from the internal carotid arteries (first recognized during TS 16) and from the vertebral arteries (first recognized during TS 19). The arterial blood is directed principally towards the grey matter of the brain because of its greater need for oxygenated blood than the white matter. The grey matter is supplied by superficial cortical arteries, while perforating vessels supply the deeper subcortical nuclei. The major arteries to the brain are all "end" arteries, there being no anastomoses between them; should occlusion of any of these vessels occur, initially ischaemia and shortly afterwards infarction (or death) of the region involved inevitably occurs.

The main branches from the internal carotid and vertebral arteries form an anastomosis in the region of the base of the brain around the optic chiasma and infundibulum (or stalk) of the pituitary to form the so-called "circle of Willis". While this is usually a complete circle, and allows the equalization of blood flow between the two sides, occasionally components of the circle are absent or, more commonly, are occluded or have an abnormally small diameter.

The arterial arrangement at the base of the brain is as follows: the two **vertebral arteries**, branches of the subclavian arteries, ascend in the foramina of the upper six cervical vertebrae, then wind around the lateral masses of the axis vertebra. They then pass through the foramen magnum by piercing the dura mater and the arachnoid, thus entering the subarachnoid space, where they amalgamate on the inferior surface of the brain stem, at the caudal border of the pons, to form the **basilar artery**. This vessel runs forward towards the middle of the pons where it divides into its two terminal branches, the **posterior cerebral arteries** (TS 21); these give off temporal branches which supply its inferior surface, as well as calcarine and parieto-occipital branches to the visual cortex. Close to its proximal and distal ends, the basilar artery gives off the **anterior inferior** and **superior cerebellar arteries**. Another branch of the posterior cerebral artery is the **posterior choroidal artery** which supplies the choroid plexus in the central part of the lateral ventricle and in the third ventricle, as well as supplying part of the thalamus, the fornix (the efferent tract of the hippocampus that arches over the thalamus and terminates in the ventral region of the hypothalamus), and the tectum of the midbrain. The basilar artery also gives off numerous pontine branches along its course.

The **internal carotid arteries** are, with the external carotid arteries, the terminal branches of the **common carotid arteries**, and pass into the middle cranial fossa through the carotid canal to enter the subarachnoid space. In this location, each divides into two terminal branches, the **anterior** and **middle cerebral arteries** (TS 22). Before branching, however, the internal carotid artery gives off the following branches: the **hypophyseal arteries**, the **ophthalmic artery**, the **posterior communicating artery** and the **anterior choroidal artery**.

The middle cerebral artery is the larger of the two terminal branches, and the direct continuation of the parent vessel; it gives off frontal, parietal and temporal branches which ramify over the lateral surface of the cerebral hemispheres. The two anterior cerebral arteries almost meet across the midline where they are joined by the **anterior communicating artery**. The anterior cerebral artery supplies the medial part of the orbital surface of the frontal lobe, including its olfactory region. It also supplies the region of the **corpus callosum** (a substantial mass of transverse fibre tracts, in the depth of the longitudinal fissure, connecting similar regions in the two cerebral hemispheres), the head of the caudate nucleus and putamen and the anterior part of the internal capsule. The terminal part of this vessel extends over the dorso-medial part of the sensori-motor strip of the frontal lobe of the brain.

The first part of the internal carotid artery is formed from the dorsal part of the third arch artery. Beyond this (and rostral to the site of the obliterated ductus caroticus, see Figure 3.3.2b), it forms from the rostral extension of the dorsal aorta. Each internal carotid artery gives off an ophthalmic branch to the developing eye, and then divides into three branches, the anterior, middle and posterior cerebral arteries. The posterior cerebral artery is initially therefore a branch of the internal carotid artery, but later becomes a terminal branch of the basilar artery. This embryological arrangement persists in about 30% of humans, where the posterior cerebral artery of one side takes origin from the internal carotid artery of the same side. For further details of the development of the intracranial vasculature in the human, see Padget (1948).

The initial venous drainage of the brain is via large numbers of unnamed **primary head veins** that later drain into the dural venous sinuses associated with the posterior cranial fossa.

The cerebral hemispheres, by contrast, have an external and an internal venous drainage system. The **external cerebral veins** lie in the subarachnoid space on all of the surfaces of the cerebral hemispheres and eventually drain into the dural venous sinuses. The central parts of the brain are drained by **internal cerebral veins** located beneath the corpus callosum in the transverse fissure. The two internal cerebral veins unite to form the **great cerebral vein** (of Galen), and this empties into the midline **straight sinus**, which is located in the midline within the fibrous tissue of the tentorium cerebelli. This in turn drains into the two transverse sinuses which then drain into the sigmoid sinuses. Both sets of veins drain to the dural venous sinuses and thence to the rostral part of the **internal jugular veins** which are located in the jugular foramina.

The spinal cord

The early stages of neural tube closure via the process of neurulation, its subsequent stratification, and the formation of the meninges have already been discussed and here we fill in some of the details of its later development that have not been covered elsewhere and point out some gross morphological differences between the spinal cord in the mouse and the human.

Morphogenesis of the caudal region

Examination of the caudal extremity of early human embryos (equivalent to TS 12) suggests that the most caudal part of the spinal cord forms by a process other than neurulation, termed **canalization, secondary neurulation** or **caudal neural tube elongation**. This region of the neural tube is believed to take origin from the **caudal eminence** (formed from the regressing primitive streak) rather than from the neural plate. Initially, a solid cord of tissue forms from the caudal eminence which then cavitates along its entire length. The lumen thus formed then joins (during the equivalent of TS 20) the neural canal that had previously formed by the process of neurulation. This is followed by a third phase of neural tube development termed **retrogressive differentiation** in which much of the neural tube in the tail region regresses and ultimately disappears to leave only the

TS	1	2	3	4	5	6	7	8	9	10	11	12	13	14	15	16	17	18	19	20	21	22	23	24	25	26
E	0	1	2	3	4	4.5	5	6	6.5	7	7.5	8	8.5	9	9.5	10	10.5	11	11.5	12	12.5	13.5	14.5	15.5	16.5	17.5

filum terminale. This is a slender fibrous band, or thread-like prolongation, that anchors the most caudal part of the neural tube (termed the conus medullaris) to the tip of the coccyx.

Detailed ultrastructural analysis (Schoenwolf, 1984) of serially sectioned mouse embryos between TS 12–20 provides some evidence that secondary neurulation also occurs in this species and involves (i) the formation of medullary rosettes (E9.5–10) or a medullary plate (E11–12), and (ii) cavitation (at E10). A medullary rosette consists of elongated tail bud cells, radially arranged around a central lumen formed by cavitation. A medullary plate also consists of elongated tail bud cells, but these cells do not surround a central cavity, but extend ventrally from the basal part of the dorsal surface ectoderm to a slit-like cavity previously formed by cavitation. A role of basement membrane and extracellular matrix components in this process has now been suggested (O'Shea, 1987). Secondary neurulation, as a phenomenon, is particularly well seen in the chick embryo (Schoenwolf and Delongo, 1980) and has also been recognized in the human (Bolli, 1966; Lemire, 1969), although its significance (for example, in relation to the development of neural tube defects) is unclear.

Retrograde differentiation with the formation of the filum terminale does, however, clearly take place in the mouse. From TS 20, the diameter of the neural tube in the distal half of the tail gradually narrows, and increasing evidence of neural luminal occlusion is seen. By TS 21, the neural lumen is completely obliterated in the distal two-thirds of the tail, and the most caudal part shows early evidence of retrograde differentiation. By TS 22, the neural lumen only extends as far as the proximal part of the tail. This region therefore represents the **conus medullaris** (TS 22), and beyond it the neural tube is replaced by the **filum terminale** (TS 22) (see above).

Cell differentiation

As far as the *histological* features of the primitive neural tube are concerned, its early wall (up to about TS 15) consists of a single layer of pseudostratified columnar epithelium consisting exclusively of neuroblasts. These cells subsequently differentiate and the definitive spinal cord is, like the brain (see above), divided into a series of concentric layers. The first (and innermost) of these is termed the **ventricular layer** (TS 16). Slightly later, during TS 16/17, the division products of these cells migrate peripherally to form an intermediate, or **mantle layer**, in which the neuroblast (or neuronal) cell bodies will be located,

and which will in due course constitute the **grey matter of the central nervous system**. Initially, but particularly up to TS 19, and even, but to a lesser extent during the early part of TS 20, the ventricular layer constitutes the majority of the tissue of the spinal cord. Even at TS 20, when the overall volume of the ventricular tissue is diminishing, the ventricular layer still comprises the majority of the dorsal half of the cord. Stratification therefore appears to take place between TS 16 and 17, and is well established by about TS 18.

Other division products of the neural precursor cells such as those of the glioblast lineage are also located within the mantle layer. These will in due course give rise to the astrocytes and oligodendrocytes that provide metabolic and structural support for the neurons. In the definitive spinal cord, additional (microglial) cells originating from the surrounding mesenchyme invade all regions of the cord. These belong to the reticuloepithelial system, and have a scavenger role, similar to that performed by the white blood cells elsewhere in the body. At about the same time that the glioblasts are first recognized, some of the division products of the neuronal precursor cells differentiate to form the specialized ependymal cells that line the central canal of the spinal cord and ventricles of the brain. These cells are largely responsible for the production and resorption of the cerebrospinal fluid (CSF, see above).

When the axonal processes migrate peripherally from the neuronal cell bodies that are located within the mantle layer of the spinal cord, an outer or **marginal layer** is established, and this constitutes the **white matter of the central nervous system**. The first evidence of a distinct marginal layer is seen during TS 17/18.[10]

A fundamental change starts to occur in the morphology of the neural tube at about TS 18 with the formation of the four discrete columns: two dorsal (or **alar**) **columns**, and two ventral (or **basal**) **columns**. Laterally, the dorsal and ventral columns are separated by a shallow groove (the **sulcus limitans**), while dorsally and ventrally the neural tissue thins in the regions of the **roof plate** and **floor plate** (see Figure 4.6.3).

Up to TS 19/20, the basal columns are considerably larger than the alar columns, but the volume of the alar columns increases considerably during TS 20/21, and now has an almost equal volume to those of the basal columns. A distinct **sulcus limitans** is also first clearly seen at this time, while, a little later (TS 21/22), the meninges start to form (see above).

The cells within the alar and basal columns have quite distinct functional roles: the **somatic motor**

---

[10] Although the changes that take place in the differentiation of the neuroepithelial elements of the primitive neural tube are, in general terms, similar in mouse and human embryos, the exact timing of the various events discussed above is not identical, with those in the mouse occurring at a slightly later period during development than that described in the corresponding (*i.e.* Carnegie stage) human embryo. This is particularly evident when an exact comparison is made between the histological/morphological features of the other organ systems which are also differentiating during this period.

cells are the first to be recognized, and become restricted to within the basal columns; those in the rostral part of the spinal cord are the first to differentiate, with the more caudal elements forming in a cranio-caudal sequence. Alkaline phosphatase enzyme activity is restricted to specific nuclei localized within a broad transverse band of cells in the future ventral horns of the spinal cord (Kwong and Tam, 1984). In the developing mouse brain at this time (between E9.5–17.5), this alkaline phosphatase-positively staining band of cells is topographically related to efferent nuclei of the cranial nerves, except those nuclei that had already migrated laterally to their final position (Tam and Kwong, 1987). Alkaline phosphatase activity may be related to the enhanced metabolism of these cells occurring during the formation of the cranial nerves and may be important in relation to the establishment of neuronal connections in the fetal brain.

Slightly later on, changes can be recognized in the cells of the alar columns as they differentiate into association neurons which have an afferent role interconnecting with axonal processes from the sensory neurons located in the **dorsal root ganglia** (Figure 4.6.2). These are neural crest-derived, and also differentiate in a cranio-caudal sequence. Within them, sensory cells relay information centrally (*i.e.* towards the central nervous system) from receptors dispersed throughout the body.

The peripheral nervous system

Aggregations of neural crest cells are first seen in association with the developing spinal cord during TS 15, while the most rostral of the dorsal root ganglia are clearly recognized by TS 16. Similar, though substantially larger aggregates of neural crest cells are also seen at about this time in the cephalic region, and are the precursors of the trigeminal and facio-acoustic ganglionic complexes (Adelmann, 1925).

At about this time (TS 16), it is also possible to recognize within the cord the presence of **ventral roots** which contain the *efferent* axonal processes of the somatic motor cells of the basal columns. These initially appear to leave the spinal cord as a broad band, but, as they grow towards and through the rostral part of the somite-derived sclerotomes, they condense to form discrete units which are located at the same "segmental" level as each pair of somites, and constitute one component of the **mixed segmental spinal nerves** (TS 17). These elements are also first evident in the cervical region of the developing spinal cord, and are directed towards the primitive "voluntary" muscle masses (myotomes) particularly those that will be differentiating within the developing limb buds.

As the emerging spinal nerve migrates from the primitive spinal cord, it passes close to the ventro-medial surface of its corresponding dorsal root ganglion (*i.e.* the ganglion which developed from the same *segmental* level). At about this time, the neurons within the dorsal root ganglion send out a *peripherally-directed* axon which joins the segmental spinal nerve as it migrates towards its appropriate end organ. Another axon from the neurons located in the dorsal root ganglion is *medially-directed*, and grows towards the dorsal column (at its own segmental level) where it synapses with the developing association neurons. The dorsal root ganglion with its connecting branches to the spinal cord (medially) and the segmental spinal nerve (laterally) are collectively termed the **dorsal root** (see Figure 4.6.2).

Within a short period (early TS 17), it is also possible to recognize the presence of **sympathetic trunks** (or chains), initially in the cervical, and shortly afterwards in the thoracic region. These are located ventro-laterally on either side of the spinal cord. A little later (late TS 17), segmental ganglia form at intervals along these trunks in the thoracic region, and, in due course (during TS 21), it is possible to recognize the superior, middle and inferior cervical sympathetic ganglia, and the para-aortic bodies (TS 22–24) and the coeliac ganglion (TS 25).

More peripherally, **parasympathetic ganglia** form in association with the abdominal aorta and its principal branches, and within the wall of the visceral organs. Finally, **vagal parasympathetic fibres** grow out from the **parasympathetic** ganglia. Some of these fibres innervate the heart, while others innervate the gastrointestinal tract. The central connections of the parasympathetic outflow are located in the brainstem (the **cranial outflow**) and reach the parasympathetic ganglia in the head and neck region and trunk viscera via the vagus (X) nerve, whereas the parasympathetic outflow in the sacral region (at the level of S2–S4) innervate the hindgut and pelvic viscera via the **pelvic splanchnic nerves**. (The development of the autonomic nervous system is discussed in more detail in Chapter 3.1 *The neural crest*.)

For a detailed review of the development of the rat spinal cord, see Altman and Bayer (1984), and for an overview of the nervous system of the rat, see Hebel and Stromberg (1986).

Arterial supply and venous drainage of the spinal cord

The arterial blood supply to the spinal cord is from three major vessels, the **anterior spinal artery**, a midline derivative of the two (*i.e.* right and left)

TS	1	2	3	4	5	6	7	8	9	10	11	12	13	14	15	16	17	18	19	20	21	22	23	24	25	26
E	0	1	2	3	4	4.5	5	6	6.5	7	7.5	8	8.5	9	9.5	10	10.5	11	11.5	12	12.5	13.5	14.5	15.5	16.5	17.5

vertebral arteries, and which runs caudally in the ventral median fissure, and the paired **posterior spinal arteries**; both the anterior and posterior spinal arteries are seen during TS 22, and are associated with an extensive vascular network related to the surface of the cord. These superficial vessels arise either from the vertebral arteries or from the posterior inferior cerebellar arteries, and runs caudally close to the origin of the dorsal spinal roots. The arterial supply to the spinal cord is supplemented by segmental spinal vessels that are branches from the vertebral vessels in the cervical region, from posterior intercostal branches of the thoracic aorta, and from lumbar arteries of the abdominal aorta. Small branches from the spinal arteries supply the spinal roots and surface of the cord, while additional branches supply the grey and white matter of the cord.

The cord is drained by pairs of **anterior** and **posterior spinal veins** (first seen during TS 20/21), and these in turn drain via **anterior** and **posterior radicular veins** into an extensive **epidural venous plexus**. This then drains into an **external vertebral plexus** before draining into vertebral, intercostal and lumbar veins. For further details of the blood supply to the brain and spinal cord of the rat, see Tokioka (1973), Hebel and Stromberg (1986), and for that of the human, see Barr (1979) and Williams (1995).

Appendix: The pituitary

The pituitary gland is interesting because it develops from two quite different sources: the first is an ectodermal diverticulum that grows upwards from the roof of the oral or buccal part of the **primitive oropharynx**, known as **Rathke's pouch**, and the second is a downgrowth from the floor of the **diencephalon**, the caudal part of the primitive forebrain. The latter is therefore also ectodermal in origin, forming from the neuroectodermal (neuroepithelial) floor of what will become the **third ventricle** (see Figure 4.6.7a).

Rathke's pouch is first seen at TS 14 as a localized midline thickening in the ectoderm of the roof of the oropharynx just in front of the **buccopharyngeal membrane** at about the time that the latter is breaking down, a process that is completed by about TS 15. The entrance to Rathke's pouch is initially quite wide, but progressively narrows, so that by about TS 20 the connection between the pouch and the roof of the oral cavity is lost. By TS 15, the anterior wall of Rathke's pouch has pushed aside the intervening mesenchyme to make direct contact with the floor of the diencephalon, but it is only by about TS 16 that the first evidence of the **infundibular recess of the third ventricle** is seen, its walls being histologically identical to the neural ectoderm elsewhere in the region of the floor of the third ventricle at this time. This recess, which becomes the **infundibulum** (proximally) and the **posterior pituitary** (distally), emerges from the median eminence of the **tuber cinereum**, a region of the diencephalon floor located between the **optic chiasma** (anteriorly) and the **mamillary bodies**[11] (posteriorly).

By TS 18, the entrance to the infundibular recess is well demarcated, but relatively narrow as compared

Figure 4.6.7. The dual origin of the pituitary gland. (a, TS 18) One component is derived from the up-growth of an oropharyngeal diverticulum (Rathke's pouch), while the other is derived from a downgrowth of diencephalic tissue from the floor of the third ventricle of the brain. Both tissues are ectodermal in origin. (b, TS 20) The derivatives of Rathke's pouch give rise to the various components of the anterior pituitary (the anterior lobe, pars intermedia and pars tuberalis), while the derivatives of the floor of the third ventricle are the infundibulum (or pituitary stalk) and the posterior lobe of the pituitary. A cavity (the residual lumen of Rathke's pouch) is often present in the adult gland separating the pars intermedia from the anterior lobe. The definitive pituitary gland is located in the pituitary fossa in the base of the skull.

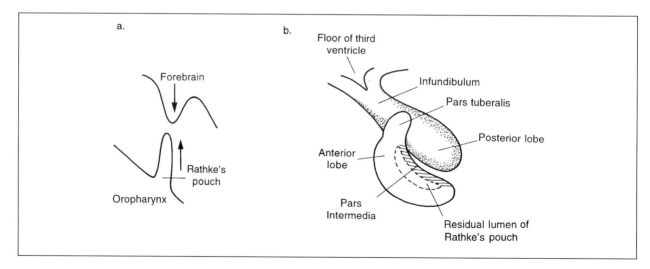

---

[11] These are a pair of smooth, side-by-side hemispherical masses of grey matter (hence their name), each enclosed in a white capsule in the floor of the interpeduncular fossa that are components of the posterior hypothalamus. They have connections with the thalamus and fornix, and are components of the limbic system that are involved in emotional and behavioural reactions, both individual and social, as well as in memory.

to the transverse diameter of the rostral part of Rathke's pouch. Initially (TS 18), only the inferior surface of the **pituitary stalk (infundibulum)** is in apposition to the rostral surface of Rathke's pouch, but, by about TS 20/21, the supero-lateral parts of Rathke's pouch (now called the **pars tuberalis**) extend upwards on either side of the stalk and may eventually surround it. At this stage, the lateral part of Rathke's pouch is close to the **internal carotid arteries** that form a plexus on the floor of the diencephalon and will later provide the pituitary blood supply (TS 21/22).

At TS 19/20, the distal part of the infundibulum forms the **neural lobe** (or **pars nervosa** or **pars posterior**) and this is connected to the median eminence of the tuber cinereum by the pituitary stalk. The neural lobe, the pituitary stalk and the median eminence together constitute the **neurohypophysis**. At about TS 20/21, Rathke's pouch differentiates into the **adenohypophysis** and its parts become clearly demarcated. These are the **pars anterior** (or **anterior lobe**, or **pars distalis**), the **pars intermedia** and the **pars tuberalis**, with the first two eventually being separated by the **hypophyseal cleft**, the remnant of the cavity of Rathke's pouch (see Figure 4.6.7b). This cavity gradually diminishes, and has often disappeared by TS 24/25 because of the increased cellularity and vascularity of the anterior pituitary.

Although Rathke's pouch seems to be in contact with the neural ectoderm of the primitive forebrain, with no intervening mesenchyme, it seems unlikely that it actually induces the differentiation of the neurohypophysis (or *vice versa*). The evidence against an inductive interaction comes from the rare instances (seen in humans) when Rathke's pouch tissue fails to "ascend" and consequently does not contact the neural tissue in the floor of the diencephalon. In these cases, the two components develop as separate entities and both appear to function quite normally, with only the posterior pituitary being found in the **hypophysial** or **pituitary fossa**, where the intact pituitary is normally located, while the anterior pituitary (termed a pharyngeal pituitary) is located close to the midline and associated with an indentation in the roof of the oral cavity. The hypophysial fossa, a hollowed-out region located in the upper part of the body of the sphenoid bone, is said to be shaped like a Turkish saddle, and hence is called the **sella turcica** (TS 25). The roof of the pituitary fossa is covered by a thin sheet of fibrous dura mater (termed the **diaphragma sellae**) and is pierced in its central part by the stalk of the pituitary gland. For information on the neural connections of the various parts of the pituitary, its hormonal output and its physiological roles, and its detailed anatomical/vascular relations with the hypothalamus, the interested reader should refer to a standard text-book of neurophysiology.

TS	1	2	3	4	5	6	7	8	9	10	11	12	13	14	15	16	17	18	19	20	21	22	23	24	25	26
E	0	1	2	3	4	4.5	5	6	6.5	7	7.5	8	8.5	9	9.5	10	10.5	11	11.5	12	12.5	13.5	14.5	15.5	16.5	17.5

4.7 The eye and the ear

The eye

Introduction

The eye is an extremely complex sense organ and, not surprisingly, it forms from a diverse range of tissues that include: (i) *neural ectoderm* (the neural and pigment layers of the retina, and the optic stalk, together with components of the peripheral nervous system that innervate the extrinsic ocular muscles and the muscles associated with the eyelids); (ii) *surface ectoderm* (the lens, the cornea and the eyelids, as well as the associated conjunctiva – the delicate membrane that lines the eyelids); (iii) *neural crest* (the sclera, and the stromal cells and Descemet's membrane in the cornea, the pigment cells associated with the choroidal region and probably the choroid itself, as well as the coverings of the peripheral nerves noted above; it may also form the intrinsic muscles of the eye); (iv) *blood vessels* (branches of the ophthalmic artery, itself a branch of the internal carotid artery); (v) *autonomic (sympathetic and parasympathetic) nerves* associated with the ciliary ganglion; and (vi) *paraxial mesoderm* (cranial somitomeres) which form the extrinsic muscles of the eye. The time course of its lengthy development extends from the early somite period (TS 12) to about 12–14 days after birth when the eyelids re-open.

The early stages

The first stage in eye development is the formation of the **optic placode**, a circumscribed central region on the neuroepithelial (inner) surface of the cephalic neural folds (where the cells tend to be columnar rather than cuboidal) that first appears when the folds start to elevate, but prior to their medial movement (early TS 12). At this stage, the placode is a flattened area of neural ectoderm in the central part of the prospective forebrain region. Shortly afterwards (TS 13), a shallow medially-directed groove, the **optic sulcus**, connects the medial part of the placode to the diencephalic region of the primitive forebrain, at the site of the future **optic chiasma** (which will form at TS 20). A slight indentation forming in the middle of the optic placode (during the latter part of this stage) is the first indication of the **optic pit**. During TS 13, this pit deepens as the cephalic neural folds enlarge, become elevated, and meet in the ventral midline region just anterior (or ventral) to the prospective forebrain. Fusion of the folds now starts (TS 13/14) and extends rapidly in both directions (*i.e.* rostrally and caudally) to form the forebrain (prosencephalic) vesicle.

While this is happening, the **optic vesicles** (sometimes termed the *primary* optic vesicles) form from the placodal tissue, with the non-placodal neuroepithelial tissue forming their peripheral margins; the optic vesicles may be considered as laterally-directed *evaginations* of the forebrain (or prosencephalic) vesicle that are connected by their **optic stalks** to the forebrain vesicle. Initially, the optic stalks and vesicles are indistinguishable in that they have a similar diameter and are also similar at the histological level, but during TS 15 the former narrow while the latter enlarge (see Kaufman, 1979). At the histological level, at this stage, all regions of the neuroepithelium appear to be multi-layered. It is only at TS 14/15, shortly after the collapse of the optic vesicle to form the optic cup, that its outer layer of cells that are destined to form the pigment layer of the retina, as well as the region that forms the optic stalk, are now seen to be present in the form of a monolayer. Once the optic cup has fully formed (TS 17/18), both the intra-retinal space and the lumen of the optic stalk become obliterated, and the diameter of the optic stalk is then seen to be at its narrowest. As the axonal processes of the neural retinal cells grow proximally within the lumen of the optic stalk, to form the definitive optic nerve, so its diameter gradually increases.

As the cephalic part of the neural tube closes, mesenchyme migrates between the optic vesicle and the overlying surface ectoderm which is raised to form a prominent protuberance on the side of the head termed the **optic eminence**. By TS 15/16, this intervening mesenchyme becomes displaced, so that the neural ectoderm of the optic vesicle directly underlies the surface ectoderm and probably induces the latter to form the **lens placode**. The central part of this placode then indents to form the lens pit (TS 16), while the outer part of the primary optic vesicle collapses inwards to form the inner layer of the (so-called *secondary*) **optic cup**, with the region on the ventral surface of the cup and optic stalk being called the **choroidal** (or fetal or optic) **fissure** (TS 18). The lens pit now deepens and eventually rounds-up to form the **lens vesicle** (TS 18) which subsequently separates from, and disappears beneath, the surface ectoderm (see Figure 4.7.1) (Wrenn and Wessells, 1969). (The fate of the fetal fissure is discusssed below in *The blood supply to the eye*.)

The lens

The cells of the lens vesicle are initially all cuboidal/columnar in shape, but those cells that constitute its posterior wall become increasingly elongated as they form lens fibres. The lumen of the lens vesicle progressively diminishes, and is eventually completely obliterated (TS 21). Further lens fibres are added from the cells at the equatorial region of the lens, so allowing the lens to enlarge as development proceeds. The nuclei of the lens fibres which are initially located towards the periphery of these cells migrate medially towards the middle of the cell and, during the final stages of lens differentiation (from TS 26 to birth), become increasingly transparent (see Figure 4.7.2). Subsequent enlargement of the lens occurs by the addition of "secondary" lens fibres from its equatorial region. In the mouse, the volume of the lens, compared to the overall volume of the globe of the eye, is considerably greater than is the case in the human[1] embryonic/fetal eye at comparable stages of development. While it is possible to recognize a primitive hyaloid vascular plexus within the hyaloid cavity of the eye from about TS 19/20, contact between these vessels and the posterior surface of the lens is only made during about TS 20/21 and these constitute the *tunica vasculosa lentis*. A well-developed lens capsule forms by about TS 25.

The retina

From about TS 15, the fates of the inner and outer layers of the optic cup are clearly different. While the outer layer remains as a monolayer of cuboidal cells that will form the **pigment layer of the retina** (pigment granules are first readily recognized in these cells in pigmented strains of mice from about TS 21), those cells in all but the periphery of the inner layer rapidly multiply, giving rise to multilayers of cells which differentiate to form the various components of the **neural retina**.

By about TS 19/20, with the "collapse" of the optic vesicle to form the **optic cup**, the space between the

Figure 4.7.1a–c. The formation of the early eye. (a, TS 16) The surface ectoderm is induced to form the lens placode by the optic vesicle, and the lateral wall of the optic vesicle then indents, leading to the formation of the so-called (secondary) optic cup. At the same time, the lens placode invaginates to form the lens pit which gradually deepens. (b, TS 17) The lens invagination separates from the surface to form the lens vesicle, while the two layers of the optic cup, which are morphologically similar at the optic vesicle stage, differentiate with the outer layer forming the pigment layer of the retina and the inner layer forms the neural retina. (c, TS 19) The transverse section through the optic stalk reveals the presence of an optic (or fetal) fissure on its ventral surface in which are located the hyaloid blood vessels (artery and vein) that supply the neural retina and lens. The hyaloid artery is a branch of the ophthalmic artery which is, in turn, a branch of the internal carotid artery.

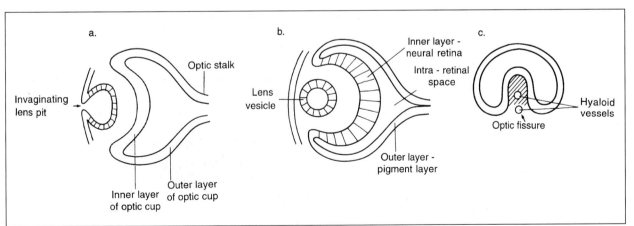

[1] Descriptions of the early stages of eye development in the human may be found in Barber (1955), Mann (1964) and O'Rahilly (1966, 1983).

TS	1	2	3	4	5	6	7	8	9	10	11	12	13	14	15	16	17	18	19	20	21	22	23	24	25	26
E	0	1	2	3	4	4.5	5	6	6.5	7	7.5	8	8.5	9	9.5	10	10.5	11	11.5	12	12.5	13.5	14.5	15.5	16.5	17.5

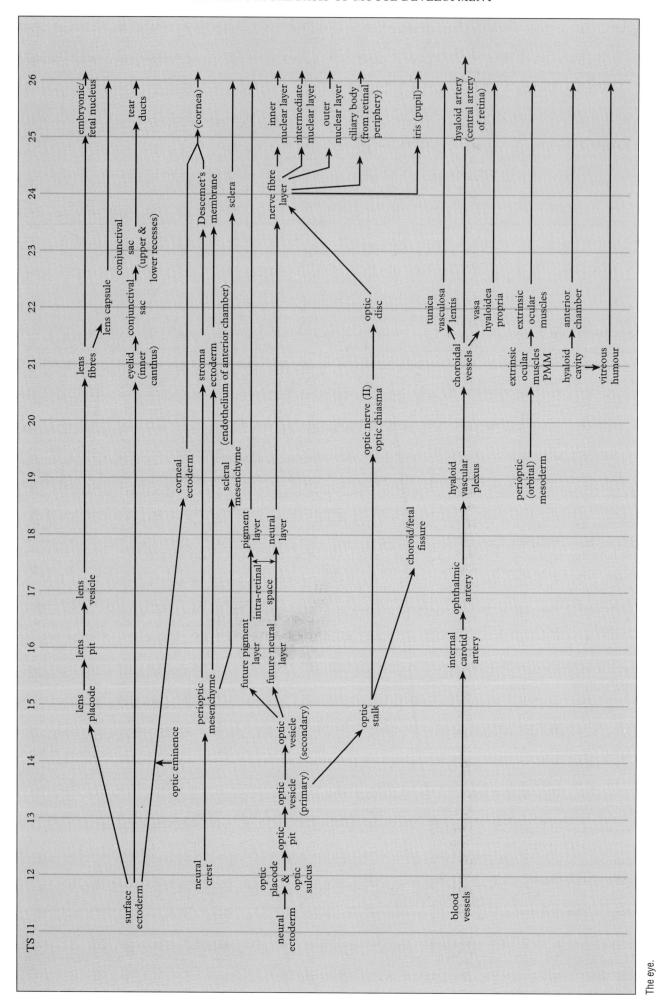

The eye.

Figure 4.7.2a–e. The differentiation of the lens of the eye. (a, TS 18/19) Soon after the lens has completely separated from the surface ectoderm, its cells have a cuboidal or columnar epithelial morphology, with peripheral nuclei; the cavity of the lens is almost as large as the diameter of the lens vesicle. (b, TS 20) Although the cells of the anterior region of the lens retain their cuboidal/columnar morphology, those near its posterior surface elongate, and the longest of these cells may now be termed *lens fibres*; their nuclei are, however, still peripherally located. (c, TS 20/21) The lens fibres at the posterior pole of the lens have elongated further, almost obliterating the last remnants of the cavity of the lens vesicle. This cavity is finally obliterated during TS 21, when most of the nuclei have migrated from the periphery towards the centre of the lens cells. (d, TS 24/25) It is only now that the centrally located nuclei within the lens fibres begin to lose their staining properties and become translucent on conventionally stained histological sections. By TS 26, the most centrally located lens fibres (its so-called *embryonic/fetal "nucleus"*) appear to lack nuclei. (e) Some 12–14 days after birth, the lens assumes its mature form. Note that the multiple nuclei seen in relation to the lens fibres in b.–d. represents what may be observed in "thick" sections through the lens. Each lens fibre has of course only a single nucleus.

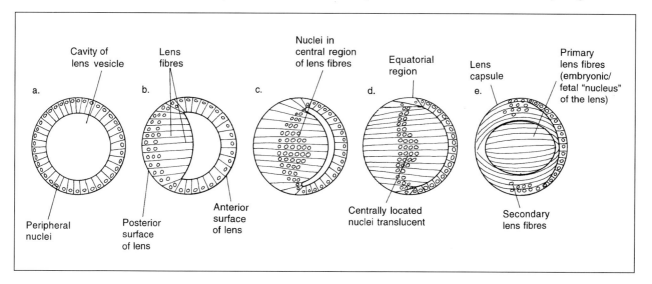

neural and pigment layers of the retina (the **intra-retinal space**), which was initially continuous with the forebrain vesicle via the lumen of the early optic stalk, and subsequently with the cavity of the third ventricle, disappears, although the degree of adhesion between the two layers is slight. The cells at the peripheral margin of the neural retina, those inaccessible to light, differentiate at about TS 25 to form the **ciliary body** (TS 25), while the very tip of the retinal cup thins out and extends over the lens to give rise to the **iris** (TS 25).

During TS 21, the nerve fibres that originate in the primitive ganglion cells of the neural retina grow towards the inner surface of the optic cup (*i.e.* on its hyaloid surface), converge on the optic pit and pass into the central region of the optic stalk, when they soon fill its lumen. These fibres, which now constitute the inner layer of the neural retina and are unmyelinated throughout their length, are directed towards the future optic chiasma. The neural layer of the retina shows increasing evidence of differentiation, but it is not until TS 25/26 that the characteristic stratification of the retina is clearly seen (an appearance considerably enhanced when histological sections are stained with silver).

At this stage, the neural retina consists of an *outer* nuclear (neuroblastic) layer, which will form the horizontal cells and the nuclei of the photoreceptor cells, a transient intermediate anuclear layer (of Chievitz), and an *inner* nuclear, or neuroblastic, layer which will mainly form the ganglion cells and supporting cells of the retina. It is from the ganglion cells that axons emerge, grow across the inner surface of the neural retina and pass into the optic stalk (for observations on cellular differentiation in the mouse retina, see Young, 1985). In the human embryo, all the cell layers of the neural retina are recognized by about the eighth month, in the mouse this degree of neural retinal differentiation is not seen until several weeks after birth.

The cornea, iris and pupil

The **corneal epithelium** differentiates from the surface epithelium overlying the lens and this process probably starts at about TS 19/20 when mesenchyme, probably of neural-crest origin, colonizes the region between the surface ectoderm and the lens (Bard and Kratochwil, 1987). By TS 21, orthogonally organized bundles of collagen fibrils are starting to form the stroma, and, a few days later (TS 24), the posterior, so-called **endothelium** of the cornea is differentiating from the posterior stromal cells with **Descemet's membrane** forming on the anterior surface of the endothelium. Development of the

TS	1	2	3	4	5	6	7	8	9	10	11	12	13	14	15	16	17	18	19	20	21	22	23	24	25	26
E	0	1	2	3	4	4.5	5		6	6.5	7	7.5	8	8.5	9	9.5	10	10.5	11	11.5	12	12.5	13.5	14.5	15.5	16.5	17.5

cornea continues through TS 25 when the eyelids have fused to enclose the conjunctival sac, and may well not be complete in the mouse until some time after birth (Haustein, 1983; see also Sevel and Isaacs, 1989).

Once the cornea has started to form, a cavity appears between the anterior surface of the lens and the overlying cornea that is termed the **aqueous chamber** (TS 24/25), while a cellular membrane that had grown out from the periphery of the eye onto the anterior surface of the lens epithelium forms the **pupillary membrane**. The peripheral margin of the optic cup grows under the deep surface of the pupillary membrane and the combined layers form the **iris**. At this stage (before the central part of the pupillary membrane breaks down), the aqueous chamber is separated into an anterior and a posterior compartment (also termed the anterior and posterior chambers, respectively). Just before the eyelids open (after birth), the central part of the pupillary membrane, which is not reinforced with optic cup tissue, breaks down to form the **pupil**. The space between the posterior surface of the lens and the inner layer of the optic cup gradually increases in size and comes to contain a jelly-like substance, the primordium of the **vitreous humour** or **body**, also termed the **hyaloid substance**. It is enclosed by a thin **hyaloid membrane** and its central part is pierced by the hyaloid canal along which the hyaloid artery passes to the posterior surface of the lens. While this vessel often regresses and disappears after birth, remnants of the presence of the hyaloid canal may still be seen in the hyaloid substance (see below). For a detailed analysis of the morphological features of the prenatal mouse eye, see Pei and Rhodin (1970).

The eyelids and other peripheral tissues

The eyelid margins are first seen in the surface ectoderm at the periphery of the eye at about TS 19 and are much more obvious by about TS 22/23. At about TS 24, epithelial outgrowths from each eyelid margin extend across the corneal surface of the eye and meet along a horizontal line where they fuse to enclose the conjunctival sac, the leading edges consisting of large rounded epithelial cells. The line of fusion between the upper and lower eyelids only breaks down at about 12 days after birth, with complete separation occurring a day or two later. Three stages in the differentiation of the eye are illustrated in Figure 4.7.3a–c, and display the features seen at about TS 20/21, at the time of birth and at about 12 days after birth, respectively. The fusion of the eyelids may allow corneal differentiation to occur in a "protected" environment *in utero* and in the early post-natal period.[2]

The eye also has peripheral components, and immediately external to the retina is the **choroidal layer** which embeds a capillary network and contains neural-crest-derived pigment cells which differentiate at about TS 21. Adjacent to (and outside) the choroid is the **sclera**, the outer layer of the eye, which forms when neural-crest-derived perioptic mesenchyme condenses and differentiates to form cells that lay down extracellular matrix (TS 24).

The nerve supply

Apart from the optic nerve that carries image data from the eye to the brain, there are also nerves to the intrinsic and extrinsic musculature and from the corneal epithelium. It is unclear exactly when the eye receives its autonomic (sympathetic and parasympathetic) nerve supply (associated with the ciliary ganglion) which enters the eye with its vascular supply. The sensory component (from the ophthalmic division of the trigeminal (V) nerve) supplies the cornea (particularly the external epithelium), iris and ciliary body, while the sympathetic supply carries vaso-constrictor fibres to the vessels of the eyeball, while the parasympathetic fibres pass to the **sphincter pupillae** and **ciliary muscle** to facilitate contraction of the pupil and allow accommodation.

By TS 20, large numbers of unmyelinated axons which originate in the ganglion cells of the neural retina pass through the region of the future optic disc into the optic stalk (and can be particularly well seen in silver-stained sections). These nerve fibres pass medially towards the **optic chiasma**, and thence to the **lateral geniculate bodies**.[3] Later (~ TS 21), the

---

[2] The time of separation of the eyelids accurately reflects the degree of immaturity of the mouse embryo at the time of birth when compared, for example, with the human fetus. In the mouse embryo, as mentioned above, the eyelids fuse at about TS 24/25 (at about day 16 of pregnancy), start to separate 3–5 days after birth and finally reopen at about 12 days after birth (see Findlater *et al.*, 1993). In the human, by contrast, the eyelids fuse at about day 60, and remain closed until the 7th month of fetal life (Hamilton and Mossman, 1972; Pearson, 1980). For a scanning micrographic and complementary histological sequence showing the detailed stages in the closure and reopening of the eyelids, see Findlater *et al.* (1993; see also Harris and MacLeod, 1982; Sevel, 1988). The timing of this event appears to be a good indicator not only of optic maturity but of overall maturity at the time of birth in all mammalian species.

[3] The nerve fibres from the lateral (temporal) half of each retina pass through the *optic tract* to the *lateral geniculate body* (or nucleus) of the thalamus on the same side; however, the fibres from the medial (*i.e.* nasal) halves of each retina cross in the optic chiasma to the other side and pass, with uncrossed temporal fibres from the opposite eye, through the *optic tract* to the *lateral geniculate body*. This partial crossing of the optic nerve fibres in the optic chiasma is an essential requirement for binocular vision.

Additional fibres from the lateral geniculate nucleus project to the *visual* (or striate) *cortex* which is located in the postero-medial part of the occipital cortex slightly above and below the calcarine sulcus. There is a detailed point-to-point projection of the retina both in the lateral geniculate nucleus and in the visual cortex, and interruption at any point along the visual pathway causes characteristic *visual field* defects. Because of the optical properties of the lens, the retinal image of an object in the visual field is inverted and reversed from right to left (as is the image on a camera film). For further details of the anatomy and physiology of the human visual system, see Barr (1979), Williams (1995); and for this information in rodents, see Hebel and Stromberg (1986).

optic stalk will form the **optic nerve**. Fairly soon after these nerve fibres are first recognized, the lumen of the optic stalk becomes completely obliterated by them.

The optic nerve at this stage is covered by a layer of mesenchyme which subsequently becomes fibrous, forming the dura mater which is continuous with that around the brain and spinal cord and with the tough fibrous coat of the eyeball.[4]

The musculature

Beyond the eyeball are **extrinsic ocular muscles** formed by the differentiation of paraxial mesoderm (cranial somitomeres) within the orbit. These voluntary muscles are first seen at about TS 20/21, and are supplied by the oculomotor (III), trochlear (IV) and abducent (VI) cranial nerves. Somewhat surprisingly, the extrinsic ocular muscles appear to be capable of differentiating within the orbit in the absence of any stimulus from the eye, and may even be present in the condition of **anophthalmia** (when the eye completely fails to develop) as, for example, occurs in homozygous *small-eye* mouse embryos, although, while the optic vesicle is initially apparently normal in these mice, it is the absence of the lens placode that inhibits subsequent development of the eye (see Hogan *et al.*, 1986). In homozygous tetraploid mice, while a morphologically abnormal optic vesicle is often present, eye development, if it occurs at all, is invariably extremely rudimentary (see Kaufman and Webb, 1990). In both of these groups of mice, eyelids form despite the absence of ocular formation within the orbit (*Sey/Sey* mice: Hogan *et al.*, 1986; Kaufman *et al.*, 1995; tetraploid mice: Kaufman and Webb, 1990).

The intrinsic ocular muscles

Controversy exists with regard to the origin of the smooth muscle associated with the iris (the **sphincter pupillae**) which, at least in birds and reptiles, is said to be of ectodermal origin, derived from the peripheral margin of the optic cup. The **ciliary muscle**, by contrast, is said to arise from the mesoderm surrounding the optic cup (Hamilton and Mossman, 1972). More recently, it has been suggested that both the pupillary muscle of the iris as well as the ciliary muscle may be derived from neural crest-derived ectomesenchymal cells in the choroid (Larsen, 1993). Both the iris and the ciliary body are first recognized during TS 25, but the time of first appearance of the smooth muscle associated with these structures to form the sphincter pupillae and ciliary muscle has yet to be determined.

The blood supply

This is first seen at TS 18 when the choroidal (or "fetal") fissure forms on the ventral aspect of the optic stalk and cup (Figure 4.7.1). The primitive **hyaloid vessels** which colonize the **hyaloid cavity** or **posterior chamber** (the space between the inner layer of the optic cup and the posterior part of the lens) at about TS 19, are branches of the **central artery of the retina** which arise from the **ophthalmic artery** (a branch of the **internal carotid artery**). The central artery enters the hyaloid cavity within the choroidal fissure, but the latter disappears as the sides of the fissure gradually meet and close so that the central vessels (the central artery with its accompanying vein) entering and leaving the optic cup eventually become completely surrounded by the optic stalk tissue (TS 22/23). These hyaloid vessels supply the neural retina and form a considerable vascular network on the posterior surface of the lens capsule (termed the **tunica vasculosa lentis**, TS 22) as well as supplying the intervening mesenchyme. When, towards the end of the fetal period, the lens loses this blood supply, the portion of the central artery beyond the retina degenerates and disappears, although the course of this vessel through the **vitreous body** is retained as the **hyaloid canal**. The proximal component of this vessel is, however, retained, as the **central artery of the retina** (TS 26). If, as occasionally happens, non-union or only partial union of the borders of the fetal fissure occurs, this results in the condition of *coloboma of the iris* (and, if the condition is more extensive, *coloboma of the iris and choroid*).

---

[4] Because the eye is an extension of the brain, the proximal part of the optic stalk is surrounded, in the adult, by the meninges. Accordingly, raised intra-cranial pressure may impair the venous drainage of the eye, and result in *papilloedema* (swelling in the region of the *optic disk*, at the site where the retinal vein enters the optic stalk).

TS	1	2	3	4	5	6	7	8	9	10	11	12	13	14	15	16	17	18	19	20	21	22	23	24	25	26
E	0	1	2	3	4	4.5	5	6	6.5	7	7.5	8	8.5	9	9.5	10	10.5	11	11.5	12	12.5	13.5	14.5	15.5	16.5	17.5

Figure 4.7.3a–c. The later development of the eye. (a, TS 20/21) The *hyaloid artery* enters the vitreous body at the site of the future *optic disc* via the *optic stalk*. At this stage, in well-fixed specimens, no *intra-retinal space* is seen, as the multilayered neural and monolayered pigment epithelia of the retina are already tightly adherent (the intraretinal spaces shown in (b) and (c) indicate its location – it is only seen as a fixation artefact at these stages of development). The eyelids are first seen at about TS 19, so that by TS 20/21 they are already reasonably well demarcated. Another feature that is clearly seen on histological sections of silver-stained material, and illustrated here, is the presence of large numbers of unmyelinated axons which originate in the cells of the neural retina pass through the region of the future optic disc into the optic stalk towards the *optic chiasma*. The epithelium overlying the eye is starting to form the *cornea*. (b, TS 25) The eyelid has closed, isolating the *conjunctival sac*, and protecting the cornea from any potentially harmful substances in the amniotic fluid. The eye now possesses an *aqueous* and a *vitreous chamber*. The vascular branches of the hyaloid artery are particularly marked. The outer (pigment) layer of the retina is now covered by layers of neural crest-derived *perioptic mesenchyme*, which will form the inner *choroid* and the outer *sclera*, the fibrous "protective" layer of the eye. Note that the lens occupies a substantial proportion of the globe of the eye. (c), At about postnatal day 12–14, the eyelids are beginning to open. The mesoderm that forms on the anterior surface of the lens epithelium (the *pupillary membrane*), possibly with additional neural crest-derived material, forms the iris, with the central part of the pupillary membrane breaking down to form the pupil. At this stage the aqueous chamber becomes subdivided into a *posterior chamber* (between the iris and the lens) and an *anterior chamber* (between the iris and the cornea). The lens is supported along its equatorial region by a *suspensory ligament* (not shown), while the choroid becomes thickened towards the margin of the optic cup to form the *ciliary body*. Branches of the *central artery of the retina* (the nominal successor to the hyaloid artery) spread over the inner surface of the neural retina; most of the branches that pass directly to the posterior surface of the lens regress leaving the *hyaloid canal* along their former pathway.

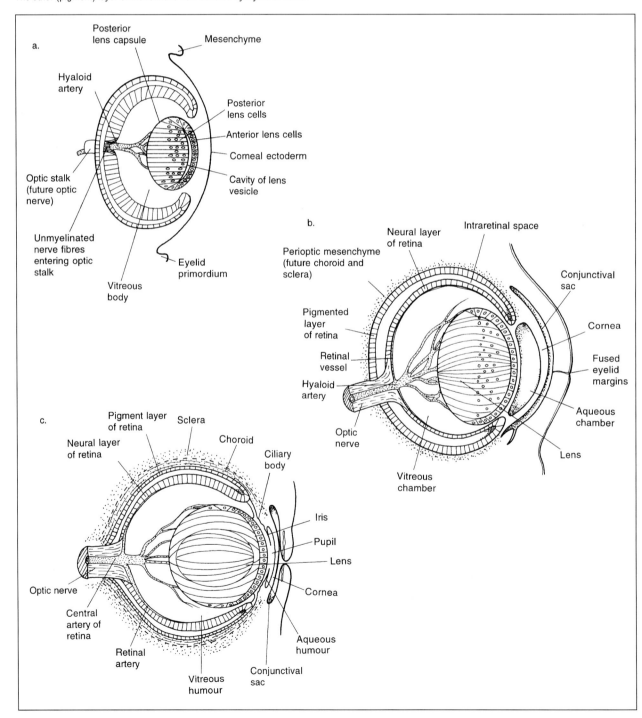

THE EYE AND THE EAR

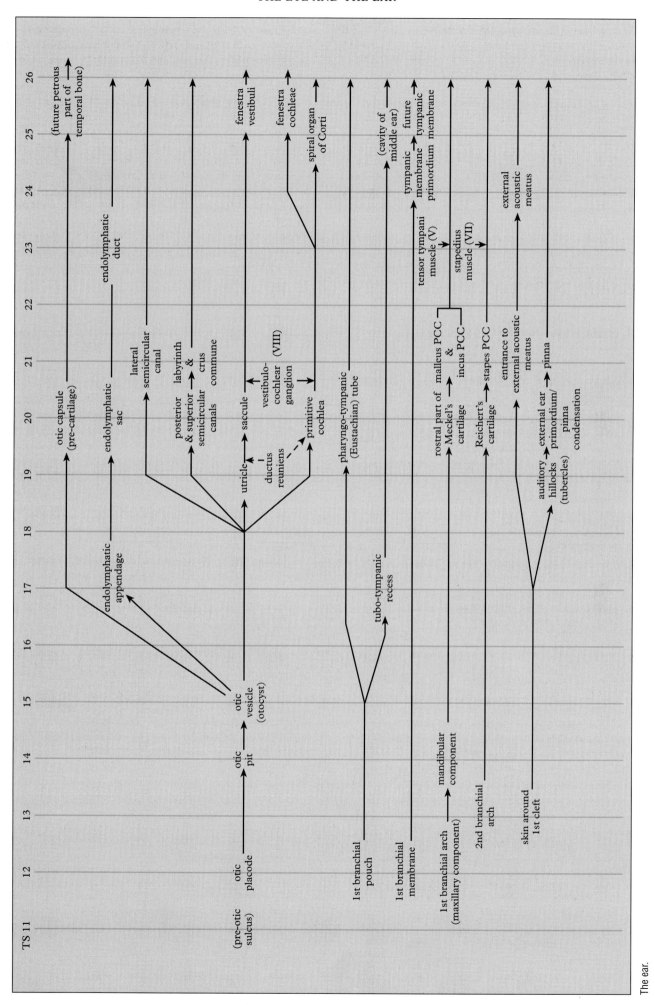

The ear.

The ear

Introduction

The auditory system has three parts, the **inner ear** apparatus, the **middle ear** apparatus and the **external ear**. While they clearly function as an integrated system, they originate from three quite different sources and are thus considered separately. As their complex organization derives from their sensory functions in hearing and balance control, this description of the ear's developmental anatomy is followed by a brief appendix detailing the functions of its various components.

The inner ear apparatus derives from the **otic** (or **auditory**) **placode** which differentiates from a specific region of surface ectoderm at the level of the future hindbrain during the early somite period (TS 12). It will form the **membranous labyrinth** which includes an **endolymphatic diverticulum**, three **semicircular canals**, the **utricle/saccule**, the **cochlear duct** and most of the neurons in the otic ganglion.

The middle ear contains three bony ossicles, the **malleus**, **incus** and **stapes**, the first two of which form from the **first (Meckel's)** and the last from the **second (Reichert's) branchial** (or **pharyngeal**) **arch cartilages**. The ossicles are located within the **tubo-tympanic recess** (*i.e.* the future middle ear cavity), and this is continuous with the **oropharynx** via the **pharyngo-tympanic (Eustachian) tube**. The cavity of the middle ear also becomes linked (after birth) with the air cells within the **mastoid process** of the **temporal** (or **mastoid**) bone. The middle ear and external ear meet at the **tympanic membrane** (the ear drum), that derives from the first pharyngeal membrane, and which transmits sounds from the external environment to the middle ear ossicles, and these in turn transmit them to the inner ear apparatus for eventual transmission via the neural system to the brain by a process of mechanical-to-neural transduction.

The external ear apparatus acts as a funnel to direct the sound from the external environment to the tympanic membrane for transmission to the middle and inner ear apparatus. The external ear forms from a series of **auditory hillocks** which develop on either side of the first pharyngeal groove. These hillocks subsequently amalgamate to form the various components of the external ear, while the first pharyngeal groove (or cleft) forms the **external acoustic** (or **auditory**) **meatus** (or **canal**).

The inner ear

The **otic placode** first appears as a thickening of the surface ectoderm in the prospective hindbrain region (TS 12) where it is closely associated with the **facio-acoustic (VII–VIII) neural crest complex**. At TS 13/14, a slight indentation forms in the central part of the otic placode to give the **otic pit** and this rapidly deepens (TS 14), while its opening narrows and finally closes (TS 15). As this happens, the pit rounds up and separates from the surface, so forming the **otic vesicle (otocyst)**.

The otic vesicle gradually changes its form, first becoming pear-shaped (TS 16/17), then starting to develop the **endolymphatic appendage** (or **diverticulum**) on its medial aspect (TS 17/18). A little later (TS 19), a number of bulges appear on the rostral surface of the otocyst, in the region of the future **utricle**, and these represent the first indication of the **semicircular canals**. The posterior and superior canals first differentiate at about TS 20, while the lateral canal appears somewhat later, at about TS 21. More caudally, in the **saccular region** of the otocyst, another bulge appears early in TS 21 and this is the first indication of the antero-medially directed **cochlear duct** whose general shape is more clearly seen towards the end of this stage. It is also at this stage that the endolymphatic appendage differentiates, while the inner ear complex becomes surrounded by the mesenchymal pre-cartilage mass of the **otic capsule** (the future **petrous part of the temporal bone**). By TS 21, all of the major components of the inner ear apparatus are present. The three semicircular canals now have their ends attached to either the utricle, or, for the common insertion of the posterior/superior canals, into the **crus commune**, while the primitive cochlea (the caudal part of the saccule) has formed the cochlear duct that will differentiate into the **spiral organ of Corti** (TS 25). For a diagrammatic sequence showing the early stages in the differentiation of the inner ear apparatus, see Figure 4.7.4, while the definitive arrangement of its gross morphology is shown in Figure 4.7.5.

The mesenchyme around the otocyst and its derivatives is initially dense (TS 20) and becomes chondrified to form the cartilaginous otic capsule (TS 22). One unusual change that now takes place is that the cartilage near the membranous labyrinth dedifferentiates to form loose periotic tissue. This is particularly evident on either side of the cochlear duct and will, in due course, form the **scala vestibuli** and **scala tympani** (see Appendix). At this stage, the cochlear duct is often termed the **scala media**, and the wall in contact with the scala vestibuli becomes the **vestibular (Reissner's) membrane**.

The middle ear

The three bony ossicles are the principal components of the middle ear and are initially located within the **tubo-tympanic recess** (TS 17), that will form the **tympanic cavity** (or cavity of the middle ear). The **malleus** and **incus** differentiate from the distal (dorsal) end of the first **branchial (Meckel's) cartilage**, while the **stapes** differentiates from the dorsal end of the **second branchial**

Figure 4.7.4. The development of the *left* inner ear apparatus of the mouse from 3-D reconstructions of serially sectioned mouse embryos. All figures are at the same magnification, and orientation. (a) At TS 16/17, the early otocyst has elongated, and, a stage later (b, TS 17/18) it has become even more elongated and slightly flattened from side to side. By TS 19 (c) a rostrally-directed *endolymphatic diverticulum* (or appendage) extends from the medial surface of the otocyst while its elongated caudal end will become the cochlear duct. A flattened region on its posterior surface is the *posterior semicircular canal* primordium, while a more rostral and less well developed area is the *superior* (anterior) *semicircular canal* primordium. By TS 20 (d), the rostral extremity of the endolymphatic duct has dilated to form the *endolymphatic sac*, while the diameter of the rest of the duct has narrowed. Both the superior and posterior semicircular ducts are well differentiated, and the cochlear duct has elongated. A stage later (e, TS 21), the *lateral semicircular duct* has formed and all three semicircular ducts have differentiated. The superior and lateral semicircular ducts are connected by the *crus commune* to the *utricle* which is now connected to the (more elongated) cochlear duct by the *saccule*.

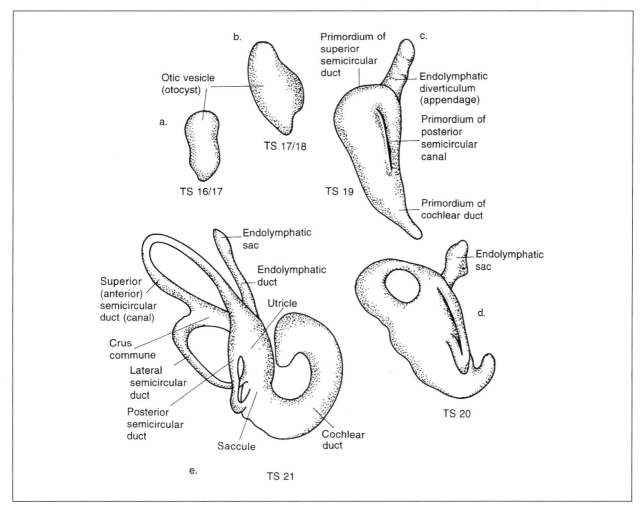

(**Reichert's**) **cartilage**. Both Meckel's and Reichert's cartilages are initially recognized as pre-cartilage condensations (TS 19/20), while the primordia of the ossicles gradually separate from the pharyngeal arch cartilages, with the incus, then the malleus and the stapes first being recognized as individual units at TS 21. The margin of the most medially directed part of the stapes, termed the **stapedial ring** eventually becomes attached to an opening (the **oval window** or **fenestra vestibuli**) which develops on the lateral wall of the otocyst (TS 26), while the long arm of the malleus becomes attached to the **tympanic membrane**. Slightly later, the ossicles ossify, and are connected together by synovial joints. The **tensor tympani** muscle is attached to the malleus, and a smaller muscle, the **stapedius**, is attached to the stapes (TS 23). Since these muscles form from first and second arch mesoderm, their respective nerve supplies are from the **mandibular division of the trigeminal (V)** and **facial (VII) cranial nerves**. Their principal function is to "dampen down" vibration, particularly when the tympanic membrane is exposed to excessive noise. Neither muscle has an opponent, and elastic recoil restores the *status quo* of the tympanic membrane and fenestra vestibuli respectively as these muscles relax.

The **pharyngo-tympanic** (or **Eustachian**) **tube** derives from the dorsal component of the **first pharyngeal pouch** (TS 20) and this extends

TS	1	2	3	4	5	6	7	8	9	10	11	12	13	14	15	16	17	18	19	20	21	22	23	24	25	26
E	0	1	2	3	4	4.5	5	6	6.5	7	7.5	8	8.5	9	9.5	10	10.5	11	11.5	12	12.5	13.5	14.5	15.5	16.5	17.5

Figure 4.7.5. A simplified diagram of the pre-term human inner ear apparatus, an arrangement similar to that of the adult mouse (based on a diagram in Hamilton and Mossman, 1972). The most prominent features are the superior, posterior and lateral semicircular canals, with the superior and posterior ducts sharing the *crus commune*. The ducts are sensory equilibration organs as, at one end, a dilated (or ampullary) region contains a *macula* whose sensitive epithelial cells can recognize fluid movement within the ducts, and so provide information on the location of the head in space and thus help control balance. Similar such cells are also located in the utricle (the *utricular macula*), in the saccule (the *saccular macula*) and along one side of the length of the cochlear canal (the primordium of the *spiral organ of Corti*). Between the utricle and saccule is the *endolymphatic duct* which terminates in the *endolymphatic sac*, while the *cochlear canal* is in continuity with the saccule via the *ductus reuniens*. The *stapes* transmits vibratory stimuli direct to the perilymphatic fluid surrounding the membranous labyrinth of the inner ear apparatus, and this is transduced via neural stimuli to the appropriate region of the brain. The whole of the inner ear apparatus is bathed in perilymphatic/cerebrospinal (CSF) fluid which is in continuity with the CSF in the subarachnoid space via the *perilymphatic duct* (or *cochlear canaliculus*). The entire membranous labyrinth of the inner ear apparatus is protected within the *petrous part of the temporal bone*. Each of the specialized regions of the membranous labyrinth receives a sensory nerve supply from the corresponding vestibular or cochlear division of the auditory (VIII) nerve. For further details of the anatomy of the bony and membranous labyrinth, see Williams (1995).

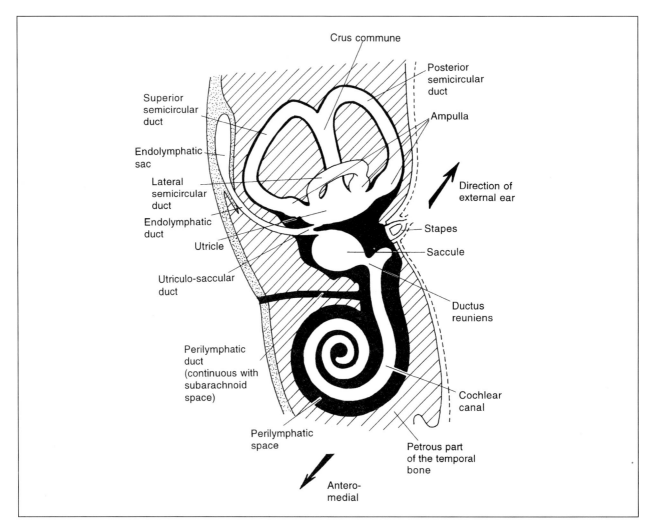

dorso-laterally to become continuous with the **tubo-tympanic recess** (about TS 22) which gradually enlarges to form the **tympanic cavity**. This cavity within the future middle ear gradually envelops the ossicles, their associated muscles and ligaments and their nerve supply, and subsequently extends dorsally, forming a much enlarged cavity within the petrous temporal bone termed the **tympanic** (or **mastoid**) **antrum** (antrum: cavity or chamber, especially within a bone). After birth, this is continuous with the **mastoid air cells** that form within the mastoid process of the temporal bone (the bony protuberance felt just behind the lower half of the ear). The postnatal growth of this process is believed to be induced by the action (or "pull") of the sternocleidomastoid muscle on its insertion on the lateral surface of the mastoid part of the temporal bone.

The external ear

The external ear develops on either side of the **first branchial groove** (or **cleft**), which separates the **first** and **second branchial arches** (TS 13). This deepens considerably (TS 14) to give rise, in its deepest part to the outer (ectodermal) component of the first pharyngeal membrane (the future *ectodermal* part of the **tympanic membrane** which differentiates after birth). This bilaminar structure has a layer of surface-derived ectoderm on its outer surface and pharyngeal endoderm, the lining of the **first branchial pouch**, on its inner aspect, there being no intervening mesoderm.

By about TS 19, a series of small bulges (the **auditory hillocks** or **tubercles**) have formed on either side of the first pharyngeal groove and these subsequently amalgamate (TS 20) to form the external ear (or **pinna**). By TS 21, a well-formed, anteriorly-directed pinna has developed, and the **external acoustic meatus** (or canal), as yet uncanalized, points antero-medially. There is as yet only minimal evidence of the tubo-tympanic recess and it is some distance from the floor of the external acoustic meatus, where the tympanic membrane will form. The expanding pinnae cover about one-half of the entrance of the external acoustic meatus by TS 22, three-quarters by TS 25, while, at the time of birth, little of the entrance is visible. It is noticeable that the external acoustic meatus, like many other components of the ear, is still immature at birth, it being largely uncanalized and filled with an epithelial plug (Figure 4.7.6).

The nerve supply

The nerve supply to the ear is complex. The principal supply is to the *inner ear* apparatus, and is via the **vestibular** and **cochlear divisions of the auditory (VIII) cranial nerve** which are principally involved in the control of balance and perception of sound, respectively. When, however, the otic pit is differentiating into the otic vesicle (TS 14/15), only the **acoustico-facial (VII–VIII) pre-ganglion complex** is present, located between it and the wall of the hindbrain. The current view is that the majority of the glial cells and some, at least, of the neurons within what will become the vestibular part of the ganglion are neural crest-derived, the otic ganglion itself is placodal in origin (Fritsch *et al.*, 1997). At TS 16/17, the facio-

Figure 4.7.6. The relationship between the various components of the external, middle and inner ear apparatus in the newborn mouse. The external ear is still immature as the majority of the pinna (which is directed forwards) has yet to separate from the surface ectoderm on the side of the head, although it does contain some elastic fibrocartilage "core" that will in due course help maintain its shape. No obvious *external acoustic meatus* is present, and its future location is indicated by the presence of a *meatal plug*. Histological examination of sections through this region show continuity between the endodermal-lining of the *pharyngo-tympanic (Eustachian) tube* and the ectodermally-derived meatal plug. In the region of the middle ear, the three ossicles (malleus, incus and stapes) are present, but the synovial joints between them (location indicated by the two arrows) have not yet differentiated. The pharyngo-tympanic tube is directed laterally from the oropharynx, but does not extend beyond the region around the proximal part of the long arm (the manubrium or handle) of the malleus. At this stage, the ossicles are immobile, and surrounded by mesenchyme, there being no evidence of a tympanic cavity. Similarly, at this stage, it is difficult to discern the location of the future tympanic membrane. The margin of the stapedial ring does, however, contact the *oval window* in the lateral wall of the *perilymphatic space*. All of the inner ear components are, however, present and seem well-differentiated. Given the relative immaturity of the external and middle ear regions, it is likely that the mouse's perception of hearing at this stage is extremely poor. However, as all of the components of the inner ear apparatus are well-formed, its 3-D awareness may be much better. Note that the nerve supply here is not shown.

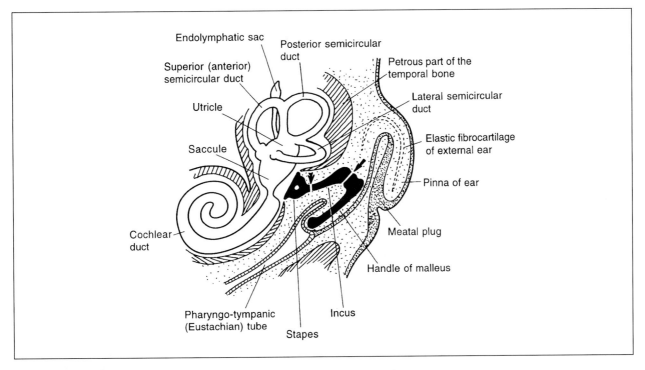

TS	1	2	3	4	5	6	7	8	9	10	11	12	13	14	15	16	17	18	19	20	21	22	23	24	25	26
E	0	1	2	3	4	4.5	5	6	6.5	7	7.5	8	8.5	9	9.5	10	10.5	11	11.5	12	12.5	13.5	14.5	15.5	16.5	17.5

acoustic (VII/VIII) neural crest complex is readily recognized and associated with the mid-ventral part of the otocyst, with the facial (VII) component being the more laterally located of the two. By TS 17/18, the facio-acoustic ganglion complex is largely associated with the rostral part of the ventral surface of the otocyst. By TS 18, this single ganglion complex is more clearly defined while, by TS 19, it is partitioned into two discrete components, the **facial (VII)** and **vestibulo-cochlear (VIII) ganglia**.

As the inner ear develops, each component of the membranous labyrinth receives its own discrete sensory nerve supply from either the corresponding vestibular or cochlear divisions of the auditory nerve. By TS 19, the rostral half of the otocyst has expanded to a greater extent than the caudal half, and the distinct facial and acoustic ganglion complexes are now mainly associated with the ventral surface of the caudal part of the developing inner ear apparatus. By TS 20, the acoustic ganglion has itself subdivided into two components, forming the vestibulocochlear (VIII) ganglion complex, and its rootlets from the basal region of the pons are readily recognized. By TS 21, the vestibular and cochlear ganglia are distinct, with the former sending branches to the semicircular canals and utricle/saccule, and the latter to the cochlear duct.

The nerve supply to the tensor tympani and stapedius muscles is via the **mandibular division of the trigeminal (V)** and **facial (VII) cranial nerves**, respectively. Both of these muscles act in unison, exerting a protective effect upon the inner ear by dampening high-intensity sound vibrations that might otherwise damage it (see above).

The nerve supply to the *external ear*, the external acoustic meatus and tympanic membrane (the ear drum) has several components. These are the **trigeminal (V) cranial nerve**, via its **auriculo-temporal branch**, the **facial (VII) cranial nerve**, via branches from the **tympanic plexus**, the **auricular branch of the vagus (X) cranial nerve** and the **second cervical (C2) nerve**, via branches from the **great auricular nerve**.

The blood supply

The blood supply to the auditory apparatus is also complex, and may be summarized as follows:

The *external ear*: from the posterior auricular and superficial temporal arteries; near the drum, from the deep auricular branch of the maxillary artery.

The *middle ear*: via branches from the external carotid artery – from the tympanic branch of the maxillary artery and stylo-mastoid branch of the posterior auricular artery, also from branches of the middle meningeal artery, the ascending pharyngeal artery and sometimes additional small branches from the internal carotid artery.

The *internal ear*: via cochlear and vestibular branches of the labyrinthine artery.

The *auditory tube*: via branches from the ascending pharyngeal and middle meningeal arteries.

Appendix: Functions of the auditory/vestibular apparatus

Given that the auditory/vestibular apparatus, with its external, internal and middle ear components is extremely complex, it seemed sensible to include here a synopsis of the physiological functions of its various parts.

While the external ear and the various components of the middle ear apparatus form critical components of the auditory system, and are exclusively involved in hearing, the inner ear apparatus has both an auditory and a vestibular function. In the latter case, it is involved in providing information on spatial orientation and movement of the head and the maintenance of a stable retinal image, as well as being involved in the maintenance of body posture and co-ordinating the movement of the limbs and body. It is easier to consider these various functions separately, though in practice they are almost invariably inter-related activities. For a simplified diagram illustrating the relationship between the components of the external and middle ear and the inner ear apparatus in the newborn mouse, see Figure 4.7.6.

Hearing

Sound energy from the external environment is converted into appropriate neural signals in the following way: the vibrations produced by external sounds which impede on the tympanic membrane are transduced through a complex (largely mechanical) chain of events into action potentials by the hair cells in the cochlear apparatus. The various signals are then transferred to the special centres in the brain where their source and nature are recognized and appropriately interpreted.

The external ear

Sound is detected by the ear as a consequence of variations in air pressure which impede on the tympanic membrane. The pinna of each external ear directs the sound waves into the external acoustic (or auditory) meatus (or canal), while the presence of two ears facilitates recognition of the origin of the source of the sound.

The middle ear

The cavity of the middle ear is filled with air, and is in continuity with the pharynx via the pharyngo-tympanic (or Eustachian) tube. Because of this arrangement, the air in the middle ear may readily be equilibrated with that of the outside pressure. Vibrations of the tympanic membrane are transferred through and amplified by the chain of middle ear ossicles to the **oval window** which represents the outer boundary of the fluid-filled inner ear apparatus (or membranous labyrinth). The **malleus** links to the tympanic membrane and is connected to the **incus** which is in turn connected to the **stapes** through a system of synovial joints. The stapes is attached to the peripheral margin of the oval window. By this system of levers, vibration energy from the air which impedes on the outer part of the tympanic membrane is amplified and transferred to the liquid environment of the inner ear.

The response to high energy sounds is mediated by two muscles: the **tensor tympani** (this arises from the cartilage of the Eustachian tube and adjoining greater wing of the sphenoid, and is inserted into the upper end of the handle of the malleus; this muscle draws the malleus and tympanic membrane medially, and makes the membrane more tense) and the **stapedius muscle** (this arises from a small conical projection termed the "pyramid" of the middle ear cavity, and is inserted into the posterior surface of the "neck" of the stapes; it tilts its anterior (inferior) end (the "foot-piece") laterally, and the posterior end medially). When the auditory system is stimulated by high energy sounds, they contract simultaneously and have a protective "dampening" effect on the energy transferred from the external environment to the inner ear apparatus.

The inner ear

The inner ear apparatus consists of the **bony labyrinth**, a series of cavities within the petrous part of the temporal bone, within which is located the **membranous labyrinth**, a series of communicating sacs (the utricle and saccule) and ducts (the semicircular ducts and cochlear duct) all of which are contained within the bony cavities. The **cochlea** is a spiral tube divided into three compartments – the centrally located **cochlear duct** (or scala media) filled with *endolymph*,[5] which is separated from the scala vestibuli above) by **Reissner's membrane**, and from the scala tympani (below) by the **basilar membrane**. The scala vestibuli and scala tympani are filled with *perilymph*. The basal region of the stapes is attached to the circumference of the oval window (or fenestra vestibuli) by the annular ligament, and allows waves to be conducted along the scala vestibuli and through the Reissner's membrane to the cochlear duct. The pressure wave passes along the scala vestibuli, depresses the basilar membrane and this causes the **round window** to bulge and thus dissipate the energy into the cavity of the middle ear.

The sound receptors are located in the **organ of Corti** which lies on the basilar membrane within the cochlear duct. Hair cells within the organ of Corti stimulate cochlear fibres which run in the auditory (VIII) cranial nerve. Sound frequencies are recognized by their stimulation of specific hair cells along the basilar membrane. All of the auditory stimuli from the auditory nerve terminate in the medulla. From there, via a series of relay nuclei (in the inferior colliculi and medial geniculate bodies), the neural stimuli eventually pass to the primary auditory cortex. The latter area is concealed in the ventral wall of the lateral fissure (corresponding to areas 41 and 42 of Brodmann, the landmarks for the auditory area are also called the anterior transverse temporal gyri, or Heschl's convolutions).

Vestibular function

As already mentioned, the various components of the inner ear apparatus that are involved in providing vestibular information (*i.e.* on spatial orientation, co-ordinating movements of the head and eye, and in regulating the maintenance of body posture) are anatomically adjacent to those parts of the inner ear apparatus (principally the cochlea) involved in hearing. The components of the membranous labyrinth that are specifically related to vestibular function are the **semicircular canals**, the **utricle** and the **saccule**. The various sacs and tubes that constitute the membranous labyrinth are filled with endolymph, and this system is surrounded by perilymph, and this in turn is protected within the bony labyrinth.

The semicircular canals

Each of the superior, posterior and lateral semicircular canals makes contact with, and their endolymphatic contents are in continuity with, the lateral part of the

---

[5] The *endolymph* is exclusively contained within the closed system of the membranous labyrinth, and is said to closely resemble intracellular fluid in its ionic composition, being rich in potassium but poor in sodium ions. While the endolymph is believed to be a secretion, its exact origin is unknown, although it has been suggested that it may be produced by specialized cells in the utricle and semicircular canals, and/or within the cochlear duct. It is believed to circulate, then pass into the endolymphatic duct to be removed by the specialized epithelial cells of the endolymphatic sac into the surrounding vascular plexus (Williams, 1995).

Perilymph, in contrast, closely resembles cerebrospinal fluid in its biochemical composition and the perilymphatic space is in continuity with the cerebrospinal-fluid filled subarachnoid space via the perilymphatic duct (or *cochlear canaliculus*). It is believed that perilymph is formed from three sources: (i) as a transudate from the blood vessels surrounding the perilymphatic spaces; (ii) from the fluid spaces surrounding the sheaths of the vestibulocochlear nerve fibres; and (iii) from a slow continuous flow of cerebrospinal fluid along the cochlear canaliculus. Its composition is said to be roughly comparable to extracellular tissue fluid. Its site of removal has yet to be established with certainty.

TS	1	2	3	4	5	6	7	8	9	10	11	12	13	14	15	16	17	18	19	20	21	22	23	24	25	26
E	0	1	2	3	4	4.5	5	6	6.5	7	7.5	8	8.5	9	9.5	10	10.5	11	11.5	12	12.5	13.5	14.5	15.5	16.5	17.5

utricle, and the sites at which this happens, known as the *ampullae* of the canals, are dilated. Each contains sense organs that respond to rotatory acceleration of the head. It is a critical feature of this system that the 3 semicircular canals lie almost at right angles to each other (*i.e.* they are in nearly orthogonal planes). The ampullae contain the *otolith organs* that are sensitive to the direction of the force of gravity and to linear acceleration of the head. The fact that the semicircular canals of the left and right sides are mirror images of each other allows the brain to receive complementary information relating to any movement.

The utricle and saccule

The utricle is a sac-like structure to which the ends of the semicircular canals are attached. Its anteriorly directed outflow passes towards and into the base of the *endolymphatic sac* via the *utriculo-saccular duct*, and then into the saccule. This is a globular-shaped sac, smaller than the utricle, whose anterior outflow is into the cochlear duct via the *ductus reuniens*. This sytem is filled with endolymphatic fluid.

Both the utricle and saccule contain special sense organs termed *maculae*; these consist of thickened regions containing hair cells associated with nerve endings that are stimulated according to the direction that the hair cells are bent. The hair cells are embedded in a gelatinous mass called the *otolith membrane* which contains calcium carbonate crystals called *otoconia* (or otoliths). The movement of the otolith membrane is gravity-dependent, as these crystals give this membrane a greater specific gravity than the endolymphatic fluid in which it is bathed. The neural output from the stimulation that occurs when the hair cells embedded in it bend, provides information on the orientation of the head in space.

Neural information from the vestibular apparatus passes to the vestibular nuclei in the brainstem, and these are connected to the nuclei of the III, IV and VI cranial nerves which control eye movement. These are an important component of the *visual fixation* reflex, that allows the eye to maintain its focus on a fixed point when the head is moving and vice versa. Input to, and feedback from, the cerebellum also plays a critical role in vestibular function.

The function of the *endolymphatic duct* and its dilated terminal part, the *endolymphatic sac*, is believed to be related to the circulation of the endolymphatic fluid, but it may also act as a "release valve" should the pressure within the endolymphatic system rise too rapidly, and need to be dissipated. The perilymphatic duct, which communicates between the perilymphatic space which surrounds the membranous labyrinth and the subarachnoid space probably also serves a similar function, although it is usually the bulging of the round window into the cavity of the middle ear that allows the pressure wave within the perilymphatic space to be dissipated.

For further details of the physiology of the auditory/vestibular system, the interested reader should refer to an appropriate specialist textbook on this topic, such as Brookhart and Mountcastle (1984), Melloni (1957), Wolfson (1966); and for a more general review, see Bray *et al.* (1994); see also Williams (1995) for detailed information on the anatomical features of this system.

4.8 The skull

Introduction

The major components of the skull are the **neurocranium** that surrounds the brain and the **viscerocranium** that constitutes the bones of the face. The **neurocranium** is subdivided into the **chondrocranium** (which represents the majority of the base of the skull), and the **vault of the skull** (or **calvarium** or skull cap).[1] All of these bones form from somitomere and neural-crest mesenchyme, apart from the occipital bone (with the exception of its interparietal part) which forms from the sclerotome of the first four occipital somites. The **mandible** is technically a component of the skull and is therefore discussed here. The mandible forms from two units that fuse together across the ventral midline (at the **symphysis menti**), unlike the other principal bony components of the skull which are either formed from discrete entities or (*e.g.* the temporal bone) by the amalgamation of viscerocranial, chondrocranial and calvarial precursors. Because the skull is such a complex structure, it is most easily understood when illustrated diagrammatically and its individual elements shown in accordance with their location in the adult skull (Figure 4.8.1).

Most of this chapter discusses the times of formation of the mesenchymal condensations and the ossification centres that will generate the bones of the skull. In general, these condensations start to form at about TS 21 and ossification then takes place from about TS 22 (E13.5) onwards, or over the last third of embryogenesis. As many of these processes are the results of events that took place some time earlier and that also reflect a wide variety of evolutionary and mechanistic interactions, this chapter starts with a brief background to these various aspects of bone morphogenesis.

The evolutionary context

The mammalian skull is a complex structure made up of a number of components of different origin, and the arrangement seen in the mouse and in other higher mammals is readily understood on the basis of the evolutionary origin of its component parts. In primitive vertebrates such as the turtle, the *chondrocranium* is well developed, having evolved to protect both the brain and the sense organs for vision, hearing and olfaction. There is also a contribution via the **pterygoquadrate** (or the **palatoquadrate**) and Meckel's cartilages (both of which are first arch derivatives) to the development of the upper and lower jaws.[2]

The chondrocranial element involved in the development of the mammalian lower jaw, Meckel's cartilage, is a transient structure, but probably acts as a guide for the membranous structure that subsequently ossifies to form the definitive mandible, as well as giving rise to the **malleus** bone, one of the

---

[1] As in other regions of the body, the terminology used in the literature is quite variable and often confusing, and that used by Thorogood (1988), for example, where the chondrocranium consists of the viscerocranial and neurocranial components is not followed here.

[2] At an early stage in the evolution of the vertebrate skull, it appears likely that the pterygo-quadrate bar and Meckel's cartilage were continuous structures, but at a slightly later stage of evolutionary development a joint cavity formed between them. The dorsal part of the second arch cartilage (Reichert's cartilage) comes to articulate with the postero-medial part of the pterygo-quadrate cartilage, being located between the latter and the otic capsule. The various parts of Meckel's cartilage are now seen to give rise to a variety of derivatives. Its *intermediate* part regresses, becomes ligamentous and forms the anterior ligament of the **malleus** and the **spheno-mandibular ligament**, its most *dorsal* part (in contact with the pterygo-quadrate cartilage) gives rise to the cartilaginous primordium of the **malleus**, while its *ventral* part is involved in the development of the lower jaw. The pterygo-quadrate cartilage itself forms the cartilage primordium of the **incus**. The anterior part of the pterygo-quadrate cartilage is believed to contribute to the greater wing of the **sphenoid bone** (the **ala temporalis**). That part of Reichert's cartilage in contact with the incus separates from the rest of the cartilaginous bar to form the **stapes**.

TS	1	2	3	4	5	6	7	8	9	10	11	12	13	14	15	16	17	18	19	20	21	22	23	24	25	26
E	0	1	2	3	4	4.5	5	6	6.5	7	7.5	8	8.5	9	9.5	10	10.5	11	11.5	12	12.5	13.5	14.5	15.5	16.5	17.5

Figure 4.8.1. The bones of the skull showing the *viscerocranium* (the facial skeleton), the *vault of the skull* (the calvarium) and the *chondrocranium* (the cartilaginous base of the skull). For most purposes, the mandible is usually considered as one of the viscerocranial elements. The location of the hyoid bone is also shown. Note that the temporal bone forms from the amalgamation of a number of calvarial and chondrocranial elements: the petromastoid part is the substantial bony element that develops around the middle and inner ear apparatus; the tympanic ring, which largely surrounds the tympanic membrane, and the squamous and zygomatic parts of the temporal bone all of which form in "membrane".

The squamous part is a large flattened plate of bone that articulates above and posteriorly with the parietal bone and the zygomatic process of the temporal bone articulates with the temporal process of the zygomatic bone (the two helping to form the zygomatic arch). The occipital bone is also made up of various components, some of which are of chondrocranial origin (the basiocciput, exocciput and supraocciput) and one of which (the interparietal part) is of calvarial origin, though rarely the interparietal part may remain as a separate bone, and is then termed the interparietal bone (the various ossification "centres" that amalgamate to form the occipital bone are shown in Figure 4.8.3).

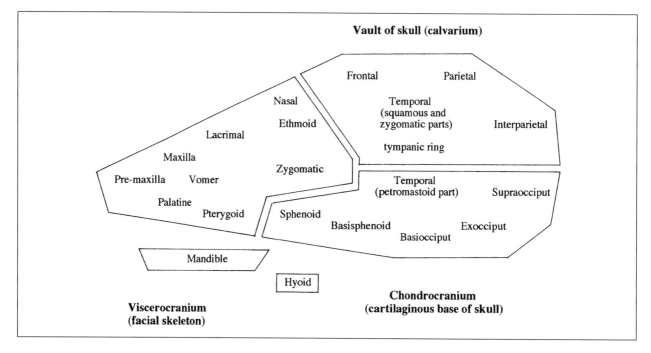

middle ear ossicles. The chondrocranial derivative that forms a component of the upper jaw is the pterygoquadrate cartilage which gives rise to the **incus**, and possibly the **greater wing of the sphenoid bone**.

These bones all form from cartilage models, while the "dermal" bones of the calvarium and the bones that constitute the viscerocranium only evolved later, so providing additional protection for the brain and sense organs. The maxilla, zygomatic bone, the squamous part of the temporal bone and other smaller bony elements and the mandible are all believed to be formed by the direct ossification of branchial arch "dermal" mesenchyme. In modern vertebrates, the chondrocranium is restricted to the base of the skull, the region around the inner ear apparatus (the **petrous part of the temporal bone**) and the **nasal capsules**.

It is worth noting the striking degree of chondrocranial conservation seen during vertebrate evolution. In lower vertebrates, three pairs of cartilaginous capsules protect the nasal, otic and optic sense organs. These fuse to form the plate-like cartilage around the margins with the olfactory, optic and otic capsules protecting the organs of smell, sight and hearing. These capsules (Figure 4.8.2) have been substantially modified as the higher vertebrates have evolved: in these, only the **otic capsule** is now readily recognized as the precursor of the **petrous part of the temporal bone**; the **nasal** (olfactory) **capsules**

unite with each other and with the anterior part of the trabecular cartilages to form the ethmoid bone, while the **optic** or **sclerotic capsule** does not chondrify in mammals, nor fuse with the basal cartilaginous complex.

The reason why the optic capsule does not chondrify in mammals and there is thus no bone directly protecting the optic region remains unclear, but one possibility is that such a bone might inhibit rapid movements of the eye. The evolutionary successor of this bone is the **sclera** of the eye (for a review of craniofacial growth in terms of the evolution of the skull, see Gans, 1988).

The region of the base of the skull that forms the **foramen magnum**, through which the spinal cord passes, is derived from the amalgamation of a series of cartilaginous units, which subsequently ossify, the largest component of which is formed from the **basiocciput**. On either side of this ventral unit are the **exoccipital bones**, while the dorsal arch is formed from the **supraoccipital bone**. The ventral part of the latter develops within a cartilaginous precursor, while its more rostral half develops within a membranous precursor (to form the **squamous part of the occipital bone**). The latter may remain as a separate unit throughout life, and is then termed the **interparietal bone** (Figure 4.8.3).

Two physiological processes may have markedly influenced the evolution of shape of the mammalian

Figure 4.8.2. The arrangement of the base of the ancestral vertebrate skull (based on diagram in Larsen, 1993). It is made up of three pairs of cartilaginous plates, the *prechordal*, *hypophyseal* and *parachordal cartilages*, with the notochord extending rostrally into the skull base where it is surrounded by the parachordal cartilages. Three cartilaginous capsules protect the olfactory apparatus (the *olfactory capsule*), the optic apparatus (the *optic capsule*) and the otocyst or otic vesicle (*otic capsule*). Evidence of this ancestral arrangement is still recognizable in the mammalian skull (Figure 4.8.4), though much modified, in that the ossified skull base develops within a cartilaginous model (the *chondrocranium*).

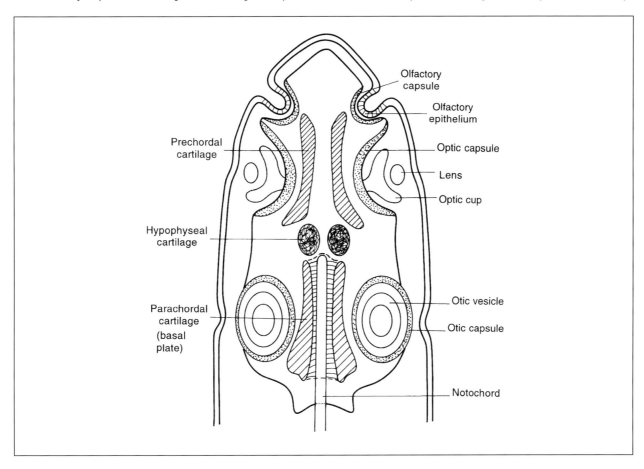

face, (i) the use of thermoregulation for cooling the brain, using the heat exchange capacity of the nasal passages, and (ii) the requirements of particular sensory systems – the face of a tarsier with primary visual orientation is markedly different from that of a canid with a primarily olfactory one. In relation to the increased size of the brain in mammals, it is noted that post-term growth leads to relatively late closure of cranial sutures (Gans, 1988).

A segmental origin of the embryonic vertebrate skull, based on the somitomeric organization of this structure, has been proposed by Meier (1981; Jacobson and Meier, 1987), despite the fact that head somites are never recognized as such. This contrasts with the view that the head is an unsegmented structure (Alberch and Kollar, 1988). A modified view now prevails, based on the current view that the occipital region is of segmental origin, as is the viscerocranium and those structures derived from the gill arches.

Experimental studies

A number of avian experimental studies have been undertaken to determine the embryological origin of the various components of the skull and these findings may well be relevant to the situation in mammals. Those craniofacial cartilages which have been studied in avian embryos include the otic and nasal capsule, Meckel's and the scleral cartilage, and the cartilaginous floor of the skull. In all cases, the mesenchyme (whether of neural crest or mesodermal origin) interacts with an epithelium (either neural or epidermal) before chondrogenesis can begin (for reviews, see Hall, 1987; Noden, 1984). Thorogood (1981) has also pointed out that the *circumstances of tissue association* are important for the differentiation of neural crest-derived cells into primary cartilage and membrane bone. Thus, for example, disturbances of the conditions for migration of these cells will generate anomalies of craniofacial morphogenesis and of skull development (see also Morriss and Thorogood, 1978).

TS	1	2	3	4	5	6	7	8	9	10	11	12	13	14	15	16	17	18	19	20	21	22	23	24	25	26
E	0	1	2	3	4	4.5	5	6	6.5	7	7.5	8	8.5	9	9.5	10	10.5	11	11.5	12	12.5	13.5	14.5	15.5	16.5	17.5

Figure 4.8.3. The bony elements that fuse together (i.e. the ossification "centres") to form the adult occipital bone. A ring of four bony elements forms around the *foramen magnum* through which the caudal part of the hindbrain is in continuity with the rostral part of the spinal cord. The basal or basilar part is formed from the basioccipital bone (or basiocciput); the lateral parts are formed from the exoccipital bones (or exocciput) on the infero-medial parts of which are located the *occipital condyles* that articulate with the superior facets of the atlas (C1) vertebra. The fourth element completes the superior part of the boundary of the foramen magnum, and is formed from the supraoccipital bone (or supraocciput). All of the ossification "centres" of the bones so far mentioned form within cartilage models and are consequently part of the chondrocranium. Above the highest nuchal line, the upper part of the squamous occipital bone forms in "membrane", within a fibrous primordium. The line of union between the upper and lower parts of the squamous occipital bone is usually recognized in the adult bone.

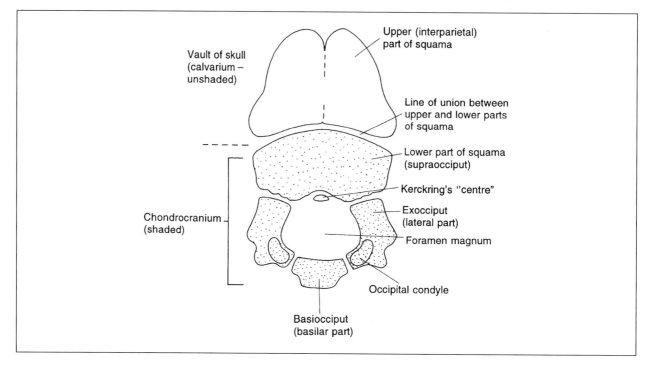

One of the most instructive studies was that undertaken by Couly *et al*. (1993) who used chick–quail grafts to investigate the origin of the various components of the bones of the skull. The skull base was divided into a prechordal part (in front of the extreme tip of the notochord – which reaches the sella turcica that surrounds the pituitary) and a part caudal to this boundary. They demonstrated that the prechordal component was derived entirely from neural crest, while the more caudal part was derived entirely from cephalic or somitic mesoderm. The parietal bone and part of the otic region was of crest origin, as was the basisphenoid, although the postsphenoid was derived from cephalic mesoderm. The occipito-otic region had a particularly complex origin, being from the paraxial mesoderm of the first five somites at the level of the hindbrain.

The origin of the frontal region of the skull of the chick embryo was studied by Tyler (1983) to determine whether epithelial influences were necessary for frontal bone development. Neural ectoderm, though required during early stages of development to induce frontal bone formation was not required during later stages (Hamburger and Hamilton (1951) stages 22–30) for osteogenesis. Epidermis was, however, required for frontal bone development during all of the stages tested: frontal mesenchyme formed bone when epidermis was present on the outer aspect of the mesenchyme, but did not when the epidermis had been removed prior to grafting, whether or not neural ectoderm was present. The presence of neural ectoderm was believed to have an inhibitory effect on chondrogenesis in the ectomeninx, though this could form cartilage in the absence of neural ectoderm.

In order to study the origin of the mandible in chick embryos between Hamburger and Hamilton stages 16–25, the mandibular processes were enzymatically separated into mesenchymal and epithelial components and grown either in organ culture or as grafts to the chorioallantoic membranes of host embryos (Tyler and Hall, 1977). Differentiation under both experimental conditions was normal, but the time of differentiation differed from that *in vivo*, and depended on the age of the embryo from which the mesenchyme had been obtained. Intramembranous ossification was observed only in the material isolated from the older embryos. It was suggested that the mandibular mesenchyme required the presence of epithelium in embryos until 4.5 days of incubation if membranous bones of the mandible were to differentiate. Mandibular epithelium was also shown to be capable of differentiating in the presence of "foreign" fibroblasts derived from the chorioallantoic membrane

Membrane and dermal bones (calvarium)

The bones of the skull form in two distinct ways: either directly from mesenchyme condensations with such "dermal" bones being said to ossify "in membrane", or through an intermediate stage where the condensed mesenchyme produces a cartilage "model", and these bones are said to ossify "in cartilage". In the skull, all of the **chondrocranium** ossifies in cartilage. The mandible largely forms in membrane, as do all of the components of the **viscerocranium** and **vault of the skull** (or **calvarium**) which are made up of a series of "dermal" bones. It is thought that the bones of the base and vault of the skull derive from the sclerotome component of the cranial paraxial mesoderm, while the mesenchyme precursors of the bones of the viscerocranium are from neural crest. There is thus no obvious lineage relationship between origin of mesenchyme and the type of bone that it forms.

Ossification of the base of the skull (the *chondrocranium*) takes place within a number of discrete "centres" in the cartilage model. These centres subsequently expand until the entire base of the skull is ossified to form a single complex unit. A particularly important growth centre is located between the **basi-occiput** and **basi-sphenoid bones**, and, as ossification proceeds here, spaces or foramina are left to allow for the passage of both cranial nerves and blood vessels whose pathways to and from the brain had already been well established at a considerably earlier stage of gestation.

The *vault of the skull* ossifies from a number of separate centres, and these eventually generate the **frontal**, **parietal**, and the **squamous parts** of the **occipital** and **temporal bones**. These membrane bones are separated by **sutures**, a special type of joint consisting of fibrous tissue that initially allows some degree of movement between the plates of bone, although they later ossify after the cessation of brain growth. In regions where the sutures are particularly wide, usually where three or more of these membrane bones are apposed, the sutural spaces are called **fontanelles**.[3]

All of the bones of the face (the *viscerocranium*) develop in membrane with ossification "centres" forming within mesenchyme precursors that derive from **cranial neural crest** of pharyngeal arch origin from the midbrain and hindbrain regions. The **mandible** also develops in membrane around, but principally lateral to **Meckel's cartilage**, a derivative of the **first branchial arch cartilage**. This cartilage also forms the **spine of the sphenoid bone**, and the **malleus** and **incus bones** of the middle ear (or **auditory ossicles**). The two bony derivatives of the second arch cartilage in the skull are the **styloid process of the temporal bone** and the **stapes** (the third of the middle ear ossicles) (for further details, see Chapter 3.3 *The branchial arch system* and Chapter 4.7 *Sensory organs: the eye and the ear*). For a comprehensive anatomical analysis, particularly as to how the skeletal elements derive from cephalic paraxial mesoderm, see de Beer (1937).

The facial bones that constitute the *viscerocranium*, are essentially modified pharyngeal cartilages that have been supplemented and largely replaced by "dermal" bones. The available evidence suggests that the viscerocranium is largely neural crest derived. The dermal bones of the face are represented by the maxillary and mandibular processes of the first pharyngeal arch. The maxillary processes are themselves derived from four pairs of ossification centres which form: the premaxilla, the maxilla, the zygomatic and squamous temporal bones (the latter fuses with the petrous part of the temporal bone).

There is also a small contribution to the skull and its associated bones from the second arch cartilages. That part of the cartilage below the stapes fuses with the otic capsule (which subsequently goes on to form the petrous part of the temporal bone) and later ossifies to form the **styloid process** (of the temporal bone). It is continuous with the **stylo-hyoid ligament** which connects the styloid process with the **lesser horn** (or cornu) of the **hyoid** bone. The caudal part of the body and the greater horn (cornu) of the hyoid bone are the sole derivatives of the third arch cartilage which plays no part in the development of the skull (see Hamilton and Mossman, 1972). For further details, see Chapter 3.3 *The branchial arch system*.

---

[3] Sutures are located where the dermal bones of the vault of the skull come together and at the sites of apposition between the dermal bones and the bones of the chondrocranium. These fibrous joints which contain non-ossified connective tissue can widen to form fontanelles, usually at the junction between three or four bony plates. The largest is the **anterior fontanelle** which forms at the junction between the two frontal and two parietal bones (located at the **bregma**, where the metopic, sagittal and coronal sutures meet). The **posterior fontanelle** is substantially smaller, and is located at the site of apposition of the posterior margins of the two parietal bones and the upper margin of the squamous part of the occipital bone, and is thus at the **lambda** (the site where the posterior part of the sagittal suture meets the two medially-directed parts of the lambdoid suture). Two equally small fontanelles are located on the side of the skull: the **sphenoidal fontanelle**, at the site of apposition of the frontal, parietal, squamous part of the temporal and sphenoid bone, and the **mastoid fontanelle** located between the squamous and mastoid parts of the temporal bone, the occipital and the parietal bones. The fontanelles narrow after birth as the dermal bones expand and eventually disappear. In relation to the increased size of the brain in mammals as compared to more primitive vertebrates, it should be noted that post-term growth leads to the relatively late closure of the cranial sutures.

TS	1	2	3	4	5	6	7	8	9	10	11	12	13	14	15	16	17	18	19	20	21	22	23	24	25	26
E	0	1	2	3	4	4.5	5	5.5	6	6.5	7	7.5	8	8.5	9	9.5	10	10.5	11	11.5	12	12.5	13.5	14.5	15.5	16.5	17.5

The chondrocranium and cranial base

In the mesenchyme between the primitive brain and the foregut, pre-cartilage and subsequently cartilaginous elements are first seen at about TS 21/22 surrounding the rostral end of the notochord. Initially two plates may be observed one on either side of the notochord, but these rapidly fuse together to form a central element, the **basal** (or **parachordal**) **plate**, and various lateral structures develop on either side of it. Both cranial nerves and blood vessels are located between these various cartilaginous elements, passing to and from the brain.

The cartilage of the basal plate then extends caudally, absorbing and amalgamating with the sclerotomes of the four **occipital somites** and surrounding the rootlets of the hypoglossal (XII) cranial nerve. The cartilaginous unit that spreads around the brainstem at its junction with the most rostral part of the spinal cord comprises various named elements which eventually fuse together to form a bony ring around the **foramen magnum** (the large opening in the occipital bone through which passes the caudal part of the medulla oblongata, where it is continuous with the rostral part of the spinal cord). These cartilaginous primordia of the occipital bone initially ossify as separate units termed the **basioccipital**, the **exoccipital** and **supraoccipital** bones. When the sutures between these elements that develop within cartilage models ossify, they amalgamate to form a single bone, the **occiput** (or occipital bone). The squamous or interparietal component of the occiput is formed in membrane, and only fuses with the other components of the occipital bone that are formed in cartilage at some time later (see below). If fusion of this membrane-derived element to the chondrocranial elements fails to occur, the squamous element is then termed the **interparietal bone**. The bony elements that amalgamate to form the occipital bone are illustrated in Figure 4.8.3.[4]

The two hypophyseal cartilages develop anterior to the basal plate one on either side of the developing pituitary gland, forming the floor of the **pituitary fossa** (or **sella turcica**). These cartilaginous precursors subsequently fuse together with other membrane-based elements (see above) to form the **sphenoid bone** (for the arrangement in the adult human, see Figure 4.8.4).

The earliest centres of ossification appear in the base of the skull at about TS 22/23 in the basioccipital (or clivus) and exoccipital bones, while ossification appears somewhat later in such other components of the chondrocranium as the sphenoid and basisphenoid bones (at TS 24 and TS 25, respectively). A small centre of ossification is first seen in the cartilaginous component (the petrous part) of the temporal bone at about TS 22/23, and the ossification at this site gradually expands to involve all of the cartilaginous precursor elements. By about TS 25, a centre appears in the lower (intra-membranous) region of the future squamous part of the temporal bone. Two distinct ossification centres appear on either side of the midline in the supraoccipital part of the future occipital bone at about E17.5 (TS 26). These centres soon amalgamate (about E18.5) to form a single ossified unit which extends across the midline.

The **otic capsule** condenses at TS 20 around the derivatives of the otocyst (which differentiate into the membranous labyrinth that gives rise to the various components of the inner ear apparatus), and soon forms the petrous part of the temporal bone. The medial walls of the nasal capsules unite to form at TS 21 the **nasal septum**. Two separate cartilaginous centres develop (~TS 22) between the otic and nasal capsules that subsequently amalgamate to form the **lesser wing** and part of the **greater wing of the sphenoid bone** (the rest of the sphenoid bone is formed from elements which ossify in membrane). The **superior orbital fissure** develops between these two "centres", and through this passes the oculomotor (III), trochlear (IV), and abducent (VI) cranial nerves that supply the extrinsic ocular muscles, and the ophthalmic division of the trigeminal (V) cranial nerve which is the principal sensory supply to the nasal and frontal regions and the anterior part of the scalp. Ophthalmic veins also pass backwards into the cranial cavity through this fissure.

At about TS 24/25, the first evidence of ossification is seen in the **tympanic ring** (os tympanicum or tympanic bone), though this centre which develops in membrane remains discrete from that in the petrous part of the temporal bone up to the time of birth, after which time it fuses with the petrous part to form the outer (or tympanic) part of the temporal bone. The tympanic "ring" only partly surrounds the tympanic membrane, as the upper part of the bony ring is absent.

In the neck, and related only from a functional point of view with the mandible (one of the tongue muscles (the hyoglossus) is attached to it), is the cartilage primordium of the body of the hyoid bone. This is first recognized at about TS 20, although an

---

[4] Occasionally in humans, the upper (interparietal) part may fail to ossify with the lower part of the squamous occipital bone (or supraocciput), and the separate element is then termed the interparietal bone. It ossifies in two "centres", one on either side of the midline and these soon unite to form a single bony element. A very small ossification "centre" often appears in the superior margin of the foramen magnum, but soon unites with the lower border of the supraoccipital bone (if present, this is termed Kerckring's "centre"). The time of union between the major parts of the occipital bone *in the human* are as follows: towards the end of the second year the supraoccipital part unites with the lateral parts, and by the sixth year, the bone consists of a single unit; between the 18th and 25th years, the basioccipital and sphenoidal bones unite to form a single bone.

Figure 4.8.4. Simplified diagram to show the "ancestral" origin of the principal components of the cartilaginous neurocranium that forms the base of the skull in the adult human (based on Larsen, 1993). This diagram should be compared with Figure 4.8.2 where the cartilaginous components of the ancestral vertebrate skull are illustrated; the same shading is given in these two diagrams to show how the primitive structures have been extensively modified and adapted in the base of the modern human skull. The human skull has been illustrated, rather than the adult mouse skull, as the comparable part of the basal region of the latter occupies only about one-third of the skull base, much of the rest of being occupied by the nasal region (anteriorly) and the zygomatic arches (laterally, through which pass the temporalis muscles to the mandible), with the most anterior part of this lateral space being occupied by the eyes and their extrinsic ocular muscles.

Note that (i) the upaired plate-like cartilaginous mass, initially located between the notochord, and the brainstem termed the *parachordal cartilage* (or *basal plate*), which originally protected the most rostral part of the notochord (see Figure 4.8.2), now forms the *basiocciput*, and extends from the caudal end of the hindbrain to the hypophysis; (ii) the *polar* or *hypophyseal cartilages* amalgamate to form the *body of the sphenoid bone* (with its associated greater and lesser wings), one of whose functions is to protect the pituitary gland; and (iii) the *ethmoid bone* forms from the amlagamation of the *prechordal cartilages*, or trabecular region (or *trabeculae cranii*) rostral to the notochord, and the supero-medial part of the nasal (olfactory) capsule.

The cartilaginous capsules that originally protected the three pairs of sense organs (for olfaction, vision and hearing/equilibrium) are present in the adult mammalian skull in the form of the nasal cartilage, the orbito-sphenoid (associated with the neural crest-derived sclera of the eye) and the petrous part of the temporal bone. The sclera of the eye does not chondrify (as it does in some lower vertebrates), nor, to allow mobility of the eye, does it fuse with the basal cartilaginous complex.

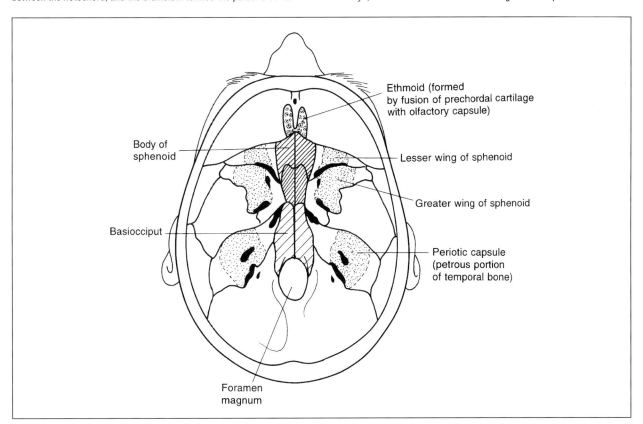

ossification centre is not observed in it until about TS 26, and this does not extend laterally to the lesser or greater horns of the hyoid bone until after birth. The laryngeal cartilages (the thyroid and cricoid cartilages) are first recognized at about TS 22/23 (and their precartilage elements one or two stages earlier) as are the cartilaginous tracheal rings.

The vault of the skull

All of the bones of the cranial vault are normally thought to ossify in membrane, although there are exceptions,[5] with most having either one or at most two centres of ossification. The **frontal bone** has two centres of ossification as early as TS 22/23, with its two precursor elements being separated by the **frontal** (or **metopic**) **suture**. The **parietal bone**

---

[5] In the mouse skull, analysis of the double-stained and cleared material indicates that, contrary to expectation, certain parts of the calvarium appear to ossify within a cartilage rather than a membranous primordium. This particularly seems to apply to the orbital part of the frontal bone, though it has to be admitted that it is difficult to distinguish this ossification centre from that of the orbito-sphenoid bone which normally ossifies in cartilage.

TS	1	2	3	4	5	6	7	8	9	10	11	12	13	14	15	16	17	18	19	20	21	22	23	24	25	26
E	0	1	2	3	4	4.5	5	5.5	6	6.5	7	7.5	8	8.5	9	9.5	10	10.5	11	11.5	12	12.5	13.5	14.5	15.5	16.5	17.5

also has two centres at this time which subsequently fuse and, by TS 23, the characteristic pattern of intramembranous ossification is especially clear in these two bones. Similarly, two centres are first seen within the **inter-parietal bone** by about TS 24, and these expand considerably, eventually fusing across the dorsal midline by about TS 26. This bone subsequently fuses along its infero-lateral border with the supraoccipital bone to form the upper (inter-parietal) part of the squamous region of the occipital bone.

The **temporal bone** also has a complex origin, in that its petrous part develops in cartilage, while its squamous and tympanic parts develop in membrane (see above). The **occipital bone** is equally complex in its origin, in that the majority of the definitive bone forms in cartilage, while its interparietal part develops in membrane (see Figure 4.8.3).

At birth, most of the skull bones, including those that comprise the facial elements, are almost completely ossified. The bones are, however, still mobile on each other to a very limited degree due to the movement allowed by the unossified sutures and fontanelles, and for this reason may readily be disarticulated. Maximum mobility is observed in the region of the cranial vault.[6]

The facial skeleton (viscerocranium)

Except for the **turbinate** (the nasal conchae) and **ethmoid bones**, which are believed to develop from the primitive cartilaginous nasal capsule, all of the other skeletal components of the face derive from "dermal bones", and consequently ossify in membrane. The major membranous components of the facial skeleton are the **pre-maxilla**, the **maxilla**, the **zygomatic** and **squamous part of the temporal bone** as well as the **palatine bone**, the **vomer**, and the **lacrimal** and **nasal bones**.

The maxilla forms from two major components, the pre-maxilla (TS 19), which is involved in the development of the primary palate (and is associated with the incisor dentition), and the maxilla proper (TS 20) which comprises the palatine shelf, probably its most important part, together with the frontal, zygomatic and alveolar components – the latter being associated with the molar dentition.

The other major component of this part of the skull is the **mandible**, a dermal bone that develops (TS 21) in close association with, and slightly lateral to, Meckel's cartilage. While the various parts of this cartilage give rise to specific structures (see Chapter 3.3 *The branchial arch system*), the part around which the mandible forms completely regresses. The mandible forms in two halves that are separated by a joint across the ventral midline termed the **symphysis menti** and its ossification within the mandible commences at about TS 23, with centres forming for the body and the inferior part of the ramus of the mandible (*i.e.* the vertical part of the mandible to which the masseter muscle is attached on its outer surface). The **condylar process** (in the region of the "head" of the mandible, and whose articular surface is involved in movement of the jaw) and **coronoid processes** develop from secondary cartilages, with the former being involved in the development of the **temporo-mandibular joint** (see Wong *et al.*, 1985), while the latter develops in association with the insertion of the temporalis muscle (one of the powerful muscles of mastication involved, with the masseter and other muscles, in closing of the jaw).

Much of the body and inferior part of the ramus of the mandible is ossified within a membranous primordium by TS 23, though the condylar region and the angle of the mandible have yet to show evidence of ossification at this time. Even at TS 24, it is possible to recognize the relationship between the ossification pattern in these sites and the right and left Meckel's cartilages. These ossification centres extend anteriorly towards the anterior tip of the lower jaw, close to the future site of the symphysis menti, and postero-medially towards the tubo-tympanic recess of the middle ear, where their relationship to the cartilage primordia of the malleus is particularly clearly seen at about TS 25/26, a day after it has separated from the posterior part of Meckel's cartilage (TS 24/25). By about TS 26, virtually all of the condylar region of the mandible has ossified.

The earliest centres of viscerocranial ossification are seen in the maxillary region, while those for the pre-maxilla and maxilla are present at about TS 22/23, with much more extensive areas of ossification being present by the end of TS 23. Additional smaller centres may also be seen in other parts of the maxilla (*e.g.* in the lateral margins of the palatal shelves, the periorbital region both lateral and infero-lateral to the nasal capsule, and on either side just deep to the first upper molar tooth primordium). Amalgamation of these maxillary centres occurs by about TS 25, and they are close to an ossification centre which develops within the zygomatic bone (first clearly seen at about TS 23), and further centres which appear within the frontal and nasal bones. Centres are also seen in the zygomatic part of the temporal bone at about TS 24.

By TS 26, it is no longer possible to distinguish between the individual components of the viscerocranium in alizarin- and Alcian blue-stained "cleared" material that respectively highlight regions of ossification and cartilage. Much of the nasal

---

[6] In the human, this ability of the dermal bones to move slightly in relation to each other at the time of birth is termed "moulding", and plays a particularly important part during parturition, particularly if there is a degree of cephalo-pelvic disproportion (*i.e.* the fetal head is slightly larger or the dimensions of the pelvis slightly smaller than normal).

capsule is cartilaginous at E18.5, and small (or accessory) cartilaginous elements are located on the lateral side of the external nares.

The blood supply

The arterial blood supply to the skull is from the numerous branches of both the internal and external carotid arteries, and it is drained by the large, thin-walled **diploic veins** located in the diploë (*i.e.* the loose spaces between the inner and outer cortical layers of the cranial bones) of most of the bones of the cranial vault. These communicate with the meningeal veins, the dural venous sinuses and pericranial veins. Many small diploic veins drain into the superior sagittal sinus. Numerous, often inconstant, emissary veins make connections between the intracranial venous sinuses and the extracranial veins, and play a significant role in the spread of infection from extracranial sites to the dural venous sinuses.

TS	1	2	3	4	5	6	7	8	9	10	11	12	13	14	15	16	17	18	19	20	21	22	23	24	25	26
E	0	1	2	3	4	4.5	5	6	6.5	7	7.5	8	8.5	9	9.5	10	10.5	11	11.5	12	12.5	13.5	14.5	15.5	16.5	17.5

5
THE INDEXES

5.1 Introduction

This section of the book is designed to provide core data about mouse embryogenesis and it has several parts: the first (5.2) is a detailed staging system based on that of Theiler, but expanded to include more recent work (from Bard et al., 1998); the second (5.3) is a fairly complete list of all the major tissues present in embryos at each of the 26 Theiler stages of development, with tissues new to a particular stage being indicated in bold type. Section 5.4 provides a list of all these tissues linked to the Theiler stage at which each can first be seen (mainly in sectioned material stained with haematoxylin and eosin). Although an earlier version of this last index has already been published in the second printing of *The Atlas of Mouse Development* (Kaufman, 1994), the version given here has some additional pointers to help readers navigate the text and a large number of corrections. The index is repeated here partly for completeness and partly for anyone who only has the first edition of the *Atlas* that lacks the index.

There is also a glossary (5.5), the references (5.6), and the author (5.7), tissue (5.8) and subject indexes (5.9). This last index provides, for example, access to details about human embryogenesis and the various congenital diseases mentioned in the text.

It should be said that indexing mouse developmental anatomy to the extent given here was not a trivial matter. Once the work had started, it soon became clear that, if the index was to be useful, we (MHK, JBLB and Duncan Davidson) would not only have to determine the stage at which each of the more than a thousand named and anatomically defined tissues in the mouse embryo first appeared, but would also have to group and categorize the tissues under appropriate organ headings. After this had been done, there were further choices at every turn: should we allow redundancy and put in items more than once (should the tibia be under limb or skeleton or both, should a particular blood vessel be categorized under the region or territory it supplies or drains or under the vascular system as a whole or all three?), should we include every tissue or just the better known ones? Should we use the more familiar anglicized version of the *nomina anatomica* (where appropriate terms are available), or supplement this with a range of other (more traditional) terms (and their synonyms) some of which are now gradually falling into disuse?

In all cases, choices were made on a pragmatic basis, with our guiding rule being that we should do what we felt would be most useful to readers and so to the field. Muscles and bones are thus indexed twice, but most veins and arteries are only included under the vascular system (branchial arch arteries are catalogued twice); moderately obscure tissues are included (the paraganglia of Zuckerkandl), very obscure ones are not (the zonule of Zinn), but the sheer amount of effort required for the enterprise meant that a line had to be drawn somewhere. As to terminology, we have tried to omit terms that we felt had outlived their usefulness while including currently used synonyms. Readers will have to decide whether we got the balance of tissues and names right.

There is another complicating factor: all embryologists know that embryos of exactly the same *temporal* age (i.e. isolated from a single female or from several females that are nominally at the same time after conception) are not as uniform as we might like, and show a degree of natural variation in their *developmental* age, while different mouse strains often have their own peculiarities (the stages given here are for (C57BL×CBA) F1 hybrids). Our solution here has been to use Theiler staging rather than embryonic age (hence the detailed staging table). But there is a more difficult problem to overcome: the stage at which a tissue is deemed to be present depends on when the viewer can first see something that was apparently not there earlier, and there are no standard criteria for establishing whether a tissue or anatomical structure is present. The reader should therefore accept that the staging of tissues still has a subjective element (while molecular markers may eventually bring some clarity to the field, a degree of fog is likely to persist) and that a particular tissue in one of their own embryos might appear up to a stage earlier or later than that described here. For all these reasons, it is unlikely that all our staging and organizational judgements are either appropriate for our strain of mouse or exactly applicable to others. Should any reader feel that there are temporal errors or logical inconsistencies in these indexes, we would appreciate their informing us.

We should also mention that the index detailing

the "contents of mice" at each Theiler stage is available over the internet up to about TS 23, although not in the format given here. This is because the internet format is designed to be part of the *Mouse Gene Expression Information Resource* (MGEIR), with the anatomical terms being used as an aid for inputting data and querying the database (for this purpose, each name can only be used once and the terminology has had to be arranged as a hierarchy). A text version of this anatomy database is included in the mouse *Gene Expression Database* (GXD, http://www.informatics.jax.org/gxd.html), the text component of MGEIR which is maintained at the Jackson Laboratory in Maine, USA, while a more sophisticated, graphical version in which all the tissues are set out as a collapsable tree is part of the *Mouse Atlas Gene Expression Database* (MAGED, http://genex.hgu.mrc.ac.uk/) which has been set up at the MRC Human Genetics Unit in Edinburgh, UK, and we express here our appreciation for its establishment to Christophe Dubreuil and Richard Baldock and our thanks to Duncan Davidson for collaborating with us on the tissue indexes.

5.2 Table of mouse and rat developmental stages

This table gives an expanded version of the Theiler system for staging embryos on the basis of their external appearance. To aid identification, the table includes both features present for the first time and those that are absent until the next stage.

Principal features to allow stage identification	Mouse[1] Theiler stage	Mouse[1] Embryonic age (E)[3]	Mouse[1] Size (mm) (approx. values)	Somites (range)	Witschi stage	Rat[2] Embryonic age (E)	Rat[2] Size (mm) (approx. values)
1-cell (fertilized) – in oviduct	1	0–1			1	1	
2-cells in oviduct	2	1–2			2	2	
Cleavage stages; early: 4–8 cells, late: morula (early to fully compacted)	3	2–2.5			3 (4 cells) 4 (8–12 cells) 5 (morula)	3 3.25 3.5	
Blastocyst, zona-intact	4	2.5–3.5			6	4	
Blastocyst, zona-free	5	3.5–4.0			7	5	
Attaching blastocyst; decidual reaction domain	6	4.0–4.5			8	6	
Early egg cylinder. Ectoplacental cone appears. Embryo implanted	7	4.5–5.5			9–10	6.75–7.25	
Differentiation of egg cylinder. Proamniotic cavity appears	8	5.5–6.0					
Advanced decidual/endometrial reaction. Primitive streak first evident at late stage. 9. Pre-streak stage (PS), ectoplacental cone invaded by blood, extraembryonic ectoderm, embryonic axis visible. 9a. Early streak (ES). Gastrulation starts, first evidence of mesoderm.	9, 9a	6.0–6.5			11		
Primitive streak and groove. Amnion formation (anterior and posterior amniotic folds – at late stage). Allantois. Ectoplacental and exocoelomic cavities – at late stage. Secondary yolk sac (late). Hensen's node. 10. Mid-streak (MS). Amniotic folds start to form. 10a. Late-streak, no allantoic bud (LSOB). Exocoelomic cavity present 10b. Late-streak, early allantoic bud (LSEB).	10, 10a, 10b	6.5–7.5			12	7.75–8.5	

TABLE OF MOUSE AND RAT DEVELOPMENT

Description	Stage	Days	CR length (mm)	Somites	Stage	Days	CR length (mm)
Late primitive steak. Neural plate. Cephalic neural (head) folds. Presomite stage. Foregut pocket and cardiogenic plate (late stage). 11 Neural plate (NP). Head process developing. Amnion complete 11a. Late neural plate (LNP). Elongated allantoic bud (late) 11b. Early head fold (EHF). 11c. Late head fold (LHF). Foregut invagination, cardiogenic plate.	11, 11a–c	7.0–7.5			13	9–9.5 }	
"Unturned", headfold stage. Neural folds begin to close in occipital/cervical region. Heart differentiation. Components of first branchial/pharyngeal arch present. Optic placode/pit (early) sulcus (late). 12. Allantois extends. First branchial arch. Heart starts to form. Foregut pocket visible. Pre-otic sulcus (2–3 somites present). Cephalic neural crest migrates 12a. Allantois contacts chorion (late stage). *Absent 2nd arch, >7 somites*	12, 12a	7.5–8.5		12 1–4 12a 5–7	14	9–9.5 }	
"Turning". Optic sulcus (well formed). Components of first and second branchial arches. Thyroid primordium. Nephrogenic cord. Early evidence of caudal neuropore. *Absent 3rd arch, >12 somites*	13	8.5–9.0		8–12	15	10	2.0
Elevation of cephalic neural folds. Formation and closure of rostral neuropore. Optic vesicle formation. Mandibular and maxillary components of first branchial arch. Components of third branchial arch. Rathke's pouch (pituitary rudiment). Liver rudiment. Pronephros. Forelimb ridge (late stage). *Absent forelimb bud*	14	8.5–9.0		13–20	16	10.5	2.2–4.0 }
Caudal neuropore diminishing in size. Components of three branchial arches. Optic cup. Otocyst (late). Forelimb bud. Hindlimb ridge. Tracheal diverticulum. Lung bud. Mesonephros. *Absent hindlimb bud*	15	9.0–9.75	1.8–3.3	21–29	17–19	11–11.75	

221

Principal features to allow stage identification	Mouse[1]				Rat[2]		
	Theiler stage	Embryonic age (E)[3]	Size (mm) (approx. values)	Somites (range)	Witschi stage	Embryonic age (E)	Size (mm) (approx. values)
Caudal neuropore closes. Components of four branchial arches. Lens placode. Otocyst (otic vesicle) separated from surface ectoderm. Olfactory placode/pit. Hindlimb bud. *Absent thin and long tail*	16	9.75–10.25	3.0–4.0	30–34	20–21	11.75–12	4–6
Deep lens indentation, optic cup. Differentiation of cephalic neural tube. Tail elongates and thins. Umbilical hernia first evident. *Absent nasal pits*	17	10–10.5	3.5–5.0	35–39	21–23	12–12.25	
Formation and closure (late) of lens vesicle, deep nasal pit. Cervical somites becoming indistinct. *Absent auditory hillocks, handplate*	18	10.5–11	5–6	40–44	24–27	12.5–13	6–8.5
Lens vesicle completely separates from surface. Cervical somites indistinct. Distal part of forelimb bud paddle-shaped (i.e. hand plate). Foot plate not yet differentiated. Auditory hillocks first evident. *Absent retinal pigmentation and sign of fingers*	19	11.0–11.5	6–7	45–47			
Earliest signs of fingers. Splayed-out digits. Foot plate present. Retinal pigmentation present (only in pigmented strains). Tongue well differentiated. Brain vesicles recognized. *Absent 5 rows of whiskers, indented anterior footplate*	20	11.5–12	7–8	48–51	28	13.5	8.5
Anterior part of footplate indented. Elbow and wrist regions evident. 5 rows of whiskers. Umbilical hernia now well defined. *Absent hair follicles, fingers separated distally*	21	12.5–13	8–9	52–55	29–30	14–14.5	9.5–10.5
Fingers separated distally. Only indentations between toes. Long bones of limbs present. Hair follicles in pectoral, pelvic and trunk regions. *Absent open eyelids, hair follicles in cephalic region*	22	13.5–14	9–10	56–>60	31	15	12
Toes separated. Hair follicles additionally in cephalic region, but not near vibrissae. Eyelids open. *Absent nail primordia, fingers 2–5 parallel*	23	14.5–15	10–11.5		32	15.5	14.2

TABLE OF MOUSE AND RAT DEVELOPMENT

Description					
Reposition of umbilical hernia (midgut loop returned to abdominal cavity). Eyelids closing. Fingers 2–5 are parallel, nail primordia seen on toes. *Absent wrinkled skin, fingers and toes joined together*	24	15.5–16	11.5–14	33	16	15.5
Skin wrinkled. Eyelids closed. No evidence of umbilical hernia. *Absent ear extending over auditory meatus, long whiskers*	25	16.5–17	13.5–16	34	17–18	16–20
Elongated whiskers. Eyes barely visible through closed eyelids. External ear covers auditory meatus.	26[4]	17.5–18	17–19	35 (ante-natal)	19–22	20–40

[1] Principally from Theiler (1989), Kaufman (1994), Bard *et al.* (1998); detailed staging of Theiler stages 9–12, courtesy of K. Lawson (personal communication). The data here are for (C57BL × CBA)F1 hybrid mice. For PO mice, see Table 2.2.1.

[2] Principally from Witschi (1962), see this reference for additional sources

[3] These figures are for *typical* embryos, although the occasional *viable* embryo may lie outside the figures and timings given here. These timings are based on matings taking place at the mid-dark point (typically 2.00a.m.) with vaginal plugs being noted the following morning and embryos being observed at about midday (*i.e.* the plug day is E0.5).

[4] Mice are typically born at about E19.5.

223

5.3 The tissues present in each stage of mouse development

(Tissues appearing for the first time are shown in bold type)

Abbreviations

E	embryonic day
CC	cartilage condensation
L	left
PCC	pre-cartilage condensation
PMM	pre-muscle mass
R	right
TS	Theiler stage

STAGE 1 – day 0 (fertilized egg)
fertilized cell
  polar body
    first
    second
zona pellucida

STAGE 2 – day 1 (two-cell stage)
two cells
  polar body (second)
zona pellucida

STAGE 3 – day 2 (cleavage stages)
early: 4–8 cells
late: morula (early to fully compacted)
  second polar body
zona pellucida

STAGE 4 – day 3 (blastocyst, zona-intact)
blastocoelic cavity
embryo
  compacted morula (early stage)
  inner cell mass (late in stage)
    second polar body
extra-embryonic tissue
  trophectoderm
    mural
    polar
zona pellucida

STAGE 5 – day 4 (blastocyst, zona-free)
blastocoelic cavity (becomes yolk-sac cavity)
embryo
  inner cell mass
extra-embryonic tissue
  trophectoderm
    mural
    polar

STAGE 6 – day 4.5 (implantation)
decidual-reaction domain
blastocoelic cavity (becomes yolk-sac cavity)
extra-embryonic tissue
  endoderm
    parietal (distal)
  trophectoderm
    mural
    polar
embryo
  epiblast (primitive ectoderm)
  primitive endoderm, (proximal, visceral)

STAGE 7 – day 5 (egg cylinder)
decidual-reaction domain
embryonic tissue
  epiblast (primitive ectoderm)
  primitive endoderm (visceral, proximal)
extra-embryonic tissue
  endoderm
    parietal (distal)
      Reichert's membrane
    visceral (proximal)
  trophectoderm
    mural
    polar
      ectoplacental cone
yolk-sac cavity (primary)

STAGE 8 – day 6 (egg cylinder)
cavity
  proamniotic canal
  yolk-sac (primary)
decidual-reaction domain
embryonic tissue
  epiblast (primitive ectoderm)
  primitive endoderm (visceral, proximal)
  proamniotic canal (emb. component)
extra-embryonic tissue
  ectoderm
  endoderm
    parietal (distal)
      Reichert's membrane
    visceral (proximal)
  proamniotic canal (extraemb. component)
  trophectoderm
    mural
      trophoblast giant cells (primary)
    polar
      ectoplacental cone
        cytotrophoblast
        syncytiotrophoblast

STAGE 9 – day 6.5 (egg cylinder)
cavity
  proamniotic canal
  yolk-sac (primary)
decidual-reaction domain
embryonic tissue
  ectoderm
    primitive streak
  endoderm
  mesoderm
  proamniotic canal (emb. component)
extra-embryonic tissue
  ectoderm
  endoderm
    parietal (distal)
      Reichert's membrane
    visceral (proximal)
  proamniotic canal (extraemb. component)
  trophectoderm
    mural
      trophoblast giant cells (primary)
    polar
      ectoplacental cone
        cytotrophoblast
        syncytiotrophoblast

STAGE 10 – day 7 (primitive streak)
cavity
  amnion (late stage)
  amniotic fold
    anterior
    posterior
  ectoplacental (late stage)
  exocoelomic (late stage)
  proamniotic canal (lost at late stage)
  yolk sac, secondary (late stage 10)
    endoderm
    mesoderm
      primordial germ cells (late stage 10)
decidual-reaction domain
embryonic tissue
  ectoderm
    neural ectoderm (neuroepithelium)
      (neural plate) (late stage 10)
    primitive streak
      Henson's node
  endoderm
    definitive
    primitive
  mesoderm
extra-embryonic tissue
  amniotic fold
  ectoderm
  endoderm
    parietal (distal)
      Reichert's membrane
    visceral (proximal)
  mesoderm
    allantois (late stage 10)
    primordial germ cells (late stage 10)
  trophectoderm
    mural
      trophoblast giant cells (primary)
    polar
      ectoplacental cone
        cytotrophoblast
        syncytiotrophoblast
      trophoblast giant cells (secondary)
  yolk sac, secondary (late stage 10)
    endoderm
    mesoderm
      primordial germ cells (late stage 10)

STAGE 11 – day 7.5 (late primitive streak)
cavity
  amniotic
  ectoplacental
  exocoelomic
  intra-embryonic coelomic
  yolk-sac
decidual-reaction domain
embryonic tissue
  ectoderm (embryonic)
    cephalic neural (head) folds
    neural crest
    neural ectoderm (neuroepithelium)
      Henson's node
  endoderm
    definitive
    primitive
  gut (primitive)
    foregut pocket or **diverticulum** (late stage)
  mesoderm
    axial
      notochordal plate
      pre-chordal plate
    cardiogenic (myocardial) plate (late stage)
    head
    paraxial/lateral
extra-embryonic tissue
  allantois
  amnion
  chorion
  ectoderm
  endoderm
    parietal (distal)
      Reichert's membrane
    visceral (proximal)
  mesoderm
    primordial germ cells
  trophectoderm
    mural
      trophoblast giant cells (primary)
    polar
      ectoplacental cone

THE TISSUES PRESENT IN EACH STAGE OF MOUSE DEVELOPMENT

STAGE 11 *continued*
    cytotrophoblast
    syncytiotrophoblast
    trophoblast giant cells (secondary)
  yolk sac
    endoderm
    mesoderm
      blood islands
      primordial germ cells

STAGE 12 – day 8 (unturned, headfold stage)
brain
  cephalic neural (head) folds
    forebrain region
    hindbrain region
      pro-rhombomeres A,B (late stage)
      pre-otic sulcus
    midbrain region
branchial arch
  1st arch (future maxillary component)
    artery
    groove
    maxillary component
    membrane
    mesenchyme
      neural crest
      head mesenchyme
    pouch
cavity
  amniotic
  exocoelomic cavity
  intra-embryonic coelomic
    pericardial component
    pericardio-peritoneal channels
    peritoneal component
  yolk-sac
decidual-reaction domain
ear, inner
  otic placode
ectoderm (surface)
  primitive streak
    Henson's node
endoderm, definitive
extra-embryonic tissue
  allantois
    primordial germ cells
  amnion
  chorion
  ectoderm
  endoderm
    parietal (distal)
      Reichert's membrane
    visceral (proximal)
  mesoderm
  trophectoderm
    mural
      trophoblast giant cells (primary)
    polar
      ectoplacental cone
        cytotrophoblast
        syncytiotrophoblast
      trophoblast giant cells (secondary)
  yolk sac
    blood islands (becoming vasculature)
    endoderm
    mesoderm
    primordial germ cells (a minority)
    vasculature (late stage)
eye, optic
  pit (evagination) (early)
  placodes (very early stage)
  sulcus
gut
  foregut pocket (diverticulum)
  hindgut pocket (diverticulum)
    wall of hindgut
      primordial germ cells (a few)
  midgut region
heart
  atrial chamber, common

bulbo-ventricular groove (sulcus)
bulbus cordis
  inferior (future right ventricle)
  superior (outflow tract – prox. part)
cardiac jelly
cardiogenic (myocardial) plate (primitive myocardium – early)
  dorsal mesocardium (dorsal mesentery of heart)
  transverse pericardial sinus
endocardial tube
outflow tract
  aortic sac (outflow tract – distal part)
primitive heart tube
sinus venosus (R & L horns)
ventricular region of primitive heart
mesoderm (embryonic)
  axial, notochord
  lateral plate
    somatopleure (lining of intraembryonic coelom adheres to ectoderm)
    splanchnopleure (lining of intraembryonic coelom adheres to endoderm)
  intermediate
  paraxial
    head mesenchyme (somitomeres)
    somites 1–8 pairs (body)
    unsegmented
  prechordal plate
mouth region
  buccopharyngeal membrane
  maxillary component of first arch
  stomatodaeum
neural crest (head), **trigeminal (V)**
neural ectoderm (neuroepithelium)
neural tube (future spinal cord)
  neural folds
  neural lumen (future spinal canal)
notochord
notochordal plate
primordial germ cells (mainly in allantois)
septum transversum
sulcus, pre-otic
vascular system
  artery
    aorta, dorsal (paired)
    vitelline
  branchial arch artery, 1st
  vein
    anterior cardinal
    primary head
    umbilical
    vitelline (omphalomesenteric)

STAGE 13 – day 8.5 (turning) (*extra-embryonic tissues omitted from now on*)
brain
  cephalic neural (head) folds
    forebrain region
    hindbrain region
      prorhombomeres A,B (early stage)
      rhombomeres 1–5 (late stage)
    midbrain region
branchial arch
  1st arch
    artery
    groove (cleft)
    maxillary component
    membrane
    mesenchyme
      neural crest
      head mesenchyme
    pouch
  2nd arch
    artery
    ectoderm
    endoderm
    membrane
    mesenchyme
      neural crest
      head mesenchyme

cavity
  intra-embryonic coelom
    pericardial component
    pericardio-peritoneal channel
    peritoneal component
ear, inner
  otic placode
ectoderm (surface)
  primitive streak
eye
  optic pit (evagination)
  optic sulcus
gland
  thyroid primordium
gut
  buccopharyngeal membrane
  foregut pocket (diverticulum)
  foregut–midgut junction
    dorsal mesentery
  hindgut pocket (diverticulum)
  midgut region
heart
  atrial chamber, common
  bulbo-ventricular groove (sulcus)
  bulbus cordis
    inferior (future right ventricle)
    superior (prox. part of outflow tract)
  cardiac jelly
  dorsal mesocardium (dorsal mesentery of heart)
    transverse pericardial sinus
  endocardial tissue
  myocardium
  outflow tract
  primitive ventricle (future left ventricle)
  sinus venosus (R & L horns)
mesenchyme
  axial, notochord
  intermediate
    nephrogenic cord
  lateral plate
    somatopleure
    splanchnopleure
  paraxial
    head mesenchyme (somitomeres)
    somites 1–13 pairs (maximum)
    unsegmented (in tail region)
mouth region
  stomatodaeum (oral pit)
    buccopharyngeal membrane
    maxillary component of first arch
neural crest (head)
  facial-acoustic (VII–VIII)
  trigeminal (V)
neural tissue (presumptive)
  luminal occlusion
  tube
    caudal neuropore
    floorplate
    neural lumen
    rostral neuropore
notochord
notochordal plate
primordial germ cells (wall of hindgut)
septum transversum
  hepatic component
  non-hepatic component
somite (see mesenchyme)
sulcus, pre-otic
  bulbo-ventricular (see heart)
tail bud
thyroid primordium
umbilicus
  artery
  vein
vascular system
  artery
    aorta
      dorsal (paired)
        midline
        internal carotid (R & L, rostral extension of dorsal aorta)

STAGE 13 continued
    umbilical (R & L)
      vitelline
      blood
      branchial arch artery
        1st
        2nd
      vein
        cardinal
          anterior
          common (Duct of Cuvier)
          posterior
        primary head
        umbilical (R & L)
        vitelline (omphalomesenteric)

STAGE 14 – day 9
brain (regions)
  forebrain (**prosencephalon**)
    3rd ventricle
    diencephalon region
      optic vesicles
    prosencephalic vesicle
      rostral neuropore
  hindbrain (**rhombencephalon**)
    4th ventricle
    rhombencephalic vesicle
      rhombomeres 1–5 (early stage)
      rhombomeres 1–8 (late stage)
  midbrain (**mesencephalon**)
    mesencephalic vesicle
branchial arch
  1st arch
    artery
    ectoderm
    endoderm
    groove (cleft)
    mandibular component
    maxillary component
    membrane
    mesenchyme
      neural crest
      head mesenchyme
    pouch
  2nd arch
    artery
    ectoderm
    endoderm
    groove
    membrane
    pouch
    mesenchyme
      neural crest
      head mesenchyme
  3rd arch
    ectoderm
    endoderm
    mesenchyme
      neural crest
      head mesenchyme
cavity
  intra-embryonic coelom
    pericardial component
    pericardio-peritoneal canals
    peritoneal component
ear, inner
  otic **pit**
  otic placode (early stage only)
ectoderm (surface)
  primitive streak
eye
  optic eminence
  optic sulcus
  optic vesicle (evagination of diencephalon)
gland
  pituitary: Rathke's pouch
  thyroid primordium
gut
  buccopharyngeal membrane
  foregut
    pharyngeal region
    thyroid primordium
  foregut–midgut junction
    biliary bud
    dorsal mesentery
  hindgut
    hindgut diverticulum
  midgut
heart
  atrial chamber, common
  atrio-ventricular canal
  bulbo-ventricular groove (sulcus)
  bulbus cordis
    inferior (future right ventricle)
    superior (prox. part of outflow tract)
  cardiac jelly
  dorsal mesocardium (dorsal mesentery of
      heart)
  transverse pericardial sinus
  endocardial tissue
  myocardium
  outflow tract
  primitive ventricle (future left ventricle)
  sinus venosus (R & L horns)
limb, **forelimb ridge** (late stage)
liver rudiment (within septum transversum)
  biliary bud (cystic diverticulum)
mesenchyme
  intermediate (see urogenital system)
  lateral
    somatopleure
    splanchnopleure
  paraxial
    head mesenchyme (somitomeres)
    somites pairs 5–20 (max.)
    unsegmented (in tail region)
mouth region
  mandibular component of first arch
  maxillary component of first arch
  Rathke's pouch (pituitary primordium)
  stomatodaeum
    buccopharyngeal membrane
neural crest (future peripheral nervous system)
  facial-acoustic (VII–VIII) + **pre-ganglion
      complex**
  trigeminal (V) + **pre-ganglion**
neural tissue (presumptive)
  folds
  tube
    caudal neuropore
    neural lumen
    luminal occlusion
    rostral neuropore (closes late stage)
nose
  olfactory placode
notochord
pituitary, Rathke's pouch
primordial germ cells (wall of hindgut)
respiratory system, pharynx
septum transversum
  hepatic component
  non-hepatic component
somite (see mesoderm)
tail bud
thyroid primordium
umbilicus
  artery (R & L)
  vein (R & L)
urogenital system (**ridge**) (intermediate mesoderm)
  nephric (Wolffian) duct
  nephrogenic cord
  pronephros
vascular system
  artery
    aorta
      dorsal (midline)
      dorsal (paired)
    branchial arch artery
      1st
      2nd
    internal carotid (R & L, rostral extension of
        dorsal aorta)
    umbilical (R & L)
    vitelline
  blood
  vein
    cardinal
      anterior
      common (Duct of Cuvier)
      posterior
    primary head
    umbilical (R & L)
    vitelline (omphalomesenteric)

STAGE 15 – day 9.5
brain
  forebrain
    diencephalon
      lamina terminalis
      third (diencephalic) ventricle
        infundibular recess
        interventricular foramen
        optic recess (lumen of optic stalk)
    telencephalon (future cerebral hemisphere)
      telencephalic vesicle (future lateral
        ventricle)
      interventricular foramen
  hindbrain (rhombencephalon)
    rhombencephalic vesicle (fourth ventricle)
      roof
    rhombomeres 1–8
  midbrain (mesencephalon)
    mesencephalic vesicle (future cerebral
      aqueduct)
branchial arch
  1st arch
    artery
    groove (cleft)
    mandibular component
    maxillary component
    membrane
    mesenchyme
      neural crest
      head mesenchyme
    pouch
  2nd arch
    artery
    groove (cleft)
    membrane
    pouch
    mesenchyme
      neural crest
      head mesenchyme
  3rd arch
    artery
    ectoderm
    endoderm
    groove
    membrane
    mesenchyme
      neural crest
      head mesenchyme
    pouch
cavity
  intra-embryonic coelom
    pericardial component
    pericardio-peritoneal canals
    peritoneal component
ear, inner
  otic pit (early stage)
  otocyst (otic vesicle) (late stage)
ectoderm (surface)
eye
  lens placode
  optic
    cup (primary)
      inner layer (future neural layer of retina)
      outer layer (future pigment layer of retina)
    eminence
    stalk
  perioptic mesenchyme
gall bladder
  (cystic) primordium

THE TISSUES PRESENT IN EACH STAGE OF MOUSE DEVELOPMENT

STAGE 15 *continued*
  cystic duct
ganglia
  dorsal (posterior) root (from neural crest)
gland
  gall bladder
    (cystic) primordium
    cystic duct
  pituitary
    infundibular recess of third ventricle
    Rathke's pouch
  thyroid primordium
gut
  foregut
    oesophageal region
    mesentery
      dorsal, ventral
    pharynx
    thyroid primordium
  foregut–midgut junction
    mesentery
      dorsal, **ventral**
    hindgut diverticulum
      dorsal mesentery
  midgut
heart
  atrial chamber
    common, **left & right components**
  atrio-ventricular canal
  bulbar cushion
  bulbo-ventricular groove (sulcus)
  bulbus cordis
    inferior (future right ventricle)
    superior (prox. part of outflow tract)
  cardiac jelly
  dorsal mesocardium (dorsal mesentery of heart)
    transverse pericardial sinus
  endocardial tissue
  myocardium
  outflow tract
  primitive ventricle (future left ventricle)
  sinus venosus (R & L horns)
limb
  hindlimb ridge
  forelimb bud
liver primordium (within septum transversum)
mesenchyme
  intermediate (see urogenital system)
  lateral
    somatopleure (+ mesothelium)
    splanchnopleure (+ mesothelium)
  paraxial
    head mesenchyme (somitomeres)
    somites 5–30 (max.) pairs (earlier ones are
      beginning to segregate into their
      dermomyotome and **sclerotome**
      components)
    presomitic (in tail region)
mouth region
  mandibular component of first arch
  maxillary component of first arch
  stomatodaeum (oral pit)
neural crest (future peripheral nervous system)
  body
    dorsal (posterior) root ganglia
  head
    facial–acoustic (VII-VIII) preganglion
      complex
    glossopharyngeal (IX) preganglion (late
      stage)
    glossopharyngeal–vagus (IX–X)
      preganglion complex (early stage)
    trigeminal (V) preganglion
    vagus (X) preganglion (late stage)
neural tube
  caudal neuropore
  lumen
  neural luminal occlusion
nose
  naso-lacrimal groove
  olfactory placodes

    process, fronto-nasal
notochord
pituitary
  infundibular recess of third ventricle
  Rathke's pouch
primordial germ cells (in wall and dorsal
  mesentery of hindgut)
respiratory system
  laryngo-tracheal groove
  lung bud
  pharynx
  tracheal diverticulum
septum transversum
  hepatic component
  non-hepatic component
somite (see mesenchyme)
tail
  neural crest
  neural tube
    neural plate
  unsegmented mesenchyme
thyroid primordium
umbilicus
  artery (R + L)
  vein (R + L)
urogenital system (from intermediate mesoderm)
  urogenital ridge
    nephric component (cord)
    mesonephros
      mesonephric tubules mesonephric
        (Wolffian) duct
vascular system
  artery
    aorta
      dorsal (midline)
      dorsal (paired)
    branchial arch artery
      1st
      2nd
      3rd
    internal carotid (R & L, rostral extension of
      dorsal aorta)
    umbilical (R & L)
    vitelline
  blood
  vein
    cardinal
      anterior
      common (Duct of Cuvier)
      posterior
    primary head
    umbilical (R & L)
    vitelline (omphalomesenteric)

STAGE 16 – day 10
brain
  forebrain
    diencephalon
      lamina terminalis
      third ventricle
        infundibular recess
        optic recess (lumen of optic stalk)
        interventricular foramen
    telencephalon
      interventricular foramen
      telencephalic vesicle
  hindbrain (rhombencephalon)
    rhombencephalic (fourth) ventricle
      roof
    rhombomeres 1–8
  midbrain (mesencephalon)
    mesencephalic vesicle (future cerebral
      aqueduct)
  layer, ventricular
branchial arch
  1st arch
    artery
    endoderm
    ectoderm
    groove (cleft)
    mandibular component

    maxillary component
    membrane
    mesenchyme
    pouch
  2nd arch
    artery
    cleft
    endoderm
    ectoderm
    membrane
    mesenchyme
    pouch
  3rd arch
    artery
    endoderm
    ectoderm
    groove
    membrane
    mesenchyme
    pouch
  4th arch
    cleft
    endoderm
    ectoderm
    membrane
    mesenchyme
    pouch
cavity
  intra-embryonic coelom
    pericardial component
    pericardio-peritoneal canal
    peritoneal component
    pleural component
ear, inner
  otocyst (otic vesicle)
ectoderm (surface)
eye
  lens pit
  optic eminence
  optic stalk
  retina (secondary optic cup)
    inner layer (future neural layer)
    intraretinal space
    outer layer (future pigment layer)
  perioptic mesenchyme
gall bladder
  (cystic) primordium
  cystic duct
ganglia (from neural crest)
  body
    dorsal (posterior) root
  cranial
    facial–acoustic (VII–VIII) ganglion complex
    glossopharyngeal (IX) ganglion (late stage)
    trigeminal (V) ganglion (semilunar,
      Gasserian)
    vagus (X) ganglion
gland
  gall bladder
    cystic duct
  pituitary
    infundibular recess of third ventricle
    Rathke's pouch
  thyroid primordium
gut
  foregut
    mesentery (dorsal, ventral)
    oesophagus
      meso-oesophagus
    pharynx
  foregut–midgut junction
    biliary system
    mesentery (dorsal, ventral)
  hindgut
    anal region
      cloacal membrane
    diverticulum
    dorsal mesentery
    primordial germ cells
  midgut
    dorsal mesentery

STAGE 16 *continued*
heart
  atrial chamber
    auricular part (left & right)
      common, left & right components
  atrio-ventricular canal
  bulbar cushion
  bulbo-ventricular groove (sulcus)
  bulbus cordis
    inferior (future right ventricle)
      superior (prox. part of outflow tract)
  cardiac jelly
  dorsal mesocardium (dorsal mesentery of heart)
    transverse pericardial sinus
  endocardial cushion tissue
  endocardial tissue
  myocardium
  outflow tract
    aortico-pulmonary spiral septum
  primitive ventricle (future left ventricle)
  interatrial septum
    septum primum (late stage)
  sinus venosus (R & L horns)
limb
  hindlimb bud
  forelimb bud
    apical ectodermal ridge
liver primordium
  cystic duct
  hepatic duct
  liver **parenchyma**
  gall bladder primordium
mesenchyme
  intermediate (see urogenital system)
  lateral plate
    somatopleure (+ mesothelium)
    splanchnopleure (+ mesothelium)
  neural-crest-derived
  paraxial
    head mesenchyme (somitomeres)
    somites 5–35 (max.) pairs (earlier ones
      segregating into their dermamyotome
      and sclerotome components)
    presomitic (unsegmented)
mouth region
  mandibular component of first arch
  maxillary component of first arch
  stomatodaeum (oral pit)
neural tube (future spinal cord)
  neural lumen
  neural luminal occlusion
  nerve root, dorsal (sensory), ventral (motor)
    [seen best in silver stain]
nose
  naso-lacrimal groove
  olfactory epithelium
  olfactory pit
  process
    fronto-nasal
      lateral-nasal
      medial-nasal
notochord
peripheral nervous system
  body
    dorsal (posterior) root ganglia
  head
    facial–acoustic (VII–VIII) ganglion complex
    glossopharyngeal (IX) ganglion (late stage)
    trigeminal (V) ganglion (semilunar, Gasserian)
    vagus (X) ganglion
pituitary
  infundibular recess of third ventricle
  Rathke's pouch
respiratory system
  laryngo-tracheal groove
  lung
    bronchus, main
      buds (L + R)
  pharynx
  trachea
septum transversum
  hepatic component
  non-hepatic component
somite (see mesoderm)
tail
  neural crest
  neural tube
    neural plate
  somites 31–35 (max.)
  unsegmented mesenchyme
thyroid primordium
umbilical **cord**
  artery (common)
  vein (common)
urogenital system (from intermediate mesoderm)
  urogenital ridge
    gonadal component
      primordial germ cells (also in hind gut)
    nephric component (cord) mesonephric
      (Wolffian) duct
    mesonephros
      tubules
      vesicles
vascular system
  artery
    aorta
      dorsal (midline)
      dorsal (paired)
    branchial arch artery
      1st
      2nd
      3rd
    internal carotid (rostral extension of dorsal aorta)
    principal (axial) to forelimb (7th cervical intersegmental)
    principal (axial) to hindlimb (5th lumbar intersegmental)
    umbilical (R & L)
    vitelline
  blood
  vein
    cardinal
      anterior
      common (Duct of Cuvier)
      posterior
      subcardinal/supracardinal
    inter-subcardinal venous anastomosis
    primary head
    umbilical (L)
    vitelline (omphalomesenteric; L & R)

STAGE 17 – day 10.5
brain
  forebrain
    diencephalon
      lamina terminalis
      third ventricle
        infundibular recess
        optic recess (lumen of optic stalk)
      interventricular foramen
    telencephalon
      interventricular foramen
      telencephalic (lateral) vesicle
  ganglia – see ganglia
  hindbrain (rhombencephalon)
    fourth ventricle
      roof
    rhombomeres 1–8
  midbrain (mesencephalon)
    mesencephalic vesicle
  layer
    ventricular
    mantle
branchial arch
  1st arch
    artery ectoderm
    endoderm
    groove (cleft)
    mandibular component
    maxillary component
    membrane
    mesenchyme
    pouch
  2nd arch
    artery
    cleft ectoderm
    endoderm
    membrane
    mesenchyme
    pouch
  3rd arch
    artery ectoderm
    endoderm
    groove
    membrane
    mesenchyme
    pouch
  4th arch
    artery
    groove
    membrane
    cervical sinus (retro-hyoid depression)
    ectoderm
    endoderm
    mesenchyme
    pouch
  6th arch
    artery
    ectoderm
    endoderm
    mesenchyme
cavity
  intra-embryonic coelom
    pericardial component
    pericardio-peritoneal canals
    peritoneal component
      greater sac
      omental bursa (lesser sac)
    pleural component
ear
  inner
    otocyst (otic vesicle)
    endolymphatic appendage (diverticulum)
  middle
    tubo-tympanic recess
  outer
    first branchial membrane
      (future tympanic membrane)
ectoderm (surface)
eye
  lens pit (vesicle at 18)
  optic ectoderm
  optic stalk
  perioptic mesenchyme
  retina
    inner (future neural) layer
    intraretinal space
    outer (future pigment) layer
gall bladder (cystic) primordium
  cystic duct
ganglia (from neural crest)
  body
    dorsal (posterior) root
    sympathetic
      thoracic
  cranial
    acoustic (VIII) (future vestibulo-cochlear ganglion)
    cranial/spinal accessory (XI)
    facial (VII) (geniculate)
    glossopharyngeal (IX)
    trigeminal (V) (semilunar, Gasserian)
    vagus (X)
gland
  gall bladder
    (cystic) primordium
    cystic duct
  pituitary
    infundibular recess of third ventricle
    Rathke's pouch
  thyroid primordium

THE TISSUES PRESENT IN EACH STAGE OF MOUSE DEVELOPMENT

STAGE 17 continued
gut
  foregut
    oesophagus
      meso-oesophagus
      mesentery (dorsal, ventral)
    pharyngeal region
    stomach
      dorsal mesentery
      thyroid primordium
  foregut–midgut junction
    biliary system
      cystic duct
      cystic (gallbladder) primordium
  hindgut
    cloaca
      cloacal membrane
      diverticulum
        preanal component
        postanal component
    dorsal mesentery
  midgut
    dosal mesentery
    loop
    physiological umbilical hernia
    dorsal mesentery
heart
  atrio-ventricular canal
  atrium
    common atrial chamber, L & R components
    auricular part (left & right)
    interatrial septum
      foramen (ostium) primum
      septum primum
    sinus venarum
      crista terminalis
  bulbar cushion
  bulbo-ventricular groove (sulcus)
  bulbus cordis
    inferior (future right ventricle)
    superior (prox. part of outflow tract)
  cardiac jelly
  dorsal mesocardium (dorsal mesentery of heart)
    transverse pericardial sinus
  endocardial cushion tissue
  endocardial tissue
  myocardium
  outflow tract
    aortico-pulmonary spiral septum
  primitive ventricle (future left ventricle)
  sinus venosus (R & L horns)
  venous valve (where R horn of sinus venosus
    enters R atrium)
limb
  forelimb bud
    apical ectodermal ridge
  hindlimb bud
    apical ectodermal ridge
liver (within septum transversum)
  hepatic duct
  hepatic primordium
  parenchyma
mesenchyme
  intermediate (*see* urogenital system)
  lateral
    body wall
  paraxial
    head mesenchyme (somitomeres)
    body
      somites pairs 5–30 (earlier ones beginning
        to segregate into dermotome,
        myotome and sclerotome)
    tail,
      pairs 31–39 (max.)
    presomitic
mouth region
  mandibular component of first arch
  maxillary component of first arch
  stomatodaeum (mouth pit)
neural tube
  nerve, segmental spinal

nerve root, dorsal (sensory), ventral (motor)
neural lumen
neural luminal occlusion
nose
  naso-lacrimal groove
  olfactory epithelium
  olfactory pit
  process
    fronto-nasal
    lateral-nasal
    medial-nasal
notochord
peripheral nervous system
  autonomic
    sympathetic (chain) trunk
  ganglia (from neural crest)
    body
      dorsal (posterior) root
      sympathetic
        thoracic
    cranial
      acoustic (VIII) (future vestibulo-cochlear
        ganglion)
      cranial/spinal accessory (XI)
      facial (VII) (geniculate)
      glossopharyngeal (IX)
      trigeminal (V) (semilunar, Gasserian)
      vagus (X)
  plexus, brachial
  segmental spinal nerve
pituitary
  infundibular recess of third ventricle
  Rathke's pouch
respiratory system
  laryngo-tracheal groove
  lung buds (L & R)
    bronchus, main
  pharynx
  trachea
    bifurcation (carina tracheae)
septum transversum
  hepatic component
  non-hepatic component
sinus, cervical (retro-hyoid depression)
somites (see mesoderm)
tail
  neural tube
  somite 31–39 (max.)
  unsegmented mesenchyme
thyroid primordium
umbilical cord
  artery (common)
  hernia, physiological umbilical
  vein (common)
urogenital system (in urogenital ridge)
  gonadal component
    primordial germ cells
  mesonephric (Wolffian) duct
  mesonephros
    tubules
    vesicles
  urorectal septum (incomplete)
vascular system
  artery
    aorta
      dorsal (midline)
      dorsal (paired)
    branchial arch artery
      1st
      2nd
      3rd
      4th
      6th
    ductus caroticus
    internal carotid
    ophthalmic
    principal (axial) to forelimb (7th cervical
      intersegmental artery)
    principal (axial) to hindlimb (5th lumbar
      intersegmental artery)
    pulmonary

    umbilical (R & L)
    vitelline
  blood
  vein
    cardinal
      anterior
      common (Duct of Cuvier)
      inter-subcardinal venous anastomosis
      posterior
      subcardinal/supracardinal
    marginal (of limb)
    primary head
    principal (axial)
      from forelimb
      from hindlimb
    umbilical (R & L)
    vitelline (omphalomesenteric)
  vitelline venous plexus

STAGE 18 – day 11
brain
  forebrain
    diencephalon
      lamina terminalis
    telencephalon
      interventricular foramen
      telencephalic vesicle
    third ventricle
      infundibular recess
      interventricular foramen
      optic recess (lumen of optic stalk)
  ganglia – *see* ganglia
  hindbrain (rhombencephalon)
    fourth ventricle
      roof
    metencephalon
      cerebellum primordium
        (rhombic lip; dorsal part of alar plate of
          metencephalon)
      pons primordium
        (basal plate of metencephalon)
      rhombomeres 1–2
    myelencephalon
      rhombomeres 3–8
  midbrain (mesencephalon)
    mesencephalic vesicle
  layer (applies to whole brain and neural tube)
    ependymal
    mantle
    marginal
branchial arch
  1st arch
    artery
    ectoderm
    endoderm
    groove (cleft)
    mandibular component
    maxillary component
    membrane
    pouch
  2nd arch
    artery
    ectoderm
    endoderm
    groove (cleft)
    membrane
    pouch
  3rd arch
    artery
    ectoderm
    endoderm
    groove (cleft)
    pouch
    membrane
  4th arch
    artery
    ectoderm
    endoderm
    groove (cleft)
    membrane
      cervical sinus (retro-hyoid depression)

STAGE 18 *continued*
    pouch
      6th arch
        artery
        ectoderm
        endoderm
        mesoderm
cavity
    intra-embryonic coelom
      pericardial component
      pericardio-peritoneal canals
      peritoneal component
        greater sac
        omental bursa (lesser sac)
      pleural component
ear
    inner
      otocyst (otic vesicle)
        endolymphatic appendage (diverticulum)
    middle
      tubo-tympanic recess
    outer (external)
      first branchial membrane (future tympanic
        membrane)
ectoderm (surface)
eye
    corneal ectoderm
    lens pit (early in stage)
    lens **vesicle** (late in stage)
    optic eminence/ectoderm
    optic stalk
    perioptic mesenchyme
    retina
      choroid (fetal) fissure
      intraretinal space
      neural (inner) **layer**
      pigment (outer) **layer**
gall bladder (cystic) primordium
    cystic duct
ganglia (from neural crest)
    body
      dorsal (posterior) root
      sympathetic
        thoracic
    cranial
      acoustic (vestibulo-cochlear) (VIII)
      cranial/spinal accessory (XI)
      facial (VII) (geniculate)
      glossopharyngeal (IX) – inferior (petrosal)
        (late stage)
      glossopharyngeal (IX) – superior (late
        stage)
      trigeminal (V) (semilunar, Gasserian)
      vagus (X) – inferior (nodose)
      vagus (X) – superior
gland
    gall bladder
      (cystic) primordium
      common bile duct
      cystic duct
    pancreas primordium
      dorsal bud
    pituitary
      infundibular recess of third ventricle
      infundibulum
      Rathke's pouch (from pharynx)
    thyroid
      foramen caecum
      thyroglossal duct
gut
    foregut
      duodenum (rostral half)
        mesoduodenum
      oesophagus
        meso-oesophagus
      mesentery (dorsal, ventral)
      pharynx
      stomach
    foregut–midgut junction
      biliary system
        cystic primordium

      mesentery (dorsal, ventral)
    hindgut
      anal membrane
      cloaca
      diverticulum
        postanal component
      dorsal mesentery
      urogenital membrane
      uro-rectal septum
    midgut
      dorsal mesentery
      duodenum (caudal half)
        mesoduodenum
      loop
      mesentery
      physiological umbilical hernia
heart
    atrio-ventricular canal
    atrium
      auricular part (left & right)
      interatrial septum
        foramen (ostium) primum
        septum primum
      sinus venarum
        crista terminalis
    bulbar cushion
    bulbo-ventricular groove (sulcus)
    bulbus cordis
      inferior (future right ventricle)
      superior (prox. part of outflow tract)
    dorsal mesocardium (mesentery)
      transverse pericardial sinus
    endocardial cushion tissue
    endocardial tissue
    myocardium
    outflow tract
      aortico-pulmonary spiral septum
    pericardium visceral (epicardium) (late)
    primitive ventricle (future left ventricle)
    sinus venosus (R & L horns)
    venous valve (where R horn of sinus venosus
      enters R atrium)
limb
    hindlimb bud
      apical ectodermal ridge
    forelimb bud
      apical ectodermal ridge
liver (in septum transversum)
    hepatic duct
    parenchyma
    mesenchyme
      body
      head
mouth region
    mandibular component of first arch
    maxillary component of first arch
      palatal shelf (secondary palate)
    oronasal cavity
    tongue
      foramen caecum
      lingual swellings
        lateral & median (tuberculum impar)
          (mandibular component of first arch)
peripheral nervous system
    ganglia
      body
        dorsal (posterior) root
        sympathetic
          thoracic
    nerve
      autonomic, sympathetic (chain) trunk
      (cranial) spinal accessory (XI)
      plexus, brachial
      segmental spinal
neural tube
    alar and basal columns
    floorplate
    layer
      ependymal, mantle, **marginal**
    nerve, segmental spinal
    nerve root, dorsal (sensory), ventral (motor)

    neural lumen
    roofplate
nose
    naso-lacrimal groove
    olfactory epithelium
    olfactory pit
    process
      fronto-nasal
        lateral-nasal
        medial-nasal
notochord
pancreas primordium
pituitary
    infundibular recess of third ventricle
    infundibulum
    Rathke's pouch
placenta
    chorionic plate
    labyrinthine part
    spongy (or basal) region (spongiotrophoblast)
respiratory system
    laryngo-tracheal groove
    lung bud
      bronchus, main
    pharynx
    trachea
      bifurcation (carina tracheae)
septum transversum
    hepatic component
    non-hepatic component
skeleton
    vertebral PCC (cervical)
somite
    trunk 12–30 pairs
    tail 31–45 (max.)
    dermatome
    myotome
    sclerotome
tail
    neural tube
    somite 31–45 (max.)
    unsegmented mesenchyme
thyroid
    foramen caecum
    thyroglossal duct
umbilical cord
    artery (common)
    hernia, physiological umbilical
    vein (common)
urogenital system
    metanephros
      metanephric mesenchyme
      ureteric bud (from **Wolffian duct**)
    urogenital ridge
      gonad primordium
        primordial germ cells
      mesonephric (Wolffian) duct
      mesonephros
        tubules
        vesicles
    urogenital sinus
    urorectal septum
vascular system
    artery
      aorta
        dorsal (midline)
        dorsal (paired)
      branchial arch artery
        1st
        2nd
        3rd
        4th
        6th
      ductus caroticus
      internal carotid
      ophthalmic
      principal (axial) to forelimb (7th cervical
        intersegmental artery)
      principal (axial) to hindlimb (5th lumbar
        intersegmental artery)
      pulmonary

THE TISSUES PRESENT IN EACH STAGE OF MOUSE DEVELOPMENT

STAGE 18 continued
- umbilical (common)
- vitelline
- blood
- vein
  - cardinal
    - anterior
    - common (Duct of Cuvier)
    - inter-subcardinal venous anastomosis
    - posterior
    - subcardinal/supracardinal
  - marginal
    - forelimb
    - hindlimb
  - primary head
  - principal (axial) from limb
    - forelimb
    - hindlimb
  - umbilical (left)
  - vitelline (omphalomesenteric)
  - vitelline venous plexus

STAGE 19 – day 11.5
- brain
  - forebrain
    - diencephalon
      - lamina terminalis
    - telencephalon
      - interventricular foramen
      - **lateral ventricle** (previously telencephalic vesicle)
      - optic recess (lumen of optic stalk)
    - third ventricle
      - floorplate
      - infundibular recess
      - interventricular foramen
      - optic recess (lumen of optic stalk)
      - roofplate
  - hindbrain (rhombencephalon)
    - fourth ventricle
      - **lateral recess**
      - roof
    - metencephalon
      - cerebellum primordium (rhombic lip; dorsal part of alar plate of metencephalon)
      - pons primordium (basal plate of metencephalon)
    - myelencephalon
  - midbrain (mesencephalon)
    - mesencephalic vesicle
  - layer (applies to whole brain & neural tube)
    - ventricular (ependymal)
    - mantle
    - marginal
- branchial arch
  - 1st arch
    - artery
    - **cartilage (Meckel's) PCC** (mandible primordium)
    - groove (cleft)
    - mandibular component
    - maxillary component
    - membrane
    - pouch
  - 2nd arch
    - artery
    - groove (cleft)
    - membrane
    - pouch
  - 3rd arch
    - artery
    - groove (cleft)
    - membrane
    - pouch
  - 4th arch
    - artery
    - cervical sinus (retro-hyoid depression)
      - pouch
  - 6th arch
    - artery
- cavity
  - intra-embryonic coelom
    - pericardial component
    - pericardio-peritoneal canals
    - peritoneal component
      - greater sac
      - omental bursa (lesser sac)
    - pleural component
- ear
  - inner
    - **ductus reuniens**
    - otocyst (otic vesicle)
      - endolymphatic appendage (diverticulum)
      - **saccule, utricle**
      - **semicircular canals (posterior, superior)**
  - middle
    - tubo-tympanic recess
  - outer (external)
    - **auditory hillocks (tubercles)**
    - first branchial membrane (future tympanic membrane)
- ectoderm (surface)
- eye
  - corneal ectoderm
  - **hyaloid vascular plexus**
  - lens vesicle
  - mesenchyme, perioptic (future sclera)
  - optic stalk
  - retina
    - choroid (fetal) fissure
    - intraretinal space
    - neural layer (inner) layer
    - pigment layer (outer) layer
- gall bladder
  - common bile duct
  - cystic duct
- ganglia (from neural crest)
  - body
    - dorsal (posterior) root
    - sympathetic
      - thoracic
  - cranial
    - cranial/spinal accessory (XI)
    - facial (VII)
    - glossopharyngeal (IX) – inferior (petrosal)
    - glossopharyngeal (IX) – superior
    - **hypoglossal (XII)**
    - trigeminal (V) (semilunar, Gasserian)
    - vagus (X) – inferior (nodose)
    - vagus (X) – superior
    - vestibulo-cochlear (VIII)
- glands
  - gall bladder
    - common bile duct
    - cystic duct
  - pancreas
    - dorsal bud
    - **ventral bud**
  - pituitary
    - infundibular recess of third ventricle
    - infundibulum
    - Rathke's pouch (from pharynx)
  - thyroid
    - foramen caecum
    - thyroglossal duct
- gut
  - foregut
    - duodenum (rostral half)
      - mesoduodenum
    - mesentery (dorsal, ventral)
    - oesophagus
      - meso-oesophagus
    - pharynx
    - **hyoid PCC**
    - stomach
      - mesentery
  - foregut–midgut junction
    - biliary system
      - cystic duct
      - common bile duct
      - gallbladder
    - mesentery (dorsal, ventral)
  - hindgut
    - cloaca
      - anal membrane
    - diverticulum
      - anal region
      - **anal pit (proctodaeum)**
      - postanal component
    - dorsal mesentery
    - urogenital membrane
    - uro-rectal septum
  - midgut
    - dorsal mesentery
    - duodenum (caudal half)
    - loop
      - dorsal mesentery
      - physiological umbilical hernia
- heart
  - atrio-ventricular canal
  - atrium
    - auricular part (left & right)
    - interatrial septum
      - foramen (ostium) primum
      - septum primum
    - sinus venarum
      - crista terminalis
  - bulbar cushion
  - **bulbar ridge**
  - dorsal mesocardium (mesentery)
  - transverse pericardial sinus
  - endocardial cushion tissue
  - endocardial tissue
  - myocardium
  - outflow tract
    - **aortic component**
    - aortico-pulmonary spiral septum
    - **pulmonary component**
  - pericardium
    - **parietal**
    - visceral (epicardium)
  - **pulmonary trunk**
    - distal part: from L sixth arch artery
    - prox. part from outflow tract
  - sinus venosus (R & L horns)
  - venous valve (where R horn of sinus venosus enters R atrium)
  - **ventricle (L & R)**
    - **interventricular groove**
    - **interventricular septum** (precursor, in bulboventricular region)
- limb
  - forelimb bud
    - apical ectodermal ridge
    - **handplate**
  - hindlimb bud
    - apical ectodermal ridge
    - **footplate**
- liver
  - common bile duct
  - hepatic duct
  - **hepatic sinusoids**
  - parenchyma
- mesenchyme
  - body
  - head
- mouth region
  - mandibular process
    - **cartilage**
      - **Meckel's PCC**
      - **Reichert's PCC**
    - **epiglottal swelling**
    - **maxillary process**
      - palatal shelf (secondary palate)
    - oronasal cavity
  - tongue
    - foramen caecum
    - **hypobranchial eminence** (3rd & 4th arch components)
    - lingual swellings
      - lateral & median (tuberculum impar; mandibular component of 1st arch)

STAGE 19 continued
    occipital myotome (gives extr. & intr.
        tongue muscles, except
        palatoglossus)
nerve
  autonomic
    sympathetic (chain) trunk
  body
    nerve trunk
      phrenic
    plexus
      brachial
        median
        radial
        ulnar
    segmental spinal
  cranial
    glossopharyngeal (IX)
    hypoglossal (XII)
    cranial/spinal accessory (XI)
  parasympathetic
    vagal trunk (X)
      recurrent laryngeal branch
neural tube (future spinal cord)
  alar and basal columns
  floorplate
  layer,
    ventricular (ependymal)
    mantle
    marginal
  nerve, segmental spinal
  nerve root, dorsal (sensory), ventral (motor)
  neural lumen (future central canal)
  roofplate
nose
  naso-lacrimal groove
  olfactory epithelium
  olfactory pit
  process
    fronto-nasal
      lateral-nasal
      medial-nasal
notochord
pancreas primordium
  dorsal bud
  ventral bud
pituitary
  infundibular recess of third ventricle
  infundibulum
  Rathke's pouch (from pharynx)
placenta
  chorionic plate
  labyrinthine part
  spongy (or basal) region (spongiotrophoblast)
renal/urinary system (in urogenital ridge)
  mesonephric (Wolffian) duct
  mesonephros
    tubules
    vesicles
  metanephros
    blastema (induced) (excretory components)
    ureteric bud (from Wolffian duct) (drainage
      component)
  urogenital
    mesentery
  urogenital sinus
  urorectal septum
reproductive system
  genital tubercle
  gonad primordium (indifferent stage; in
    urogenital ridge)
    "germinal" epithelium
    primordial germ cells
  mesonephric (Wolffian) duct (future ductus
    deferens in male)
  urogenital mesentery
  urogenital sinus
respiratory system
  arytenoid swelling PCC
  laryngo-tracheal groove
  lung bud
    bronchus
      lobar
      main
  pharynx
  trachea
    bifurcation (carina tracheae)
septum transversum (non-hepatic component)
skeleton
  vertebral axis
    rib PCC
    vertebral CC (cervical)
    vertebral sclerotomal condensations (lumbar,
      thoracic)
somite
  body pairs 26–30
  tail: pairs 31–48 (max.)
  dermatome, myotome, sclerotome
tail
  neural tube
  somites pairs 31–48 (max.)
  unsegmented mesenchyme
thyroid
  foramen caecum
  thyroglossal duct
umbilical cord
  artery (R & L)
  hernia, physiological umbilical
  vein (R & L)
vascular system
  artery
    aorta
      ascending thoracic (from outflow tract)
      dorsal (midline)
      dorsal (paired)
    basilar
    branchial arch artery
      3rd
      4th
      6th
    common iliac
    ductus caroticus
    hyoid
    iliac
    internal carotid
    maxillary
    mesenteric
    ophthalmic
    principal (axial) to hindlimb
    principal (axial) to forelimb
    pulmonary
    pulmonary trunk
      distal part from L 6th arch artery
      prox. part from outflow tract
    stapedial
    umbilical (R & L)
    vertebral
  blood
  vein
    cardinal
      anterior
      common (Duct of Cuvier)
      subcardinal/supracardinal
    ductus venosus
    hepato-cardiac
    marginal (of fore- and hindlimb)
    mesenteric
    portal
    primary head
    principal (axial) from forelimb
    principal (axial) from hindlimb
    umbilical (R & L)
    vena cava
      inferior, post-hepatic part (from posterior
        cardinal)

STAGE 20 – day 12
brain
  forebrain
    diencephalon
      epithalamus
      hypothalamus
      lamina terminalis
      thalamus
    telencephalon
      corpus striatum
      interventricular foramen
      lateral ventricle
        choroid invagination (tela choroidea)
      neopallial cortex (future cerebral cortex)
    third ventricle
      infundibular recess
      interventricular foramen
      optic recess (lumen of optic stalk)
  hindbrain (rhombencephalon)
    fourth ventricle
      lateral recess
      roof
        choroid plexus
    metencephalon
      cerebellum primordium (alar plate)
      pons (basal plate)
      sulcus limitans
    myelencephalon
      medulla oblongata
      sulcus limitans
  midbrain (mesencephalon)
    mesencephalic vesicle
  layer, ependymal, mantle, marginal
  primitive ectomeninx
cavity
  intra-embryonic coelom
    pericardial component
    pericardio-peritoneal canals
    peritoneal component
      greater sac
      omental bursa (lesser sac)
    pleural component
ear
  ganglia
    cochlear (VIII) (late stage)
    vestibular (VIII) (late stage)
    vestibulo-cochlear (VIII) (early stage only)
  inner
    cochlea
    ductus reuniens
    endolymphatic appendage (diverticulum)
    endolymphatic sac (future duct)
    saccule (utricle/saccule)
    otic capsule (pre-cartilage)
      semicircular canals (posterior, superior)
  middle
    pharyngo-tympanic tube (future Eustachian
      tube)
    tubo-tympanic recess
  outer (external)
    external acoustic meatus (entrance only)
    external ear primordium (pinna)
    future tympanic membrane (1st branchial
      membrane)
    pinna condensation
    temporal bone (petrous part) PCC
eye
  cornea
    ectoderm
    stroma
  hyaloid vascular plexus
  lens vesicle
  mesenchyme, perioptic (future sclera)
  optic
    (II) nerve
    chiasma
    fissure
    stalk
  retina
    choroid (fetal) fissure
    neural (inner) layer
    pigment (outer) layer
gall bladder
  cystic duct
ganglia
  body
    dorsal (posterior) root

THE TISSUES PRESENT IN EACH STAGE OF MOUSE DEVELOPMENT

STAGE 20 continued
  sympathetic
    thoracic
  cranial
    acoustic (vestibulo-cochlear) (VIII) (early
        stage only)
    cochlear (VIII)
    cranial/spinal accessory
    facial (VII)
    glossopharyngeal (IX) – inferior
    glossopharyngeal (IX) – superior
    trigeminal (V) (semilunar, Gasserian)
    vagus (X) – inferior (nodose)
    vagus (X) – superior
    vestibular (VIII)
    vestibulo-cochlear (VIII)
glands
  gall bladder
    common bile duct
    cystic duct
  pancreas
    body, tail, left & right lobes
    pancreatic duct
    uncinate process
  pituitary
    adenohypophysis
      lumen (remnant of Rathke's pouch, now
        detached)
      pars anterior (anterior lobe, pars distalis)
      pars intermedia
      pars tuberalis
    neurohypophysis
      infundibulum (pituitary stalk)
      median eminence
      pars nervosa (pars posterior)
  thyroid
    foramen caecum
    thyroglossal duct
gut
  foregut
    duodenum (rostral half)
      mesoduodenum
    oesophagus
      meso-oesophagus
    mesentery (dorsal, ventral)
    pharynx
      hyoid CC
    stomach
      **glandular, proventricular ("cutaneous",
        non-glandular), pyloric regions**
      dorsal, ventral mesogastrium
  foregut–midgut junction
    biliary system
      common bile duct
      gallbladder
    mesentery (dorsal, ventral)
  hindgut
    cloaca
    diverticulum
    anal region
      anal membrane
      anal pit (proctodaeum)
      post-anal component
    dorsal mesentery
    urogenital membrane
    uro-rectal septum
  midgut
    dorsal mesentery
    duodenum (caudal half)
      dorsal mesoduodenum
    loop
      dorsal mesentery
      physiological umbilical hernia
heart
  aorta
    ascending thoracic (from outflow tract)
  aortic arch (from left fourth arch artery)
  atrio-ventricular canal
  atrio-ventricular cushion tissue
  atrium
    auricular part (L & R)
    interatrial septum
      foramen (ostium) primum
      septum primum
    sinus venarum
      crista terminalis
  bulbar cushion
  bulbar ridge
  dorsal mesocardium (mesentery)
    transverse pericardial sinus
  endocardial cushion tissue
  endocardial tissue
  myocardium
  outflow tract
    aortic component
    aortico-pulmonary spiral septum
    pulmonary component
  pericardium
    parietal
    visceral (epicardium)
  pulmonary trunk
    distal part, from L sixth arch artery
    prox. part from outflow tract
  sinus venosus (L & R horns)
  venous valve (where R horn of sinus venosus
    enters R atrium)
  ventricle (L & R)
    interventricular groove
    interventricular septum
      muscular part
limb
  forelimb
    handplate
      digit condensations
      digital interzones
      humeral PCC
      radio-ulnar PCC
  hindlimb
    femoral PCC
    footplate
      digit condensations (late stage)
      digital interzones
liver
  common bile duct
  hepatic duct
  hepatic sinusoids
  parenchyma
mesenchyme
  body
  head
mouth region
  epiglottis primordium
  mandibular process
    cartilage (Meckel's, primordium of mandible)
    mandible primordium
  maxillary process
    palate
      palatal (secondary) shelf
      **primary (premaxilla, intermaxillary
        segment)**
  oral cavity
  tongue
    foramen caecum
    hypobranchial eminence (3rd & 4th arch
      components)
    lingual swellings
      lateral & median (tuberculum impar;
        mandibular component of 1st arch)
    occipital myotome
nerve
  body
    autonomic
      parasympathetic
        vagus (X)
        vagal trunk (X)
          recurrent laryngeal branch
      sympathetic (chain) trunk
    nerve trunk
      phrenic
    plexus
      brachial
        median, radial, ulnar
      lumbo-sacral
    segmental spinal
  cranial
    accessory (XI)
      cranial, spinal components
    glossopharyngeal (IX)
    hypoglossal (XII)
    oculomotor (III)
    optic (II)
    trigeminal (V)
nose
  nasal cavity
  olfactory epithelium
  olfactory pit
  primitive nasal septum
  vomeronasal (Jacobson's) organ
  process
    fronto-nasal
      lateral-nasal
      medial-nasal
notochord
pancreas
  body
  duct
  left lobe
  right lobe
  tail
  uncinate process
pituitary
  Rathke's pouch (now detached from pharynx)
  adenohypophysis
    **pars anterior (anterior lobe, pars
      distalis)**
    pars intermedia
    pars tuberalis
  neurohypophysis
    infundibulum (pituitary stalk)
    median eminence
    pars nervosa (pars posterior)
placenta
  chorionic plate
  labyrinthine part
  spongy (or basal) region (spongiotrophoblast)
renal/urinary system
  mesonephros (regressing)
    tubules
    vesicles
  mesonephric (Wolffian) duct
  metanephros
    excretory components
      blastemal cells
    drainage component (ureteric bud
      derivatives)
      primitive collecting ducts
      primitive ureter
    sinus
  urogenital
    mesentery
  urogenital sinus
  urorectal septum
reproductive system
  genital tubercle
  gonad primordium (indifferent stage)
    "germinal" epithelium
    primordial germ cells
  mesonephric (Wolffian) duct (future ductus
    deferens in male)
  mesonephric component (future rete testis in
    male)
  urogenital
    membrane
    mesentery
  urogenital sinus
respiratory system
  arytenoid swelling PCC
  laryngo-tracheal groove
  lung bud
    bronchus
      lobar
      main
  pharynx

STAGE 20 *continued*
  trachea
    bifurcation (carina tracheae)
  upper respiratory tract
    nasopharynx
septum transversum (non-hepatic component)
skeleton
  cranium
    1st arch, Meckel's cartilage CC
    2nd arch, Reichert's cartilage CC
    chondrocranium
      temporal bone
        petrous part PCC
  limb, forelimb
    humerus PCC
    radio-ulnar PCC
  limb, hindlimb
    femoral PCC
  vertebral axis
    hyoid bone CC
    rib PCC
    vertebrae
      atlas PCC
      axis PCC
      cervical vertebrae PCC
skin
  cervical sinus (retro-hyoid depression)
  vibrissae elevations
somites
  tail: pairs 35–52 (max.)
  dermatome, myotome, sclerotome
spinal cord (previously neural tube)
  central canal (previously neural lumen)
  floorplate
  layer
    ependymal
    mantle (alar & basal columns)
    marginal
  nerve, segmental spinal
  nerve root, dorsal (sensory), ventral (motor)
  roofplate
  sulcus limitans
tail
  neural tube
  somites: pairs 35–52 (max.)
  unsegmented mesenchyme
thyroid
  foramen caecum
  thyroglossal duct
umbilical cord
  artery
  hernia, physiological umbilical
  vein
vascular system
  artery
    aorta
      aortic arch (pre-ductal, from L 4th arch artery)
      ascending thoracic (from outflow tract)
      dorsal
    basilar
    branchial arch artery
      6th
    common iliac
      external iliac
    ductus caroticus (remnant)
    hyoid
    iliac
    innominate (brachiocephalic)
    internal carotid
    intersegmental (e.g. to limbs)
    maxillary
    mesenteric
    ophthalmic
    principal (axial) to hindlimb (7th cervical intersegmental)
    principal (axial) to forelimb (5th lumbar intersegmental)
    pulmonary
    pulmonary trunk
      proximal: from outflow tract
      distal: from L 6th arch artery
    stapedial
    subclavian
    umbilical (R & L)
    vertebral
  blood
  vein
    cardinal
      common (Duct of Cuvier)
      subcardinal/supracardinal
    ductus venosus
    internal jugular (from anterior cardinal)
    marginal (of fore- and hindlimb)
    mesenteric
    portal
    primary head
    principal (axial, of fore- and hindlimb)
    subclavian
    umbilical (R & L)
    vena cava
      inferior, post-hepatic part

STAGE 21 – day 12.5
brain
  forebrain
    diencephalon
      epithalamus
      hypothalamus
      lamina terminalis
      thalamus
    telencephalon
      choroid fissure
      choroid plexus
      corpus striatum
      interventricular foramen
      lateral ventricle
        anterior, inferior, posterior & superior horns
        choroid invagination (tela choroidea)
      neopallial cortex (future cerebral cortex)
      olfactory lobe
      olfactory cortex
    third ventricle
      choroid plexus
      infundibular recess
      interventricular foramen
      optic recess (lumen of optic stalk)
  hindbrain (rhombencephalon)
    fourth ventricle
      lateral recess
        choroid plexus
      roof
        choroid plexus
    metencephalon
      cerebellum
      pons
      sulcus limitans
    myelencephalon
      medulla oblongata
      sulcus limitans
  midbrain (mesencephalon)
    cerebral aqueduct (derived from mesencephalic vesicle)
  covering (brain and spinal cord)
    meninges
      arachnoid mater
      dura mater
      pia mater
  layer
    ependymal, mantle, marginal
cavity
  diaphragm
    central tendon (from septum transversum)
    crus
    pleuro-pericardial canal (channel)
    pleuro-pericardial folds
    pleuro-peritoneal canal (channel)
    pleuro-peritoneal folds
  intra-embryonic coelom
    pericardial component
    pericardio-peritoneal canals
    peritoneal component
      greater sac
      omental bursa (lesser sac)
    pleural component
ear
  ganglia
    cochlear (VIII)
    vestibular (VIII)
  inner
    cochlea
    cochlear canal (duct)
    ductus reuniens
    endolymphatic appendage (diverticulum)
    endolymphatic sac (future duct)
    labyrinth
      crus commune
    otic capsule (pre-cartilage)
    saccule (utricle/saccule)
    semicircular canals (posterior, superior, **lateral**)
  middle
    ossicles (malleus & incus PCC (1st arch), stapes PCC (2nd arch))
    pharyngo-tympanic (Eustachian) tube
    tubo-tympanic recess
  outer (external)
    external acoustic meatus (entrance only)
    first branchial membrane (future tympanic membrane)
    pinna
  temporal bone (petrous part) PCC
eye
  canthus, inner (nasal)
  choroidal vessels
  cornea
    ectoderm
    stroma
  eyelid (upper & lower)
  hyaloid cavity
    hyaloid vascular plexus
    vitreous humour
  lens
    fibres
  muscle
    extrinsic ocular PMM
  optic
    (II) nerve
    chiasma
    fissure
    stalk
  perioptic mesenchyme (future sclera)
  retina
    choroid (fetal) fissure
    neural (inner) layer
    pigment (outer) layer
gall bladder
  common bile duct
  cystic duct
ganglia
  body
    dorsal (posterior) root
    sympathetic
      inferior cervical (cervico-thoracic, stellate)
      middle cervical
      superior cervical
      thoracic
  cranial
    cochlear (VIII)
    cranial/spinal accessory (XI)
    facial (VII)
    glossopharyngeal (IX) – inferior
    glossopharyngeal (IX) – superior
    trigeminal (V) (semilunar, Gasserian)
    vagus (X) – inferior (nodose)
    vagus (X) – superior
    vestibular (VIII)
gland
  adrenal primordium
  gall bladder
    common bile duct
    cystic duct

THE TISSUES PRESENT IN EACH STAGE OF MOUSE DEVELOPMENT

STAGE 21 *continued*
- **mammary gland**
- **mammary ridge**
- pancreas
  - body, tail, left & right lobes
  - pancreatic duct
  - uncinate process
- pituitary
  - adenohypophysis
    - lumen (remnant of Rathke's pouch, now detached)
    - pars anterior (anterior lobe, pars distalis)
    - pars intermedia
    - pars tuberalis
  - neurohypophysis
    - infundibulum (pituitary stalk)
    - median eminence
    - pars nervosa (pars posterior)
- **sublingual gland primordium** (late)
- **submandibular gland primordium** (late)
- **thymus primordium**
- thyroid
  - foramen caecum
  - thyroglossal duct
- gut
  - foregut
    - duodenum (rostral half)
      - dorsal mesoduodenum
    - oesophagus
      - meso-oesophagus
    - pharynx
    - stomach
      - **fundus**
      - glandular, proventricular & pyloric regions
      - **mesentery**
        - dorsal mesogastrium
        - ventral mesogatrium
          - **lesser omentum**
      - **pyloric antrum**
  - foregut–midgut junction
    - biliary system
      - common bile duct
      - gallbladder
  - hindgut
    - cloaca
      - anal pit (proctodaeum)
      - anal membrane
    - dorsal mesentery
    - **perineal "body"**
    - **rectum**
    - urogenital membrane
    - uro-rectal septum
  - midgut
    - dorsal mesentery
    - duodenum (caudal half)
      - dorsal mesoduodenum
    - loop
    - mesentery
    - physiological umbilical hernia
- haemolymphoid system
  - blood
  - **thymus primordium**
- heart
  - aorta, ascending thoracic (from outflow tract)
  - aortic arch (from left fourth arch artery)
  - atrio-ventricular canal
  - atrio-ventricular cushion tissue
  - atrium
    - auricular part (L & R)
    - interatrial septum
      - **foramen (ostium) secundum**
      - septum primum
    - sinus venarum
      - crista terminalis
  - bulbar cushion
  - bulbar ridge
  - dorsal mesocardium
    - transverse pericardial sinus
  - **ductus arteriosus** (from L 6th arch artery)
  - endocardial cushion tissue
  - endocardial tissue
  - myocardium
  - outflow tract
    - aortic component
    - aortico-pulmonary spiral septum
    - pulmonary component
  - pericardium
    - parietal
    - visceral (epicardium)
  - pulmonary trunk
    - proximal: from outflow tract
    - distal: from L sixth arch artery
  - **valve**
    - **aortic**
    - **mitral (bicuspid)**
    - **pulmonary**
    - **tricuspid**
  - venous valve (leaflets, entrance of inferior vena cava)
  - ventricle (L & R)
    - interventricular groove
    - interventricular septum
      - muscular part
- limb
  - forelimb
    - bones
      - **carpal PCC**
      - humeral CC
      - **metacarpal PCC**
      - radius CC
      - ulnar CC
    - handplate
      - **digit primordia**
      - digital interzones
    - **joints**
      - **elbow primordium**
      - **shoulder primordium**
      - **wrist region**
    - mesenchyme
  - hindlimb
    - bones
      - femoral CC
      - **metatarsal PCC**
      - **tarsal PCC**
      - **tibial-fibular PCC**
    - footplate
      - digit primordia
      - digital interzones
      - **calcaneum PCC**
      - **talus PCC**
    - joints
      - ankle region
      - hip region
      - knee joint primordium
    - mesenchyme
- liver
  - **"bare" area**
  - common bile duct
  - cystic duct
  - hepatic duct
  - hepatic sinusoids
  - parenchyma
- **mammary gland primordium**
- **mammary ridge**
- mesenchyme
  - body
  - head
- mouth region
  - epiglottis primordium
  - jaw, lower
    - cartilage (Meckel's; mandible)
    - **mandible**
    - **tooth primordium**
      - **bell stage**
      - **bud stage**
      - **cap stage**
      - **dental laminae**
        - **incisors & molars**
      - **enamel organ** (outer epithelial layer)
  - **lips (upper & lower)**
    - philtrum
  - **mandible**
  - maxilla
    - premaxilla
  - palate
    - primary (intermaxillary segment)
    - secondary (palatal shelf of maxilla)
  - **sublingual gland primordium** (late)
  - **submandibular gland primordium** (late)
  - tongue
    - foramen caecum
    - muscle
      - **extrinsic PMM**
      - **intrinsic PMM**
- muscle
  - cranial
    - **extrinsic ocular PMM**
  - girdle, pectoral
    - **deltoid PMM**
    - **subscapularis PMM**
  - **lingual PMM**
  - vertebral axis
    - **thoracic/abdominal wall PMMs**
- nerve
  - body
    - autonomic
      - parasympathetic
        - vagus (X)
          - **recurrent laryngeal branch**
          - **trunk** R & L above diaphragm
          - post. & ant. below diaphragm
      - sympathetic (chain) trunk
    - nerve trunk
      - **intercostal**
      - phrenic
      - **sciatic**
    - plexus
      - lumbo-sacral
      - brachial
        - median, radial, ulnar
    - segmental spinal
      - **intercostal**
  - cranial,
    - accessory (XI)
      - cranial and spinal components
    - **cochlear (VIII)**
    - **facial (VII)**
    - glossopharyngeal (IX)
    - hypoglossal (XII)
    - oculomotor (III)
    - **olfactory (I)**
    - optic (II)
    - trigeminal (V)
      - **mandibular division**
      - **maxillary division**
      - **ophthalmic division**
    - vagal (X) trunks, L & R
    - **vestibular (VIII)**
- nose
  - naris
    - external
    - **primary choana** (primitive posterior)
  - nasal (olfactory) cavity
  - **nasal capsule (cartilaginous)**
  - **nasal septum**
    - vomeronasal (Jacobson's) organ
  - olfactory epithelium
- **nucleus pulposus** (previously notochord)
- pancreas
  - body, left lobe, right lobe, tail
  - pancreatic duct
  - uncinate process
- pituitary
  - adenohypophysis
    - lumen (remnant of Rathke's pouch, now detached)
    - pars anterior (anterior lobe, pars distalis)
    - pars intermedia
    - pars tuberalis
  - neurohypophysis
    - infundibulum (pituitary stalk)
    - median eminence
    - pars nervosa (pars posterior)

STAGE 21 continued
placenta
    chorionic plate
    labyrinthine part
    spongy (or basal) region (spongiotrophoblast)
renal/urinary system
    mesonephric (Wolffian) duct
    mesonephros
        tubules (regressing remnants only)
        vesicles (regressing remnants only)
    metanephros
        excretory components
            cortical layer
                blastemal cells
                early nephrons
            medullary zone
        drainage component
            pelvis
            primitive collecting ducts
            ureter
    urogenital membrane
    urogenital mesentery – see reproductive system
    urogenital sinus
        phallic part
        urachus
        vesical part (future bladder)
    urorectal septum
reproductive system
    female
        genital tubercle
        gonad (ovary)
            "germinal" epithelium
            primordial germ cells
            rete ovarii
        paramesonephric (Müllerian) duct
        urogenital mesentery
            mesovarium (female)
        urogenital sinus
    male
        genital tubercle
        gonad (testis)
            "germinal" epithelium
            primordial germ cells
            testicular cords (future seminiferous tubules)
        mesonephric (Wolffian) duct (future ductus deferens)
        urogenital mesentery
            mesorchium (male)
        urogenital sinus
respiratory system
    arytenoid swelling CC
    hyoid bone PCC
    laryngeal aditus
    laryngeal cartilages
    lung
        bronchus
            lobar
            main
            segmental
        hilus (root)
        left lobe
        right lobes
            accessory, caudal, cranial, middle
    pharynx
    trachea
        bifurcation (carina tracheae)
    upper respiratory tract
        nasopharynx
skeleton
    cranium
        chondrocranium
            basioccipital PCC
            basisphenoid PCC
            exoccipital PCC
            temporal bone
                petrous part CC
            Meckel's cartilage CC (central element of **mandible**)
        vault of skull
            frontal bone primordium
            inter-parietal bone primordium
            parietal bone primordium
            temporal bone
                squamous part PCC
        viscerocranium
            ethmoid bone primordium
            facial bones primordia
            sphenoid bone PCC
    girdle
        pectoral
            clavicle PCC
            scapula PCC
            shoulder joint primordium
        pelvic
            hip joint primordium
            iliac PCC
    limb
        forelimb
            bones
                carpal PCC
                humeral CC
                radius CC
                ulnar CC
            handplate
                digit primordia
                digital interzones
            joints
                elbow primordium
                shoulder primordium
                wrist region
        hindlimb
            bones
                femoral CC
                metatarsal bone PCC
                tarsal PCC
                      calcaneum PCC
                      talus PCC
                tibial-fibular PCC
            digit primordia (late stage)
            digital interzones
            joints
                ankle region
                hip primordium
                knee primordium
    vertebral axis
        hyoid bone CC
        rib
        vertebrae
            atlas
            axis
                dens (odontoid process of C2)
            cervical vertebrae
                centrum
                intervertebral disc primordium (surrounds nucleus pulposus)
                neural arch
                nucleus pulposus
            lumbar, sacral, thoracic vertebrae
                neural arch, centrum etc.
skin
    dermis
    eyelids (upper & lower) – see eye
    mammary gland
    mammary ridge
    vibrissae elevations
    vibrissae precursors
somites
    tail: pairs 40–55
    dermotome, myotome, sclerotome
spinal cord
    central canal
    floorplate
    layer
        ependymal, mantle (alar & basal columns), marginal
    meninges – see brain
    nerve, segmental spinal
    nerve root, dorsal (sensory), ventral (motor)
    roofplate
    sulcus limitans
tail
    neural tube
    somites: pairs 40–55
thymus primordium (from ventral part of third branchial pouch)
thyroid
    foramen caecum
    thyroglossal duct
umbilical cord
    artery
    hernia, physiological umbilical
    Wharton's jelly
    vein
vascular system
    artery
        anterior cerebral
        aorta
            abdominal (post-ductal, from L dorsal aorta)
            aortic arch (pre-ductal, from left fourth arch artery)
            ascending thoracic (from outflow tract)
            dorsal
            post-ductal part of thoracic (from L dorsal aorta)
        basilar
        common carotid
        common iliac
            external iliac
            internal iliac
        ductus arteriosus (from left sixth arch artery)
        external carotid
        hepatic
        hyoid
        iliac
        innominate (brachiocephalic)
        internal carotid
        intersegmental
        maxillary
        mesenteric
        ophthalmic
        posterior cerebral
        principal (axial) to hindlimb (7th cervical intersegmental)
        principal (axial) to forelimb (5th lumbar intersegmental)
        pulmonary
        pulmonary trunk
            proximal: from outflow tract
            distal: from L 6th arch artery
        renal
        subclavian
        stapedial
        umbilical (R & L)
        vertebral
    blood
    vein
        common iliac
        ductus venosus
        femoral
        jugular (internal, **external**)
        mesenteric
        portal
        principal (axial) from fore- and hindlimb
        subclavian
        umbilical (R & L)
        vena cava
            inferior
                post hepatic part
                pre-hepatic part
            superior (cranial) vena cava

STAGE 22 – day 13.5
adrenal gland
    cortex (mesodermal, coelomic epithelium)
    medulla (ectodermal, neural crest)
brain
    forebrain
        diencephalon
            epithalamus
            hypothalamus
            lamina terminalis

THE TISSUES PRESENT IN EACH STAGE OF MOUSE DEVELOPMENT

STAGE 22 *continued*
    internal capsule
        thalamus
    ganglionic eminence (lateral & medial
        aspects)
    telencephalon
        corpus striatum
            caudate nucleus
            lentiform nucleus
        choroid fissure
        choroid invagination (tela choroidea)
        choroid plexus
        interventricular foramen
        lateral ventricle
            anterior, inferior, posterior and superior
                horns
        neopallial cortex (future cerebral cortex)
        olfactory lobe
            olfactory cortex
    third ventricle
        choroid plexus
        infundibular recess
        interventricular foramen
        optic recess (lumen of optic stalk)
    hindbrain (rhombencephalon)
        fourth ventricle
            lateral recess
                choroid plexus
            roof
                choroid plexus
        metencephalon
            cerebellum
            pons
            sulcus limitans
        myelencephalon
            medulla oblongata
            sulcus limitans
    midbrain (mesencephalon)
        cerebral aqueduct
        tegmentum
    coverings
        meninges (includes arachnoid, dura & pia
            mater)
        dura
            falx cerebri
            tentorium cerebelli
        pontine cistern
        subarachnoid space
    layers
        ependymal, mantle, marginal
    sinus
        venous dural
cavity
    diaphragm
        central tendon
        crus
        dome (R & L)
        pleuro-pericardial canal (channel)
        pleuro-pericardial folds
        pleuro-peritoneal canal (channel)
        pleuro-peritoneal folds
    intraembryonic coelom
        pericardial cavity
        pericardio-peritoneal canals (future pleural
            cavity)
        peritoneal cavity
            greater sac
            hepatic recess (R & L)
            omental bursa (lesser sac) **superior**
                recess
        pleural cavity
            pleura, parietal (late stage)
diaphragm – *see* cavity
ear
    ganglia
        cochlear (VIII)
        vestibular (VIII)
    inner
        cochlea
        cochlear canal (duct)
        ductus reuniens
        endolymphatic appendage (diverticulum)
        endolymphatic sac (future duct)
        labyrinth
            crus commune
        otic capsule (pre-cartilage)
        saccule (utricle/saccule)
        semicircular canals (posterior, superior,
            lateral)
    middle
        ossicles (malleus & incus CC (1st arch),
            stapes CC (2nd arch))
        pharyngo-tympanic (Eustachian) tube
        tubo-tympanic recess
    outer (external)
        external acoustic meatus (entrance only)
        first branchial membrane (future tympanic
            membrane)
        pinna
    temporal bone (petrous part) CC
eye
    anterior chamber
    choroidal vessels
    conjunctival sac
    cornea
        ectoderm
        stroma
    eyelid (upper & lower)
        canthus, inner (nasal)
    hyaloid cavity
        hyaloid vascular plexus
            tunica vasculosa lentis
            vasa hyaloidea propria
        vitreous humour
    lens
        capsule
        fibres
    muscle
        extrinsic ocular
            **oblique (inf., sup.), rectus
                (lateral/temporal, medial/ nasal,
                inferior, superior)**
    optic
        (II) nerve
        chiasma
        fissure
        stalk
    perioptic mesenchyme (future sclera)
    retina
        choroid (fetal) fissure
        optic disc
        neural (inner) layer
        pigment (outer) layer
gall bladder
    common bile duct
    cystic duct
ganglia
    body
        **autonomic
            ciliary
            paraganglia of Zuckerkandl** (para-aortic
                "bodies")
        dorsal (posterior) root
        sympathetic
            cervical
                inferior
                middle
                superior (cervico-thoracic, stellate)
            paraganglia of Zuckerkandl (para-aortic
                "bodies")
            thoracic
    cranial
        cochlear (VIII)
        cranial/spinal accessory (XI)
        facial (VII)
        glossopharyngeal (IX) – inferior
        glossopharyngeal (IX) – superior
        trigeminal (V) (semilunar, Gasserian)
        vagus (X) – inferior (nodose)
        vagus (X) – superior
        vestibular (VIII)
glands
    **adrenal
        cortex** (mesodermal, coelomic epithelium)
        medulla (ectodermal, neural crest)
    gall bladder
        common bile duct
        cystic duct
    mammary gland
    mammary ridge
    pancreas
        body, tail, left & right lobes
        pancreatic duct
        uncinate process
    parathyroid primordium
    pituitary
        adenohypophysis
            lumen (remnant of Rathke's pouch, now
                detached)
            pars anterior (anterior lobe, pars distalis)
            pars intermedia
            pars tuberalis
        neurohypophysis
            infundibulum (pituitary stalk)
            median eminence
            pars nervosa (pars posterior)
    sublingual gland
    spleen primordium
    submandibular gland
    thymus
    thyroid
        foramen caecum
        **isthmus
        lobes (L & R)**
        thyroglossal duct
gut
    foregut
        duodenum (rostral half)
            mesentery (dorsal mesoduodenum)
        gastro-oesophageal junction
        oesophagus
            dorsal meso-oesophagus
        pharynx
        stomach
            fundic glandular mucous membrane
            fundus
            glandular, proventricular & pyloric
                regions
            mesogastrium, dorsal
                **gastro-splenic ligament
                lieno-renal ligament
                splenic primordium**
            mesogastrium, ventral
                falciform ligament
                lesser omentum
            pyloric antrum
    foregut–midgut junction
        biliary system
            common bile duct
            gallbladder
            hepatic duct (L & R)
        dorsal mesentery
        pancreatic duct
    hindgut
        **anal region
            anal canal
                rostral part** (hindgut)
            sphincter (external & internal)
        cloaca
            anal pit (proctodaeum)
            urogenital membrane
        dorsal mesentery
        perineal "body"
        rectum
        uro-rectal septum
    midgut
        dorsal mesentery
        duodenum (caudal half)
            mesentery (dorsal mesoduodenum)
        loop
        physiological umbilical hernia
haemolymphoid system
    blood

STAGE 22 continued
lymph node
  jugular lymph sac
  spleen primordium
  thymus primordium
hair – *see* skin
heart
  aorta
    ascending thoracic (from outflow tract)
  aortic arch (from left fourth arch artery)
  atrio-ventricular canal
  atrio-ventricular cushion tissue
  atrium (L & R)
    auricular part (L & R)
    bulbar cushion
    interatrial septum
      foramen (ostium) secundum
      foramen ovale
      septum primum
      septum secundum (crista dividens)
    sinus venarum
      crista terminalis
  bulbar cushion
  dorsal mesocardium
    transverse pericardial sinus
  ductus arteriosus (from L sixth arch artery)
  endocardial cushion tissue
  endocardial tissue
  epicardium (visceral pericardium)
  myocardium
  pericardium
    parietal
    visceral (epicardium)
  pulmonary trunk
    distal: from L sixth arch artery
    proximal: from outflow tract
  valve
    aortic, mitral (bicuspid), pulmonary, tricuspid
  venous valve
    leaflets (entrance of inferior vena cava)
  ventricle (L & R)
    interventricular groove
    interventricular septum
      muscular part
limb
  forelimb (for bones, *see* skeleton)
    digits
    mesenchyme
  hindlimb (for bones, *see* skeleton)
    digits
    mesenchyme
liver
  "bare" area
  hepatic duct
  hepatic sinusoids
  lobes, L & R (including caudate & quadrate)
  parenchyma
mammary gland
mesenchyme
  body
  head
mouth region
  epiglottis
  lips
    philtrum
    upper & lower
  lower jaw
    mandible
      Meckel's cartilage
  maxilla
  oral cavity
  palate
    primary (intermaxillary segment)
    secondary (palatal shelf of maxilla)
  premaxilla
  sublingual gland
  submandibular gland
  tongue
    circumvallate papilla
      foramen caecum
    fungiform papillae

intermolar eminence
median fibrous septum
muscle
  extrinsic
    genioglossus
    hyoglossus
    palatoglossus
    styloglossus
  intrinsic
    transverse & vertical component
tooth primordium
  dental laminae
    incisors & molars
  dental papilla
  enamel organ
  sulcus, alveolar
muscle
  anterior body wall
    ext. & int. oblique,
    rectus abdominus
    transversus abdominis
  axial
    erector spinae
    psoas major
    psoas minor
  cranial
    extrinsic ocular (*see* eye)
    masseter
    mylohyoid
    sterno-mastoid
    temporalis
  girdle, pectoral and thoracic body wall
    deltoid
    latissimus dorsi
    pectoralis major (pectoralis profundus)
    pectoralis minor
    subscapularis
    supraspinatus
    trapezius
  girdle, pelvic
    gluteus maximus (superficialis)
  lingual – *see* tongue
  skin (panniculus carnosus)
    platysma
  thoracic/abdominal wall
nerve
  body (*see also* ganglia)
    autonomic
      parasympathetic
        vagus (X)
          recurrent laryngeal branch
          R & L above diaphragm
          post. & ant. below diaphragm
      sympathetic (chain) trunk
  nerve trunk
    intercostal
    phrenic
    sciatic
  plexus
    autonomic, hypogastric
    brachial
      median, radial, ulnar
    lumbar
    lumbo-sacral
    segmental spinal
  cranial
    abducent (VI)
    accessory (XI)
      cranial component
      spinal component
    cochlear (VIII)
    facial (VII)
      chorda tympani branch
    glossopharyngeal (IX)
    hypoglossal (XII)
    oculomotor (III)
    olfactory (I)
    optic (II)
    trigeminal (V)
      mandibular division
      maxillary division

      ophthalmic division
    trochlear (IV)
    vagal (X) trunks
    vestibular (VIII)
nose
  naris
    anterior
    external
    primary choana (primitive posterior)
  nasal (olfactory) cavity
  nasal capsule (cartilaginous)
  nasal septum
    vomeronasal (Jacobson's) organ
  olfactory epithelium
nucleus pulposus
pancreas
  body, left lobe, right lobe, tail
  duct
  uncinate process
parathyroid primordium (derived from dorsal part
  of third branchial pouch)
pituitary
  adenohypophysis
    lumen (remnant of Rathke's pouch)
    pars anterior (anterior lobe, pars distalis)
    pars intermedia
    pars tuberalis
  neurohypophysis
    infundibulum (pituitary stalk)
    median eminence
    pars nervosa (pars posterior)
placenta
  chorionic plate
  labyrinthine part
  spongy (or basal) region (spongiotrophoblast)
renal/urinary system
  bladder
    fundus region
    urachus
  mesonephros
    tubules (regressing remnants only)
    vesicles (regressing remnants only)
  metanephros
    excretory components
      cortex
        early nephrons
        glomerulus
        stem cells
      medulla
        stromal cells
    ureteric bud (drainage) derivatives
      pelvis
      primitive collecting ducts
      ureter
  urogenital membrane
  urogenital sinus
    pelvic part
    phallic part
  urogenital mesentery – *see* reproductive system
  urorectal septum
reproductive system
  female
    genital tubercle (becomes phallus)
    gonad (ovary)
      "germinal" epithelium
      primordial germ cells
      rete ovarii
    labial swelling
    Müllerian tubercle
    paramesonephric (Müllerian) duct
    urogenital mesentery
      mesovarium (female)
    urogenital sinus
  male
    genital tubercle (becomes phallus)
    gonad (testis)
      "germinal" epithelium
      mediastinum testis
      primitive seminiferous tubules
      primordial germ cells
      rete testis (from mesonephric duct)

THE TISSUES PRESENT IN EACH STAGE OF MOUSE DEVELOPMENT

STAGE 22 *continued*
    mesonephric (Wolffian) duct (future ductus
        deferens)
    urogenital mesentery
        mesorchium (male)
    urogenital sinus
respiratory system
    arytenoid CC
    hyoid CC
    laryngeal aditus
    laryngeal cartilages
    larynx
        mucous membrane
    lung
        bronchus
            lobar
            main
            segmental
            terminal
        hilus (root)
        left lobe
        pleura
            parietal (late)
        right lobe (accessory, caudal, cranial, middle)
    pharynx
    trachea
        bifurcation (carina tracheae)
    upper respiratory tract
        nasopharynx
skeleton
    cranium
        1st arch derivative
            Meckel's cartilage
                (central element of mandible; malleus,
                incus)
        2nd arch derivative
            temporal bone
                Reichert's cartilage
                stapes
        chondrocranium
            basioccipital CC
            basisphenoid CC
            exoccipital CC
                squamous part (mes)
            occipital PCC
                squamous part
            sphenoid bone CC
            temporal bone
                petrous part
        mandible
        vault of skull
            frontal bone primordium
            inter-parietal bone
            parietal bone
            supraoccipital bone (**tectum posterius**)
            temporal bone
                squamous part CC
        viscerocranium
            ethmoid bone primordium
            facial bones primordia
            maxilla
                palatal shelf
    girdle
        pectoral
            clavicle CC
            scapula CC
            shoulder joint
        sternum
            sternebral bone PCC
        pelvic
            acetabular region
            hip joint
            iliac CC
            ischial PCC
            pubic PCC
    limb
        forelimb
            bones
                carpal CC
                humerus
                metacarpal CC
                phalangeal PCC
                radius CC
                ulna CC
            joints
                elbow
                gleno-humeral (shoulder)
                wrist
        hindlimb
            bones
                femur CC
                fibula CC
                metatarsal CC
                phalangeal PCC
                future tarsus
                    calcaneus CC
                    talus CC
                    tibia CC
            joints
                ankle primordium
                hip
                knee
    vertebral axis
        hyoid bone CC
        rib
        vertebrae
            atlas
            axis
                dens (odontoid process of C2)
            cervical vertebrae
                centrum
                foramen transversarium
                intervertebral disc (surrounds nucleus
                    pulposus)
                neural arch
                nucleus pulposus
            lumbar, sacral, thoracic vertebrae
                neural arch, centrum etc.
skin
    dermis
    eyelid (upper & lower) – see eye
    hair
        tactile (**sinus**) **hair follicle primordium**
        vibrissae
            elevations
            precursors
    mammary gland
        mammary ridge
    muscle (panniculus carnosus)
        platysma
somites,
    tail: pairs 46–60 (max.)
spinal cord
    central canal
    conus medullaris
    filum terminale
    floorplate
    layer
        ependyma
        mantle
            alar & basal columns
        marginal
    meninges – see brain
    nerve, segmented spinal
    nerve root, dorsal (sensory), ventral (motor)
    roofplate
    sulcus limitans
spleen primordium
tail
    spinal cord
    somites: pairs 46–60 (max.)
thymus (from ventral part of third branchial
    pouch)
thyroid
    foramen caecum
    isthmus
    lobes (left & right)
    thyroglossal duct
umbilical cord
    artery
    hernia, physiological umbilical
    Wharton's jelly
    vein
vascular system
    artery
        anterior cerebral
        aorta
            abdominal (from L dorsal aorta)
            aortic arch (from L 4th arch artery)
            ascending thoracic (from outflow tract)
            post-ductal part of thoracic (from left
                dorsal aorta)
        basilar
        bronchial
        carotid sinus
        common carotid (from third arch artery)
        common iliac
            external iliac
            internal iliac
        ductus arteriosus (from left sixth arch artery)
        external carotid
        hepatic
        hyoid (from 2nd arch artery)
        iliac
        innominate (brachiocephalic)
        internal carotid
        intersegmental
        maxillary (from 1st arch artery)
        mesenteric
        middle cerebral
        ophthalmic
        posterior cerebral
        principal (axial) to hindlimb (7th cervical
            intersegmental)
        principal (axial) to forelimb (5th lumbar
            intersegmental)
        pulmonary
        pulmonary trunk
            proximal: from outflow tract
            distal: from L 6th arch artery
        renal
        spinal (anterior, posterior)
        stapedial (from 2nd arch artery)
        subclavian
        umbilical (R & L)
        vertebral
    blood
lymphatic system
    jugular lymph sac
vein
    azygos
    common iliac
        external iliac
        internal iliac
    ductus venosus
    femoral
    hemiazygos
    hepatico-cardiac
    jugular (internal, external)
    mesenteric
    portal
    principal (axial) from fore- and hindlimb
    spinal (anterior, posterior)
    subclavian
    superior (cranial) vena cava
    umbilical (R & L)
    vena cava
        inferior
            post-hepatic part
            pre-hepatic part
        superior (cranial) vena cava

STAGE 23 – day 14.5
adrenal gland
    cortex (mesodermal, coelomic epithelium)
    medulla (ectodermal, neural crest)
brain
    forebrain
        diencephalon
            epithalamus
                pineal primordium (epiphysis)
            hypothalamus
            internal capsule

STAGE 23 *continued*
- thalamus
  - **interthalamic adhesion (massa intermedia)**
- ganglionic eminence (lateral & medial aspects)
- telencephalon
  - choroid fissure
  - choroid invagination (tela choroidea)
  - choroid plexus
  - corpus striatum
    - caudate nucleus
      - **head & tail**
      - **caudate–putamen**
    - lentiform nucleus
  - interventricular foramen
  - lateral ventricle
    - anterior, inferior, posterior and superior horns
  - neopallial cortex (future cerebral cortex)
  - olfactory lobe
    - olfactory cortex
  - third ventricle
    - **epithalamic (pineal) recess**
    - **choroid plexus**
    - infundibular recess
    - interventricular foramen
    - lamina terminalis
    - optic recess (lumen of optic stalk)
- hindbrain (rhombencephalon)
  - fourth ventricle
    - lateral recess
      - choroid plexus
    - roof
      - choroid plexus
  - metencephalon
    - cerebellum
      - **intraventricular portion**
    - pons
    - sulcus limitans
  - myelencephalon
    - medulla oblongata
      - **medullary raphe**
    - sulcus limitans
- midbrain (mesencephalon)
  - cerebral aqueduct
  - tegmentum
- coverings
  - meninges (includes arachnoid, dura & pia mater)
    - dura
      - falx cerebri
      - tentorium cerebelli
    - pontine cistern
    - subarachnoid space
  - layers
    - ependymal, mantle, marginal
  - sinus
    - venous dural
      - **inferior sagittal**
      - **superior sagittal**
      - **transverse**
- cavity
  - diaphragm
    - central tendon
    - crus
    - dome (R & L)
  - intra-embryonic coelom
    - pericardial cavity
    - peritoneal cavity
      - **greater sac**
        - hepatic recess (R & L)
        - omental bursa (lesser sac) (superior recess)
      - **pelvic recess (recto-uterine pouch of Douglas)**
    - pleural cavity
      - pleura, parietal
      - **pleura, visceral**
- ear
  - ganglia
    - cochlear (VIII)
    - vestibular (VIII)
  - inner
    - cochlea
    - cochlear canal (duct)
    - ductus reuniens
    - **endolymphatic duct**
    - labyrinth
      - crus commune
    - otic capsule (cartilage)
    - saccule (utricle/saccule)
    - semicircular canals (posterior, superior, lateral)
  - middle
    - ossicles (malleus & incus bone (1st arch), stapes bone (2nd arch))
    - pharyngo-tympanic (Eustachian) tube
    - **stapedius muscle**
    - **tensor tympani muscle**
    - tubo-tympanic recess
  - outer (external)
    - external acoustic meatus (entrance only)
    - first branchial membrane (future tympanic membrane)
    - pinna
  - temporal bone (petrous part) cartilage
- eye
  - anterior chamber
  - choroidal vessels
  - conjunctival sac
    - **upper & lower recesses**
  - cornea
    - ectoderm
    - stroma
  - eyelid (upper & lower)
    - canthus, inner (nasal)
  - hyaloid cavity
    - hyaloid vascular plexus
      - tunica vasculosa lentis
      - vasa hyaloidea propria
    - vitreous humour
  - lens
    - capsule
    - fibres
  - mesenchyme, periopic (future sclera)
  - muscle
    - extrinsic ocular
      - oblique (inf., sup.), rectus (lateral/temporal, medial/nasal, inferior, superior)
  - optic
    - (II) nerve
    - chiasma
    - stalk
  - retina
    - choroid (fetal) fissure
    - neural (inner) layer
    - optic disc
    - pigment (outer) layer
- gall bladder
  - common bile duct
  - cystic duct
- ganglia
  - body
    - autonomic
      - ciliary
      - paraganglia of Zuckerkandl (para-aortic "bodies")
    - dorsal (posterior) root
    - sympathetic
      - cervical
        - inferior
        - middle
        - superior (cervico-thoracic, stellate)
      - paraganglia of Zuckerkandl
        - para-aortic "bodies"
      - thoracic
  - cranial
    - cochlear (VIII)
    - cranial/spinal accessory (XI)
    - facial (VII)
    - glossopharyngeal (IX) – inferior
    - glossopharyngeal (IX) – superior
    - trigeminal (V) (semilunar, Gasserian)
    - vagus (X) – inferior (nodose)
    - vagus (X) – superior
    - vestibular (VIII)
- glands
  - gall bladder
    - common bile duct
    - cystic duct
  - mammary gland
  - pancreas
    - pancreatic duct
    - body, tail, left & right lobes, uncinate process
  - **pineal primordium**
  - pituitary
    - adenohypophysis
      - lumen (remnant of Rathke's pouch, now detached)
      - pars anterior (anterior lobe, pars distalis)
      - pars intermedia
      - pars tuberalis
    - neurohypophysis
      - infundibulum (pituitary stalk)
      - median eminence
      - pars nervosa (pars posterior)
  - spleen
  - sublingual gland
  - submandibular gland
  - thymus
  - thyroid
    - foramen caecum
    - isthmus
    - lobes (L & R)
    - thyroglossal duct
- gut
  - foregut
    - duodenum (rostral half)
      - mesentery (dorsal mesoduodenum)
    - gastro-oesophageal junction
    - oesophagus
      - dorsal meso-oesophagus
    - pharynx
      - hyoid bone
    - stomach
      - fundic glandular mucous membrane
      - fundus
      - glandular, proventricular & pyloric regions
      - mesentery (dorsal mesogastrium)
        - gastro-splenic ligament
        - lieno-renal ligament
        - splenic primordium
      - mesentery (ventral)
        - falciform ligament
        - lesser omentum
      - omentum
        - **greater**
      - pyloric antrum
  - foregut–midgut junction
    - biliary system
      - common bile duct
      - gallbladder
      - hepatic duct (L & R)
    - duodenum
      - dorsal mesoduodenum
      - **duodenal papilla (ampulla of Vater)**
      - pancreatic duct
      - **sphincter of Oddi**
  - hindgut
    - anal region
      - anal canal
        - rostral part (from hindgut)
        - **caudal part** (from anal pit, after membrane perforates)
      - sphincter (external & internal)
    - cloaca
    - dorsal mesentery
    - perineal "body"
    - rectum
    - uro-rectal septum

THE TISSUES PRESENT IN EACH STAGE OF MOUSE DEVELOPMENT

STAGE 23 continued
- midgut
  - dorsal mesentery
  - duodenum (caudal half)
    - mesentery (dorsal mesoduodenum)
  - **jejunum**
  - loop
  - physiological umbilical hernia
- haemolymphoid system
  - blood
  - lymph node
    - jugular lymph sac
  - spleen primordium
  - thymus primordium
- hair – see skin
- heart
  - aorta, ascending thoracic (from outflow tract)
  - aortic arch (from L fourth arch artery)
  - **aortic sinus**
  - atrio-ventricular canal
  - atrio-ventricular cushion tissue
  - atrium (L & R)
    - auricular part (L & R)
    - bulbar cushion
    - interatrial septum
      - foramen (ostium) secundum
      - foramen ovale
      - septum primum
      - septum secundum (crista dividens)
    - sinus venarum
      - crista terminalis
  - bulbar ridge
  - dorsal mesocardium
    - transverse pericardial sinus
  - ductus arteriosus (from L sixth arch artery)
  - endocardial cushion tissue
  - endocardial tissue
  - epicardium (visceral pericardium)
  - myocardium
  - pericardium
    - parietal
    - visceral (epicardium)
  - pulmonary trunk
    - distal part: from L sixth arch artery
    - prox. part: from outflow tract
  - **trabeculae carneae**
  - valve
    - aortic, mitral (bicuspid), pulmonary, tricuspid
    - **leaflets (aortic, mitral, pulmonary, tricuspid)**
    - venous valve (leaflets, entrance of inferior vena cava)
  - ventricle (L & R)
    - interventricular groove
    - interventricular septum
      - muscular part
- limb – for bones and joints, see skeleton; for muscles, see muscle
  - forelimb
    - digits
    - mesenchyme
    - **palmar (digital) "pads"** (early stage)
  - hindlimb – for bones and joints, see skeleton
    - digits
    - mesenchyme
    - **plantar (digital) "pads"** (late stage)
- liver
  - "bare" area
  - hepatic ducts
  - hepatic sinusoids
  - lobes, L & R (including caudate & quadrate)
  - parenchyma
- mammary gland
- mesenchyme
  - body
  - head
- mouth region
  - epiglottis
  - lips
    - philtrum
    - upper & lower
  - lower jaw
    - Meckel's cartilage
    - mandible
  - maxilla
    - premaxilla
  - oral cavity
  - palate
    - primary (intermaxillary segment)
    - secondary (palatal shelf of maxilla)
  - **parotid gland** (late)
  - sublingual gland
  - submandibular gland
  - tongue
    - circumvallate papilla
    - **filiform papillae**
    - foramen caecum
    - **frenulum**
    - fungiform papillae
    - intermolar eminence
    - median fibrous septum
    - muscle
      - extrinsic
        - genioglossus
        - hyoglossus
        - palatoglossus
        - styloglossus
      - intrinsic
        - transverse & vertical component
  - tooth primordium
    - dental laminae
      - incisors & molars
    - dental papilla
    - outer epithelial layer (enamel organ)
    - sulcus, alveolar
- muscle
  - anterior body wall
    - ext. & int. oblique
    - rectus abdominus
    - transversus abdominis
  - axial
    - erector spinae
    - **ilio-psoas**
    - **intercostal (external & internal layers)**
    - psoas major
    - psoas minor
    - **quadratus lumborum**
    - **serratus anterior**
  - cranial
    - extrinsic ocular – see eye
    - masseter
    - mylohyoid
    - sterno-mastoid
    - temporalis
  - girdle, pectoral and thoracic body wall
    - deltoid
    - latissimus dorsi
    - pectoralis major (pectoralis profundus)
    - pectoralis minor
    - subscapularis
    - supraspinatus
    - trapezius
  - girdle, pelvic
    - gluteus maximus (superficialis)
  - limb
    - forelimb
      - **biceps (brachii)**
      - **teres major**
      - **triceps (brachii)**
    - hindlimb
      - **hamstring (flexor) group**
      - **quadriceps (extensor) group**
  - lingual – see tongue
  - skin (panniculus carnosus)
  - platysma
- nerve – see also ganglia
  - body
    - autonomic
      - **hypogastric plexus**
      - parasympathetic
        - vagus (X) trunks
          - (R & L above diaphragm)
          - post. & ant. below diaphragm)
      - recurrent laryngeal branch
      - sympathetic (chain) trunk
    - nerve trunk
      - intercostal
      - phrenic
      - sciatic
    - plexus
      - autonomic, hypogastric
      - brachial
        - median, radial, ulnar
      - lumbar
      - lumbo-sacral
      - segmental spinal
        - intercostal
  - cranial
    - abducent (VI)
    - accessory (XI)
      - cranial component
      - spinal component
    - cochlear (VIII)
    - facial (VII)
      - chorda tympani branch
    - glossopharyngeal (IX)
    - hypoglossal (XII)
    - oculomotor (III)
    - olfactory (I)
    - optic (II)
    - trigeminal (V)
      - mandibular division
      - maxillary division
      - ophthalmic division
    - trochlear (IV)
    - vagal (X) trunks
    - vestibular (VIII)
- nose
  - **meatus**
  - naris
    - anterior
    - external
    - primary choana (primitive posterior)
  - nasal capsule (cartilaginous)
  - nasal (olfactory) cavity
  - nasal septum
    - vomeronasal (Jacobson's) organ
  - olfactory epithelium
  - **turbinate bones (conchae)**
- nucleus pulposus
- pancreas
  - body, left & right lobes, tail, uncinate process
  - duct
- parathyroid (derived from dorsal part of third branchial pouch)
- pituitary
  - adenohypophysis
    - lumen (remnant of Rathke's pouch)
    - pars anterior (anterior lobe, pars distalis)
    - pars intermedia
    - pars tuberalis
  - neurohypophysis
    - infundibulum (pituitary stalk)
    - median eminence
    - pars nervosa (pars posterior)
- placenta
  - chorionic plate
  - labyrinthine part
  - spongy (or basal) region (spongiotrophoblast)
- renal/urinary system
  - bladder
    - fundus region
    - urachus
  - mesonephros
    - **duct becomes ductus deferens** (male, see reproductive system)
    - tubules (regressing remnants only)
    - vesicles (regressing remnants only)
  - metanephros
    - **capsule**
    - excretory components

STAGE 23 continued
    cortex
        early nephrons
        glomerulus
        stem cells
    medulla
        stromal cells
    ureteric bud derivatives
        calyces, minor & major
        pelvis
        primitive collecting ducts
        ureter
    urogenital mesentery – *see* reproductive system
    urogenital sinus
        pelvic part
        phallic part
    urorectal septum
reproductive system
    female
        clitoris
        gonad (ovary)
            "germinal" epithelium
            primordial germ cells
            rete ovarii
        labial swelling
        Müllerian tubercle
        paramesonephric (Müllerian) duct
        urethral groove
        urogenital sinus
    male
        gonad (testis)
            cortical region ("germinal" epithelium)
            gubernaculum
            mediastinum testis
            medullary region
            primordial germ cells
            primitive seminiferous tubules
        mesonephric (Wolffian) duct
            ductus deferens
            rete testis
        penis
        urethra
        urethral groove
    urogenital mesentery
        mesorchium (male)
        mesovarium (female)
    uroregenital sinus
respiratory system
    arytenoid cartilage (4th & ?5th branchial arch cartilages)
    cricoid cartilage (?6th branchial arch cartilage)
    hyoid bone (2nd & 3rd branchial arch cartilages)
    laryngeal aditus
    laryngeal cartilages
    thyroid cartilage (4th & ?5th branchial arch cartilages)
    larynx
        mucous membrane
    lung
        bronchus
            lobar
            main
            segmental
            terminal
        hilus (root)
        left
        pleura
            parietal
            visceral
        right (accessory, caudal, cranial, middle lobes)
    pharynx
    trachea
        bifurcation (carina tracheae)
        "rings" (from somatopleuric mesenchyme)
    upper respiratory tract
        nasopharyngeal cavity
skeleton
    cranium
        1st arch derivative
            Meckel's cartilage (central element of mandible; malleus, incus)
        2nd arch derivative
            temporal bone
            Reichert's cartilage
            stapes
            styloid process
        chondrocranium
            basioccipital bone
            basisphenoid bone
            carotid canal
            exoccipital CC
                squamous part (mes)
            foramen ovale
            foramen rotundum
            jugular foramen
            occipital CC
                squamous part
            sphenoid bone
            temporal bone
                petrous part
        mandible
        vault of skull
            frontal bone primordium
            inter-parietal bone
            parietal bone
            supraoccipital bone (tectum posterius)
            temporal bone
                squamous part
        viscerocranium
            ethmoid bone
            facial bone primordia
            intermaxillary (premaxilla)
            maxilla
            palatal shelf
            optic foramen (canal)
            orbital fissure, superior
            orbito-sphenoid
            turbinate (conchae)
    girdle
        pectoral
            clavicle
            scapula
            shoulder joint
            sternum
                manubrium sterni
                sternebral bone CC
                xiphisternum (xiphoid process)
        pelvic
            acetabular region
            hip joint
            pelvis (innominate bone) CC
                iliac bone
                ischial CC
                pubic CC
    limb
        forelimb
            bone
                carpal
                humerus
                metacarpal CC
                phalangeal CC
                radius
                ulna
            joint
                elbow
                gleno-humeral (shoulder)
                wrist
        hindlimb
            bone
                femur
                fibula CC
                metatarsal
                phalangeal CC
                tarsus
                    calcaneus, talus
                tibia CC
            joint
                ankle
                hip
                knee
                fabella PCC
                ligamentum patellae
                patella PCC
    vertebral axis
        hyoid bone
            body, greater horn PCC
        intervertebral disc (nucleus pulposus)
        rib
            chondro-sternal joint primordium
            costal cartilage
            costal margin
        vertebrae
            atlas
            axis
            cervical vertebrae
                dens (odontoid process of C2)
                foramen transversarium
                intervertebral disc (surrounds nucleus pulposus)
                neural arch
                nucleus pulposus
            coccygeal (tail), lumbar, sacral, thoracic vertebrae
                neural arch, centrum etc.
skin
    dermis
    eyelid (upper & lower)
    hair
        tactile (sinus) hair follicle
        vibrissae elevations
        vibrissae precursors
    mammary gland
    muscle (panniculus carnosus)
        cutaneous muscle of trunk
        platysma
spinal cord
    central canal
    conus medullaris
    filum terminale
    grey horns (dorsal, ventral & lateral)
    floorplate
    layer
        ependymal
        mantle
            alar & basal columns
        marginal
    meninges – *see* brain
    nerve, segmental spinal
    nerve root, dorsal (sensory), ventral (motor)
    roofplate
    sulcus limitans
spleen primordium
tail
    spinal cord
    vertebrae
thymus
thyroid
    foramen caecum
    isthmus
    lobes (left & right)
    thyroglossal duct
umbilical cord
    artery
    hernia, physiological umbilical
    Wharton's jelly
    vein
vascular system
    artery
        aorta
            abdominal (from L dorsal aorta)
            aortic arch (L 4th arch artery)
            ascending thoracic (from outflow tract)
            post-ductal part of thoracic (from left dorsal aorta)
        basilar
        bronchial
        carotid sinus
        cerebral (anterior, middle, posterior)
        common carotid (3rd arch artery)

THE TISSUES PRESENT IN EACH STAGE OF MOUSE DEVELOPMENT

STAGE 23 *continued*
    common iliac
        external iliac
        internal iliac
    ductus arteriosus (L 6th arch artery)
    external carotid
    hepatic
    hyoid
    iliac
    innominate (brachiocephalic)
    internal carotid
    intersegmental
    internal thoracic (mammary)
    maxillary
    mesenteric (**superior, inferior**)
    ophthalmic
    posterior communicating
    principal (axial) to hindlimb (7th cervical intersegmental)
    principal (axial) to forelimb (5th lumbar intersegmental)
    pulmonary
    pulmonary trunk
        distal: from L 6th arch artery
        proximal: from outflow tract
    renal
    spinal (anterior, posterior)
    stapedial
    subclavian
    superior vesical
    testicular
    thyroid (superior, inferior)
    umbilical (R & L)
    vertebral
blood
lymphatic system
    jugular lymph sac
vein
    azygos
    common iliac
        external iliac
        internal iliac
    ductus venosus
    femoral
    hemiazygos
    hepatico-cardiac
    internal thoracic (mammary)
    jugular (internal, external)
    mesenteric (superior, inferior)
    portal
    pulmonary
    subclavian
    testicular (pampiniform plexus)
    umbilical (R & L)
    vena cava
        inferior
            post-hepatic part
            pre-hepatic part
        superior (cranial) vena cava

STAGE 24 – day 15.5
adrenal
    cortex (mesodermal, coelomic epithelium)
    medulla (ectodermal, neural crest)
brain
    forebrain
        diencephalon
            epithalamus
                pineal gland (epiphysis)
            hypothalamus
            internal capsule
            thalamus
                interthalamic adhesion (massa intermedia)
            ganglionic eminence (lateral & medial aspects)
        telencephalon
            choroid fissure
            choroid invagination (tela choroidea)
            choroid plexus
            corpus striatum
                caudate nucleus
                    head & tail
                caudate-putamen
                lentiform nucleus
            interventricular foramen
            lateral ventricle
                anterior, inferior, posterior and superior horns
            neopallial cortex (future cerebral cortex)
            olfactory lobe
            olfactory cortex
        temporal lobe
        third ventricle
            epithalamic (pineal) recess
                choroid plexus
            infundibular recess
            interventricular foramen
            lamina terminalis
            optic recess (lumen of optic stalk)
    hindbrain (rhombencephalon)
        fourth ventricle
            lateral recess
                choroid plexus
            roof
                choroid plexus
        metencephalon
            cerebellum
                intraventricular portion
            pons
            sulcus limitans
        myelencephalon
            medulla oblongata
                medullary raphe
            sulcus limitans
    midbrain (mesencephalon)
        cerebral aqueduct
        tegmentum
    coverings
        meninges (includes arachnoid, dura & pia mater)
        dura
            falx cerebri
            tentorium cerebelli
        pontine cistern
        subarachnoid space
    layers
        ependymal, mantle & marginal
    sinus
        venous dural
            inferior sagittal
            superior sagittal
            transverse
cavity
    diaphragm
        central tendon
        crus
        dome (R & L)
    intra-embryonic coelom
        pericardial cavity
        peritoneal cavity
            greater sac
            hepatic recess (R & L)
            omental bursa (lesser sac)
                superior recess
            pelvic recess (recto-uterine pouch of Douglas)
        pleural cavity
ear
    ganglia
        cochlear (VIII)
        vestibular (VIII)
    inner
        cochlea
        cochlear canal (duct)
        ductus reuniens
        endolymphatic duct
        labyrinth
            crus commune
        saccule (utricle/saccule)
        semicircular canals (posterior, superior, lateral)
    middle
        ossicles (malleus & incus bone (1st arch), stapes bone (2nd arch))
        pharyngo-tympanic (Eustachian) tube
        stapedius muscle
        tensor tympani muscle
        tubo-tympanic recess
    outer (external)
        external acoustic meatus (entrance only)
        first branchial membrane (future tympanic membrane)
        pinna
    temporal bone (petrous part) cartilage
eye
    anterior chamber
    choroidal vessels
    conjunctival sac
        upper & lower recesses
    cornea
        Descemet's membrane
        ectoderm
        stroma
    eyelid (upper & lower)
        canthus, inner (nasal)
    hyaloid cavity
        hyaloid vascular plexus
            tunica vasculosa lentis
            vasa hyaloidea propria
        vitreous humour
    lens
        capsule
        fibres
    muscle
        extrinsic ocular
            oblique (inf., sup.), rectus (lateral/temporal, medial/nasal, inferior, superior)
    optic
        (II) nerve
        chiasma
        stalk
    retina
        choroid (fetal) fissure
        neural (inner) layer
            nerve fibre layer
            optic disc
        pigment (outer) layer
    sclera
gall bladder
    common bile duct
    cystic duct
ganglia
    body
        autonomic
        ciliary
        paraganglia of Zuckerkandl
            para-aortic "bodies"
    dorsal (posterior) root
    sympathetic
        cervical
            inferior
            middle
            superior (cervico-thoracic, stellate)
        paraganglia of Zuckerkandl (para-aortic "bodies")
        thoracic
    cranial
        cochlear (VIII)
        cranial/spinal accessory (XI)
        facial (VII)
        glossopharyngeal (IX) – inferior
        glossopharyngeal (IX) – superior
        trigeminal (V) (semilunar, Gasserian)
        vagus (X) – inferior (nodose)
        vagus (X) – superior
        vestibular (VIII)
gland
    gall bladder
        common bile duct
        cystic duct
    mammary gland

STAGE 24 *continued*
- pancreas
  - body, tail, left & right lobes
  - pancreatic duct
  - uncinate process
- parathyroid
- **pineal**
- pituitary
  - adenohypophysis
    - lumen (remnant of Rathke's pouch, now detached)
    - pars anterior (anterior lobe, pars distalis)
    - pars intermedia
    - pars tuberalis
  - neurohypophysis
    - infundibulum (pituitary stalk)
    - median eminence
    - pars nervosa (pars posterior)
- sublingual gland
- submandibular gland
- spleen
- thymus
- thyroid
  - foramen caecum
  - isthmus
  - lobes (L & R)
  - thyroglossal duct
- gut
  - foregut
    - duodenum (rostral half)
      - mesentery (dorsal mesoduodenum)
    - gastro-oesophageal junction
    - oesophagus
      - dorsal meso-oesophagus
    - pharynx
      - **constrictor muscle (superior & inferior)**
      - hyoid bone
    - stomach
      - fundic glandular mucous membrane
      - fundus
      - glandular, proventricular & pyloric regions
      - mesentery (dorsal mesogastrium)
        - gastro-splenic ligament
        - lieno-renal ligament
        - **spleen**
      - mesentery (ventral)
        - falciform ligament
        - lesser omentum
      - omentum, greater
      - pyloric antrum
  - foregut–midgut junction
    - biliary system
      - common bile duct
      - gallbladder
      - hepatic duct (L & R)
    - duodenum
      - dorsal mesoduodenum
      - duodenal papilla (ampulla of Vater)
      - pancreatic duct
      - sphincter of Oddi
  - hindgut
    - anal region
      - anal canal
        - caudal part: anal pit
        - remnant of anal membrane (pectinate line will mark site)
        - rostral part: hindgut
      - sphincter (external & internal)
    - cloaca
    - dorsal mesentery
    - perineal "body"
    - rectum
    - uro-rectal septum
  - midgut
    - dorsal mesentery
    - duodenum (caudal half)
    - jejunum
    - loop
    - physiological umbilical hernia
- haemolymphoid system
  - blood
  - lymph node
    - jugular lymph sac
  - **Peyer's patches**
  - **spleen**
  - **thymus**
- hair – *see* skin
- heart
  - aorta
    - ascending thoracic (from outflow tract)
    - aortic arch (from left fourth arch artery)
  - aortic sinus
  - atrio-ventricular canal
  - atrio-ventricular cushion tissue
  - atrium
    - auricular part (L & R)
    - bulbar cushion
    - interatrial septum
      - foramen (ostium) secundum
      - foramen ovale
      - septum primum
      - septum secundum (crista dividens)
    - sinus venarum
      - crista terminalis
  - bulbar cushion
  - dorsal mesocardium
    - transverse pericardial sinus
  - ductus arteriosus (from left sixth arch artery)
  - endocardial cushion tissue
  - endocardial tissue
  - epicardium (visceral pericardium)
  - myocardium
  - pericardium
    - **fibrous**
    - parietal
    - visceral (epicardium)
  - pulmonary trunk
    - distal part: from L sixth arch artery
    - proximal part: from outflow tract
  - trabeculae carneae
  - valve
    - aortic, mitral (bicuspid), pulmonary, tricuspid
    - leaflets (aortic, mitral, pulmonary, tricuspid)
    - venous valve (leaflets, entrance of inferior vena cava)
  - ventricle (L & R)
    - interventricular groove
    - interventricular septum
      - muscular part
- limb – for bones and joints, *see* skeleton; for muscles, *see* muscle)
  - forelimb
    - **claw primordium**
    - digits
    - mesenchyme
    - palmar (digital) "pads"
  - hindlimb
    - **claw primordium**
    - digits
    - mesenchyme
    - plantar digital "pads"
- liver
  - "bare" area
  - hepatic ducts
  - hepatic sinusoids
  - lobes, L & R (including caudate & quadrate)
  - parenchyma
- mammary gland
- mesenchyme
  - body
  - head
- mouth region
  - epiglottis
  - **gum, primitive**
  - lips
    - philtrum
    - upper & lower
  - lower jaw
    - mandible
    - Meckel's cartilage
  - maxilla
    - premaxilla
- oral cavity
- palate
  - **canal, naso-palatine**
  - primary (intermaxillary segment)
  - **secondary** (definitive)
  - **soft**
- parotid gland
- **sublingual caruncle**
- sublingual gland
- submandibular gland
- tongue
  - circumvallate papilla
  - filiform papillae
  - foramen caecum
  - frenulum
  - fungiform papillae
  - intermolar eminence
  - median fibrous septum
  - muscle
    - extrinsic
      - genioglossus
      - hyoglossus
      - palatoglossus
      - styloglossus
    - intrinsic
      - transverse & vertical component
  - **sulcus terminalis**
- tooth primordium
  - dental laminae
    - incisors & molars
  - dental papilla
  - enamel organ
  - sulcus, alveolar
- muscle
  - anterior body wall
    - ext. & int. oblique,
    - rectus abdominus
    - transversus abdominis
  - axial
    - erector spinae
    - ilio-psoas
    - intercostal (external & internal layers)
    - psoas major
    - psoas minor
    - quadratus lumborum
    - serratus anterior
  - cranial
    - extrinsic ocular – *see* eye
    - masseter
    - mylohyoid
    - **pterygoid (medial, lateral)**
    - sterno-mastoid
    - temporalis
  - girdle, pectoral and thoracic body wall
    - deltoid
    - **infraspinatus**
    - lassimus dorsi
    - pectoralis major (pectoralis profundus)
    - pectoralis minor
    - subscapularis
    - supraspinatus
    - trapezius
  - girdle, pelvic
    - gluteus maximus (superficialis)
  - limb
    - forelimb
      - biceps (brachii)
      - teres major
      - triceps (brachii)
    - hindlimb
      - hamstring (flexor) group
      - quadriceps (extensor) group
  - lingual – *see* tongue
  - pharyngeal region
    - **constrictor (superior & inferior)**
  - skin (panniculus carnosus)
    - platysma
- nerve – *see also* ganglia
  - body
    - autonomic
      - hypogastric plexus

THE TISSUES PRESENT IN EACH STAGE OF MOUSE DEVELOPMENT

STAGE 24 *continued*
- parasympathetic
  - vagus (X) trunks
    - (R & L above diaphragm. post & ant. below diaphragm)
    - recurrent laryngeal branch
  - sympathetic (chain) trunk
- nerve trunk
  - intercostal
  - phrenic
  - sciatic
- plexus
  - autonomic, hypogastric
  - brachial
    - median, radial, ulnar
    - **axillary (circumflex)**
  - lumbar
  - lumbo-sacral
  - segmental spinal
  - intercostal
- cranial
  - abducent (VI)
  - cochlear (VIII)
  - accessory (XI)
    - cranial component
    - spinal component
  - facial (VII)
    - chorda tympani branch
  - glossopharyngeal (IX)
  - hypoglossal (XII)
  - oculomotor (III)
  - olfactory (I)
  - optic (II)
  - trigeminal (V)
    - mandibular division
    - maxillary division
    - ophthalmic division
  - trochlear (IV)
  - vagal (X) trunks
  - vestibular (VIII)
- cranial/spinal
  - accessory (XI)
- nose
  - meatus
  - naris
    - anterior
    - external
    - primary choana (primitive posterior)
  - nasal capsule (cartilaginous)
  - nasal (olfactory) cavity
  - nasal septum
    - vomeronasal (Jacobson's) organ
  - olfactory epithelium
  - turbinate bones (conchae)
- nucleus pulposus
- pancreas
  - body, left & right lobes, tail
  - duct
  - uncinate process
- parathyroid (derived from dorsal part of third branchial pouch)
- pituitary
  - adenohypophysis
    - lumen (remnant of Rathke's pouch)
    - pars anterior (anterior lobe, pars distalis)
    - pars intermedia
    - pars tuberalis
  - neurohypophysis
    - infundibulum (pituitary stalk)
    - median eminence
    - pars nervosa (pars posterior)
- placenta
  - chorionic plate
  - labyrinthine part
  - spongy (or basal) region (spongiotrophoblast)
- renal/urinary system
  - bladder
    - fundus region
    - **trigone region**
    - urachus
  - mesonephros (degenerating)
    - duct becomes ductus deferens, efferent ducts, ejaculatory duct (male)
    - tubule (regressing remnants only)
    - vesicles (regressing remnants only)
  - metanephros
    - capsule
    - excretory components
      - cortex
        - early nephrons
        - **convoluted tubules**
        - glomerulus
          - **Bowman's (glomerular) capsule**
        - stem cells
      - medulla
        - stromal cells
    - ureteric bud derivatives
      - calyces, minor & major
      - **collecting tubules**
      - pelvis
      - ureter
  - urethra – *see also* male and female reproductive sytems
  - urogenital mesentery – *see* reproductive system
  - **urogenital sinus**
    - vagina (lower two thirds)
  - urorectal septum
- reproductive system
  - female
    - clitoris
    - **cystic vesicular appendage** (prob. mesonephric/paramesonephric duct origin)
    - gonad (ovary)
      - **germinal cells (oogonia)**
      - "germinal" epithelium
      - **ovarian capsule**
      - **ovigerous cords**
      - **primary oocytes**
      - rete ovarii
      - **suspensory ligament**
    - labial swelling
    - Müllerian tubercle
    - paramesonephric (Müllerian) duct
      - **oviduct**
      - **uterine horn**
      - **vagina** (upper one third)
    - **urethra**
    - urogenital sinus
  - male
    - gonad (testis)
      - **appendix epididymis** (mesonephric duct)
      - **appendix testis** (paramesonephric duct)
      - cortical region ("germinal" epithelium)
      - gubernaculum
      - mediastinum testis
      - medullary region
      - primitive seminiferous tubules
      - **suspensory ligament**
    - mesonephric (Wolffian) duct
      - ductus deferens
      - **efferent ducts**
      - **ejaculatory duct**
      - rete testis
    - penis
      - **glans penis**
      - **prepuce**
    - urethra
      - **prostatic region**
      - **phallic region**
      - urethral groove
  - urogenital mesentery
    - mesorchium (male)
    - mesovarium (female)
  - urogenital sinus
- respiratory system
  - hyoid bone – *see* skeleton, vertebral axis
  - laryngeal aditus
  - laryngeal cartilages
    - arytenoid, cricoid, thyroid
  - larynx
    - mucous membrane
  - lung
    - **bronchiole**
    - bronchus
      - lobar, main, segmental, terminal
    - hilus (root)
    - left
    - pleura
      - parietal
      - visceral
    - right (accessory, caudal, cranial, middle lobes)
  - pharynx
    - hyoid bone
  - trachea
    - bifurcation (carina tracheae)
    - "rings" (from somatopleuric mesenchyme)
  - upper respiratory tract
    - nasopharyngeal cavity
- skeleton
  - cranium
    - 1st arch derivative
      - Meckel's cartilage (central element of mandible; malleus, incus)
    - 2nd arch derivative
      - temporal bone
        - Reichert's cartilage
        - stapes
        - styloid process
    - chondrocranium
      - basioccipital bone
      - basisphenoid bone
      - carotid canal
      - exoccipital
        - squamous part (mes)
      - foramen ovale
      - foramen rotundum
      - **hypoglossal (anterior condylar) canal**
      - jugular foramen
      - occipital
        - squamous part
      - "**sphenoidal**" **canal**
      - sphenoid
      - temporal bone
        - petrous part
        - squamous part
      - **zygomatic process**
    - mandible
    - vault of skull
      - frontal bone
      - inter-parietal bone
      - parietal bone
      - supraoccipital bone (tectum posterius)
      - temporal bone
        - squamous part
    - viscerocranium
      - ethmoid bone
      - facial bone
      - intermaxillary (premaxilla)
      - maxilla
        - palatal shelf
      - optic foramen (canal)
      - orbital fissure, superior
      - orbito-sphenoid
      - turbinate (conchae)
  - girdle
    - pectoral
      - clavicle
      - **costo-vertebral joint primordium**
      - scapula
      - shoulder joint
      - sternum
        - manubrium sterni
        - sternebral bone
        - xiphisternum (xiphoid process)
    - pelvic
      - acetabular region
      - hip joint
      - **synovial cavity**
      - pelvis (innominate bone)
        - iliac bone
        - ischial bone

STAGE 21 *continued*
    pubic bone
  ligament
    cruciate (C1)
  limb
    forelimb
      bone
        carpal
        humerus
        metacarpal
        phalangeal
        radius
        ulna
      joint
        elbow
        gleno-humeral (shoulder)
        wrist
    hindlimb
      bone
        femur
        fibula
        metatarsal
        phalangeal
        tarsus
          calcaneus, talus
        tibia
      joint
        ankle
        knee
          fabella CC
          ligamentum patellae
          patella CC
  vertebral axis
    hyoid bone
      body PCC
      horn (**lesser**, greater) PCC
    intervertebral disc (nucleus pulposus)
    rib
      chondro-sternal joint
      costal cartilage
      costal margin
    vertebrae
      atlas
        cruciate (triradiate) ligament
        hypochordal arch (bow)
      axis
      cervical vertebrae
        dens (odontoid process of C2)
        foramen transversarium
        intervertebral disc (surrounds nucleus pulposus)
        neural arch
        nucleus pulposus
      coccygeal (tail), lumbar, sacral, thoracic vertebrae
        neural arch, centrum etc.
    costo-vertebral joint primordium
skin
  dermis
  eyelid (upper & lower)
  hair
    bulb
    follicle (ordinary)
    root sheath
    tactile (sinus) hair follicle
    vibrissae, **definitive**
  mammary gland
  muscle (panniculus carnosus)
    cutaneous muscle of trunk
    platysma
spinal cord
  central canal
  conus medullaris
  filum terminale
  floorplate
  grey horns (dorsal, ventral & lateral)
  layer
    ependymal
    mantle
      alar & basal columns
    marginal

  meninges – *see* brain
  nerve, segmental spinal
  nerve root, dorsal (sensory), ventral (motor)
  roofplate
  sulcus limitans
spleen
tail
  spinal cord
  vertebrae
thymus
thyroid
  foramen caecum
  isthmus
  lobes (left & right)
  thyroglossal duct
umbilical cord
  artery
  hernia, physiological umbilical
  Wharton's jelly
  vein
vascular system
  artery
    aorta
      abdominal (from L dorsal aorta)
      aortic arch (from left fourth arch artery)
      ascending thoracic (from outflow tract)
      post-ductal part of thoracic (from left dorsal aorta)
    axillary
    basilar
    bronchial
    carotid sinus
    cerebral (anterior, middle, posterior)
    common carotid (from third arch artery)
    common iliac
      external iliac
      internal iliac
    ductus arteriosus (left 6th arch artery)
    epigastric (superior, inferior)
    external carotid
    hepatic
    hyoid
    iliac
    innominate (brachiocephalic)
    intercostal
    intersegmental
    internal carotid
    internal thoracic (mammary)
    maxillary
    mesenteric (superior, inferior)
    ophthalmic
    palatine, greater
    posterior communicating
    principal (axial) to hindlimb (7th cervical intersegmental)
    principal (axial) to forelimb (5th lumbar intersegmental)
    pulmonary
    pulmonary trunk
      distal: from L 6th arch artery
      proximal part from outflow tract
    renal
    spinal (anterior, posterior)
    stapedial
    subclavian
    superior vesical
    testicular
    thyroid (superior, inferior)
    umbilical (R & L)
    vertebral
  blood
  lymphatic system
    jugular lymph sac
  vein
    axillary
    azygos
    common iliac
      external iliac
      internal iliac
    ductus venosus
    epigastric (superior, inferior)

    femoral
    hemiazygos
    hepatico-cardiac
    intercostal
    internal thoracic (mammary)
    jugular (internal, external)
    mesenteric (superior, inferior)
    portal
    pulmonary
    spinal (anterior, posterior)
    subclavian
    testicular (pampiniform plexus)
    umbilical (R & L)
    vena cava
      inferior
        post-hepatic part
        pre-hepatic part
      superior (cranial) vena cava

STAGE 25 – day 16.5
adrenal
  cortex (mesodermal, coelomic epithelium)
  medulla (ectodermal, neural crest)
brain
  forebrain
    diencephalon
      epithalamus
        pineal gland (epiphysis)
          stalk (**peduncle**)
      hypothalamus
        nucleus, lenticular
      internal capsule
      thalamus
        interthalamic adhesion (massa intermedia)
      ganglionic eminence (lateral & medial aspects)
      hippocampus
    telencephalon
      choroid fissure
      choroid invagination (tela choroidea)
      choroid plexus
      corpus striatum
        caudate nucleus
          head & tail
          caudate-putamen
        dentate gyrus
        lentiform nucleus
      interventricular foramen
      lateral ventricle
        anterior, inferior, posterior and superior horns
      neopallial cortex (future cerebral cortex)
      olfactory lobe
      olfactory cortex
    temporal lobe
    third ventricle
      anterior commissure
      epithalamic (pineal) recess
        choroid plexus
      infundibular recess
      interventricular foramen
      lamina terminalis
      optic recess (lumen of optic stalk)
  hindbrain (rhombencephalon)
    fourth ventricle
      lateral recess
        choroid plexus
      roof
        choroid plexus
    metencephalon
      cerebellum
        cerebellar plate
          intraventricular portion
      pons
      sulcus limitans
    myelencephalon
      medulla oblongata
        medullary raphe
      sulcus limitans

THE TISSUES PRESENT IN EACH STAGE OF MOUSE DEVELOPMENT

STAGE 25 *continued*
- midbrain (mesencephalon)
  - cerebral aqueduct
  - **roof (rostral part, tectum)**
  - tegmentum
- coverings
  - meninges (includes arachnoid, dura & pia mater)
    - dura
      - falx cerebri
      - tentorium cerebelli
    - pontine cistern
    - subarachnoid space
      - **basal (interpeduncular) cistern (cisterna interpeduncularis)**
      - **cisterna chiasmatica**
- layers
  - ependymal, mantle & marginal
- sinus
  - venous dural
    - **cavernous**
    - inferior sagittal
    - **sigmoid**
    - superior sagittal
    - transverse
- cavity
  - diaphragm
    - central tendon
    - crus
    - dome (R & L)
  - intra-embryonic coelom
    - pericardial cavity
    - peritoneal cavity
      - greater sac
        - hepatic recess (R & L)
        - omental bursa (lesser sac)
          - superior recess
        - pelvic recess (recto-uterine pouch of Douglas)
  - pleural cavity
- ear
  - ganglia
    - cochlear (VIII)
    - vestibular (VIII)
  - inner
    - cochlea
    - cochlear canal (duct)
    - ductus reuniens
    - endolymphatic duct
    - labyrinth
      - crus commune
    - saccule (utricle/saccule)
    - semicircular canals (posterior, superior, lateral)
    - **spiral organ of Corti**
  - middle
    - ossicles (malleus & incus bone (1st arch), stapes bone (2nd arch))
    - pharyngo-tympanic (Eustachian) tube
    - stapedius muscle
    - tensor tympani muscle
    - tubo-tympanic recess
  - outer (external)
    - external acoustic meatus (entrance only)
    - pinna
  - **tympanic membrane primordium**
  - temporal bone (petrous part) cartilage
- eye
  - anterior chamber
  - choroidal vessels
  - **ciliary body**
  - conjunctival sac
    - upper & lower recesses
  - cornea
    - Descemet's membrane
    - ectoderm
    - stroma
  - eyelid (upper & lower)
    - canthus, inner (nasal)
  - **fat pad, orbital**
  - **Harderian gland (within orbit)**
  - hyaloid cavity
    - vitreous humour
  - hyaloid vascular plexus
    - tunica vasculosa lentis
    - vasa hyaloidea propria
  - **iris (boundary of pupil)**
  - lens
    - capsule
    - fibres
  - muscle
    - extrinsic ocular
      - oblique (inf., sup.), rectus (lateral/temporal, medial/nasal, inferior, superior)
  - optic
    - (II) nerve
    - chiasma
    - stalk
  - retina
    - choroid (fetal) fissure
    - neural (inner) layer
      - **inner nuclear layer**
      - **intermediate anucleate layer (of Chievitz)**
      - nerve fibre layer
      - optic disc
      - **outer nuclear (neuroblastic) layer**
    - pigment (outer) layer
  - sclera
- gall bladder
  - common bile duct
  - cystic duct
- ganglia
  - body
    - autonomic
      - ciliary
      - paraganglia of Zuckerkandl
        - para-aortic "bodies"
    - dorsal (posterior) root
    - sympathetic
      - cervical
        - inferior
        - middle
        - superior (cervico-thoracic, stellate)
      - **coeliac (pre-aortic, abdominal sympathetic)**
      - paraganglia of Zuckerkandl (para-aortic "bodies")
      - thoracic
  - cranial
    - cochlear (VIII)
    - cranial/spinal accessory (XI)
    - facial (VII)
    - glossopharyngeal (IX) – inferior
    - glossopharyngeal (IX) – superior
    - trigeminal (V) (semilunar, Gasserian)
    - vagus (X) – inferior (nodose)
    - vagus (X) – superior
    - vestibular (VIII)
- glands
  - gall bladder
    - common bile duct
    - cystic duct
  - mammary gland
  - pancreas
    - pancreatic duct
    - body, tail, left & right lobes
    - uncinate process
  - pituitary
    - adenohypophysis
      - lumen (remnant of Rathke's pouch, now detached)
      - pars anterior (anterior lobe, pars distalis)
      - pars intermedia
      - pars tuberalis
    - neurohypophysis
      - infundibulum (pituitary stalk)
      - median eminence
      - pars nervosa (pars posterior)
  - spleen
  - sublingual gland
  - submandibular gland
  - thymus
  - thyroid
    - foramen caecum
    - isthmus
    - lobes (L & R)
    - thyroglossal duct
- gut
  - foregut
    - duodenum (rostral half)
      - mesentery (dorsal mesoduodenum)
    - gastro-oesophageal junction
    - oesophagus
      - dorsal meso-oesophagus
    - pharynx
      - constrictor muscle (superior & inferior)
      - hyoid bone
    - stomach
      - fundic glandular mucous membrane
      - fundus
      - glandular, proventricular & pyloric regions
      - mesentery (dorsal mesogastrium)
        - gastro-splenic ligament
        - lieno-renal ligament
        - spleen
      - mesentery (ventral)
        - falciform ligament
        - lesser omentum
      - omentum, greater
      - pyloric antrum
  - foregut–midgut junction
    - biliary system
      - common bile duct
      - gallbladder
      - hepatic duct (L & R)
    - duodenum (caudal part)
      - dorsal mesoduodenum
      - duodenal papilla (ampulla of Vater)
      - pancreatic duct
      - sphincter of Oddi
  - hindgut
    - anal region
      - anal canal
        - rostral part: hindgut
        - caudal part: anal pit
      - **pectinate line** (site of anal membrane)
      - sphincter (external & internal)
    - dorsal mesentery
    - perineal "body"
    - rectum
    - uro-rectal septum
  - midgut
    - dorsal mesentery
    - duodenum (caudal half)
    - jejunum
- haemolymphoid system
  - blood
  - lymph node
    - jugular lymph sac
    - Peyer's patches
  - spleen
  - thymus
- hair – *see* skin
- **Harderian gland** – *see* eye
- heart
  - aorta
    - ascending thoracic (from outflow tract)
    - aortic arch (from left fourth arch artery)
    - aortic sinus
  - atrio-ventricular canal
  - atrio-ventricular cushion tissue
  - atrium
    - auricular part (L & R)
    - interatrial septum
      - foramen (ostium) secundum
      - foramen ovale
      - septum primum
      - septum secundum (crista dividens)
    - sinus venarum
      - crista terminalis
  - bulbar cushion
  - **coronary artery**

STAGE 25 continued
    dorsal mesocardium
        transverse pericardial sinus
    ductus arteriosus (from left sixth arch artery)
    endocardial cushion tissue
    endocardial tissue
    epicardium (visceral pericardium)
    myocardium
    pericardium
        fibrous
        parietal
        visceral (epicardium)
    pulmonary trunk
        distal part: from L sixth arch artery
        proximal part: from outflow tract
    trabeculae carneae
    valve
        aortic, mitral (bicuspid), pulmonary, tricuspid
        leaflets (aortic, mitral, pulmonary, tricuspid)
        venous valve (leaflets, entrance of inferior vena cava)
    ventricle (L & R)
        interventricular groove
        interventricular septum
            muscular part
limb – for bones & joints, see skeleton; for muscles, see muscle
    forelimb
        claw primordium
        digit
        mesenchyme
        plantar (digital) "pads"
    hindlimb
        claw primordium
        digit
        mesenchyme
        plantar (digital) "pads"
liver
    "bare" area
    cystic duct
    hepatic ducts
    hepatic sinusoids
    lobes, L & R (including caudate & quadrate)
    parenchyma
mammary gland
mouth region
    epiglottis
    gum, primitive
    lips
        philtrum
        upper & lower
    lower jaw
        mandible
        canal, mandibular (inferior alveolar)
        Meckel's cartilage
    maxilla
        premaxilla
    oral cavity
    palate
        canal
        incisive
            naso-palatine
        primary (intermaxillary segment)
        secondary (definitive)
        soft
    parotid gland
    sublingual caruncle
    sublingual gland
    submandibular gland
    tongue
        circumvallate papilla
        filiform papillae
        foramen caecum
        frenulum
        fungiform papillae
        intermolar eminence
        median fibrous septum
        muscle
            extrinsic
                genioglossus
                hyoglossus
                palatoglossus
                styloglossus
            intrinsic
                transverse & vertical component
        sulcus terminalis
    tooth primordium
        dental laminae
            incisors & molars
        dental papilla
        enamel organ
        sulcus, alveolar
muscle
    anterior body wall
        ext. & int. oblique,
        rectus abdominus
        transversus abdominis
    axial
        erector spinae
        ilio-psoas
        intercostal (external & internal layers)
        psoas major
        psoas minor
        quadratus lumborum
        serratus anterior
    cranial
        digastric
        extrinsic ocular – see eye
        masseter
        mylohyoid
        pterygoid (medial, lateral)
        sterno-mastoid
        temporalis
    girdle, pectoral and thoracic body wall
        deltoid
        infraspinatus
        lassimus dorsi
        pectoralis major (pectoralis profundus)
        pectoralis minor
        subscapularis
        supraspinatus
        trapezius
    girdle, pelvic
        gluteus maximus (superficialis)
        gluteus medius
        iliacus
        levator ani
    limb
        forelimb
            biceps (brachii)
            teres major
            triceps (brachii)
        hindlimb
            adductor group
            flexor digitorum profundus (sole of foot)
            hamstring (flexor) group
            quadriceps (extensor) group
    lingual – see tongue
    pharyngeal region
        constrictor (superior & inferior)
    skin (panniculus carnosus)
        platysma
nerve – see also ganglia
    body
        autonomic
            hypogastric plexus
            parasympathetic
                vagus (X) trunk
                    (R & L above diaphragm
                    post. & ant. below diaphragm)
                    recurrent laryngeal branch
            sympathetic (chain) trunk
        nerve trunk
            intercostal
            phrenic
            sciatic
        plexus
            autonomic, hypogastric
            brachial
                median, radial, ulnar
                    axillary (circumflex)
            lumbar
            lumbo-sacral
        segmental spinal
            intercostal
    cranial
        abducent (VI)
        cochlear (VIII)
        accessory (XI)
            cranial component
            spinal component
        facial (VII)
            chorda tympani branch
        glossopharyngeal (IX)
        hypoglossal (XII)
        oculomotor (III)
        olfactory (I)
        optic (II)
        trigeminal (V)
            mandibular division
            lingual branch
            maxillary division
            ophthalmic division
        trochlear (IV)
        vagal (X) trunks
        vestibular (VIII)
    cranial/spinal
        accessory (XI)
nose
    nasal bone
    meatus
    naris
        anterior
        external
        primary choana (primitive posterior)
    nasal capsule (cartilaginous)
    nasal (olfactory) cavity
    nasal septum
        vomeronasal (Jacobson's) organ
    olfactory epithelium
    turbinate bones (conchae)
nucleus pulposus
pancreas
    body, left & right lobes, tail,
    duct
    uncinate process
parathyroid (derived from dorsal part of third branchial pouch)
pituitary
    adenohypophysis
        lumen (remnant of Rathke's pouch)
        pars anterior (anterior lobe, pars distalis)
        pars intermedia
        pars tuberalis
    neurohypophysis
        infundibulum (pituitary stalk)
        median eminence
        pars nervosa
placenta
    chorionic plate
    labyrinthine part
    spongy (or basal) region (spongiotrophoblast)
renal/urinary system
    bladder
        fundus region
        trigone region
        urachus
    mesonephros (degenerating)
        duct has now become
            ductus deferens (male)
            efferent ducts
            ejaculatory ducts (male)
        tubules (regressing remnants only)
        vesicles (regressing remnants only)
    metanephros
        capsule
        excretory components
            cortex
                developing nephrons
                convoluted tubules
                glomerulus
                    Bowman's (glomerular) capsule
                stem cells

THE TISSUES PRESENT IN EACH STAGE OF MOUSE DEVELOPMENT

STAGE 25 *continued*
- medulla
  - stromal cells
- ureteric bud derivatives
  - calyces, major & minor
  - collecting tubules
  - pelvis
  - ureter
- urethra – *see also* reproductive system, male
- urogenital mesentery – *see* reproductive system
- urogenital sinus
  - vagina (lower two thirds)
- urorectal septum

reproductive system
- female
  - clitoris
    - **crus of clitoris**
    - **glans clitoridis**
  - cystic vesicular appendage (prob. mesonephric/paramesonephric duct origin)
  - gonad (ovary)
    - germinal cells (oogonia)
    - ovarian capsule
    - ovigerous cords
    - primary oocytes
    - rete ovarii
    - suspensory ligament
  - labial swelling
  - Müllerian tubercle
  - paramesonephric (Müllerian) duct
    - **cervix uteri**
    - oviduct
    - uterine horn
      - **mesometrium**
  - vagina (upper one third)
  - urogenital sinus
- male
  - gonad (testis)
    - appendix epididymis (mesonephric duct)
    - appendix testis (paramesonephric duct)
    - gubernaculum
    - **interstitial cells (includes future Leydig cells)**
    - mediastinum testis
    - medullary region
    - **primitive Sertoli cells**
    - primitive seminiferous tubules
    - **spermatogonia**
    - suspensory ligament
    - **tunica vaginalis testis** (coelomic epithelial covering of testis)
  - **inguinal canal**
  - **inguinal ring (internal (deep) external (superficial))**
  - mesonephric (Wolffian) duct
    - ductus deferens
      - ejaculatory duct
      - rete testis
    - **seminal vesicle**
  - penis
    - glans penis
    - prepuce
  - urethra
    - phallic region
    - prostatic region
  - urogenital mesentery
    - mesorchium (male)
    - mesovarium (female)
  - urogenital sinus

respiratory system
- hyoid bone – *see* skeleton, vertebral axis
- laryngeal aditus
- laryngeal cartilages
  - arytenoid, cricoid, thyroid
- larynx
  - mucous membrane
  - **vocal folds**
- lung
  - **alveolar duct**
  - bronchiole
  - bronchus
    - lobar, main, segmental, terminal
  - hilus (root)
  - pleura
    - parietal
    - visceral
  - right (accessory, caudal, cranial, middle lobes)
- pharynx
- **smooth muscle**
  - **trachealis**
- trachea
  - bifurcation (carina tracheae)
  - "rings" (from somatopleuric mesenchyme)
- upper respiratory tract
  - nasopharyngeal cavity

skeleton
- cranium
  - 1st arch derivative
    - Meckel's cartilage
      - (central element of mandible; malleus, incus)
  - 2nd arch derivative
    - temporal bone
      - Reichert's cartilage
      - stapes
      - styloid process
  - chondrocranium
    - basioccipital bone
    - basisphenoid bone
    - carotid canal
    - **craniopharyngeal canal**
    - exoccipital
      - squamous part (mes)
    - foramen ovale
    - foramen rotundum
    - hypoglossal (anterior condylar) canal
    - jugular foramen
    - occipital
      - squamous part
    - **pituitary** (hypophyseal) **fossa (sella turcica)**
    - "sphenoidal" canal
    - sphenoid
    - temporal bone
      - petrous part
      - squamous part bone
    - **temporo-mandibular joint**
    - **tympanic ring (os tympanicum)**
    - zygomatic process
  - mandible
  - vault of skull
    - frontal bone
    - inter-parietal bone
    - parietal bone
    - supraoccipital bone (tectum posterius)
    - **sutures**
      - **fontanelle (anterior, posterior)**
  - viscerocranium
    - ethmoid bone
      - **cribriform plate**
    - facial bone
    - intermaxillary (premaxilla)
    - maxilla
      - palatal shelf
    - **nasal bone**
    - optic foramen (canal)
    - orbital fissure, superior
    - orbito-sphenoid
    - turbinate (conchae)
- girdle
  - pectoral
    - clavicle
    - **costo-vertebral joint**
    - scapula
    - shoulder joint
  - sternum
    - manubrium sterni
    - sternebral bone
    - **sterno-clavicular joint**
    - xiphisternum (xiphoid process)
  - pelvic
    - acetabular region
    - hip joint
      - synovial cavity
    - pelvis (innominate bone)
      - iliac bone
      - ischial bone
      - pubic bone
    - ligament
      - cruciate (C1)
- limb
  - forelimb
    - bone
      - carpal
      - humerus
      - metacarpal
      - phalangeal
      - radius
      - ulna
    - joint
      - elbow
      - gleno-humeral (shoulder)
      - **radio-carpal (wrist)**
      - **radio-ulnar (proximal, distal)**
  - hindlimb
    - bone
      - femur
      - fibula
      - metatarsal
      - phalangeal
      - tarsus
        - calcaneus, talus
      - tibia
    - joint
      - ankle
      - knee
        - fabella
        - ligamentum patellae
        - patella
- vertebral axis
  - hyoid bone
    - body PCC
    - horn (lesser, greater) CC
  - intervertebral disc (nucleus pulposus)
  - **ligamentum nuchae**
  - rib
    - chondro-sternal joint
    - costal cartilage
    - costal margin
  - vertebrae
    - atlas
      - cruciate (triradiate) ligament
      - hypochordal arch (bow)
    - axis
    - cervical vertebrae
      - dens (odontoid process of C2)
      - foramen transversarium
      - intervertebral disc (surrounds nucleus pulposus)
      - neural arch
      - nucleus pulposus
    - coccygeal (tail), lumbar, sacral, thoracic vertebrae
      - neural arch, centrum etc.
    - costo-vertebral joint

skin
- **brown fat**
- dermis
- **epicranial aponeurosis (galea aponeurotica)**
- eyelid (upper & lower)
- hair
  - bulb
  - follicle (ordinary)
  - root sheath
  - tactile (sinus) hair follicle
  - vibrissae, definitive
- mammary gland
- muscle (panniculus carnosus)
  - cutaneous muscle of trunk
  - platysma

STAGE 25 continued
spinal cord
  central canal
  conus medullaris
  filum terminale
  floorplate
  grey horns (dorsal, ventral & lateral)
  layer
    ependymal
    mantle
      alar & basal columns
    marginal
  meninges – see brain
  nerve, segmental spinal
  nerve root, dorsal (sensory), ventral (motor)
  roofplate
  sulcus limitans
spleen
tail
  spinal cord
  vertebrae
thymus
thyroid
  foramen caecum
  isthmus
  lobes (left & right)
  thyroglossal duct
umbilicus cord
  artery
  Wharton's jelly
  vein
vascular system
  artery
    aorta
      abdominal (from L dorsal aorta)
      aortic arch (from L 4th arch artery)
      ascending thoracic (from outflow tract)
        post-ductal part of thoracic (from L dorsal aorta)
    axillary
    basilar
    bronchial
    carotid "body"
    carotid sinus
    cerebral (anterior, middle, posterior)
    coeliac trunk
    common carotid (from third arch artery)
    common iliac
      external iliac
      internal iliac
    coronary
    ductus arteriosus (L 6th arch artery)
    epigastric (superior, inferior)
    external carotid
    facial
    hepatic
    hyoid
    iliac
    innominate (brachiocephalic)
    intercostal
    internal carotid
    internal thoracic (mammary)
    intersegmental
    maxillary
    median sacral
    mesenteric (superior, inferior)
    musculo-phrenic
    ophthalmic
    palatine, greater
    posterior communicating
    principal (axial) to hindlimb (7th cervical intersegmental)
    principal (axial) to forelimb (5th lumbar intersegmental)
    pulmonary
    pulmonary trunk
      distal: from L 6th arch artery
      proximal: from outflow tract
    renal
    spinal (anterior, posterior)
    splenic
    stapedial
    subclavian
    superior epigastric
    superior vesical
    testicular
    thyroid (superior, inferior)
    umbilical (R & L)
    vertebral
  blood
  lymphatic system
    jugular lymph sac
  vein
    accessory hemiazygos
    axillary
    azygos
    common iliac
      external iliac
      internal iliac
    deep dorsal of clitoris/penis
    ductus venosus
    epigastric (superior, inferior)
    facial
    femoral
    hemiazygos
    hepatic
    hepatico-cardiac
    intercostal
    internal thoracic (mammary)
    jugular (internal, external)
    mesenteric (superior, inferior)
    musculo-phrenic
    portal
    pulmonary
    renal
    spinal (anterior, posterior)
    subclavian
    superior epigastric
    testicular (pampiniform plexus)
    umbilical (R & L)
    vena cava
      inferior
        post-hepatic part
        pre-hepatic part
      superior (cranial) vena cava

STAGE 26 – day 17.5
adrenal
  cortex (mesodermal, coelomic epithelium)
  medulla (ectodermal, neural crest)
brain
  forebrain
    diencephalon
      epithalamus
        pineal gland (epiphysis)
          stalk (peduncle)
      hypothalamus
        nucleus, lenticular
      internal capsule
      thalamus
        interthalamic adhesion (massa intermedia)
      ganglionic eminence (lateral & medial aspects)
      hippocampus
    telencephalon
      choroid fissure
      choroid invagination (tela choroidea)
      choroid plexus
      corpus striatum
        caudate nucleus
          head & tail
        caudate-putamen
      dentate gyrus
      lentiform nucleus
      interventricular foramen
      lateral ventricle
        anterior, inferior, posterior & superior horns
      neopallial cortex (future cerebral cortex)
      olfactory lobe
      olfactory cortex
      temporal lobe
    third ventricle
      anterior commissure
      epithalamic (pineal) recess
        choroid plexus
      infundibular recess
      interventricular foramen
      lamina terminalis
      optic recess (lumen of optic stalk)
  hindbrain (rhombencephalon)
    fourth ventricle
      lateral recess
        choroid plexus
      roof
        choroid plexus
    metencephalon
      cerebellum
        cerebellar plate
        dorsal part (vermis)
        intraventricular portion
      pons
      sulcus limitans
    myelencephalon
      medulla oblongata
        medullary raphe
      inferior olivary nucleus
      sulcus limitans
  midbrain (mesencephalon)
    cerebral aqueduct
    floor (tegmentum)
    roof (rostral part, tectum)
    tegmentum
  covering
    meninges (includes arachnoid, dura & pia mater)
    dura
      falx cerebri
      tentorium cerebelli
    pontine cistern
    subarachnoid space
      basal (interpeduncular) cistern (cisterna interpeduncularis)
      cisterna chiasmatica
  layer
    ependymal, mantle, marginal
  sinus
    venous dural
      cavernous
      inferior petrosal
      inferior sagittal
      pineal venous plexus
      sigmoid
      superior sagittal
      transverse
cavity
  diaphragm
    arcuate ligaments (medial & lateral)
    central tendon
    crus
    dome (R & L)
  intra-embryonic coelom
    pericardial cavity
      pericardium
        parietal
        visceral
    peritoneal cavity
      greater sac
      hepatic recess (R & L)
      omental bursa (lesser sac)
        superior recess
      pelvic recess (recto-uterine pouch of Douglas)
    pleural cavity
ear
  ganglia
    cochlear (VIII)
    vestibular (VIII)
  inner
    cochlea
    cochlear canal (duct)

THE TISSUES PRESENT IN EACH STAGE OF MOUSE DEVELOPMENT

STAGE 26 *continued*
- ductus reuniens
- endolymphatic duct
- **fenestra cochleae**
- **fenestra vestibuli**
- labyrinth
  - crus commune
  - saccule (utricle/saccule)
  - semicircular canals (posterior, superior, lateral)
  - spiral organ of Corti
- middle
  - ossicles (malleus & incus bone (1st arch), stapes bone (2nd arch))
  - pharyngo-tympanic (Eustachian) tube
  - stapedius muscle
  - tensor tympani muscle
  - tubo-tympanic recess
- outer (external)
  - external acoustic meatus (entrance only)
  - pinna
  - tympanic membrane
- temporal bone (petrous part) cartilage

eye
- anterior chamber
- choroidal vessels
- ciliary body
- conjunctival sac
  - upper & lower recesses
- cornea
  - Descemet's membrane
  - ectoderm
  - stroma
- eyelid (upper & lower)
  - canthus, inner (nasal)
- fat pad, orbital
- Harderian gland (within orbit)
- hyaloid cavity
  - vitreous humour
- hyaloid vascular plexus
  - **hyaloid artery (central artery of retina)**
  - tunica vasculosa lentis
  - vasa hyaloidea propria
- iris (boundary of pupil)
- lens
  - capsule
  - **embryonic/fetal "nucleus"**
  - fibres
  - **posterior suture**
- muscle
  - extrinsic ocular
    - oblique (inf., sup.), rectus (lateral/temporal, medial/nasal, inferior, superior)
- optic
  - (II) nerve
  - chiasma
  - stalk
- retina
  - choroid (fetal) fissure
  - neural (inner) layer
    - inner nuclear layer
    - intermediate anucleate layer (of Chievitz)
    - nerve fibre layer
    - optic disc
    - outer nuclear (neuroblastic) layer
  - pigment (outer) layer
- sclera
- **tear ducts**

gall bladder
- common bile duct
- cystic duct

ganglia
- body
  - autonomic
    - ciliary
    - paraganglia of Zuckerkandl
      - para-aortic "bodies"
  - dorsal (posterior) root
  - sympathetic
    - cervical
      - inferior
      - middle
      - superior (cervico-thoracic, stellate),
    - coeliac (pre-aortic, abdominal sympathetic)
      - paraganglia of Zuckerkandl (para-aortic "bodies")
    - thoracic
- cranial
  - cochlear (VIII)
  - cranial/spinal accessory (XI)
  - facial (VII)
  - glossopharyngeal (IX) – inferior
  - glossopharyngeal (IX) – superior
  - trigeminal (V) (semilunar, Gasserian)
  - vagus (X) – inferior (nodose)
  - vagus (X) – superior
  - vestibular (VIII)

gland
- adrenal
  - cortex (mesodermal, coelomic epithelium)
  - medulla (ectodermal, neural crest)
- **bulbo-urethral**
- gall bladder
  - common bile duct
  - cystic duct
- mammary gland
- pancreas
  - body, tail, left & right lobes
  - **islets of Langerhans**
  - pancreatic duct
  - uncinate process
- pituitary
  - adenohypophysis
    - lumen (remnant of Rathke's pouch, now detached)
    - pars anterior (anterior lobe, pars distalis)
    - pars intermedia
    - pars tuberalis
  - neurohypophysis
    - infundibulum (pituitary stalk)
    - median eminence
    - pars nervosa (pars posterior)
- **prostate**
- spleen
- sublingual gland
- submandibular gland
- thymus
  - **thin fibrous capsule**
  - **medullary core**
- thyroid
  - **colloid-filled follicles**
  - foramen caecum
  - isthmus
  - lobes (L & R)
  - thyroglossal duct

gut
- foregut
  - duodenum (rostral half)
    - mesentery (dorsal mesoduodenum)
  - gastro-oesophageal junction
  - oesophagus
    - dorsal meso-oesophagus
  - pharynx
    - constrictor muscle (superior & inferior)
    - hyoid bone
    - **pharyngeal ligament**
    - **pharyngeal tubercle**
  - stomach
    - **"cutaneous" non-glandular (proventricular) portion**
    - fundic glandular mucous membrane
    - fundus
    - glandular, proventricular & pyloric regions
    - mesentery (dorsal mesogastrium)
      - gastro-splenic ligament
      - lieno-renal ligament
      - spleen
    - mesentery (ventral)
      - falciform ligament
      - lesser omentum
      - omentum, greater
    - pyloric antrum
    - **pyloric sphincter**
  - foregut–midgut junction
    - biliary system
      - common bile duct
      - gallbladder
      - hepatic duct (L & R)
    - duodenum
      - dorsal mesoduodenum
      - duodenal papilla (ampulla of Vater)
      - pancreatic duct
      - sphincter of Oddi
- hindgut
  - anal region
    - anal canal (rostral: hindgut)
    - anal canal (caudal: anal pit)
    - pectinate line (site of anal membrane)
    - sphincter (external & internal)
  - dorsal mesentery
  - **large intestine**
  - perineal "body"
  - rectum
  - urorectal septum
- midgut
  - dorsal mesentery
  - duodenum (caudal half)
    - dorsal mesoduodenum
  - jejunum

haemolymphoid system
- blood
- lymph node
  - jugular lymph sac
  - Peyer's patches
  - **thoracic duct**
- spleen
  - **hilum**
  - **medullary region**
- thymus
  - **cortex**
  - **medulla**

hair – *see* skin

Harderian gland – *see* eye

heart
- aorta
  - ascending thoracic (from outflow tract)
  - aortic arch (from left fourth arch artery)
  - aortic sinus
- atrio-ventricular canal
- atrio-ventricular cushion tissue
- atrium
  - auricular part (L & R)
  - interatrial septum
    - foramen (ostium) secundum
    - foramen ovale
    - septum primum
    - septum secundum (crista dividens)
  - sinus venarum
    - crista terminalis
- bulbar cushion
- coronary artery
- dorsal mesocardium
  - transverse pericardial sinus
- ductus arteriosus (from left sixth arch artery)
- endocardial cushion tissue
- endocardial tissue
- epicardium (visceral pericardium)
- myocardium
- pericardium
  - fibrous
  - parietal
  - visceral (epicardium)
- pulmonary trunk
  - distal part: from L sixth arch artery
  - proximal part: from outflow tract
- trabeculae carneae
- valve
  - aortic, mitral (bicuspid), pulmonary, tricuspid
  - leaflets (aortic, mitral, pulmonary, tricuspid)
- venous valve (leaflets, entrance of inferior vena cava)

STAGE 26 continued
  ventricle (L & R)
    interventricular groove
    interventricular septum
      muscular part
limb – for bones & joints, see skeleton; for muscles, see muscle
  forelimb
    claw primordium
    digit
    mesenchyme
    palmar (digital) "pads"
  hindlimb
    claw primordium
    digit
    mesenchyme
    palmar (digital) "pads"
liver
  "bare" area
  cystic duct
  hepatic ducts
  hepatic sinusoids
  lobes, L & R (including caudate & quadrate)
  parenchyma
mammary gland
mouth region
  epiglottis
  gum, primitive
  lips
    philtrum
    upper & lower
  lower jaw
    mandible
      mandibular canal (inferior alveolar)
    Meckel's cartilage
  maxilla
    premaxilla
  oral cavity
  palate
    canal
      incisive
      naso-palatine
    primary (intermaxillary segment)
    secondary (definitive)
    soft
  parotid gland
  sublingual caruncle
  sublingual gland
  submandibular gland
  tongue
    circumvallate papilla
    filiform papillae
    foramen caecum
    frenulum
    fungiform papillae
    intermolar eminence
    median fibrous septum
    muscle
      extrinsic
        genioglossus
        hyoglossus
        palatoglossus
        styloglossus
      intrinsic
        transverse & vertical component
    sulcus terminalis
  tooth primordium
    dental laminae
      incisors & molars
    dental papilla
    enamel organ
    sulcus, alveolar
muscle
  anterior body wall
    ext. & int. oblique,
    rectus abdominus
    transversus abdominis
  axial
    erector spinae
    ilio-psoas
    intercostal (external & internal layers)
    psoas major
    psoas minor
    quadratus lumborum
    serratus anterior
  cranial
    digastric
    extrinsic ocular (see eye)
    masseter
    mylohyoid
    pterygoid (medial, lateral)
    sterno-mastoid
    temporalis
  girdle, pectoral and thoracic body wall
    deltoid
    infraspinatus
    lassimus dorsi
    pectoralis major (pectoralis profundus)
    pectoralis minor
    subscapularis
    supraspinatus
    transversus thoracis
    trapezius
  girdle, pelvic
    gluteus maximus (superficialis)
    gluteus medius
    iliacus
    levator ani
  limb
    forelimb
      biceps (brachii)
      teres major
      triceps (brachii)
    hindlimb
      adductor group
      flexor digitorum profundus (sole of foot)
      hamstring (flexor) group
      quadriceps (extensor) group
  lingual – see tongue
  pharyngeal region
    constrictor (superior & inferior)
    pretracheal
      cricothyroid, sternohyoid, sternothyroid, thyrohyoid
  skin (panniculus carnosus)
  platysma
nerve – see also ganglia
  body
    autonomic
      hypogastric plexus
        superior & inferior
      parasympathetic
        vagus (X) trunk
          R & L above diaphragm
          post. & ant. below diaphragm
          recurrent laryngeal branch
      sympathetic (chain) trunk
  nerve trunk
    intercostal
    phrenic
    sciatic
  plexus
    autonomic, hypogastric
    brachial
      median, radial, ulnar
        axillary (circumflex)
    lumbar
    lumbo-sacral
  segmental spinal
    intercostal
  cranial
    abducent (VI)
    cochlear (VIII)
    accessory (XI)
      cranial component
      spinal component
    facial (VII)
      chorda tympani branch
    glossopharyngeal (IX)
    hypoglossal (XII)
    oculomotor (III)
    olfactory (I)
    optic (II)
    trigeminal (V)
      mandibular division
        lingual branch
      maxillary division
      ophthalmic division
    trochlear (IV)
    vagal (X) trunks
    vestibular (VIII)
  cranial/spinal
    accessory (XI)
nose
  meatus
    middle
  naris
    anterior
    external
    primary choana (primitive posterior)
  nasal bone
  nasal capsule (cartilaginous)
  nasal (olfactory) cavity
  nasal septum
    vomeronasal (Jacobson's) organ
  olfactory epithelium
  turbinate bones (conchae)
nucleus pulposus
pancreas
  body, left & right lobes, tail, uncinate process
  duct
  islets of Langerhans
parathyroid (derived from dorsal part of third branchial pouch)
pituitary
  adenohypophysis
    lumen (remnant of Rathke's pouch)
    pars anterior (anterior lobe, pars distalis)
    pars intermedia
    pars tuberalis
  neurohypophysis
    infundibulum (pituitary stalk)
    median eminence
    pars nervosa
placenta
  chorionic plate
  labyrinthine part
  spongy (or basal) region (spongiotrophoblast)
renal/urinary system
  bladder
    detrusor muscle
    fundus region
    trigone region
    urachus
  mesonephros
    duct – see reproductive system, male
      ductus deferens
      efferent ducts
      ejaculatory ducts
      vesicles (aberrant remnants only)
  metanephros
    capsule
    excretory components
      cortex
        immature nephrons
        nephrons
          Bowman's glomerular capsule
          convoluted tubules (proximal & distal)
          glomerulus
          stem cells
      medulla
        loop of Henle
        stromal cells
    ureteric bud derivatives
      calyces, minor & major
      collecting tubules
      pelvis
      ureter
  urethra – see also reproductive system, male
  urogenital mesentery – see reproductive system
  urogenital sinus
    vagina (lower two thirds)

THE TISSUES PRESENT IN EACH STAGE OF MOUSE DEVELOPMENT

STAGE 26 *continued*
  urorectal septum
  reproductive system
    female
      clitoris
        crus of clitoris
        glans clitoridis
      cystic vesicular appendage (prob.
          mesonephric/paramesonephric duct
          origin)
      gonad (ovary)
        capsule
        germinal cells (oogonia)
        ovigerous cords
        primary follicles, follicle cells
        primary oocytes
        rete ovarii
        suspensory ligament
      labial swelling
      Müllerian tubercle
      paramesonephric (Müllerian) duct
        cervix uteri
        oviduct
        uterine horn
          mesometrium
        vagina (upper one third)
      urogenital sinus
    inguinal canal
    inguinal ring (internal (deep) external
      (superficial))
    male
      ductus deferens (from Wolffian duct)
        ejaculatory duct
      epididymis (caput (head); cauda (tail);
        corpus (body))
        rete testis
        seminal vesicle
      gland
        bulbo-urethral
        prostate
      gonad (testis)
        appendix epididymis (mesonephric duct)
        appendix testis (paramesonephric duct)
        gubernaculum
        interstitial cells (includes **Leydig cells**)
        mediastinum testis
        medullary region
        seminiferous tubules
        spermatogonia
        suspensory ligament
        Sertoli (sustentacular) cells
        tunica albuginea (fibrous capsule of
          testis)
        tunica vaginalis testis (coelomic epithelial
          covering of testis)
        vasa efferentia (efferent ducts)
      penis
        crus penis
        glans penis
        prepuce
      scrotum
      urethra
        bulbar part
        bulbo-urethral glands
        penile part
        phallic region
        prostatic region
        prostatic utricle (from paramesonephric
          duct)
    urogenital mesentery
      mesorchium (male)
      mesovarium (female)
    urogenital sinus
  respiratory system
    hyoid bone – *see* skeleton, vertebral axis
    laryngeal aditus
    laryngeal cartilages: arytenoid, cricoid,
      hyroid
    larynx
      mucous membrane
      vocal folds
    lung
      alveolar duct
      alveolus
      bronchiole
      bronchus
        lobar, main, segmental, terminal
      hilus (root)
      left
      pleura
        parietal
        visceral
      right (accessory, caudal, cranial, middle
        lobes)
    pharynx
    trachea
      bifurcation (carina tracheae)
      "rings" (from somatopleuric mesenchyme)
    trachealis (smooth) muscle
    upper respiratory tract
      nasopharyngeal cavity
  skeleton
    cranium
      1st arch derivative
        Meckel's cartilage
          (central element of mandible; malleus,
          incus)
      2nd arch derivative
        temporal bone
          Reichert's cartilage
          stapes
          styloid process
      chondrocranium
        basioccipital bone
        basisphenoid bone
        carotid canal
        craniopharyngeal canal
        exoccipital
          squamous part (mes)
        foramen ovale
        foramen rotundum
        hypoglossal (anterior condylar) canal
        jugular foramen
        occipital
          squamous part
        pituitary (hypophyseal) fossa (sella
          turcica)
        sphenoid bone
        "sphenoidal" canal
        temporal bone
          petrous part
          squamous part
          temporo-mandibular joint
          tympanic ring (os tympanicum)
          zygomatic process
      mandible
      vault of skull
        frontal bone
        inter-parietal bone
        parietal bone
        supraoccipital bone (tectum posterius)
        sutures
          fontanelle (anterior, posterior)
      viscerocranium
        ethmoid bone
          cribriform plate
        facial bone
        intermaxillary (premaxilla)
        maxilla
          palatal shelf
        nasal bone
        optic foramen (canal)
        orbital fissure, superior
        orbito-sphenoid
        turbinate (conchae)
    girdle
      pectoral
        clavicle
        costo-vertebral joint
        scapula
        shoulder joint
        sternum
          manubrium sterni
          sternebral bone
          sterno-clavicular joint
          xiphisternum (xiphoid process)
      pelvic
        acetabular region
        hip joint
          synovial cavity
        pelvis
          iliac bone
          ischial bone
          pubic bone
        sacro-iliac joint
    ligament
      cruciate (C1)
    limb
      forelimb
        bone
          carpal
          humerus
          metacarpal
          phalangeal bone
          radius
          ulna
        joint
          annular ligament (prox. radio-ulnar)
          elbow
          gleno-humeral (shoulder)
          metacarpo-phalangeal
          radio-carpal (wrist)
          radio-ulnar (proximal, distal)
      hindlimb
        bone
          femur
          fibula
          metatarsal
          phalangeal bone
          tarsus
            calcaneus, talus
          tibia
        joint
          ankle
          knee
            fabella
            ligamentum patellae
            patella
    vertebral axis
      hyoid bone
        body, horn (lesser, greater)
      intervertebral disc (nucleus pulposus)
      ligamentum nuchae
      rib
        chondro-sternal joint
        costo-vertebral joint
        costal cartilage
        costal margin
      vertebrae
        atlas
          cruciate (triradiate) ligament
          hypochordal arch (bow)
        axis
        cervical vertebrae
          dens (odontoid process of C2)
          foramen transversarium
          intervertebral disc (surrounds nucleus
            pulposus)
          neural arch
          nucleus pulposus
        coccygeal (tail), lumbar, sacral, thoracic
          vertebrae
          neural arch, centrum etc.
        costo-vertebral joint
  skin
    brown fat
    dermis
    epicranial aponeurosis (galea aponeurotica)
    eyelid (upper & lower)
    hair
      bulb
      follicle (ordinary)
      root sheath

STAGE 26 continued
    tactile (sinus) hair follicle
      vibrissae, definitive
    mammary gland
    muscle (panniculus carnosus)
      cutaneous muscle of trunk
      platysma
spinal cord
    central canal
    conus medullaris
    filum terminale
    floorplate, roofplate
    grey horns (dorsal, ventral & lateral)
    layer: ependymal, mantle, marginal
    meninges – *see* brain
    nerve, segmental spinal
    nerve root, dorsal (sensory), ventral (motor)
    roofplate
    sulcus limitans
spleen
    hilum
    medullary region
tail
    spinal cord
    vertebrae
thymus
    medullary core
    thin fibrous capsule
thyroid
    colloid-filled follicles
    foramen caecum
    isthmus
    lobes (L & R)
    thyroglossal duct
umbilicus
    artery
    cord
    Wharton's jelly
    vein
vascular system
    artery
      aorta
        abdominal
        aortic arch (from L dorsal aorta)
        ascending thoracic (from outflow tract)
        post-ductal part of thoracic (from left dorsal aorta)
      axillary

basilar
bronchial
carotid "body"
carotid sinus
cerebral (anterior, middle, posterior)
coeliac trunk
common carotid (from third arch artery)
common iliac
    external iliac
    internal iliac
communicating (anterior/posterior)
coronary
ductus arteriosus (L 6th arch artery)
epigastric (superior, inferior)
external carotid
facial
hepatic
hyaloid
hyoid
iliac
innominate (brachiocephalic)
intercostal
internal carotid
internal thoracic (mammary)
intersegmental
maxillary
median sacral
mesenteric (superior, inferior)
musculo-phrenic
ophthalmic
palatine, greater
posterior communicating
principal (axial) to hindlimb (7th cervical intersegmental)
principal (axial) to forelimb (5th lumbar intersegmental)
pulmonary
pulmonary trunk
    distal: from L 6th arch artery
    proximal: from outflow tract
renal
spinal (anterior, posterior)
splenic
stapedial
subclavian
superior cerebellar
superior epigastric
superior vesical

testicular
thyroid (superior, inferior)
umbilical (R & L)
vertebral
blood
carotid sheath
lymphatic system
    jugular lymph sac
thoracic duct
vein
    accessory hemiazygos
    axillary
    azygos
    common iliac
      external iliac
      internal iliac
    deep dorsal of clitoris/penis
    ductus venosus
    epigastric (superior, inferior)
    facial
    femoral
    great cerebral vein of Galen
    hemiazygos
    hepatic
    hepatico-cardiac
    intercostal
    internal thoracic (mammary)
    jugular (internal, external)
    lingual
    mesenteric (superior, inferior)
    musculo-phrenic
    ophthalmic
    palatine, greater
    portal
    pulmonary
    renal
    spinal (anterior, posterior)
    subclavian
    superficial temporal
    superior epigastric
    testicular (pampiniform plexus)
    umbilical (R & L)
    vena cava
      inferior
        post-hepatic part
        pre-hepatic part
      superior (cranial) vena cava

5.4 Index of first occurrences of tissues

Abbreviations
E   embryonic day
CC   cartilage condensation
L   left
PCC   pre-cartilage condensation
PMM   pre-muscle mass
R   right
TS   Theiler stage

adrenal gland
   primordium – TS 21
   cortex (mesodermal, coelomic epithelium) – TS 22
   medulla (ectodermal, neural crest) – TS 22
alkaline phosphatase (intracellular) enzyme activity
   apical ectodermal ridge – TS 16
   mantle zone of basal plate of neural tube – TS 16/17
   primordial germ cells – TS 10
aortic arches – *see* branchial arches
arteries, named – *see* vascular system

blastocoelic cavity – TS 4
blastocyst
   zona-free – TS 5
   zona-intact – TS 4
body cavity
   diaphragm
     body wall derivatives
       arcuate ligaments (medial & lateral) – TS 26
       crus – TS 21
       pleuro-pericardial folds (later membrane) – TS 21
         (separates lateral part of pericardial cavity from upper part of pleural cavity)
       pleuro-peritoneal folds (later membrane) – TS 20/21
         (separates caudal part of pleural cavity from peritoneal cavity)
       dome (left & right) – TS 21
     canals (channels)
       pleuro-pericardial – TS 16/17
       pleuro-peritoneal – TS 14
     mesenteries
       oesophagus, dorsal and ventral – TS 19/20
     septum transversum derivative
       central tendon – TS 21
   intra-embryonic coelomic cavity – TS 11
     channels – TS 12
     pericardial component – TS 12
       pericardium
         parietal – TS 19
         visceral – TS 18 (late)
     pericardio-peritoneal canals (channels) – TS 13
     pleural component – TS 16
     peritoneal component – TS 12
       greater sac – TS 17
       hepatic recess (left & right) – TS 22
       lesser sac (omental bursa) – TS 17
         superior recess – TS 22
   pericardial cavity – TS 21 (late)
   peritoneal cavity – TS 22
     pelvic recess (recto-uterine pouch of Douglas) – TS 23
   pleural cavity – TS 22
     pleura, parietal – TS 22 (late)
     pleura, visceral – TS 23
brain
   cephalic neural folds – TS 11

layer
   ependymal (ventricular) – TS 16
   mantle – TS 17
   marginal – TS 17
covering
   primitive ectomeninx – TS 20
     meninges (includes arachnoid, dura & pia mater) – TS 21
dura
   falx cerebri – TS 22
   subarachnoid space – TS 22
   tentorium cerebelli – TS 22
forebrain region – TS 12
   caudate nucleus – TS 22
     head & tail – TS 23
   choroid plexus – TS 21
   corpus striatum – TS 20
     caudate-putamen – TS 23
     ganglionic eminence (lateral & medial aspects) – TS 22
   dentate gyrus – TS 25
   diencephalon – TS 14
     epithalamus – TS 20
       pineal gland – TS 24
         primordium (epiphysis) – TS 23
         stalk (peduncle) – TS 25
     hypothalamus – TS 20
       nucleus, lenticular – TS 25
     thalamus – TS 21
       interthalamic adhesion (massa intermedia) – TS 23
   hippocampus – TS 25
   internal capsule – TS 22
   lamina terminalis – TS 15
   layer
     ependymal (ventricular) – TS 16
     mantle – TS 17
     marginal – TS 17
   neural folds – TS 12
   pituitary gland – *see* pituitary
   prosencephalic region – TS 12
   prosencephalon – TS 13
   prosencephalic vesicle – TS 14
   striatum – TS 20 (*see* "corpus striatum")
   telencephalon (future cerebral hemisphere) – TS 15
     choroid fissure – TS 21
     choroid invagination – TS 20
     interventricular foramen – TS 16
     lateral ventricle – TS 19
       anterior horn – TS 21
       choroid plexus – TS 21
       inferior horn – TS 21
       interventricular foramen – TS 16
       posterior horn – TS 21
       superior horn – TS 21
     neopallial cortex (future cerebral cortex) – TS 20
     olfactory lobe – TS 21
       olfactory cortex – TS 21
     telencephalic vesicle (future lateral ventricle) – TS 15
   temporal lobe – TS 24
   third ventricle – TS 15
     anterior commissure – TS 25
     epithalamic (pineal) recess – TS 23
     choroid plexus – TS 23
     infundibular recess – TS 15
     interventricular foramen – TS 16
     optic recess – TS 17
     optic vesicle – TS 13/14
     telencephalic vesicle – TS 15
headfold
   cephalic neural folds – TS 11

region – TS 11
hindbrain region – TS 12
   cerebellum – TS 21
     cerebellar plate – TS 25
     dorsal part (vermis) – TS 26
     intraventricular portion – TS 23
     primordium (rhombic lip; dorsal part of alar plate of metencephalon) – TS 18
   fourth ventricle – TS 15
     lateral recess – TS 19
       choroid plexus – TS 22
     roof – TS 15
       choroid plexus – TS 21
   metencephalon – TS 17
     alar plate (cerebellar primordium) – TS 18
     basal plate (pons primordium) – TS 18
   myelencephalon – TS 18
     medulla oblongata – TS 20
       inferior olivary nucleus – TS 26
       medullary raphe – TS 23
   pons primordium (basal plate of metencephalon) – TS 18
   sulcus, pre-otic – TS 11
   rhombencephalon – TS 14
   rhombencephalic vesicle – TS 14
   rhombomeres (A & B) – TS 12
   rhombomeres (1–5) – TS 13 (late)
   rhombomeres (6–8) – TS 14 (late)
midbrain region – TS 12
   mesencephalic vesicle (future cerebral aqueduct) – TS 15
   midbrain (mesencephalon) – TS 15
     cerebral aqueduct – TS 21
     mesencephalic vesicle – TS 14
       floor (tegmentum) – TS 22
       roof (rostral part, tectum) – TS 22
     tectum – TS 25
     tegmentum – TS 22
neural folds – TS 12
pia mater – TS 21
primitive ectomeninx – TS 20
sinus
   cervical – TS 17
   venous dural – TS 22
     cavernous – TS 25
     inferior petrosal – TS 26
     inferior sagittal – TS 23
     pineal venous plexus – TS 26
     sigmoid – TS 25
     superior sagittal – TS 23
     transverse – TS 23
subarachnoid space
   basal (interpeduncular) cistern (cisterna interpeduncularis) – TS 25
   cisterna chiasmatica – TS 25
   pontine cistern – TS 22
branchial arch
(ectoderm, endoderm, mesenchyme from head mesoderm and neural crest)
1st arch – TS 12
   artery – TS 12
     maxillary – TS 22
   cartilage
     Meckel's cartilage PCC – TS 19
     incus PCC – TS 21
     malleus PCC – TS 21
   groove (cleft) – TS 12
     (external acoustic meatus – after birth)
   mandibular component – TS 14
   maxillary component – TS 12
   membrane – TS 12
     (future tympanic membrane – TS 17)
   pouch – TS 12
     (tubo-tympanic recess – TS 21)

2nd arch – TS 13
    artery – TS 13
        hyoid – TS 22
        stapedial – TS 22
    cartilage
        Reichert's cartilage PCC – TS 19
        hyoid bone: lesser horn & upper part of body – TS 24
        stapes PCC – TS 20
    groove (cleft) – TS 13
    membrane – TS 13
    pouch (future tonsillar cleft) – TS 13
3rd arch – TS 13
    artery – TS 15
        (common carotid artery – TS 21)
    cartilage
        hyoid bone: lower part of body & greater horn – TS 23
    groove (cleft) – TS 15
    membrane – TS 15
    pouch – TS 15
        (parathyroid primordium from dorsal part – TS 22)
        (thymus primordium from ventral part – TS 21)
4th arch – TS 16
    artery – TS 17
        (arch of aorta from L arch artery – stage 21)
    cartilage
        components of laryngeal cartilages – TS 23
    cervical sinus (retro-hyoid depression) – TS 17
    pouch (future ultimobranchial body) – TS 15
6th arch – TS 17
    artery – TS 17
        (ductus arteriosus, from L arch – TS 21)
    cartilage
        cricoid cartilage – TS 23

canal
    proamniotic (ectoplacental duct) – TS 8
    yolk sac – TS 10
cavity, embryonic – *see* body cavity
cavity, extra-embryonic (*for embryonic cavities, see* body cavity)
    amniotic – TS 10 (late)
    amniotic fold
        within anterior fold – TS 10
        within posterior fold – TS 10
    blastocoelic – TS 4
    ectoplacental – TS 10
    exocoelomic (extra-embryonic coelomic) – TS 12
    proamniotic (canal) – TS 8
    yolk sac
        primary – TS 7
        secondary – TS 10
cleavage stages – TS 3
cystic (gall bladder) primordium – TS 15

decidual reaction – TS 6
diaphragm
    body wall derivatives
        arcuate ligaments (medial & lateral) – TS 26
        crus – TS 21
        pleuro-pericardial folds (later membrane) – TS 21
        pleuro-peritoneal folds (later membrane) – TS 20/21
            dome (left & right) – TS 21
    canals (channels)
        pleuro-pericardial – TS 16/17
        pleuro-peritoneal – TS 14
    septum transversum derivative
        central tendon – TS 21
duct
    common bile – TS 20
    cystic – TS 17
    hepatic – TS 16

inner ear
    cochlear – TS 21
    endolymphatic – TS 23
pancreatic – TS 20
renal/urinary system
    mesonephric – TS 16
    nephric – TS 14
    paramesonephric (Müllerian) – TS 21
thoracic – TS 26
thyroglossal – TS 18

ear
    cochlear (VIII) ganglion – TS 20 (late)
    inner
        otic
            capsule (pre-cartilage) – TS 20
            pit – TS 14
            placode – TS 12
            vesicle (otocyst) – TS 15 (late)
        otocyst – TS 15 (late)
            cochlea – TS 20
            cochlear canal (duct, scala media) – TS 21
            ductus reuniens – TS 19
            endolymphatic appendage (diverticulum) – TS 17
            endolymphatic sac (future duct) – TS 20
                endolymphatic duct – TS 23
            fenestra cochleae – TS 26
            fenestra vestibuli (oval window) – TS 26
            labyrinth – TS 21
                crus commune – TS 21
            saccule (utricle/saccule) – TS 19
            semicircular canals
                lateral – TS 21
                posterior – TS 19
                superior – TS 19
            spiral organ of Corti – TS 25
    middle
        ossicles (malleus & incus PCC (1st arch), stapes PCC (2nd arch)) – TS 21
        pharyngo-tympanic (Eustachian) tube – TS 20
        stapedius muscle – TS 23
        tensor tympani muscle – TS 23
        tubo-tympanic recess – TS 17
    outer
        auditory hillocks (tubercles) – TS 19
        external acoustic meatus (from first arch cleft) – TS 24
            entrance – TS 21
        external ear primordium – TS 20
        first branchial membrane (future tympanic membrane [forms after birth]) – TS 17
        pinna – TS 21
        pinna condensation – TS 20
        tympanic membrane primordium – TS 25
    temporal bone (petrous part) PCC – TS 20
    vestibular (VIII) ganglion – TS 20 (late)
    vestibulo-cochlear (VIII) ganglion – TS 20
ectoderm (embryonic)
    primitive (epiblast) – TS 7
    definitive – TS 8
        primitive streak – TS 9
        neural ectoderm – TS 10
egg cylinder – TS 7
endoderm
    definitive – TS 10
    parietal (distal, extra-embryonic) – TS 6
    visceral (proximal, embryonic) – TS 6
epiglottis – *see* "mouth region"
extra-embryonic tissue
    allantois – TS 10
    amnion – TS 10 (late)
    amniotic fold
        anterior – TS 10
        posterior – TS 10
    chorion – TS 11
    ectoderm – stage 8
    ectoplacental cone – TS 7
        cytotrophoblast – TS 8
        syncytiotrophoblast – TS 8

endoderm
    distal, parietal, primary – TS 6
    visceral layer – TS 7
exocoelomic cavity – TS 11
mesoderm – TS 9
proamniotic canal – TS 9
proamniotic cavity – TS 8
Reichert's membrane – TS 7
trophectoderm – TS 7
trophoblast giant cells
    primary – TS 8
    secondary – TS 10
umbilicus
    artery (common) – TS 13
    cord – TS 16
    hernia, physiological – TS 17
    vein
        left, right – TS 13
        common only – TS 15
    Wharton's jelly – TS 21
visceral yolk sac – TS 11
yolk sac "canal" – TS 10
yolk sac cavity
    primary – TS 7
    secondary – TS 10
yok-sac blood islands – TS 11
yolk sac endoderm – TS 12
yolk sac mesoderm – TS 12
yolk sac vasculature – TS 12
extremitätenleiste (lateral ridge from which limb buds develop) – TS 14 (late)
eye
    aqueous chamber – TS 22
    anterior chamber – TS25
    posterior chamber – TS25
    choroid (fetal) fissure – TS 18
    ciliary body – TS 25
    conjunctival sac – TS 22
        upper & lower recesses – TS 23
    cornea – TS 19
        Descemet's membrane – TS 24
        ectoderm – TS 17
        stroma – TS 20
    eyelid (upper & lower) – TS 21
        canthus, inner (nasal) – TS 21
    fat pad, orbital – TS 25
    gland, Harderian (within orbit) – TS 25
    hyaloid cavity – TS 21
        vitreous humour – TS 21
    intraretinal space – TS 16
    iris (boundary of pupil) – TS 25
    lens
        capsule – TS 22
        embryonic/fetal "nucleus" – TS 26
        fibres – TS 21
        pit – TS 16
        placode – TS 15
        posterior suture – TS 26
        vesicle – TS 17
    mesenchyme, perioptic (future sclera) – TS 15
    muscle
        extrinsic ocular
            PMM – TS 21
            oblique (inferior, superior) – TS 22
            rectus (lateral/temporal, medial/nasal, inferior, superior) – TS 22
    optic derivative of neural ectoderm
        chiasma – TS 20
        cup (secondary) – TS 15
        eminence – TS 14
        pit (evagination) – TS 12 (early)
        placode – TS 12
        stalk – TS 15
        sulcus – TS 12
        vesicle (primary) – TS 13/14
    retina
        future neural layer (inner layer of optic cup) – TS 15
        future pigment layer (outer layer of optic cup) – TS 15
        pigment layer – TS 18

INDEX OF FIRST OCCURRENCES OF TISSUES

intraretinal space – TS 16
neural layer – TS 18
  inner nuclear layer – TS 25
  intermediate anucleate layer (of Chievitz) – TS 25
  nerve fibre layer – TS 24
  optic disc – TS 22
  outer nuclear (neuroblastic) layer – TS 25
optic (II) nerve – TS 20
sclera – TS 24
tear ducts – TS 26
vascular system
  choroidal vessels – TS 21
  hyaloid vascular plexus – TS 19
    hyaloid artery (central artery of retina) – TS 26
    tunica vasculosa lentis – TS 22
    vasa hyaloidea propria – TS 22
  ophthalmic artery – TS 17

first polar body – TS 1
flexure, cervical – TS 18
fluid, cerebrospinal – TS 14

gall bladder
  (cystic) primordium – TS 15
  cystic duct – TS 15
  definitive – TS 19
ganglia
  autonomic
    ciliary – TS 22
    para-aortic "bodies" (paraganglia of Zuckerkandl) – TS 22
  cranial
    acoustic (VIII) (future vestibulo-cochlear) – TS 16
    cochlear (VIII) – TS 20 (late)
    cranial/spinal accessory (XI) – TS 17
    facial-acoustic (VII–VIII) pre-ganglion complex – TS 14
    facial-acoustic (VII–VIII) ganglion complex – TS 16
    glossopharyngeal (IX) – inferior (petrosal) – TS 18 (late)
    glossopharyngeal–vagal preganglion (IX–X) – TS 15 (early stage)
    glossopharyngeal (IX) preganglion – TS 15 (late)
    glossopharyngeal (IX) – superior – TS 18 (late)
    hypoglossal (XII) – TS 19
    trigeminal (V) (semilunar, Gasserian) preganglion – TS 14
    trigeminal (V) (semilunar, Gasserian) ganglion – TS 16
    vagus (X) preganglion – TS 15 (late)
    vagus (X) ganglion – TS 16
    vagus (X) – inferior (nodose) – TS 18
    vagus (X) – superior – TS 18
    vestibular (VIII) – TS 20 (late)
    vestibulo-cochlear (VIII) – TS 20
  dorsal (posterior) root ganglia – TS 15
  sympathetic
    coeliac (pre-aortic, abdominal sympathetic) – TS 25
    inferior cervical (cervico-thoracic, stellate) – TS 21
    middle cervical – TS 21
    paraganglia of Zuckerkandl (para-aortic "bodies") – TS 22
    superior cervical – TS 21
    thoracic – TS 17
germ cell, primordial – TS 12
gland
  adrenal
    primordium – TS 21
      cortex (mesodermal, coelomic epithelium) – TS 22
      medulla (ectodermal, neural crest) – TS 22
  bulbo-urethral – TS 26
  gall bladder
    (cystic) primordium – TS 15
    cystic duct – TS 15
    definitive – TS 19
  Harderian (within orbit) – TS 25
  mammary
    gland – TS 21
    ridge – TS 21
  pancreas – TS 18
    ventral bud – TS 19
    islets of Langerhans – TS 26
    pancreatic duct – TS 20
    primordium (dorsal bud) – TS 18
    body, tail, left & right lobes – TS 20
    uncinate process – TS 20
  parathyroid (from dorsal part of third branchial pouch) – TS 22
  parotid – TS 23 (late)
  pituitary
    adenohypophysis – TS 20
      pars anterior (anterior lobe, pars distalis) – TS 20
      pars intermedia – TS 20
      pars tuberalis – TS 20
    hypophyseal fossa – TS 22
    infundibular recess of third ventricle – TS 15
    neurohypophysis (pars nervosa, pars posterior) – TS 20
    infundibulum (pituitary stalk) – TS 18
    median eminence – TS 20
    Rathke's pouch – TS 14
  prostate – TS 26
  spleen – TS 24
    primordium – TS 22
  sublingual – TS 21 (late)
  submandibular – TS 21 (late)
  thymus
    thin fibrous capsule – TS 26
    medullary core – TS 26
    primordium (from ventral part of third branchial pouch) – TS 21
  thyroid
    colloid-filled follicles – TS 26
    foramen caecum – TS 18
    isthmus – TS 22
    lobes (left & right) – TS 22
    primordium (first appearance) – TS 13
    thyroglossal duct – TS 18
gonad – see reproductive system
gut
  biliary bud (hepatic diverticulum) – TS 14
  buccopharyngeal membrane – TS 12
  cloaca
    anal membrane – TS 19
      perforates – TS 23
    anal pit (proctodaeum) – TS 19
    cloacal membrane – TS 17
    urogenital membrane – TS 19
  foregut
    dorsal meso-oesophagus – TS 22
    duodenum (rostral half) – TS 18
    gastro-oesophageal junction – TS 22
    lower respiratory tract – see respiratory tract
    oesophageal region – TS 15
    oesophagus – TS 16
      dorsal mesentery – TS 22
    pharyngeal region – TS 14
      constrictor muscle (superior & inferior) – TS 24
      hyoid PCC – TS 19
      pharyngeal ligament – TS 26
      pharyngeal tubercle – TS 26
      pocket (diverticulum) – TS 11
    stomach – TS 17
      "cutaneous" non-glandular (proventricular) portion – TS 26
      fundic glandular mucous membrane – TS 21
      fundus – TS 21
      glandular, proventricular & pyloric regions – TS 20
      mesentery
        dorsal mesogastrium – TS 20
          gastro-splenic ligament – TS 22
          lieno-renal ligament – TS 22
          splenic primordium – TS 22
        ventral mesentery – TS 20
          falciform ligament – TS 22
          lesser omentum – TS 21
      omentum greater – TS 23
      pyloric antrum – TS 21
      pyloric sphincter – TS 26
  foregut–midgut junction – TS 13
    biliary system
      cystic (gall bladder) primordium – TS 15
      common bile duct – TS 18
      sphincter of Oddi – TS 23
    dorsal mesoduodenum – TS 21
    duodenum
      duodenal papilla (ampulla of Vater) – TS 23
    pancreas – see pancreas
    septum transversum – TS 12
      hepatic primordium – TS 13
      hepatic duct (left & right) – TS 16
  hindgut – TS 16
    anus
      anal canal (caudal part from anal pit) – TS 23
      anal canal (rostral part from hindgut) – TS 22
      anal pit (proctodaeum) – TS 19
      ano-rectal junction – TS 25
        pectinate line – TS 25
      sphincter (external & internal anal) – TS 22
    cloaca – TS 16
      membrane – TS 17
      region – TS 17
    diverticulum
      hindgut pocket – TS 12
      preanal component – TS 17
      postanal component – TS 17
    dorsal mesentery – TS 17
    large intestine – TS 26
    pectinate line – TS 25
    perineal "body" – TS 21
    rectum – TS 21
    uro-rectal septum – TS 18
  midgut
    dorsal mesentery ("the" mesentery) – TS 17
    duodenum (caudal half) – TS 18
    jejunum – TS 23
    loop – TS 17
    physiological umbilical hernia
      forms – TS 17
      disappears – TS 24
    region – TS 12
  primitive gut – TS 11
  proctodaeum (anal pit) – TS 19

haemolymphoid system
  blood – TS 13
  lymph node
    jugular lymph sac – TS 22
    Peyer's patches – TS 24
    thoracic duct – TS 26
  spleen – TS 24
    hilum – TS 26
    medullary region – TS 26
    primordium – TS 22
  thymus
    cortex – TS 26
    medulla – TS 26
    primordium – TS 21
hair – see skin
heart
  aortic arch (from left fourth arch artery) – TS 21
  aortic sac (distal outflow tract) – TS 12
  aortic sac, continuous with truncus arteriosus – TS 13
  aortic sinus – TS 23
  ascending thoracic component of aorta (from outflow tract) – TS 19

atrial chamber
  common – TS 12
  left & right components – TS 16 (late)
  sinus venarum – TS 17
    crista terminalis – TS 17
atrio-ventricular canal – TS 14
atrio-ventricular cushion tissue – TS 20
atrium (left & right) – TS 19
  auricular part – TS 16
  interatrial septum
    foramen (ostium) primum – TS 17
    foramen (ostium) secundum – TS 21
    foramen ovale – TS 22
    septum primum – TS 16 (late)
    septum secundum (crista dividens) – TS 22
  sinus venarum – TS 17
    crista terminalis – TS 17
blood – TS 13
bulbar cushion – TS 15
bulbar ridge – TS 19
bulbo-ventricular groove (sulcus) (becomes interventricular groove) – TS 12
bulbus cordis (inferior: future R ventricle; superior: prox. part of outflow tract) – TS 12
cardiac jelly – TS 12
cardiogenic plate – TS 11 (late)
  myocardium – TS 12
coronary artery – TS 25
dorsal mesocardium (dorsal mesentery of heart) – TS 12
  transverse pericardial sinus – TS 12
ductus arteriosus (from L sixth arch artery) – TS 21
endocardial cells – TS 12
endocardial cushion tissue – TS 16
endocardial tissue – TS 13
epicardium (visceral pericardium) – TS 18 (late)
interventricular groove – TS 20
myocardium – TS 12
ostium (foramen)
  primum – TS 17
  secundum – TS 21
outflow tract – TS 12
  aortic component – TS 19
  aortico-pulmonary spiral septum
    initiation – TS 16
    complete – TS 18/19
  pulmonary component – TS 19
pericardium
  fibrous – TS 24
  parietal – TS 19
  visceral (epicardium) – TS 18 (late)
primitive heart tube – TS 12
primitive right ventricle (previously bulbus cordis) – TS 20
primitive ventricle (future left ventricle) – TS 13
pulmonary trunk (proximal part from pulmonary component of outflow tract) – TS 19
pulmonary trunk (distal part from prox. part of L sixth arch artery) – TS 19
sinus venosus (left & right horns) – TS 12
thoracic aorta, ascending – TS 20
trabeculae carneae – TS 23
truncus arteriosus – TS 13
valve
  aortic – TS 21
  mitral (bicuspid) – TS 21
  pulmonary – TS 21
  tricuspid – TS 21
    leaflets (aortic, mitral, pulmonary, tricuspid) – TS 23
  venous valve – TS 17
    leaflets (entrance of inferior vena cava) – TS 21
ventricle, left, right – TS 19
  interventricular septum (precursor, in bulbo-ventricular region) – TS 19
    muscular part – TS 20
ventricular region of primitive heart – TS 12

inguinal canal – *see* reproductive system
inner cell mass – TS 4

joints – *see* appropriate location under "skeleton" and/or "limbs"

kidney *see under* renal/urinary system

limb
  forelimb
    apical ectodermal ridge – TS 16
    claw primordium – TS 24
    digit primordia – TS 21 (early)
    digital interzones – TS 20
    elbow joint primordium – TS 21
    forelimb – TS 19
    forelimb bud – TS 15
    forelimb ridge – TS 14 (late)
    handplate – TS 19
      palmar (digital) "pads" – TS 23 (early stage)
    humeral PCC – TS 20
    metacarpo-phalangeal joint – TS 26
    muscle groups, various – *see* muscle, forelimb
    radio-ulnar PCC – TS 20
    wrist joint primordium – TS 22
    wrist region – TS 21
  hindlimb
    ankle region – TS 21
    ankle joint – TS 23
    ankle joint primordium – TS 22
    apical ectodermal ridge – TS 17
    calcaneum PCC – TS 21
    claw primordium – TS 24
    digit primordia – TS 21 (late)
    digital interzones – TS 20
    femoral PCC – TS 20
    fibula PCC – TS 22
    footplate – TS 19
      plantar (digital) "pads" – TS 23 (late)
    muscle groups, various – *see* muscle, hindlimb
    hindlimb ridge – TS 15
    hindlimb bud – TS 16
    hip primordium – TS 22
    talus PCC – TS 21
    tibial-fibular PCC – TS 21
    tibia PCC – TS 22
liver rudiment TS 14
  "bare" area – TS 21
  biliary bud (hepatic diverticulum) – TS 14
  hepatic component of septum transversum – TS 13
  hepatic duct – TS 16
  hepatic primordium – TS 15
  hepatic sinusoids – TS 19
  lobes, left & right (including caudate & quadrate) – TS 22
  parenchyma – TS 16
  septum transversum – TS 12
lung – *see under* respiratory system

mesoderm (embryonic) – TS 9
  axial
    notochordal plate – TS 11
    notochord – TS 12
    nucleus pulposus – TS 21
  intermediate plate – TS 12
    nephrogenic cord – TS 13
  lateral plate mesoderm – TS 12
    somatopleure (lining of intra-embryonic coelom) – TS 12
      mesothelium – TS 14
    splanchnopleure (lining of intra-embryonic coelom) – TS 12
      mesothelium – TS 14
  paraxial – approximate figures for somites, (allowing for partition into dermomyotome & sclerotome after ~ 20 h)
    somites 1–4 – TS 12

somites 1–10 – TS 13
somites 1–17 – TS 14
somites 5–25 – TS 15
somites 5–32 – TS 16
somites 5–38 – TS 17
somites 11–42 – TS 18
somites 18–46 – TS 19
somites 25–50 – TS 20
morula
  early to fully compacted – TS 3
mouth region
  cartilage (Meckel's, Reichert's [hyoid]) – TS 19
  epiglottis – TS 22
    swelling (primordium) – TS 19
  gum, primitive – TS 24
  lips
    philtrum – TS 21
    upper and lower – TS 21
  mandible – TS 21
    canal, mandibular (inferior alveolar) – TS 25
  mandibular component of first arch – TS 14
  maxilla
    maxillary component of first arch – TS 12 (late)
    palatal shelf – TS 19
    primordium – TS 19
    pre-maxilla – TS 20
  oral cavity – TS 20
  oronasal cavity – TS 18
  palatal shelf – TS 19
  palate
    canal
      incisive – TS 25
      naso-palatine – TS 24
    primary (intermaxillary segment) – TS 20
    secondary (shelf of maxilla) – TS 18
    secondary (definitive) – TS 24
    soft – TS 24
  parotid gland – TS 23 (late)
  philtrum – TS 21
  primary palate – TS 20
  stomatodaeum – TS 12
    buccopharyngeal membrane – TS 12
  sublingual caruncle – TS 24
  sublingual gland – TS 21 (late)
  submandibular gland – TS 21 (late)
  tongue
    circumvallate papilla – TS 22
    filiform papillae – TS 23
    foramen caecum – TS 18
    fungiform papillae – TS 22
    frenulum – TS 23
    intermolar eminence – TS 22
    lingual swelling (lateral) – TS 18
    median fibrous septum – TS 22
    muscle
      extrinsic PMM – TS 21
        genioglossus – TS 22
        hyoglossus – TS 22
        palatoglossus – TS 22
        styloglossus – TS 22
        thyroglossus – TS 22
      intrinsic PMM – TS 21
        transverse (longtitudinal) & vertical component – TS 22
      occipital myotome – TS 19
      (gives rise to extrinsic & intrinsic tongue muscles [not palatoglossus])
      lingual PMM – TS 21
    primordium
      hypobranchial eminence (3rd & 4th arch components) – TS 19
      lingual swellings (mandibular component of 1st arch)
        lateral & median (tuberculum impar) – TS 18
    sulcus terminalis – TS 24
  tooth primordium
    bud stage – TS 21
    bell stage – TS 21
    cap stage – TS 21

INDEX OF FIRST OCCURRENCES OF TISSUES

dental laminae – TS 21
dental papilla (neural crest) – TS 22
incisors & molars – TS 21
enamel organ – TS 21
sulcus, alveolar – TS 22
muscle
  axial – see *vertebral axis*
  bladder, detrusor muscle – TS 26
  cranial
    constrictor (inferior & superior) – TS 24
    digastric – TS 25
    extrinsic ocular – TS 22
    extrinsic ocular PMM – TS 21
    extrinsic of tongue – *see* tongue
    intrinsic of tongue – *see* tongue
    masseter – TS 23
    mylohyoid – TS 22
    pterygoid (medial, lateral) – TS 24
    sterno-mastoid – TS 22
    temporalis – TS 22
  ear
    stapedius muscle – TS 23
    tensor tympani muscle – TS 23
  girdle
    pectoral
      deltoid – TS 22
      deltoid PMM – TS 21
      infraspinatus – TS 24
      latissimus dorsi – TS 22
      subscapularis PMM – TS 21
      subscapularis – TS 22
      supraspinatus – TS 22
      transversus thoracis – TS 26
      trapezius – TS 22
    pelvic – TS 22
      gluteus maximus (superficialis) – TS 23
      gluteus medius – TS 25
      iliacus – TS 25
      ilio-psoas – TS 25
      levator ani – TS 25
      psoas major – TS 22
      psoas minor – TS 22
  limb
    forelimb
      biceps (brachii) – TS 23
      teres major – TS 23
      triceps (brachii) – TS 23
    hindlimb
      adductor group – TS 25
      flexor digitorum profundus (sole of foot) – TS 25
      hamstring (flexor) group – TS 23
      quadriceps (extensor) group – TS 23
  occipital myotome – TS 19
    (gives rise to extrinsic & intrinsic tongue muscles, except palatoglossus)
    lingual PMM – TS 21
  pharyngeal region
    constrictor (superior & inferior) – TS 24
  pretracheal
    cricothyroid – TS 26
    sternohyoid – TS 26
    sternothyroid – TS 26
    thyrohyoid – TS 26
  skin (panniculus carnosus)
    cutaneous muscle of trunk – TS 23
    platysma – TS 22
  smooth
    respiratory system
      trachealis – TS 25
  tongue
    extrinsic PMM – TS 21
      genioglossus – TS 22
      hyoglossus – TS 22
      palatoglossus – TS 22
      styloglossus – TS 22
      thyroglossus – TS 22
    intrinsic PMM – TS 21
      transverse (longitudinal) & vertical component – TS 22
    occipital myotome – TS 19
      (gives rise to extrinsic & intrinsic tongue muscles, except palatoglossus)
    lingual PMM – TS 21
vertebral axis
  body wall – TS 22
  gluteus maximus (superficialis) – TS 23
  gluteus medius – TS 25
  ilio-psoas – TS 23
  intercostal (exterior & interior layers) – TS 23
  pectoralis major (pectoralis profundus) – TS 22
  psoas major – TS 22
  psoas minor – TS 22
  quadratus lumborum – TS 23
  serratus anterior – TS 23
  thoracic/abdominal wall PMM – TS 21
  thoracic/abdominal wall
    erector spinae – TS 22
    external oblique – TS 22
    internal oblique – TS 22
    rectus abdominis – TS 22
    transversus abdominis – TS 22
    transversus thoracis – TS 26
    trapezius – TS 22

nerve
  autonomic
    hypogastric plexus – TS 23
      superior & inferior – TS 26
  parasympathetic
    vagus (X)
      nerve – TS 20
      recurrent laryngeal branch – TS 21
      trunk (left becomes anterior) – TS 21
      trunk (right becomes posterior) – TS 21
  sympathetic (chain) trunk – TS 17
  cranial
    abducent (VI) – TS 22
    cochlear (VIII) – TS 21
    cranial/spinal accessory (XI) nerve – TS 18
    facial (VII) – TS 21
      chorda tympani branch – TS 22
    glossopharyngeal (IX) – TS 19
    hypoglossal (XII) – TS 19
    left & right vagal (X) trunks – TS 21
    oculomotor (III) nerve – TS 20
    olfactory (I) – TS 21
    optic (II) – TS 20
    trigeminal (V)
      lingual branch – TS 25
      mandibular division – TS 21
      maxillary division – TS 21
      nerve – TS 20
      ophthalmic division – TS 21
    trochlear (IV) – TS 22
    vagal (X) trunks (L & R) – TS 21
    vestibular – TS 21
  cranial/spinal
    accessory (XI) – TS 18
  nerve trunk
    axillary (circumflex) – TS 24
    intercostal – TS 21
    phrenic – TS 19
    sciatic – TS 21
    segmental spinal – TS 17
  plexus
    autonomic, hypogastric – TS 22
    brachial – TS 17
      median – TS 19
      radial – TS 19
      ulnar – TS 19
    lumbar – TS 22
    lumbo-sacral – TS 20
neural crest – TS 12
  dorsal root ganglia – TS 15
  facio-acoustic (VII–VIII) – TS 13 (early)
  glossopharyngeal (IX) – TS 15
  glossopharyngeal–vagal (IX–X) – TS 15
  periotic mesenchyme – TS 15
  trigeminal (V) – TS 12
  vagal (X) – TS 15

neural tube
  (future spinal cord) – TS 12
  caudal neuropore – TS 13
  conus medullaris – TS 22
  filum terminale – TS 22
  floorplate – TS 19
  grey horns (dorsal, ventral & lateral) – TS 23
  layer
    ependymal (ventricular) – TS 16
    mantle – TS 17
      alar & basal columns – TS 18
    marginal – TS 17
  neural ectoderm (neuroepithelium) – TS 10
  neural lumen (future spinal canal) – TS 12
  neural luminal occlusion – TS 13
  neural fold – TS 12
  neural plate – TS 11
  roofplate – TS 19
  rostral neuropore – TS 13
  segmental spinal nerves – TS 17
  spinal cord – TS 20
    central canal (previously "neural lumen") – TS 20
  sulcus limitans – TS 20
nose
  choana – TS 22
  meatus – TS 23
    middle – TS 26
  naris
    anterior naris – TS 22
    external naris – TS 21
    primary choana (primitive posterior naris) – TS 21
  nasal bone – TS 25
  nasal capsule (cartilaginous) – TS 21
  nasal (olfactory) cavity – TS 21
  nasal cavity (upper respiratory tract) – TS 20
  naso-lacrimal duct – TS 21
  naso-lacrimal groove – TS 15
  olfactory epithelium – TS 16
  olfactory pit – TS 16
  olfactory placodes – TS 15
  oronasal cavity – TS 18
  primitive nasal septum – TS 20
    nasal septum – TS 21
  processes
    fronto-nasal – TS 15
    lateral-nasal – TS 16
    medial-nasal – TS 16
  turbinate bones (conchae) – TS 23
  vomeronasal (Jacobson's) organ – TS 20
notochord – TS 12
  nucleus pulposus – TS 21
  notochordal plate – TS 11

one-cell stage – TS 1

palate – *see* "mouth region"
pancreas
  ventral bud – TS 19
  islets of Langerhans – TS 26
  pancreatic duct – TS 20
  primordium (dorsal bud) – TS 18
  body, tail, left & right lobes – TS 20
  uncinate process – TS 20
parathyroid gland (from dorsal part of third branchial pouch) – TS 22
parotid gland – TS 23 (late)
pharyngeal arches – *see* branchial arches
pituitary
  adenohypophysis – TS 20
    pars anterior (anterior lobe, pars distalis) – TS 20
    pars intermedia – TS 20
    pars tuberalis – TS 20
  hypophyseal fossa – TS 22
  infundibular recess of third ventricle – TS 15
  neurohypophysis (pars nervosa, pars posterior) – TS 20
    infundibulum (pituitary stalk) – TS 18
    median eminence – TS 20

Rathke's pouch – TS 14
placenta
  chorionic plate – TS 18
  labyrinthine part – TS 18
  spongy (or basal) region
      (spongiotrophoblast) – TS 18
  umbilical cord
    artery (common) – TS 13
    cord – TS 16
    hernia, physiological – TS 17
    vein
      left, right – TS 13
      common only – TS 15
    Wharton's jelly – TS 21
polar body
  first – TS 1
  second – TS 1
pouch, Rathke's – *see* pituitary
primitive groove – TS 9
primitive streak – TS 9
  Henson's node – TS 10
primordium
  adrenal – TS 21
  cerebellum – TS 19
  gall bladder (cystic) primordium – TS 15
    cystic duct – TS 16
  liver
    hepatic component of septum transversum –
      TS 15
    hepatic duct – TS 16
    lobes, right (+ caudate & quadrate) – TS 22
  mammary gland – TS 21
  pancreas – TS 18
  parathyroid (from dorsal part of 3rd branchial
    pouch) – TS 22
  parotid – TS 23 (late)
  reproductive system, gonad (indifferent stage) –
    TS 19
  spleen – TS 22
  sublingual gland – TS 21
  submandibular gland – TS 21
  tactile (sinus) hair follicle – TS 22
  thymic rudiment (left & right, ventral part of 3rd
    pouch) – TS 23
  thyroid – TS 13
  tongue (see mouth) – TS 18
  tooth (see mouth) – TS 21
proctodaeum (anal pit) – TS 19

renal/urinary system
  bladder – TS 22
    detrusor muscle – TS 26
    fundus region – TS 22
      urachus – TS 21
    trigone region – TS 24
  gonadal ridges – TS 16
  mesonephric (Wolffian) duct – TS 16
    becomes ductus deferens (male) – TS 23
  mesonephros – TS 15
    degenerating – TS 23
    tubule – TS 16
    vesicles – TS 15
  metanephros
    excretory components
      blastema (future definitive kidney) – TS 18
      blastemal cells (induced) – TS 19
      Bowman's (glomerular) capsule – TS 24
      capsule – TS 23
      convoluted tubules (proximal & distal) –
        TS 24
      cortex – TS 22
      cortical layer – TS 20
      early nephrons – TS 21
      glomerulus – TS 22
      loop of Henle – TS 26
      medulla – TS 22
      medullary zone – TS 20
      stromal cells – TS 22
    ureteric bud derivatives
      calyces (major & minor) – TS 24
      collecting tubules – TS 24

      pelvis – TS 21
      primitive collecting ducts – TS 20
      ureter – TS 20
      ureteric bud – TS 18
  nephric (Wolffian) duct – TS 14
  nephrogenic cord – TS 14
  paramesonephric (Müllerian) duct – TS 21
    female derivatives
      oviduct – TS 24
      uterine horn – TS 24
      vagina (upper one third) – TS 24
    male derivatives
      appendix testis – TS 24
      prostatic utricle – TS 26
  pronephros – TS 14
  urachus – TS 21
  ureter – TS 20
  urethra – *see* reproductive system
  urogenital sinus – TS 18
    pelvic part – TS 22
    phallic part – TS 21
    vagina (lower two thirds) – TS 24
    vesical part (future bladder) – TS 21
  urogenital
    mesentery – TS 19
      mesorchium (male) – TS 21
      mesovarium (female) – TS 21
    ridge – TS 14
      gonadal component – TS 16
      mesonephric component – TS 16
  urorectal septum
    incomplete – TS 17
    complete – TS 18
reproductive system
  indifferent stage (up to TS 20)
    genital tubercle – TS 19
    "germinal" epithelium (peritoneal mesothelial
      covering of gonad) – TS 19
    gonadal component of urogenital ridge – TS
      16
    gonad primordium – TS 18
      germ cells apparent – TS 18/19
    primordial germ cells – TS 10
    urogenital mesentery – TS 19
  female
    clitoris – TS 23
      crus of clitoris – TS 25
      glans clitoridis – TS 25
    cystic vesicular appendage
      (? mesonephric/paramesonephric duct
      origin) – TS 24
    labial swelling – TS 22
    Müllerian tubercle – TS 22
    ovary – TS 21
      capsule – TS 24
      germinal cells (oogonia) – TS 24
      primary follicles, follicle cells – TS 26
      primary oocytes – TS 24
      mesovarium – TS 21
      ovigerous cords – TS 24
      rete ovarii – TS 21
      suspensory ligament – TS 24
    ovarian capsule – TS 24
    oviduct – TS 24
    paramesonephric (Müllerian) duct – TS 21
      cervix uteri – TS 25
      oviduct – TS 24
      uterine horn – TS 24
      vagina (upper third) – TS 24
    urethra – TS 24
    urethral groove – TS 23
    uterine horn – TS 24
      cervix uteri (paramesonephric duct
        origin) – 25
      mesometrium – TS 25
    vagina – TS 24
      lower two thirds from urogenital sinus
      upper one third from paramesonephric
        (Müllerian) duct
  male
    gland

      bulbo-urethral – TS 26
      prostate – TS 26
    inguinal canal – TS 25
    inguinal ring (internal (deep) & external
      (superficial)) – TS 25
    mesonephric (Wolffian) duct derivatives
      appendix epididymis – TS 24
      ductus deferens – TS 23
      efferent ducts – TS 24
      ejaculatory duct – TS 24
      rete testis – TS 22
      seminal vesicle – TS 25
      vasa efferentia (from efferent ducts) – TS
        26
    mesonephros derivative
      epididymis (caput (head); cauda (tail);
        corpus (body)) – TS 26
    penis – TS 23
      crus penis – TS 26
      glans penis – TS 24
      prepuce – TS 24
    prostatic utricle (from paramesonephric
      duct) – TS 26
    scrotum – TS 26
    seminal vesicle – TS 25
    testis – TS 21
      appendix epididymis (mesonephric duct) –
        TS 24
      appendix testis (paramesonephric duct) –
        TS 24
      cortical region ("germinal" epithelium) –
        TS 23
      gubernaculum – TS 23
      interstitial cells – TS 25
        primitive Leydig cells – TS 26
      mediastinum testis – TS 22
      medullary region – TS 23
      mesorchium – TS 21
      primitive seminiferous tubules – TS 22
      primitive Sertoli cells – TS 25
      rete testis (from mesonephric duct) – TS
        22
      Sertoli (sustentacular) cells – TS 26
      spermatogonia – TS 25
      suspensory ligament – TS 24
      testicular cords (future seminiferous
        tubules) – TS 21
      tunica albuginea (fibrous capsule of
        testis) – TS 26
      tunica vaginalis testis (coelomic epithelial
        covering of testis) – TS 25
    urethra
      bulbar part – TS 26
      bulbo-urethral glands – TS 26
      penile part – TS 26
      prostatic region – TS 24
      urethral groove – TS 23
respiratory system
  alveolar duct – TS 25
  alveolus – TS 26
  arytenoid cartilage (4th branchial arch
    cartilage) – TS 23
  arytenoid swelling PCC – TS 19
  cricoid cartilage (6th branchial arch cartilage) –
    TS 23
  diaphragm – *see* body cavity or diaphragm
  hyoid bone PCC – TS 19
    body – TS 23
    greater horn – TS 23
    lesser horn – TS 24
  laryngeal aditus – TS 21
  laryngeal cartilages – TS 21
    thyroid cartilage (4th branchial arch
      cartilage) – TS 23
  laryngo-tracheal groove – TS 15
  larynx – TS 22
    mucous membrane – TS 22
    vocal folds – TS 25
  lung
    alveolar duct – TS 25
    alveolus – TS 26

INDEX OF FIRST OCCURRENCES OF TISSUES

bronchiole – TS 24
bronchus
  lobar – TS 19
  main – TS 16
  segmental – TS 21
  terminal – TS 22
bud – TS 15
hilus (root) – TS 21
lobe
  left – TS 21
  right (accessory, caudal, cranial, middle) – TS 21
pleura
  parietal – TS 22 (late)
  visceral – TS 23
Meckel's cartilage – TS 19
upper respiratory tract
  oronasopharynx – TS 14/15
  pharyngeal region – TS 14
  pharynx – TS 14
  nasopharynx (nasal cavity) – TS 20
pleural cavity (*see* "body cavity")
smooth muscle
  trachealis – TS 25
thyroid cartilage (4th & branchial arch cartilage) – TS 23
trachea – TS 16
  "rings" – TS 23
  bifurcation (carina tracheae) – TS 17
tracheal diverticulum – TS 15

second polar body – TS 1
septum transversum – TS 12
  hepatic component – TS 13
  non-hepatic component (part of future diaphragm) – TS 13
sinus, *see* brain
skeleton
  cranium
    1st arch
      Meckel's cartilage PCC – TS 19
        incus PCC – TS 21
        malleus PCC – TS 21
    2nd arch
      temporal bone
        Reichert's cartilage PCC – TS 19
          stapes PCC – TS 21
          styloid process – TS 23
    chondrocranium
      basioccipital PCC – TS 21
      basisphenoid PCC – TS 21
      carotid canal – TS 23
      craniopharyngeal canal – TS 25
      exoccipital PCC – TS 21
        squamous part (mesenchyme) – TS 22
      foramen ovale – TS 23
      foramen rotundum – TS 23
      hypoglossal (anterior condylar) canal – TS 24
      jugular foramen – TS 23
      occipital PCC – TS 22
        squamous part (mesenchyme) – TS 22
      pituitary fossa (sella turcica) – TS 25
      "sphenoidal" canal – TS 24
      temporal bone
        petrous part PCC – TS 20
        squamous part PCC – TS 21
        temporo-mandibular joint – TS 25
        tympanic ring (os tympanicum) – TS 25
        zygomatic process – TS 24
    vault of skull
      frontal bone – TS 21
      inter-parietal bone – TS 21
      parietal bone (mes) – TS 21
      supraoccipital bone (tectum posterius) – TS 22
      sutures, fontanelles (anterior, posterior) – TS 25
    viscerocranium
      ethmoid bone primordium – TS 21
        cribriform plate – TS 25

facial bones primordia – TS 21
  intermaxillary (premaxilla) – TS 23
  maxilla – TS 22
    palatal shelf – TS 22
  nasal bone – TS 25
  optic foramen (canal) – TS 23
  orbital fissure, superior – TS 23
  orbito-sphenoid – TS 23
  sphenoid bone PCC – TS 21
  turbinate – TS 23
mandible – TS 21
girdle
  pectoral
    PCC – TS 21
    clavicle – TS 23
    clavicle PCC – TS 21
    costo-sternal joint – TS 25
    scapula PCC – TS 21
    shoulder joint primordium – TS 21
    sternum
      manubrium sterni – TS 23
      sternebral bone PCC – TS 22
      sterno-clavicular joint – TS 25
      xiphisternum (xiphoid process) – TS 23
  pelvic
    PCC – TS 21
    acetabular region – TS 22
    hip joint primordium – TS 21
    hip joint
      synovial cavity – TS 24
    iliac PCC – TS 21
    ischial PCC – TS 22
    pelvis (innominate bone) – TS 23
    pubic PCC – TS 22
    sacro-iliac joint – TS 26
limb
  forelimb
    bones
      carpal PCC – TS 21
      humerus PCC – TS 20
      metacarpal PCC – TS 21
      phalangeal PCC – TS 22
      radio-ulnar PCC – TS 20
      radius PCC – TS 21
      ulna PCC – TS 21
    joints
      annular ligament (proximal radio-ulnar joint) – TS 26
      elbow joint primordium – TS 21
      gleno-humeral joint (shoulder) primordium – TS 21
      gleno-humeral joint – TS 25
      metacarpo-phalangeal joint – TS 26
      radio-carpal (wrist) joint – TS 25
      radio-humeral (proximal, distal) joints – TS 25
      wrist joint primordium – TS 22
  hindlimb
    bones
      femur PCC – TS 21
      fibula – TS 22
      metatarsal bone PCC – TS 21
      phalangeal bone PCC – TS 22
      tarsal bone PCC – TS 21
        calcaneus (calcaneum) – TS 22
        talus – TS 22
      tibia – TS 22
      tibial-fibular PCC – TS 21
    joints
      ankle joint – TS 22
      knee joint
        fabella PCC – TS 23
        ligamentum patellae – TS 23
        patella PCC – TS 23
        primordium – TS 21
vertebral axis
  atlas – TS 21
  axis – TS 21
  dens (odontoid process of C2) PCC – TS 21

cervical vertebrae PCC – TS 21
  foramen transversarium – TS 21
hyoid bone PCC – TS 19
  body – TS 23
  horn (greater) – TS 23
  horn (lesser) – TS 24
intervertebral disc primordium (nucleus pulposus) – TS 21
ligamentum nuchae – TS 25
neural arch PCC – TS 21
rib primordium, PCC – TS 19
  chondro-sternal joint primordium – TS 23
  costal cartilage – TS 23
  costal margin – TS 23
  thoracic vertebral bodies PCC – TS 22
sclerotome – TS 18
sternum – *see* skeleton, pectoral girdle
vertebrae (sclerotome-derived) – TS 18–21
  centrum
  costal process
  hypochordal arch (bow)
  intervertebral disc (surrounds nucleus pulposus)
  nucleus pulposus (of notochordal origin)
vertebrae (regions) PCC – TS 19–21
  (sclerotome condensations)
  (cervical, coccygeal (tail), lumbar, sacral, thoracic)
  costo-vertebral joint primordium – TS 24
  cruciate ligament (C1) – TS 24
skin
  brown fat – TS 25
  dermis – TS 21
  epicranial aponeurosis (galea aponeurotica) – TS 25
  eyelid (upper & lower) – TS 21
  hair
    bulb – TS 24
    follicle (ordinary) – TS 24
    root sheath – TS 24
    tactile (sinus) hair follicle primordium – TS 22
  muscle (panniculus carnosus)
    cutaneous muscle of trunk – TS 23
    platysma – TS 22
  vibrissae elevations – TS 20
  vibrissae precursors – TS 21
  vibrissae
    definitive – TS 24
somite – approximate figures
  (allowing for partition into dermomyotome & sclerotome after ~ 20 h and disappearance after ~ 60 h)
  somites 1–4 – TS 12
  somites 1–10 – TS 13
  somites 1–17 – TS 14
  somites 5–25 – TS 15
  somites 5–32 – TS 16
  somites 5–38 – TS 17
  somites 11–42 – TS 18
  somites 18–46 – TS 19
  somites 25–50 – TS 20
  condensations (for vertebrae) – TS 19–21
sphincter
  anal (external & internal) – TS 22
  pyloric – TS 26
spinal cord – *see* neural tube
spleen
  hilum – TS 26
  medullary region – TS 26
  primordium – TS 22
sulcus, pre-otic – TS 11
stomatodaeum – TS 12

tail – TS 13
thymus
  medullary core – TS 26
  primordium (from ventral part of third branchial pouch) – TS 21
  thin fibrous capsule – TS 26

thyroid
 colloid-filled follicles – TS 26
 foramen caecum – TS 18
 isthmus – TS 22
 lobes (left & right) – TS 22
 primordium (first appearance) – TS 13
 thyroglossal duct – TS 18
tongue – *see* mouth region
tooth – *see* mouth region
trophectoderm
 mural – TS 4
 polar – TS 4
trophoblast giant cells
 primary – TS 8
 secondary – TS 10

umbilicus
 artery (common) – TS 13
 cord – TS 16
 hernia, physiological – TS 17
 vein
  common only – TS 15
  left, right – TS 13
 Wharton's jelly – TS 21
urinary system – *see* renal/urinary system

vascular system
 artery
  aorta
   aortic arch (from L fourth arch artery) – TS 21
   ascending thoracic (from outflow tract) – TS 19
   descending abdominal (from L dorsal aorta) – TS 21
   dorsal midline – TS 13
   dorsal (paired) – TS 12
   post-ductal part of descending – TS 21
  axillary – TS 24
  basilar – TS 19
  bronchial – TS 22
  carotid
   artery
    common carotid – TS 21
    external carotid – TS 21
    internal carotid – TS 13
   body – TS 25
   sheath – TS 26
   sinus – TS 22
  cerebral
   anterior – stage 21
   middle – TS 22
   posterior – TS 21
  coeliac trunk – TS 25
  common iliac – TS 19
  communicating (anterior/posterior) – TS 26
  coronary – TS 25
  ductus arteriosus (from L sixth arch artery) – TS 21
  ductus caroticus – TS 17
  epigastric (superior, inferior) – TS 24
  external iliac – TS 20
  facial – TS 25
  hepatic – TS 21
  hyaloid – TS 26
  hyoid – TS 19
  iliac – TS 19
   external TS 21
   internal TS 21
  innominate (brachiocephalic) – TS 20
  intercostal – TS 24
  internal iliac – TS 23
  internal thoracic (mammary) – TS 23
  intersegmental – TS 16
  maxillary – TS 19
  median sacral – TS 25
  mesenteric – TS 19
   superior, inferior – TS 23
  musculo-phrenic – TS 25
  ophthalmic – TS 17
  palatine, greater – TS 24
  posterior communicating – TS 23
  principal (axial) to forelimb (7th cervical intersegmental) – TS 16
  principal (axial) to forelimb (subclavian) – TS 20
  principal (axial) to hindlimb (5th lumbar intersegmental) – TS 16
  principal (axial) to hindlimb (external iliac) – TS 20
  pulmonary – TS 17
  pulmonary trunk (proximal part from pulmonary component of outflow tract) – TS 19
  pulmonary trunk (distal part from prox. part of L sixth arch artery) – TS 19
  renal – TS 21
  spinal (anterior & posterior) – TS 22
  splenic – TS 25
  stapedial – TS 19
  subclavian – TS 20
  superior cerebellar – TS 26
  superior epigastric – TS 25
  superior vesical – TS 23
  testicular – TS 23
  thyroid (superior, inferior) – TS 23
  umbilical – TS 13
  vertebral – TS 19
  vitelline – TS 12
 blood – TS 13
 branchial arch artery
  1st – TS 12 (maxillary – TS 22)
  2nd – TS 13 (hyoid, stapedial – TS 22)
  3rd – TS 15 (common carotid – TS 21)
  4th – TS 17 (arch of aorta – TS 20/21)
  6th – TS 17 (distal part of pulmonary trunk, ductus arteriosus – TS 21)
 lymphatic system
  jugular lymph sac – TS 22
  thoracic duct – TS 26
 vein
  accessory hemiazygos – TS 25
  axillary – TS 24
  azygos – TS 22
  cardinal
   anterior – TS 12
   anterior (internal jugular) – TS 20
   common (Duct of Cuvier) – TS 13
   posterior – TS 13
    becomes inferior vena cava – TS 19
   subcardinal/supracardinal – TS 16
  common iliac – TS 21
  deep dorsal of clitoris/penis – TS 25
  ductus venosus – TS 19
  epigastric (superior, inferior) – TS 24
  facial – TS 25
  femoral – TS 21
  great cerebral vein of Galen – TS 26
  hemiazygos – TS 22
  hepatic – TS 25
  hepato-cardiac – TS 19
  inter-subcardinal venous anastomosis – TS 16
  intercostal – TS 24
  internal thoracic (mammary) – TS 23
  jugular (external) – TS 21
  jugular (internal) – TS 20
  lingual – TS 26
  marginal (of limb) – TS 17
  mesenteric – TS 19
   superior, inferior – TS 23
  musculo-phrenic – TS 25
  ophthalmic – TS 26
  palatine, greater – TS 26
  portal – TS 19
  principal (axial) from limb – TS 17
  primary head vein – TS 12
  pulmonary – TS 23
  renal – TS 25
  spinal (anterior, posterior) – TS 24
  subcardinal/supracardinal – TS 16
  subclavian – TS 23
  superficial temporal – TS 26
  superior epigastric – TS 25
  testicular (pampiniform plexus) – TS 23
  umbilical – TS 12
  vena cava – TS 19
   inferior
    post-hepatic part – TS 19
    pre-hepatic part – TS 21
   superior (cranial, from anterior cardinal) – TS 21
  vitelline (omphalomesenteric) – TS 12
  vitelline venous plexus – TS 17

yolk sac – *see* "extra-embryonic tissue"
yolk sac cavity – TS 7

zona pellucida – TS 1

5.5 Glossary of terms

aditus an entrance, opening or approach into a cavity or organ e.g. *laryngeal aditus* – the entrance into the larynx, where the pharynx communicates with the laryngeal cavity

alar winglike e.g. *alar plate* one of the pair of dorsolateral longitudinally-running grey horns of the neural tube/spinal cord where the sensory or receptive part of the grey matter is located and where most of the dorsal root fibres terminate

antrum a cavity or chamber, particularly within a bone e.g. *maxillary antrum* – the maxillary air sinus

arcuate shaped like an arc or curved bow e.g. *arcuate line* (syn. linea semicircularis) a crescentic-shaped line marking the lower border of the posterior sheath of the rectus abdominis muscle

arytenoid shaped like a jug or ladle e.g. *arytenoid cartilage*, one of the cartilages of the larynx so-named because of its supposed resemblance to a ladle

axillary pertaining to the axilla, the hollow or armpit where the arm joins the body at the shoulder

blastema the (usually) small group of cells that, following cellular migration or differentiation, give rise to either a part of an organ or a complete organ

bursa a sac or pouch-like structure e.g. the *prepatellar bursa(e)*- the sac (or occasionally several communicating sacs) lined by synovial membrane and containing synovial fluid that is located in front of the patella and which permits the skin overlying the patella to move freely over it

canthus the angle at either end of the palpebral fissure, the slit between the eyelids; these are the outer (or temporal), and inner (or nasal) cathi

caudate possessing or shaped like a tail e.g. *caudate nucleus*, a mass of grey matter that forms one of the basal ganglia: it possesses a pear-shaped head, a long slender body and a long curved tail

choana a funnel-shaped cavity e.g. the *primary choana* – the indentation in the embryo made by the olfactory pit, being initially a blind-ending sac which overlies the front part of the primitive oral cavity which subsequently breaks down allowing communication between the nasal and oral cavities

choroid resembling the chorion or other vascular or villous membrane e.g. *choroid plexus* – the delicate network of vascular villi derived from the pia mater that protrudes from the walls of the ventricles of the brain, being the site of cerebrospinal fluid production

cistern a reservoir or (often) closed space e.g. *basal cistern* – a part of the subarachnoid space located on the basal surface of the brain that acts as a reservoir, or site of accumulation, of cerebrospinal fluid

commissure in neuroanatomy: a midline region, or bridge, where nerve fibres from similar structures on the two sides of the brain or spinal cord are connected: in the spinal cord, these are the "bridges" of grey matter that cross above and below the central canal

diverticulum a pouch or pocket leading off from a tube or cavity e.g. *thyroid diverticulum* – an indentation on the dorsum of the tongue, at the foramen caecum, and located at the apex of the sulcus terminalis, from which the thyroglossal duct grows caudally to form the thyroid gland

dorsum the back; in relation to the hand – it refers to the back of the hand; in relation to the foot – it refers to the top of the foot when the individual is in the standing position

ductus a passage (or duct) through which passes blood, lymph, excretions or secretions e.g. *ductus arteriosus*- a channel which is usually only patent in the fetus through which deoxygenated blood flows from the pulmonary trunk to the aorta, and is derived from the left sixth aortic (pharyngeal) arch artery: it normally closes shortly after birth

ectomeninx a dense mesenchymatous layer present in the embryo that surrounds the central nervous system and which gives rise to the dura mater and is believed also to contribute to the cartilage and bone of the neurocranium

eminence a projection, ridge or prominence, particularly on the surface of a bone e.g. *hypobranchial eminence* – a large median swelling on the floor of the pharynx formed from components of the third and fourth pharyngeal arches: it gives rise to the posterior two-thirds, or pharyngeal component of the tongue and the epiglottis

ependymal composed of ependyma: the cellular layer that lines both the ventricles of the brain and the central canal of the spinal cord

fenestra a window-like opening in the wall between two chambers or spaces e.g. *fenestra vestibuli* – (syn: oval window) a small oval window on the medial wall of the tympanic cavity where it communicates with the vestibule of the inner ear

filiform threadlike, or like a filament e.g. *filiform papillae* – the tall conical papillae located on the dorsum of the tongue just anterior to the V-shaped sulcus terminalis which approximately corresponds to the junction between the anterior two-thirds and the posterior one-third of the tongue

filum a threadlike or stringlike structure or part e.g. *filum terminale* – the slender threadlike caudal extension of the conus medullaris of the spinal cord which anchors the latter to the coccyx

fistula an abnormal connection, sinus or tract which allows communication between two normally separate structures e.g. *urachal fistula* – a persistent communication between the distal part of the urachus and the umbilicus that allows the leakage of urine at the umbilicus

foramen an opening or hole into or through a structure, particularly a bone e.g. *foramen ovale* – an obliquely directed opening which allows communication between the right and left atria of the fetal heart, and which normally closes shortly after birth

fossa a trenchlike depression, hollow or recess e.g. *acetabular fossa* – the hollow roughened *non-articular* region in the central part of the concavity on the lateral part of the innominate (or hip) bone around but principally above which the head of the femur articulates

fungiform shaped like a mushroom or fungus e.g. *fungiform papillae* – the round mushroomlike papillae found on the dorsum of principally the oral part of the tongue, but particularly near its lateral margins and tip

ganglion (plural: ganglia) in neuroanatomy: a collection or mass of nerve cell bodies, usually associated with the peripheral nervous system e.g. the *cranial ganglia*, *dorsal root ganglia* or *autonomic ganglia*

gubernaculum a term usually used for a fibrous cord that directs the course (usually descent) of a structure attached to it during development e.g. *gubernaculum testis* – the fibromuscular cord that is initially attached to the lower pole of the mesonephros, and later the testis, that directs its descent through the coverings of the abdominal wall to the scrotum

gyrus in neuroanatomy: an elevation or convolution of the cerebral cortex (the surface of the brain) resulting from the infolding of adjacent sulci (or fissures)

hernia the protrusion of either an organ or a tissue through a weakness or abnormal opening e.g. *physiological umbilical hernia* – during the normal development of the gut, a loop of midgut protrudes through the umbilicus enabling this region of the gut to both increase in length and differentiate: at a later stage, and when space within the abdominal cavity permits, this loop of bowel returns to the abdominal cavity and the site of weakness at the umbilicus is repaired

hilum (or hilus) a small hollow or depression, usually in an organ, where the vessels, and nerves and ducts enter or leave it e.g. *hilum of the lung* – the depression on the medial surface of the lung where the main bronchi, vessels and nerves enter

hyaloid (or hyaline) having a glassy or glasslike appearance e.g. *hyaline cartilage* – the semitranslucent cartilage characteristically present on articular surfaces that has a low coefficient of friction

hypogastric (of the hypogastrium) – the lower midline region of the abdomen, below the region of the stomach

intercostal the region between the ribs

isthmus a narrow passage or strip of tissue connecting two larger parts (usually of an organ) e.g. *isthmus of the thyroid* – the strip of glandular tissue between the two lobes of the thyroid gland

ligament usually a tough fibrous band connecting two bones (often in association with a joint), or a fibrous band or fold of peritoneum that supports an organ e.g. *gastrosplenic* and *lienorenal ligaments* – the two components of the dorsal mesentery of the stomach (the dorsal mesogastrium) in which the spleen is suspended (between the posterior abdominal wall and the greater curvature of the stomach)

mantle a covering or surrounding layer e.g. *mantle layer* – the middle layer of the neural tube, located between the inner *ependymal* layer and the outer *marginal* layer

manubrium a hilt, or handle-like structure, usually used in relation to the upper part of the sternum (the *manubrium sterni*) that articulates (laterally) with the clavicles and the first costal cartilages and inferiorly with the body of the sternum

meatus an opening (usually) into a canal or passage e.g. *external acoustic* (or *auditory*) *meatus* – the external auditory canal which is lined by skin, and in the depth of which is the tympanic membrane (or ear drum)

mediastinum a term usually used for the space between the two pleural sacs: it contains the heart and its great vessels and the other thoracic viscera located in this region (e.g. the thymus gland)

medulla (usually) the central region of an organ, in contrast to its outer part, or cortical region; also used to describe any tissue that resembles marrow or pulp e.g. *adrenal medulla* – the central region of the adrenal gland

membrane a thin layer of tissue that covers a surface, or divides a space or an organ e.g. *extra-embryonic membranes* – the coverings of the embryo, and later the fetus, that provide for its protection and nutrition as well as facilitates respiration and excretion

naris one of the openings or orifices into the nasal cavity e.g. the *anterior* (or *external*) and *posterior* (or *internal*) *nares* (or *nasal apertures*)

nucleus (usually) the central, usually spherical, membrane-bound component within a cell that contains its genetic material (the DNA), or the inner part (or core) of any structure e.g. *sensory nucleus of the trigeminal nerve* – the collection of nerve cells (or neurons) forming the sensory component of the trigeminal (V) cranial nerve

omentum a double layered region of the peritoneum that passes from the stomach to an adjacent organ e.g. *lesser omentum* – part of the ventral mesentery of the stomach located between the posterior (or dorsal) region of the liver and the lesser curvature of the stomach

ostium an orifice or small opening, particularly into a tubular organ e.g. *ostium primum* – an opening between the two atria in the embryonic heart located below the lower border of the septum primum

papilla a small nipple-like elevation or projection e.g. *greater duodenal papilla* – a small elevation into the postero-medial region of the lumen of the second part of the duodenum that transmits (usually separately) both the common bile duct and the pancreatic duct

paraganglion one of the (usually oval) paired masses of chromaffin tissue, derived from the neural crest, usually a component of the sympathetic system, but may also be associated with the aorta and its major branches, and in various other parts of the body

parasympathetic system the craniosacral part of the autonomic (involuntary) nervous system, whose principal function is to restore the energy sources of the body e.g. by decreasing the rate and force of the heart, lowering the blood pressure, and augmenting the activity of the digestive system in contrast to the activity of the *sympathetic system* (see below)

parietal (usually) associated with the walls of an organ or body cavity e.g. *parietal pleura* – the serous membrane associated with the lining of the pleural cavity

pia mater the innermost of the three meninges (the membranous coverings) (also *arachnoid mater* and *dura mater*): this layer is particularly closely adherent to the surface of the brain and spinal cord, and forms the inner boundary of the subarachnoid space

pleura the serous membrane that covers the lungs (the *visceral pleura*) and lines the pleural cavity (the *parietal pleura*)

plexus an aggregation of nerves, veins or lymphatics e.g. *brachial plexus* – the plexus of nerves in the axilla formed (usually) by the union of the anterior branches (or divisions) of the lower four cervical and the first thoracic nerves, being the nerve supply to the forelimb

process (in anatomy) a slender projection, prominence or outgrowth of tissue e.g. *transverse process of a vertebra* – the laterally-directed outgrowth of a vertebral body

putamen the larger and lateral (or outer) part of the lenticular nucleus of the corpus striatum

pyloric associated with the pylorus: the terminal (distal) part of the stomach which opens into the duodenum; the *pyloric sphincter* is located in this region of the stomach

ramus a primary division of a nerve or blood vessel, also used for a principal division of a bone e.g. *ramus of the* *mandible* – the posterior, ascending part of the lower jawbone

raphe a (usually midline) seam, ridge or line of union of various bilaterally symmetrical structures e.g. *anococcygeal raphe* – a median fibrous septum extending from the anus to the coccyx

recess a small space or hollowed out region or cavity e.g. *lateral recess of the fourth ventricle* – the most lateral (and widest) part of the cavity of the hindbrain

reflection a turning or bending back, or fold e.g. the region where the visceral and parietal layers of peritoneum meet, for example, in relation to the upper surface of the liver, where the peritoneal reflections form the boundary of the "bare" area of the liver

rest (in anatomy) a small amount of embryonic (or undifferentiated) tissue that has been retained (usually) at its site of origin, rather than migrating elsewhere e.g. *embryonic rest* – a small collection of embryonic cells, or embryonic tissue, which fails to develop in the normal way and is retained in its undifferentiated state in the adult

rete a meshwork of nerve fibres, vessels or ducts e.g. *rete testis* – the network of cords or ducts in the hilar region of the testis, of mesonephric duct origin, through which the seminiferous tubules drain to the head of the epididymis

septum a dividing wall or partition e.g. *septum primum* – the first and relatively thin component of the partition that separates the primitive atrium into two parts: it is crescentic in shape, and grows caudally from the cranial wall of the atrium towards the atrio-ventricular cushion tissue

sinus (usually) a cavity or hollow space that may be filled with air (e.g. *frontal* and *maxillary air sinuses*, in the cranial bones), venous blood (e.g. *cavernous sinus* or *dural venous sinus*) or lymph (an example of which is the *cisterna chyli*): (also) a recess or hollow cavity or space e.g. *urogenital sinus* – the anterior (or ventral) subdivision of the cloaca, separated from the hindgut by the downgrowth of the urorectal septum

striatum (or neostriatum) phylogenetically the more recent part of the corpus striatum, consisting of the caudate nucleus and the putamen or lateral part of the lenticular nucleus

subarachnoid the region or space between the pia mater and the arachnoid mater that contains the cerebrospinal fluid

sulcus in neuroanatomy: linear grooves or furrows on the surface of the cerebral hemispheres separating the convolutions or gyri of the brain: also, any groove or furrow e.g. the *intertubercular sulcus* – the groove on the anterior surface of the humerus between the greater and lesser tubercules along which runs the tendon of the long head of the biceps muscle

sympathetic system the autonomic outflow from the intermediolateral column of all of the thoracic and the upper two or three lumbar segments of the spinal cord: this system stimulates activities that are accompanied by an expenditure of energy stores, including acceleration of the heart rate and increased efficiency in the force of its contraction, rise in arterial blood pressure and blood sugar level and direction of the blood flow towards the skeletal muscles and away from the viscera and the skin – the basis of the "fight and flight" response

symphysis a growing together or natural site of union; (in anatomy) a fibrocartilaginous joint with special features – each of the two joint surfaces is covered by a thin layer of hyaline cartilage, and these are separated by an intervening disk or pad of fibrocartilage; such joints are usually located in the median plane e.g. *pubic symphysis* – the joint between the two pubic bones; *symphysis menti* – the midline joint between the two halves of the lower jaw

tegmentum in neuroanatomy: the dorsal portion of the pons including the major part of the cerebral peduncle (or crus cerebri) of the midbrain

terminalis situated at an end or boundary e.g. *sulcus terminalis of the tongue* – a v-shaped furrow on the dorsum of the tongue separating the anterior two-thirds (or oral part) from the posterior one-third (or pharyngeal part): its apex is at the foramen caecum, and its two arms spread anterolaterally to the sides of the tongue

truncus the major undivided part of a blood or lymphatic vessel or nerve from which branches may arise e.g. *truncus arteriosus* – the distal part of the outflow tract of the primitive heart tube from which the first pair of aortic arch arteries arise

trunk (see truncus): used to describe the major part of the body *excluding* the head and neck and the extremities

tuberosity a broad bump, prominence or swelling situated on a bone e.g. *deltoid tuberosity of the humerus* – the prominent site of insertion of the deltoid muscle onto the antero-lateral surface of the upper part of the shaft of the humerus

tunica a membrane or coat that lines or covers an organ e.g. *tunica media* of a blood vessel – the middle and usually thickest coat of the wall of a blood vessel containing circularly arranged smooth muscle fibres with some elastic or collagenous fibres

ventricular pertaining to a ventricle e.g. *ventricular myocardium* – the muscular wall of the ventricle of the heart

visceral (usually) situated near to an organ (or viscus) e.g. *visceral pleura* – the membranous covering of the lungs, by contrast with the *parietal pleura* – the membranous lining of the pleural cavity

5.6 References

Abramovich, D.R. (1968) The volume of amniotic fluid in early pregnancy. *J. Obstet. Gynaecol. Br. Commonw.* **75**: 728–731.

Abramovich, D.R. (1973) The volume of amniotic fluid and factors affecting or regulating this. In: *Amniotic Fluid. Research and Clinical Application*, eds D.V.I. Fairweather and T.K.A.B. Eskes, pp. 29–51. Amsterdam: Excerpta Medica.

Adams, F.H., Fujiwara, T. and Rowshan, G. (1963) The nature and origin of the fluid in the fetal lamb lung. *J. Pediatr.* **63**: 881–888.

Addison, W.H.F. and Appleton, J.L. (1915) The structure and growth of the incisor teeth of the albino rat. *J. Morphol.* **26**: 42–96.

Adelmann, H.B. (1925) The development of the neural folds and cranial ganglia in the rat. *J. Comp. Neurol.* **39**: 19–171.

Adesanya, T., Grillo, I. and Shima, K. (1966) Insulin content and enzyme histochemistry of the human foetal pancreatic islet. *J. Endocrinol.* **36**: 151–158.

AhPin, P., Ellis, S., Arnott, C. and Kaufman, M.H. (1989) Prenatal development and innervation of the circumvallate papilla in the mouse. *J. Anat.* **162**: 33–42.

Aikawa, E. and Kawano, J. (1982) Formation of coronary arteries sprouting from the primitive aortic sinus wall of the chick embryo. *Experientia* **38**: 816–818.

Alberch, P. and Kollar, E. (1988) Strategies of head development: workshop report. *Development* **103**: Suppl. 25–30.

Alden, R.H. (1948) Implantation of the rat egg. III. Origin and development of primary trophoblast giant cells. *Am. J. Anat.* **83**: 143–181.

Aldridge, R.T. and Campbell, P.E. (1968) Ganglion cell distribution in the normal rectum and anal canal. A basis for the diagnosis of Hirschsprung's disease by anorectal biopsy. *J. Pediatr. Surg.* **3**: 475–490.

Alexander, D.P. and Nixon, D.A. (1961) The foetal kidney. *Br. Med. Bull.* **17**, 112–117.

Allen, B. (1904) The embryonic development of the ovary and testis in mammals. *Am. J. Anat.* **3**: 88–153.

Altman, J. and Bayer, S.A. (1982) Development of the cranial nerve ganglia and the nuclei in the rat. *Adv. Anat. Embryol. Cell Biol.* **74**: 1–90.

Altman, J. and Bayer, S.A. (1984) The development of the rat spinal cord. *Adv. Anat. Embryol. Cell Biol.* **85**: 1–116.

Altman, J. and Bayer, S.A. (1995) *Atlas of Prenatal Rat Brain Development*. Boca Raton: CRC Press.

Alvarez-Bolado, G. and Swanson L.W. (1996) *Developmental Brain Maps: Structure of the Embryonic Rat Brain*. Amsterdam: Elsevier.

Amoroso, E.C. (1952) Placentation. In: *Marshall's Physiology of Reproduction*, ed. A.S. Parkes, vol. 2, pp. 127–311. London: Longmans Green.

Ansell, J.D. (1975) The differentiation and development of mouse trophoblast. In: *The Early Development of Mammals*, eds M. Balls and A.E. Wild, pp. 133–144. Cambridge: Cambridge University Press.

Ansell, J.D. and Snow, M.H.L. (1975) The development of trophoblast *in vitro* from blastocysts containing varying amounts of inner cell mass. *J. Embryol. Exp. Morphol.* **33**, 177–185.

Arey, L.B. (1974) *Developmental Anatomy*. Philadelphia: W.B. Saunders.

Austin, C.R. (1965) *Fertilization*. New Jersey: Prentice-Hall.

Avery, B., Bak, A. and Schmidt, M. (1989) Differential cleavage rates and sex determination in bovine embryos. *Theriogenology* **32**: 139–147.

Backhouse, K.M. and Butler, H. (1960) The gubernaculum testis of the pig (*Sus scrofa*). *J. Anat.* **94**: 107–120.

Baker, T.G. and Franchi, L.L. (1967) The fine structure of oogonia and oocytes in human ovaries. *J. Cell Sci.* **2**: 213–224.

Barber, A.N. (1955) *Embryology of the Human Eye*. St Louis: C.V. Mosby.

Bard, J.B.L. (1992) *Morphogenesis: the Cellular and Molecular Processes of Developmental Anatomy*. Cambridge: Cambridge University Press.

Bard, J.B.L. and Kaufman, M.H. (1994) The mouse. In *Embryos: Color Atlas of Development*, ed. J.B.L. Bard. pp. 183–206. London: Wolfe Publishing.

Bard, J.B.L. and Kratochwil, K. (1987) Corneal morphogenesis in the Mov13, mutant mouse is characterised by normal cellular organization but disordered and thin collagen. *Development* **101**: 547–555.

Bard, J.B.L., Kaufman, M.H., Dubreuil, C., Brune. R.M., Burger, A., Baldock, R.A. and Davidson, D.R. (1998) An internet-accessible database of mouse developmental anatomy based on a systematic nomenclature. *Mech. Dev.* **74**: 111–120.

Barr, M.L. (1979) *The Human Nervous System: An Anatomical Viewpoint*, 3rd edn. Hagerstown, Maryland: Harper and Row.

Barron, D.H. (1944) The changes in the fetal circulation at birth. *Physiol. Rev.* **24**: 277–295.

Bartelings, M.M. and Gittenberger-de Groot, A.C. (1989) The outflow tract of the heart – embryologic and morphologic considerations. *Int. J. Cardiol.* **22**: 289–300.

Bates, M.N. (1948) The early development of the hypoglossal musculature in the cat. *Am. J. Anat.* **83**: 329–356.

Beatty, R.A. (1957) *Parthenogenesis and Polyploidy in Mammalian Development*. Cambridge: Cambridge University Press.

Bee, J. and Thorogood, P. (1980) The role of tissue interactions in the skeletogenic differentiation of avian neural crest cells. *Dev. Biol.* **78**: 47–62.

Bell, L. and Williams, L. (1982) A scanning and transmission electron microscopical study of the morphogenesis of human colonic villi. *Anat. Embryol.* **165**: 437–455.

Bell, S.C. (1985) Comparative aspects of decidualization in rodents and human: cell types, secreted products and associated function. In: *Implantation of the Human Embryo*, eds R.G. Edwards, J.M. Purdy and P.C. Steptoe, pp. 71–122. London: Academic Press.

Bellairs, R. (1963) The development of somites in the chick embryo. *J. Embryol. Exp. Morphol.* **11**: 697–714.

Bellairs, R. and Osmond, M. (1998) *The Atlas of Chick Development*. San Diego: Academic Press.

Bellairs, R., Ede, D.A. and Lash, J.W. (1986) *Somites in Developing Embryos*. New York: Plenum Press.

Bernfield, M.R., Bannerjee, S.D., Koda, J.E. and

Rapraeger, A.C. (1984) Remodelling of the basement membrane and mechanism of morphogenetic tissue interactions. In: *The Role of the Extracellular Matrix in Development*, ed. R.L. Trelstad, pp. 542–572. New York: Plenum Press.

Berry, J.M. (1900) On the development of the villi of the human intestine. *Anat. Anz.* **17**: 242–249.

Bloom, W. and Bartelmez, G.W. (1940) Hematopoiesis in young human embryos. *Am. J. Anat.* **67**: 21–52.

Bolli, V.P. (1966) Sekundäre Lumenbildungen im Neuralrohr und Rückenmark Menschlicher Embryonen. *Acta Anat.* **64**: 48–81.

Bonneville, K. (1950) New facts on mesoderm formation and proamnion derivatives in the normal mouse embryo. *J. Morphol.* **86**: 495–545.

Borysenko, M. and Beringer, T. (1989) *Functional Histology*, 3rd edn. Boston: Little, Brown.

Bourne, G. (1962) *The Human Amnion and Chorion*. London: Lloyd-Luke.

Boyd, J.D. (1933) The classification of the upper lip in mammals. *J. Anat.* **67**: 409–416.

Boyd, J.D. and Hamilton, W.J. (1970) *The Human Placenta*. Cambridge: W. Heffer and Sons Ltd.

Brambell, F.W.R. (1927) The development and morphology of the gonads of the mouse. II. The development of the Wolffian body and ducts. *Proc. R. Soc. Lond. [Biol.]* **102**: 206–221.

Bray, J.J., Cragg, P.A., Macknight, A.D.C., Mills, R.G. and Taylor, D.W. (eds) (1994) *Lecture Notes on Human Physiology*, 3rd edn. Oxford: Blackwell Scientific Publications.

Bremner, C.G. (1968) Studies on the pyloric muscle. I. The embryology of the pyloric muscle. *S. Afr. J. Surg.* **6**: 79–85.

Brinkley, L.L. and Morris-Wiman, J. (1987) Effects of chlorcyclizine-induced glycosaminoglycan alterations on patterns of hyaluronate distribution during morphogenesis of the mouse secondary palate. *Development* **100**: 637–640.

Brökelmann, J. and Biggers, J.D. (1979) Studies on the development of cell contacts and of the intercellular matrix during decidualization in the rat. *Arch. Gynecol.* **227**: 103–117.

Bronner-Fraser, M. (1982) Distribution of latex beads and retinal pigment epithelial cells along the ventral neural crest pathway. *Dev. Biol.* **91**: 50–63.

Brookhart, J.M. and Mountcastle, V.B. (eds) (1984) *Handbook of Physiology*, Section 1. *The Nervous System*, vol III. Washington: American Physiological Society.

Bucher, U. and Reid, L. (1961a) Development of the intrasegmental bronchial tree: the pattern of branching and development of cartilage at various stages of intrauterine life. *Thorax* **16**: 207–218.

Bucher, U. and Reid, L. (1961b) Development of the mucus-secreting elements in human lung. *Thorax* **16**: 219–225.

Buckingham, M. (1996) The formation and maturation of skeletal muscle during mammalian development. In: *Mammalian Development*, ed. P. Lonai, pp. 81–99. Amsterdam: Harwood Academic.

Bulfone, A., Puelles, L., Porteus, M.H., Frohman, M.A., Martin, G.R. and Rubenstein, J.L.R. (1993) Spatially restricted expression of Dlx-1, Dlx-2 (Tes-1), Gbx-2, and Wnt-3 in the embryonic day 12.5 mouse forebrain defines potential transverse and longitudinal segmental boundaries. *J. Neurosci.* **13**: 3155–3172.

Bulmer, D. (1957) The development of the human vagina. *J. Anat.* **91**: 490–509.

Burgess, D.R. (1975) Morphogenesis of intestinal villi. II. Mechanism of formation of previllous ridges. *J. Embryol. Exp. Morphol.* **34**: 723–740.

Burgoyne, P.S. (1993) A Y-chromosomal effect on blastocyst cell number in mice. *Development* **117**: 341–345.

Burke, A.C., Nelson, C.E., Morgan, B.A. and Tabin, C. (1995) Hox genes and the evolution of vertebrate axial morphology. *Development* **121**: 333–346.

Butler, H. and Juurlink, B.H.J. (1987) *An Atlas for Staging Mammalian and Chick Embryos*. Boca Raton, Florida: CRC Press.

Byskov, A.G. (1978) The anatomy and ultrastructure of the rete system in the fetal mouse ovary. *Biol. Reprod.* **19**: 720–735.

Byskov, A.G. (1986) Differentiation of mammalian embryonic gonad. *Phys. Rev.* **66**: 71–117.

Byskov, A.G. and Lintern-Moore, S. (1973) Follicle formation in the immature mouse ovary: the role of the rete ovarii. *J. Anat.* **116**: 207–217.

Carlson, B.M. (1996) *Patten's Foundations of Embryology*, 6th edn. New York: McGraw-Hill.

Chacko, A.W. and Reynolds, S.R.M. (1954) Architecture of distended and nondistended human umbilical cord tissues, with special reference to the arteries and veins. *Contrib. Embryol. Carnegie Inst.* **35**: 135–150.

Challice, C.E. and Virágh, S. (1973) The embryonic development of the mammalian heart. In: *Ultrastructure of the Mammalian Heart*, eds C.E. Challice and S. Virágh, pp. 91–126. New York: Academic Press.

Chan, S.T.H. (1970) Natural sex reversal in vertebrates. *Philos. Trans. R. Soc. Lond. [Biol.]* **259**: 59–71.

Chang, H.H., Schwartz, Z. and Kaufman, M.H. (1996) Limb and other postcranial skeletal defects induced by amniotic sac puncture in the mouse. *J. Anat.* **189**: 37–49.

Chen, J.M. (1952a) Studies on the morphogenesis of the mouse sternum. I. Normal embryonic development. *J. Anat.* **86**: 373–386.

Chen, J.M. (1952b) Studies on the morphogenesis of the mouse sternum. II. Experiments on the origin of the sternum and its capacity for self-differentiation *in vitro*. *J. Anat.* **86**: 387–401.

Chiquoine, A.D. (1954) The identification, origin, and migration of the primordial germ cells in the mouse embryo. *Anat. Rec.* **118**: 135–146.

Chiquoine, A.D. (1960) The development of the zona pellucida of the mammalian ovum. *Am. J. Anat.* **106**: 149–169.

Christie, G.A. (1964) Developmental stages in somite and post-somite rat embryos, based on external appearance, and including some features of the macroscopic development of the oral cavity. *J. Morphol.* **114**: 263–286.

Clark, E.B. (1990) Growth, morphogenesis, and function. The dynamics of cardiac development. In: *Fetal, Neonatal, and Infant Cardiac Disease*, eds J.H. Moller and W.A. Neal. Norwalk, Connecticut: Appleton and Lange.

Clark, J.M. and Eddy, E.M. (1975) Fine structural observations on the origin and associations of primordial germ cells of the mouse. *Dev. Biol.* **47**: 136–155.

Clermont, Y. and Huckins, C. (1961) Microscopic anatomy of the sex cords and seminiferous tubules in growing and adult male albino rats. *Am. J. Anat.* **108**: 79–97.

Cole, R.J. (1967) Cinemicrographic observations on the trophoblast and zona pellucida of the mouse blastocyst. *J. Embryol. Exp. Morphol.* **17**: 481–490.

Comline, R.S., Cross, K.W., Dawes, G.S. and Nathanielz, P.W. (eds) (1973) *Foetal and Neonatal Physiology*. Cambridge: Cambridge University Press.

Congdon, E.D. (1922) Transformation of the aortic-arch system during the development of the human embryo. *Contrib. Embryol. Carnegie Inst.* **14**: 47–110.

Conklin, J.L. (1962) Cytogenesis of the human fetal pancreas. *Am. J. Anat.* **111**: 181–204.

Copp, A.J. (1978) Interaction between inner cell mass and trophectoderm of the mouse blastocyst. I. A study of cellular proliferation. *J. Embryol. Exp. Morphol.* **48**: 109–125.

Copp, A.J. (1979) Interaction between inner cell mass and trophectoderm of the mouse blastocyst. II. The fate of the polar trophectoderm. *J.Embryol. Exp. Morphol.* **51**: 109–120.

Cordier, A.C. and Haumoont, S.M. (1980) Development of thymus, parathyroids, and ultimo-branchial bodies in NMRI and nude mice. *Am. J. Anat.* **157**: 227–263.

Cornes, J.S. (1965) Number, size, and distribution of Peyer's patches in the human small intestine. Part 1. The development of Peyer's patches. *Gut* **6**: 225–233.

Coulombre, A.J. (1956) The role of intraocular pressure in the development of the chick eye. I. Control of eye size. *J. Exp. Zool.* **133**: 211–225.

Coulombre, A.J. and Coulombre, J.L. (1958a) Intestinal development. I. Morphogenesis of the villi and musculature. *J. Embryol. Exp. Morphol.* **6**: 403–411.

Coulombre, A.J. and Coulombre, J.L. (1958b) The role of mechanical factors in brain morphogenesis. *Anat. Rec.* **130**: 289–290 (Abstract).

Couly, G.F., Coltey, P.M. and Le Douarin, N.M. (1993) The triple origin of skull in higher vertebrates: a study in quail–chick chimeras. *Development* **117**: 409–429.

Coupland, R.E. (1958) The innervation of the pancreas of the rat, cat and rabbit as revealed by the cholinesterase technique. *J. Anat.* **92**: 143–149.

Cunha, G.R. (1975) The dual origin of vaginal epithelium. I. *Am. J. Anat.* **143**: 387–392.

Cuthbertson, K.S.R., Cobbold, P.H. and Whittingham, D.G. (1981) Free Ca^{2+} increases in exponential phases during mouse oocyte activation. *Nature* **294**: 754–757.

Dalcq, A. (1957) *Introduction to General Embryology*. Oxford: Oxford University Press.

David, K.M., McLachlan, J.C., Aiton, J.F., Whiten, S.C., Smart, S.D., Thorogood, P.V. and Crockard, H.A. (1998) Cartilaginous development of the human craniovertebral junction as visualised by a new three-dimensional computer reconstruction technique. *J. Anat.* **192**: 269–277.

Davies, J.A. and Bard, J.B.L. (1998) The develcpment of the kidney. In: *Current Topics in Developmental Biology*, eds R.A. Pedersen and G.P. Schatten, pp. 245–301. San Diego: Academic Press.

Davis, C.L. (1927) Development of the human heart from its first appearance to the stage found in embryos of twenty paired somites. *Contrib. Embryol. Carnegie Inst.* **19**: 245–284.

Dawes, G.S. (1961) Changes in the circulation at birth. *Br. Med. Bull.* **17**: 148–153.

de Beer, G.R. (1937) *The Development of the Vertebrate Skull*. Oxford: Oxford University Press (reissued 1985, Chicago: University of Chicago Press).

de Beer, G.R. (1947) The differentiation of neural crest cells into visceral cartilages and odontoblasts in *Amblystoma*, and re-examination of the germ-layer theory. *Proc. R. Soc. Lond. B.* **134**: 377–398.

De Haan, R.L. (1959) Cardia bifida and the development of pacemaker function in the early chicken heart. *Dev. Biol.* **1**: 586–602.

De Haan, R.L. (1965) Morphogenesis of the vertebrate heart. In: *Organogenesis*, eds R.L. De Haan and H. Ursprung, pp. 377–419. New York: Holt, Rinehart and Winston.

De Martino, C. and Zamboni, L. (1966) A morphologic study of the mesonephros of the human embryo. *J. Ultrastruct. Res.* **16**: 399–427.

Desmond, M.E. and Jacobson, A.G. (1977) Embryonic brain enlargement requires cerebrospinal fluid pressure. *Dev. Biol.* **57**: 188–198.

Deuchar, E.M. (1958) Regional differences in catheptic activity in *Xenopus laevis* embryos. *J. Embryol. Exp. Morphol.* **6**: 223–237.

De Vries, P.A. and Saunders, J.B. de C.M. (1962) Development of ventricles and spiral outflow tract in human heart. *Contrib. Embryol. Carnegie Inst.* **37**: 87–114.

Dickson, A.D. (1966) The form of the mouse blastocyst. *J. Anat.* **100**: 335–348.

Didier, E. (1973) Recherches sur la morphogénèse du canal de Müller chez les oiseaux. II. Étude expérimentale. *Roux Archiv.* **172**: 287–302.

Downs, K.M. (1998) The murine allantois. In: *Current Topics in Developmental Biology*, vol 39, eds R.A. Pedersen and G.P. Schatten, pp. 1–33. San Diego: Academic Press.

Downs, K.M. and Davies, T. (1993) Staging of gastrulating mouse embryos by morphological landmarks in the dissecting microscope. *Development* **118**: 1255–1266.

Downs K.M. and Gardner, R.L. (1995) An investigation into early placental ontogeny: allantoic attachment to the chorion is selective and developmentally regulated. *Development* **121**: 407–416.

Downs, K.M. and Harmann, C. (1997) Developmental potency of the murine allantois. *Development* **124**: 2769–2780.

Drachman, D.B. and Sokoloff, L. (1966) The role of movement in embryonic joint development. *Dev. Biol.* **14**: 401–420.

Ducibella, T. (1977) Surface changes of the developing trophoblast cell. In: *Development in Mammals*, ed. M.H. Johnson, vol. 1, pp. 5–30. Amsterdam: North-Holland.

Durbec, P.L., Larsson-Blomberg, L.B., Schuchardt, A., Constantini, F. and Pachnis, V. (1996) Common origin and developmental dependence on *c-ret* of subsets of enteric and sympathetic neuroblasts. *Development* **122**: 349–358.

Duthie, G.M. (1925) An investigation of the occurrence, distribution and histological structure of the embryonic remains in the human broad ligament. *J. Anat.* **59**: 410–431.

Duval, M. (1892) Le Placenta des Rongeurs. Extrait du *Journal de l'Anatomie et de la Physiologie*. Années 1889–1892, ed. F. Alcan. Paris: Ancienne Librairie Germer Baillière (cited by Gardner and Papaioannou, 1975).

Dziadek, M. (1979) Cell differentiation in isolated inner cell masses of mouse blastocysts *in vitro*: onset of specific gene expression. *J. Embryol. Exp. Morphol.* **53**: 367–379.

Ebers, D.W., Smith, D.I. and Gibbs, G.E. (1956) Gastric

acidity on the first day of life. *Pediatrics* **18**: 800–802.

Edwards, R.G. (1980a) Sexual differentiation, infancy and puberty. In: *Conception in the Human Female*, pp. 23–98. London: Academic Press.

Edwards, R.G. (1980b) *Conception in the Human Female*. London: Academic Press.

Edwards, R.G. and Gates, A.H. (1959) Timing of the stages of the maturation divisions, fertilization and the first cleavage of eggs of adult mice treated with gonadotrophins. *J. Endocrinol.* **18**: 292–304.

Edwards, R.G. and Steptoe, P.C. (1975) Induction of follicular growth, ovulation and luteinization in the human ovary. *J. Reprod. Fertil. (Suppl.)* **22**: 121–163.

Eggermont, E. (1966) Enzymic activities in meconium from human foetuses and newborns. *Biol. Neonate* **10**: 266–280.

Eik-Nes, K.B. (1969) Patterns of steroidogenesis in the vertebrate gonads. *Gen. Comp. Endocrinol. Suppl.* **2**: 87–100.

Ellington, S.K.L. (1985) A morphological study of the development of the allantois of rat embryos in vivo. *J. Anat.* **142**: 1–11.

Ellington, S.K.L. (1987) A morphological study of the development of the chorion of rat embryos. *J. Anat.* **150**: 247–263.

Emery, J. (1969) Embryogenesis. In: *The Anatomy of the Developing Lung*, ed. J. Emery, pp. 1–7. London: Heinemann Medical.

Enders, A.C. (1965) Comparative study of the fine structure of the trophoblast in several hemochorial placentas. *Am. J. Anat.* **116**: 29–68.

Enders, A.C. (1971) The fine structure of the blastocyst. In: *The Biology of the Blastocyst*, ed. R.J. Blandau, pp. 71–94. Chicago: Chicago University Press.

Enders, A.C. and Schlafke, S.J. (1965) The fine structure of the blastocyst: some comparative studies. In: *Preimplantation Stages of Pregnancy: A Ciba Foundation Symposium*, ed. G.E.W. Wolstenholme and M. O'Connor, pp. 29–54. London: J. & A. Churchill.

Erickson, R.A. (1968) Inductive interactions in the development of the mouse metanephros. *J. Exp. Zool.* **169**: 33–42.

Evans, M.J. and Kaufman, M.H. (1981) Establishment in culture of pluripotential cells from mouse embryos. *Nature* **292**: 154–156.

Evans, M.J. and Martin, G.R. (1975) The differentiation of clonal teratocarcinoma cell cultures in vitro. In: *Roche Symposium on Teratocarcinomas and Differentiation*, eds. M. Sherman and D. Solter, pp. 237–250. New York: Academic Press.

Everett, N.B. (1943) Observational and experimental evidences relating to the origin and differentiation of the definitive germ cells in mice. *J. Exp. Zool.* **92**: 49–91.

Fairweather, D.V.I. and Eskes, T.K.A.B. (eds) (1973) *Amniotic Fluid. Research and Clinical Application.* Amsterdam: Excerpta Medica.

Fananapazir, K. and Kaufman, M.H. (1988) Observations on the development of the aortico-pulmonary spiral septum in the mouse. *J. Anat.* **158**: 157–172.

Fawcett, D.W. (1975) Ultrastructure and function of the Sertoli cell. In: *Handbook of Physiology. Male Reproductive System*, eds. R.O. Greep and E.B. Astwood, pp. 21–55. Washington, DC: American Philosophical Society.

Ferguson, M.W.J. (1987) Palate development: mechanisms and malformations. *Irish J. Med. Sci.* **156**: 309–315.

Ferguson, M.W.J. (1988) Palate development. *Development, Suppl.* **103**: 41–60.

Findlater, G.S., McDougall, R.D. and Kaufman, M.H. (1993) Eyelid development, fusion and subsequent reopening in the mouse. *J. Anat.* **183**: 121–129.

Fishel, S.B. (1985) Uterine histology, biochemistry and secretions. In: *Implantation of the Human Embryo*, eds. R.G. Edwards, J.M. Purdy and P.C. Steptoe, pp. 47–63. London: Academic Press.

Fitzgerald, M.J., Nolan, J.P. and O'Neill, M.N. (1971a) The position of the human caecum in fetal life. *J. Anat.* **109**: 71–74.

Fitzgerald, M.J., Nolan, J.P. and O'Neill, M.N. (1971b) The formation of the ascending colon. *Irish J. Med. Sci.* **140**: 258–262.

Fitzharris, T.P. (1981) Origin and migration of cushion tissue in the developing heart. *Scand. Electron Microsc.* **2**: 255–260.

Fontaine, J. (1979) Multistep migration of calcitonin cell precursors during ontogeny of the mouse pharynx. *Gen. Comp. Endocrinol.* **37**: 81–92.

Forest, M.G. (1983) Role of androgens in fetal and pubertal development. *Horm. Res.* **18**: 69–83.

Forsberg, J-G. (1965) Origin of vaginal epithelium. *Obstet. Gynecol.* **25**: 787–791.

Fowler, R.E., Kaufman, M.H. and Grainge, C. (1986) The secretions of the cumulus–oocyte complex in relation to fertilization and early embryonic development: a histochemical study. *Histochem. J.* **18**: 541–550.

Franchi, L.L. and Mandl, A.M. (1962) The ultrastructure of oogonia and oocytes in the foetal and neonatal rat. *Proc. R. Soc. Lond. [Biol.]* **157**: 99–114.

Franchi, L.L., Mandl, A.M. and Zuckerman, S. (1962) The development of the ovary and the process of oogenesis. In: *The Ovary*, ed. S. Zuckerman, vol. 1, pp. 1–88. New York: Academic Press.

Franklin, K.B.J. (1996) *The Mouse Brain: Stereotactic Coordinates.* San Diego: Academic Press.

Fraser, E.A. (1950) The development of the vertebrate excretory system. *Biol. Rev.* **25**: 159–187.

Frazer, J.E. (1926) The disappearance of the pre-cervical sinus. *J. Anat.* **61**: 132–143.

Frazer, J.E. (1931) *A Manual of Embryology.* London: Baillière, Tindall and Cox.

Frazer, J.E. (1935) The terminal part of the Wolffian duct. *J. Anat.* **69**: 455–468.

Frazer, J.E. and Robbins, R.H. (1916) On the factors concerned in causing rotation of the intestine in man. *J. Anat. Physiol.* **50**: 75–110.

Fredericks, C.M., Azzam, M.E.A. and Hafez, E.S.E. (1977) The motility in vitro of the rabbit uterovarian ligament. *J. Reprod. Fertil.* **49**: 387–389.

Friedberg, J. (1989) Pharyngeal cleft sinuses and cysts, and other benign neck lesions. *Pediatr. Clin. North Am.* **36**: 1451–1469.

Fritsch, B., Barald, K.F. and Lomax, M.I. (1997) Early embryology of the vertebrate ear. In: *Development of the Auditory System*, eds E.W. Rubel, A.N. Popper and R.R. Fay, pp. 80–145. New York: Springer-Verlag.

Fujinaga, M., Brown, N.A. and Baden, J.M. (1992) Comparison of staging systems for the gastrulation and early neurulation period in rodents: a proposed new system. *Teratology* **46**: 183–190.

Fujiwara, T., Adams, F.H. and Scudder, A. (1964) Fetal lamb amniotic fluid: relationship of lipid composition to surface tension. *J. Pediatr.* **65**: 824–830.

Fukiishi, Y. and Morriss-Kay, G.M. (1992) Migration of cranial neural crest cells to the pharyngeal arches and heart in rat embryos. Cell Tissue Res. 268: 1–8.

Fyfe, D.M. and Hall, B.K. (1983) The origin of the ectomesenchymal condensations which precede the development of the bony scleral ossicles in the eyes of embryonic chicks. *J. Embryol. Exp. Morphol.* 73: 69–86.

Gadd, R.L. (1966) The volume of the liquor amnii in normal and abnormal pregnancies. *J. Obstet. Gynaecol. Br. Commonw.* 73: 11–22.

Gans, C. (1988) Craniofacial growth. *Development Suppl.* 103: 3–15.

Gardner, R.L. and Johnson, M.H. (1973) Investigation of early mammalian development using interspecific chimaeras between rat and mouse. *Nature (New Biol.)* 246: 86–89.

Gardner, R.L. and Papaioannou, V.E. (1975) Differentiation in the trophectoderm and inner cell mass. In: *The Early Development of Mammals*, eds M. Balls and A.E. Wild, pp. 107–132, Cambridge: Cambridge University Press.

Gardner, R.L., Papaioannou, V.E. and Barton, S.C. (1973) Origin of the ectoplacental cone and secondary giant cells in mouse blastocysts reconstituted from isolated trophoblast and inner cell mass. *J. Embryol. Exp. Morphol.* 30: 561–572.

Gartler, S.M., Rivest, M. and Cole, R.E. (1980) Cytological evidence for an inactive X chromosome in murine oogonia. *Cytogenet. Cell Genet.* 28: 203–207.

Gasser, R.F. (1975) *Atlas of Human Embryos*. Hagerstown, Maryland: Harper and Row

Geelan, J.A.G. and Langman, J. (1977) Closure of the neural tube in the cephalic region of the mouse embryo. *Anat. Rec.* 189: 625–640.

Gersh, I. (1937) The correlation of structure and function in the developing mesonephros and metanephros. *Contrib. Embryol. Carnegie Inst.* 26: 35–58.

Geubelle, F., Karlberg, P., Koch, G., Lind, J., Wallgren, G. and Wegelius, C. (1959) Aeration of the lung in the newborn infant. *Biol. Neonate* 1: 169–210.

Gilbert, S.F. (1997) *Developmental Biology*, 5th edn. Sunderland, Massachusetts: Sinauer Associates Inc.

Ginsburg, M., Snow, M.H.L. and McLaren, A. (1990) Primordial germ cells in the mouse embryo during gastrulation. *Development* 110: 521–528.

Gladstone, R.J. and Hamilton, W.J. (1941) A presomite human embryo (Shaw) with primitive streak and chorda canal, with special reference to the development of the vascular system. *J. Anat.* 76: 9–44.

Glenister, T.W. (1962) The development of the utricle and of the so-called 'median' lobe of the human prostate. *J. Anat.* 96: 443–455.

Glucksmann, A., Ooka-Souda, S., Miura-Yasugi, E. and Mizuno, T. (1976) The effect of neonatal treatment of male mice with antiandrogens and of females with androgens on the development of the os penis and os clitoridis. *J. Anat.* 121: 363–370.

Godin, I.E., Garcia-Porrero, J.A., Coutinho, A., Dieterlen-Lièvre, F and Marcos, M.A.R. (1993) Para-aortic splanchnopleura from early mouse embryos contains B1a cell progenitors. *Nature*, 364: 67–70.

Gondos, B. (1980) Development and differentiation of the testis and male reproductive tract. In: *Testicular Development, Structure and Function*, eds. A. Steinberger and E. Steinberger, pp. 3–20. New York: Raven Press.

Goodrich, E.S. (1930) *Studies on the Structure and Development of Vertebrates*. London: Macmillan.

Gosden, R., Krapez, J. and Briggs, D. (1997) Growth and development of the mammalian oocyte. *BioEssays* 19: 875–882.

Goss, C.M. (1938) The first contractions of the heart in rat embryos. *Anat. Rec.* 70: 505–524.

Grand, R.J., Watkins, J.B. and Torti, F.M. (1976) Progress in gastroenterology: development of the human gastrointestinal tract. A review. *Gastroenterology* 70: 790–810.

Green, E.L. (ed.) (1966) *Biology of the Laboratory Mouse*, 2nd edn. New York: McGraw-Hill.

Green, M.N., Clarke, J.T. and Shwachman, H. (1958) Studies in cystic fibrosis of the pancreas: protein pattern in meconium ileus. *Pediatrics* 21: 635–641.

Gruenwald, P. (1941) The relation of the growing Müllerian duct to the Wolffian duct and its importance for the genesis of malformations. *Anat. Rec.* 81: 1–19.

Grüneberg, H. (1963) *The Pathology of Development. A Study of Inherited Skeletal Disorders in Animals*. Oxford: Blackwell Scientific Publications.

Gupta, M., Gulamhusein, A.P. and Beck, F. (1982) Morphometric analysis of the visceral yolk sac endoderm in the rat *in vivo* and *in vitro*. *J. Reprod. Fertil.* 65: 239–245.

Halbert, S.A. and Conrad, J.T. (1975) *In vitro* contractile activity of the mesotubarium superius from the rabbit oviduct in various endocrine states. *Fertil. Steril.* 26: 248–256.

Hall, B.K. (1987) Initiation of chondrogenesis from somitic, limb and craniofacial mesenchyme: search for a common mechanism. In: *Somites in Developing Embryos*, eds. R. Bellairs, D.A. Ede and J.W. Lash, pp. 247–259. New York: Plenum.

Hall, B.K. (1988a) *The Neural Crest. Including a Facsimile Reprint of the Neural Crest by Sven Hörstadius*. London: Oxford University Press.

Hall, B.K. (1988b) The embryonic development of bone. *Am. Sci.* 76: 174–181.

Halpern, M.H. (1953) The azygos vein system in the rat. *Anat. Rec.* 116: 83–93.

Hamburger, V. and Hamilton, H.L. (1951) A series of normal stages in development of the chick embryo. *J. Morphol.* 88: 49–92.

Hamilton, W.J. and Mossman, H.W. (1972) *Hamilton, Boyd and Mossman's Human Embryology. Prenatal Development of Form and Function*, 4th edn. Cambridge: W. Heffer and Sons Ltd.

Hammett, F.S. and Justice, E.S. (1923) The geometrical symmetry of growth of the upper incisors of the albino rat. *Anat. Rec.* 26: 141–144.

Harary, I. and Farley, B. (1963) *In vitro* studies on single beating rat heart cells. II. Intercellular communication. *Exp. Cell Res.* 29: 466–474.

Harris, M.J. and McLeod, M.J. (1982) Eyelid growth and fusion in fetal mice. A scanning electron microscope study. *Anat. Embryol.* 164: 207–220.

Harvey, S.C. and Burr, H.S. (1926) The development of the meninges. *Arch. Neurol. Psychiatr.* 15: 545–565.

Harvey, S.C., Burr, H.S. and Van Campenhout, E. (1933) Development of the meninges. Further experiments. *Arch. Neurol. Psychiatr.* 29: 683–690.

Haustein, J. (1983) On the ultrastructure of the developing and adult mouse corneal stroma. *Anat. Embryol.* 168: 291–305.

Hay, D.A., Markwald, R.R. and Fitzharris, T.P. (1984) Selected views of early heart development by scanning electron microscopy. *Scanning Electron Micros.* **4**: 1983–1993.

Hebel, R and Stromberg, M.W. (1986) *Anatomy and Embryology of the Laboratory Rat*. Worthsee: BioMed Verlag.

Hess, A.F. (1913) The gastric secretion of infants at birth. *Am. J. Dis. Child.* **6**: 264–276.

Heuser, C.H. and Streeter, G.L. (1941) Development of the macaque embryo. *Contrib. Embryol.* **29**: 15–55.

Hilfer, S.R., Rayner, R.M. and Brown J.W. (1985) Mesenchymal control of branching pattern in the fetal mouse lung. *Tissue Cell* **17**: 523–538.

Hillman, N. and Tasca, R. (1969) Ultrastructural and autoradiographic studies of mouse cleavage stages. *Am. J. Anat.* **126**: 151–174.

Hislop, A.A., Wigglesworth, J.S. and Desai, R. (1986) Alveolar development in the human fetus and infant. *Early Hum. Dev.* **13**: 1–11.

Ho, E. and Shimada, Y. (1978) Formation of the epicardium studied with the scanning electron microscope. *Dev. Biol.* **66**: 579–585.

Hogan, B.L.M., Horsburgh, G., Cohen, J., Hetherington, C.M., Fisher, G. and Lyon, M.F. (1986) *Small eyes (Sey)*: a homozygous lethal mutation on chromosome 2 which affects the differentiation of both lens and nasal placodes in the mouse. *J. Embryol. Exp. Morphol.* **97**: 95–110.

Holland, P.W.H. and Hogan, B.L.M. (1988) Spatially restricted patterns of the homeobox-containing gene *Hox 2.1* during mouse embryogenesis. *Development* **102**: 159–174.

Hörstadius, S. and Sellman, S. (1941) Experimental studies on the determination of the chondrocranium in *Amblystoma mexicanum. Ark. Zool. Stockholm* 33A, no. 13, 1–8 (cited by Le Douarin, 1982).

Hörstadius, S. and Sellman, S. (1946) Experimentelle über die Determination des Knorpeligen Kopfskelettes bei Urodelen. *Nova Acta Soc. Scient. Uppsaliensis Ser.* 4, **13**: 1–170 (cited by Le Douarin, 1982).

Hoshino, K. (1967) Comparative study of the skeletal development in the fetus of rat and mouse. *Congenital Anomalies (Japan)* **7**: 32–38.

Huang, R., Zhi, Q., Wilting, J. and Christ, B. (1994) The fate of somitocoele cells in avian embryos. *Anat. Embryol.* **190**: 243–250.

Huang, R., Zhi, Q., Neubüser, A., Müller, T.S., Brand-Saberi, B., Christ, B. and Wilting, J. (1996) Function of somite and somitocoele cells in the formation of the vertebral motion segment in avian embryos. *Acta Anat.* **155**: 231–241.

Hummel, K.P., Richardson, F.L. and Fekete, E. (1966) Anatomy. In: *Biology of the Laboratory Mouse*, 2nd edn, ed. E.L. Green, pp. 247–307. New York: McGraw-Hill.

Humphrey, T. (1969) The relation between human fetal mouth opening reflexes and closure of the palate. *J. Anat.* **125**: 317–344.

Hunter, R.H.F. (1988) *The Fallopian Tubes: Their Role in Fertility and Infertility*. Berlin: Springer-Verlag.

Hutchins, G.M., Kessler-Hanna, A. and Moore, G.W. (1988) Development of the coronary arteries in the embryonic human heart. *Circulation* **77**: 1250–1257.

Hutchinson, D.L., Hunter, C.B., Neslen, E.D. and Plentl, A.A. (1955) The exchange of water and electrolytes in the mechanism of amniotic fluid formation and the relationship to hydramnios. *Surg. Gynecol. Obstet.* **100**: 391–396.

Ingalls, N.W. (1921) A human embryo at the beginning of segmentation, with special reference to the vascular system. *Contrib. Embryol. Carnegie Inst.* **11**: 61–90.

Jacobson, A. and Meier, S. (1987) Somitomeres: the primordial body segments. In: *Somites in Developing Embryos*, eds R. Bellairs, D.A. Ede and J.W. Lash, pp. 1–16. New York: Plenum.

Jacobson, A.G. and Tam, P.P.L. (1982) Cephalic neurulation in the mouse embryo analyzed by SEM and morphometry. *Anat. Rec.* **203**: 375–396.

Jeffcoate, T.N.A. and Scott, J.S. (1959) Polyhydramnios and oligohydramnios. *Can. Med. Assoc. J.* **80**: 77–86.

Jenkinson, J.W. (1902) Observations on the histology and physiology of the placenta in the mouse. *Tijdschr. Ned. Dierkd. Vereeniging* **2**: 124–198 (cited by Gardner and Papaioannou, 1975).

Jeon, U.W. and Kennedy, J.R. (1973) The primordial germ cells in early mouse embryos: light and electron microscopic studies. *Dev. Biol.* **31**: 275–284.

Johnson, F.P. (1910) The development of the mucous membrane of the oesophagus, stomach and small intestine of the human embryo. *Am. J. Anat.* **10**: 521–561.

Johnson, F.P. (1913) The development of the mucous membrane of the large intestine and vermiform process in the human embryo. *Am. J. Anat.* **14**: 187–233.

Johnson, F.P. (1914) The development of the rectum in the human embryo. *Am. J. Anat.* **16**: 1–58.

Johnson, M.H. and Ziomek, C.A. (1981) The foundation of two distinct cell lineages within the mouse morula. *Cell* **24**: 71–80.

Johnson, M.L. (1933) The time and order of appearance of ossification centers in the albino mouse. *Am. J. Anat.* **52**: 241–271.

Jollie, W.P. (1964) Fine structural changes in placental labyrinth of the rat with increasing gestational age. *J. Ultrastruct. Res.* **10**: 27–47.

Jones, H.W. and Scott, W.W. (1958) *Hermaphroditism, Genital Anomalies and Related Endocrine Disorders*. Baltimore: Williams and Wilkins.

Jost, A. (1947) Recherches sur la différenciation sexuelle de l'embryon de lapin. IV. Organogénèse sexuelle masculine après décapitation du foetus. *Arch. Anat. Microsc. Morphol. Exp.* **40**: 247–281 (cited by Byskov, 1986).

Jost, A. (1965) Gonadal hormones in the sex differentiation of the mammalian fetus. In: *Organogenesis*, eds R.L. De Haan and H. Ursprung, pp. 611–628. New York: Holt, Rinehart and Winston.

Jost, A. and Magre, S. (1988) Control mechanisms of testicular differentiation. *Philos. Trans. R. Soc. B* **322**: 55–61.

Jost, A., Vigier, B., Prépin, J. and Perchellet, J.P. (1973) Studies on sex differentiation in mammals. *Recent Prog. Horm. Res.* **29**: 1–41.

Kalia, M. and Richter, D. (1985) Morphology of physiologically identified slowly adapting lung stretch receptor afferents stained with intra-axonal horseradish peroxidase in the nucleus of the tractus solitarius of the cat. I. A light microscopic analysis. *J. Comp. Neurol.* **241**: 503–520.

Kanagasuntheram, R. (1957) Development of the human lesser sac. *J. Anat.* **91**: 188–206.

Kanagasuntheram, R. (1960) Some observations on the development of the human duodenum. *J. Anat.* **94**: 231–240.

Kaufman, M.H. (1973) Timing of the first cleavage division of the mouse and the duration of its component stages: a study of living and fixed eggs. *J. Cell Sci.* **12**: 799–808.

Kaufman, M.H. (1979) Cephalic neurulation and optic vesicle formation in the early mouse embryo. *Am. J. Anat.* **155**: 425–444.

Kaufman, M.H. (1981a) Parthenogenesis: a system facilitating understanding of factors that influence early mammalian development. In: *Progress in Anatomy*, eds R.J. Harrison and R.L. Holmes, vol 1, pp. 1–34. Cambridge: Cambridge University Press.

Kaufman, M.H. (1981b) The role of embryology in teratological research, with particular reference to the development of the neural tube and heart. *J. Reprod. Fertil.* **62**: 607–623.

Kaufman, M.H. (1983a) *Early Mammalian Development: Parthenogenetic Studies*. Cambridge: Cambridge University Press.

Kaufman, M.H. (1983b) The origin, properties and fate of trophoblast in the mouse. In: *Biology of Trophoblast*, eds. Y.W. Loke and A Whyte, pp. 23–68. Amsterdam:Elsevier/North Holland Biomedical Press.

Kaufman, M.H. (1983c) Occlusion of the neural lumen in early mouse embryos analysed by light and electron microscopy. *J. Embryol. Exp. Morphol.* **78**: 211–228.

Kaufman, M.H. (1986) Occlusion of the lumen of the neural tube, and its role in the early morphogenesis of the brain. In: *Spina Bifida – Neural Tube Defects*, eds D. Voth and P. Glees, pp. 29–46. Berlin: Walter de Gruyter.

Kaufman, M.H. (1988) The development of the female genital tract. In: *Genital Tract Infection in Women*, ed. M.J. Hare, pp. 3–25. Edinburgh: Churchill Livingstone.

Kaufman, M.H. (1990) Morphological stages of postimplantation embryonic development. In: *Postimplantation Mammalian Embryos: A Practical Approach*, eds A.J. Copp and D.L. Cockroft, pp. 81–91. Oxford: IRL Press.

Kaufman, M.H. (1991) Critical role of the yolk sac in erythropoiesis and in the formation of the primordial germ cells in mammals. *Int. J. Radiat. Biol.* **60**: 544–547.

Kaufman, M.H. (1992) *The Atlas of Mouse Development*. London: Academic Press.

Kaufman, M.H. (1994) *The Atlas of Mouse Development*, 2nd printing with index. London: Academic Press.

Kaufman, M.H. (1997) Mouse and human embryonic development: a comparative overview. In: *Molecular Genetics of Early Human Development*, eds T. Strachan, S. Lindsay and D.I. Wilson, pp. 77–110. Oxford: BIOS Scientific Publishers Ltd.

Kaufman, M.H., Brune, R.M., Baldock, R.A., Bard, J.B.L. & Davidson, D. (1997) Computer-aided 3-D reconstruction of serially sectioned mouse embryos: its use in integrating anatomical organization. *International Journal of Developmental Biology*, **41**, 223–233.

Kaufman, M.H., Chang, H-H. and Shaw, J.P. (1995) Craniofacial abnormalities in homozygous *Small eye (Sey/Sey)* embryos and newborn mice. *J. Anat.* **186**: 607–617.

Kaufman, M.H., Fowler, R.E., Barratt, E. and McDougall, R.D. (1989) Ultrastructural and histochemical changes in the murine zona pellucida during the final stages of oocyte maturation prior to ovulation. *Gamete Res.* **24**: 35–48.

Kaufman, M.H. and Navaratnam, V. (1981) Early differentiation of the heart in mouse embryos. *J. Anat.* **133**: 235–246.

Kaufman, M.H. and Webb, S. (1990) Postimplantation development of tetraploid mouse embryos produced by electrofusion. *Development* **110**: 1121–1132.

Keene, M.F.L. and Hewer, E.E. (1929) Digestive enzymes of the human foetus. *Lancet* **i**, 767–769.

Keibel, F. (1937) *Normentafl zur Entwicklungsgeschichte der Wanderratte (Rattus norvegicus Erxleben)*. Jena: Fischer.

Keibel, F. and Mall, F.P. (1910) *Manual of Human Embryology*, vol. 1. Philadelphia: Lippincott.

Keibel, F. and Mall, F.P. (1912) *Manual of Human Embryology*, vol. 2. Philadelphia: Lippincott.

Keiffer, H. (1926) Recherches sur la physiologie de l'amnios humain. *Gynéc. Obstét.* **14**: 1–20.

Kelly, S.J. (1975) Studies of the potency of the early cleavage blastomeres of the mouse. In: *The Early Development of Mammals*, eds M. Balls and A.E. Wild, pp. 97–107. Cambridge: Cambridge University Press.

Kelly, S.J. (1979) Studies of the developmental potential of 4- and 8-cell stage mouse blastomeres. *J. Exp. Zool.* **200**: 365–376.

Kingery, H.M. (1917) Oogenesis in the white mouse. *J. Morphol.* **30**: 261–316.

Kirby, D.R.S., Billington, W.D., Bradbury, S. and Goldstein, D.J. (1964) Antigen barrier of the mouse placenta. *Nature* **204**: 548–549.

Kirby, D.R.S. and Bradbury, S. (1965) The hemochorial mouse placenta. *Anat. Rec.* **152**: 279–282.

Kirby, M.L. (1987) Cardiac morphogenesis – recent research advances. *Pediatr. Res.* **21**: 219–224.

Kirby, M.L. and Waldo, K.L. (1990) Role of neural crest in congenital heart disease. *Circulation* **82**: 332–340.

Kirby, M.L., Gale, T.F. and Stewart, D.E. (1983) Neural crest cells contribute to normal aorticopulmonary septation. *Science* **220**: 1059–1061.

Kline, D. (1993) Cell signalling and regulation of exocytosis at fertilization of the egg. In: *Signal Transduction during Biomembrane Fusion*, ed. D.H. O'Day, pp. 75–102. San Diego: Academic Press.

Kline, D. and Kline, J.T. (1992) Repetitive calcium transients and the role of calcium in exocytosis and cell cycle activation in the mouse egg. *Dev. Biol.* **149**: 80–89.

Koff, A.K. (1933) Development of the vagina in the human fetus. *Contrib. Embryol. Carnegie Inst.* **24**: 61–90.

Kramer, T.C. (1942) The partitioning of the truncus and conus and the formation of the membranous portion of the interventricular septum in the human heart. *Am. J. Anat.* **71**: 343–370.

Kwong, W.H. and Tam, P.P.L. (1984) The pattern of alkaline phosphatase activity in the developing mouse spinal cord. *J. Embryol. Exp. Morphol.* **82**: 241–251.

Larsen, W.J. (1993) *Human Embryology*. New York: Churchill Livingstone.

Lawson, K.A. and Pedersen, R.A. (1992) Clonal analysis of cell fate during gastrulation and early neurulation in the mouse. In: *Post-implantation Development in the Mouse. CIBA Symposium* no. 165, pp. 3–20. Chichester: John Wiley.

Lawson, K.A., Meneses, J.J. and Pedersen, R.A. (1991) Clonal analysis of epiblast fate during germ layer formation in the mouse embryo. *Development* **113**: 891–911.

Layton, W.M. (1985) The biology of asymmetry and the development of the cardiac loop. In: *Cardiac Morphogenesis*, eds V.J. Ferrans, G.C. Rosenquist and C. Weinstein, pp. 134–140. New York: Elsevier Science.

Le Douarin, N.M. (1974) Cell recognition based on natural morphological nuclear markers. *Med. Biol.* **52**: 281–319.

Le Douarin, N.M. (1976) Cell migration in early vertebrate development studied in interspecific chimeras. In:

Embryogenesis in Mammals. CIBA Symposium **40**: pp. 71–101. Amsterdam: Excerpta Medica.

Le Douarin, N.M. (1982) *The Neural Crest*. Cambridge: Cambridge University Press.

Le Lièvre, C. (1976) Contribution des crêtes neurales à la genèse des structures céphaliques et cervicales chez les Oiseaux. Thèse d'Etat. Nantes. (cited in Le Douarin, 1982).

Le Lièvre, C.S. (1978) Participation of neural crest-derived cells in the genesis of the skull in birds. *J. Embryol. Exp. Morphol.* **47**: 17–37.

Leeson, C.R. and Leeson, T.S. (1965) The fine structure of the rat umbilical cord at various times of gestation. *Anat. Rec.* **151**: 183–197.

Lemire, R.J. (1969) Variations in development of the caudal neural tube in human embryos (Horizons XIV–XXI). *Teratology* **2**: 361–370.

Lewis, O.J. (1956) The development of the circulation in the spleen of the foetal rabbit. *J. Anat.* **90**: 282–289.

Lieberman, M. (1985) Initiation of cardiac contractions. In: *Cardiac Morphogenesis*, eds V.J. Ferrans, G.C. Rosenquist and C. Weinstein, pp. 141–145. New York: Elsevier Science.

Lillie, F.R. (1917) The free-martin; a study of the action of sex hormones in the foetal life of cattle. *J. Exp. Zool.* **23**: 371–452.

Lind, J. (1969) Normal perinatal circulation. In: *The Anatomy of the Developing Lung*, ed. J. Emery, pp. 116–146. London: Heinemann Medical.

Liu, H.M. and Potter, E.L. (1962) Development of the human pancreas. *Arch. Pathol.* **74**: 439–452.

Lloyd, J.R. and Clatworthy, H.W. (1958) Hydramnios as an aid to the early diagnosis of congenital obstruction of the alimentary tract: a study of the maternal and fetal factors. *Pediatrics* **21**: 903–909.

Lo, C.W. (1984) Gap junctional communication compartments and the regulation of development. *J. Embryol. Exp. Morphol.* **82** suppl. 88 (Abstract).

Lo. C.W. and Gilula, N.B. (1979) Gap junctional communication in the preimplantation mouse embryo. *Cell* **18**: 399–409.

Mackay, S., Strachan, L. and McDonald, S.W. (1993) Increased vascularity is the first sign of testicular differentiation in the mouse. *J. Anat.* **183**: 171 (Abstract).

Manasek, F.J. and Monroe, R.G. (1972) Early cardiac morphogenesis is independent of function. *Dev. Biol.* **27**: 584–588.

Manasek, F.J., Burnside, B. and Waterman, R.E. (1972) Myocardial cell shape change as a mechanism of embryonic heart looping. *Dev. Biol.* **29**: 349–371.

Manasek, F.J., Kulikowski, R.R. and Fitzpatrick, L. (1978) Cytodifferentiation: a causal antecedent of looping? *Birth Defects* **14**: 161–178.

Mann, I. (1964) *The Development of the Human Eye*, 3rd edn. London: British Medical Association.

Marieb, E.N. (1995) *Human Anatomy and Physiology*, 3rd edn. Redwood City, California: Benjamin/Cummings.

Massion, J. (1967) The mammalian red nucleus. *Physiol. Rev.* **47**: 383–436.

Mathan, M., Moxey, P.C. and Trier, J.S. (1976) Morphogenesis of fetal rat duodenal villi. *Am. J. Anat.* **146**: 73–92.

McBride, R.E., Moore, G.W. and Hutchins, G.M. (1981) Development of the outflow tract and closure of the interventricular septum in the normal human heart. *Am. J. Anat.* **160**: 309–331.

McIlhinney, R.A.J. (1983) The biology of human germ cell tumours: experimental approaches. In: *Current Problems in Germ Cell Differentiation*, eds A. McLaren and C.C. Wylie, pp. 175–197. Cambridge: Cambridge University Press.

McKay, D.C., Hertig, A.T., Adams, E.C. and Danziger, S. (1953) Histochemical observations on the germ cells of human embryos. *Anat. Rec.* **117**: 201–219.

McKelvey, J.L. and Baxter, J.S. (1935) Abnormal development of the vagina and genitourinary tract. *Am. J. Obstet. Gynecol.* **29**: 267–271.

McMinn, R.M.H. (1990) *Last's Anatomy. Regional and Applied*, 8th edn. Edinburgh: Churchill Livingstone.

Meier, S. (1981) Development of the chick embryo mesoblast: morphogenesis of the prechordal plate and cranial segments. *Dev. Biol.* **83**: 49–61.

Meier, S. and Tam, P.P.L. (1982) Metameric pattern development in the embryonic axis of the mouse. I. Differentiation of the cranial segments. *Differentiation* **21**: 95–108.

Melloni, B.J. (1957) *The Internal Ear, An Atlas of Some Pathological Conditions of the Eye, Ear and Throat*. Chicago: Abbott Laboratories.

Mérida-Velasco, J.A., García-García, J.D., Espín-Ferra, J. and Linares, J. (1989) Origin of the ultimobranchial body and its colonizing cells in human embryos. *Acta Anat.* **136**: 325–330.

Merklin, R.J. and Michels, N.A. (1958) The variant renal and suprarenal blood supply with data on the inferior phrenic, ureteral and gonadal arteries: a statistical analysis based on 185 dissections and a review of the literature. *J. Int. Coll. Surg.* **29**: 41–76.

Meyer, R. (1909) Zur Kenntnis des Gartnerschen (oder Wolffschen) Ganges besonders in der Vagina und dem Hymen des Menschen. *Archiv. Mikrosk. Anat.* **73**: 751–792.

Miyazaki, S. and Igusa, Y. (1981) Fertilization potential in golden hamster eggs consists of recurring hyperpolarizations. *Nature* **290**: 706–707.

Moffat, D.B. (1959) Developmental changes in the aortic arch system in the rat. *Am. J. Anat.* **105**: 1–35.

Moore, K.L. and Persaud, T.V.N. (1993) *The Developing Human*, 4th edn. Philadelphia: W.B. Saunders.

Morriss, G.M. and Thorogood, P.V. (1978) An approach to neural crest cell migration and differentiation in mammalian embryos. In: *Development in Mammals*, vol. 3, ed. M.H. Johnson, pp. 363–412. Amsterdam: Elsevier/North-Holland.

Morriss-Kay, G. and Tan, S.-S. (1987) Mapping cranial neural crest cell migration pathways in mammalian embryos. *Trends Genet.* **3**: 257–261.

Mossman, H.W. (1937) Comparative morphogenesis of the fetal membranes and accessory uterine structures. *Contrib. Embryol. Carnegie Inst.* **26**: 133–246.

Moxey, P.C. and Trier, J.S. (1978) Specialized cell types in the human fetal small intestine. *Anat. Rec.* **191**: 269–286.

Mullen, R.J. and Whitten, W.K. (1971) Relationship of genotype and degree of chimerism in coat colour to sex ratios and gametogenesis in chimeric mice. *J. Exp. Zool.* **178**: 165–176.

Müller, F. and O'Rahilly, R. (1980) The human chondrocranium at the end of the embryonic period proper, with particular reference to the nervous system. *Am. J. Anat.* **159**: 33–58.

Müller, F. and O'Rahilly, R. (1994) Occipitocervical

segmentation in staged human embryos. *J. Anat.* **185**: 251–258.

Munson, P.L., Hirsch, P.F., Brewer, H.B., Reisfeld, R.A., Cooper, C.W., Wasthed, B., Orimo, H. and Potts, J.T. (1968) Thyrocalcitonin. *Recent Prog. Horm. Res.* **24**: 589–650.

Murakami, R. (1987) A histological study of the development of the penis of wild-type and androgen-insensitive mice. *J. Anat.* **153**: 223–231.

Murakami, R. and Mizuno, T. (1984) Histogenesis of the os penis and os clitoridis in rats. *Dev. Growth Diffn.* **26**: 419–426.

Nadijcka, M and Hillman, N. (1974) Ultrastructural studies of the mouse blastocyst substages. *J. Embryol. Exp. Morphol.* **32**: 675–695.

Navaratnam, V. (1965) Development of the nerve supply to the human heart. *Br. Heart J.* **27**: 640–650.

Navaratnam, V., Kaufman, M.H., Skepper, J.N., Barton, S. and Guttridge, K.M. (1986) Differentiation of the myocardial rudiment of mouse embryos; an ultrastructural study including freeze–fracture replication. *J. Anat.* **146**: 65–85.

Niswander, L. (1996) Molecular control of growth and pattern formation in the developing vertebrate limb. In: *Mammalian Development*, ed. P. Lonai, pp. 135–150. Amsterdam: Harwood Academic.

Noback, C.R., Barnett, J.C. and Kupperman, H.S. (1949) The time of appearance of ossification centers in the rat is influenced by injections of thyroxin, thiouracil, estradiol and testosterone propionate. *Anat. Rec.* **103**: 49–68.

Noden, D.M. (1983) The role of the neural crest in patterning of avian cranial skeleton, connective, and muscle tissues. *Dev. Biol.* **96**: 144–165.

Noden, D.M. (1984) The use of chimeras in analyses of craniofacial development. In: *Chimeras in Developmental Biology*, eds N.M. Le Douarin and A. McLaren, pp. 241–280. London: Academic Press.

Noden, D.M. (1988) Interactions and fates of avian craniofacial mesenchyme. *Dev.* **103** suppl. 121–140.

Norris, E.H. (1937) The parathyroid glands and the lateral thyroid in Man: their morphogenesis, histogenesis, topographic anatomy and prenatal growth. *Contrib. Embryol. Carnegie Inst.* **26**: 247–294.

Norris, E.H. (1946) Anatomical evidence of prenatal function of the human parathyroid glands. *Anat. Rec.* **96**: 129–141.

Odgers, P.N.B. (1935) The formation of the venous valves, the foramen secundum and the septum secundum in the human heart. *J. Anat.* **69**: 412–422.

Odor, D.L. and Blandau, R.J. (1969) Ultrastructural studies on fetal and early postnatal mouse ovaries. II. Cytodifferentiation. *Am. J. Anat.* **125**: 177–216.

Ohno, S. (1963) Life history of female germ cells in mammals. In: *Proc. Int. Conf. Congenital Malformations*, 2nd edn, pp. 36–42. (cited by Byskov, 1986).

O'Rahilly, R. (1966) The early development of the eye in staged human embryos. *Contrib. Embryol.* **38**: 1–42.

O'Rahilly, R. (1971) The timing and sequence of events in human cardiogenesis. *Acta Anat.* **79**: 70–75.

O'Rahilly. R. (1977) Prenatal human development. In: *Biology of the Uterus*, 2nd edn, ed. R.A. Wynn, pp. 35–57. New York: Plenum.

O'Rahilly, R. (1978) The timing and sequence of events in the development of the human digestive system and associated structures during the embryonic period proper. *Anat. Embryol.* **153**: 123–136.

O'Rahilly, R. (1983) The timing and sequence of events in the development of the human eye and ear during the embryonic period proper. *Anat. Embryol.* **168**: 87–99.

O'Rahilly, R. and Boyden, E.A. (1973) The timing and sequence of events in the development of the human respiratory system during the embryonic period proper. *Z. Anat. Entwickl.* **141**: 237–250.

O'Rahilly, R. and Müller, F. (1987) *Developmental Stages in Human Embryos*. Washington, DC: Carnegie Institute Publication no. 637.

O'Rahilly, R. and Müller, F. (1994) *The Embryonic Human Brain: An Atlas of Developmental Stages*. New York: Wiley–Liss.

O'Rahilly, R., Müller, F. and Meyer, D.B. (1983) The human vertebral column at the end of the embryonic period proper. 2. The occipitocervical region. *J. Anat.* **136**: 181–195.

O'Rahilly, R. and Tucker, J.A. (1973) The early development of the larynx in staged human embryos. I. Embryos of the first five weeks (to stage 15). *Ann. Otol. Rhinol. Laryngol.* **82**: 1–27.

Orts Llorca, F., Collardo, J.J. and Gil, D.R. (1960) La fase plexiforme del desarrollo cardiaco en el hombre. *Anal. Desarr.* **8**: 79–98 (cited by Hamilton and Mossman, 1972).

Osathanondh, V. and Potter, E.L. (1964a) Pathogenesis of polycystic kidneys. Type I. Due to hyperplasia of interstitial portions of collecting tubules. *Arch. Pathol.* **77**: 466–473.

Osathanondh, V. and Potter, E.L. (1964b) Pathogenesis of polycystic kidneys. Type II. Due to inhibition of ampullary activity. *Arch. Pathol.* **77**: 474–484.

Osathanondh, V. and Potter, E.L. (1964c) Pathogenesis of polycystic kidneys. Type III. Due to multiple abnormalities of development. *Arch. Pathol.* **77**: 485–501.

Osathanondh, V. and Potter, E.L. (1964d) Pathogenesis of polycystic kidneys. Type IV. Due to urethral obstruction. *Arch. Pathol.* **77**: 502–509.

O'Shea, K.S. (1987) Differential deposition of basement membrane components during formation of the caudal neural tube in the mouse embryo. *Development*, **99**: 509–519.

Otis, E.M. and Brent, R. (1954) Equivalent ages in mouse and human embryos. *Anat. Rec.* **120**: 33–64.

Ozdzeński, W. (1967) Observations on the origin of primordial germ cells in the mouse. *Zool. Polon.* **17**: 367–379.

Padget, D.H. (1948) The development of the cranial arteries in the human embryo. *Contrib. Embryol. Carnegie Institute* **32**: 205–261.

Parrington, J., Lai, F.A. and Swann, K. (1998) A novel protein for Ca^{2+} signaling at fertilization. In: *Current Topics in Developmental Biology*, vol. 39, eds. R.A. Pedersen and C.P. Schatten. pp. 215–243. Dan Diego: Academic Press.

Patten, B.M. (1960) The development of the heart. In: *Pathology of the Heart*, 2nd edn, ed. S.E. Gould, pp. 24–92. Springfield, Illinois: C.C. Thomas.

Pattle, R.E. (1969) The development of the foetal lung. In: *Foetal Autonomy*, eds. G.E.W. Wolstenholme and M. O'Connor, pp. 132–146 (Includes discussion). London: J. and A. Churchill.

Patton, J.T. and Kaufman, M.H. (1995) The timing of ossification of the limb bones, and growth rates of various long bones of the fore and hind limbs of the prenatal and early postnatal laboratory mouse. *J. Anat.* **186**: 175–185.

Pauerstein, C.J. (1974) *The Fallopian Tube* Baltimore: Lea and Fabiger.

Paxinos, G., Törk, I., Tecott, L.H. and Valentino, K.L. (1991) *Atlas of the Developing Rat Brain*. San Diego: Academic Press.

Peacock, A. (1951) Observations on the prenatal development of the intervertebral disc in Man. *J. Anat.* **85**: 260–274.

Pearce, R.M. (1903) The development of the islands of Langerhans in the human embryo. *Am. J. Anat.* **2**: 445–455.

Pearson, A.A. (1980) The development of the eyelids. Part I. External features. *J. Anat.* **130**: 33–42.

Pei, Y.F. and Rhodin, J.A.G. (1970) The prenatal development of the mouse eye. *Anat. Rec.* **168**: 105–126.

Pelliniemi, L.J. (1975) Ultrastructure of gonadal ridge in male and female pig embryos. *Anat. Embryol.* **147**: 19–34.

Penefsky, Z.J. (1984) Regulation of contractility in developing heart. In: *The Developing Heart*, ed. M.J. Legato, pp. 113–148. Boston: Martinus-Nijhoff.

Peters, H. (1970) Migration of gonocytes into the mammalian gonad and their differentiation. *Philos. Trans. R. Soc. Lond.* [*Biol.*] **9**: 91–101.

Pexieder, T. (1975) Cell death in the morphogenesis and teratogenesis of the heart. *Adv. Anat. Embryol. Cell Biol.* **51**: 3–99.

Pexieder, T. (1978) Development of the outflow tract of the embryonic heart. In: *Morphogenesis and Malformation of the Cardiovascular System*, eds G. Rosenquist and D. Bergsma, pp. 29–68. New York: Alan R. Liss.

Platzer, A.C. (1978) The ultrastructure of normal myogenesis in the limb of the mouse. *Anat. Rec.* **190**: 639–658.

Poelmann, R.E. (1981a) The formation of the embryonic mesoderm in the early post-implantation mouse embryo. *Anat. Embryol.* **162**: 29–40.

Poelmann, R.E. (1981b) The head-process and the formation of the definitive endoderm in the mouse embryo. *Anat. Embryol.* **162**: 41–49.

Polani, P.E. (1962) Sex chromosome anomalies in man. In: *Chromosomes in Medicine*, ed. J.L. Hamerton, pp. 73–139. London: Heinemann.

Pomeranz, H.D., Rothman, T.P. and Gershon, M.D. (1991) Colonization of the post-umbilical bowel by cells derived from the sacral neural crest: direct tracing of cell migration using an intercalating probe and a replication-deficient retrovirus. *Development* **111**: 647–655.

Potter, E.L. (1965) Development of the human glomerulus. *Arch. Pathol.* **80**: 241–255.

Prader, A. (1947a) Die frühembryonale Entwicklung der meschlichen Zwischenwirbelscheibe. *Acta Anat.* **3**: 68–83.

Prader, A. (1947b) Die Entwicklung der Zwischenwirbelscheibe beim menschlichen Keimling. *Acta Anat.* **3**: 115–152.

Pritchard, J.A. (1966) Fetal swallowing and amniotic fluid volume. *Obstet. Gynecol.* **28**: 606–610.

Puschel, A.W., Westerfield, M. and Dressler, G.R. (1992) Comparative analysis of *Pax-2* protein distributions during neurulation in mice and zebrafish. *Mech. Dev.* **38**: 197–208.

Read, J.B. and Burnstock, G. (1970) Development of the adrenergic innervation and chromaffin cells in human fetal gut. *Dev. Biol.* **22**: 513–534.

Reinius, S. (1965) Morphology of the mouse embryo from the time of implantation to mesoderm formation. *Z. Zellforsch. Mikrosk. Anat.* **68**: 711–723.

Rodriguez, E.R. and Ferrans, V.J. (1985) The assembly of myofibrils in the developing heart. In: *Cardiac Morphogenesis*, eds V.J. Ferrans, G.C. Rosenquist and C. Weinstein, pp. 86–93. New York: Elsevier Science.

Rohwedel, J., Sehlmeyer, U., Shan, J., Meister, A. and Wobus, A.M. (1996) Primordial germ cell-derived mouse embryonic germ (EG) cells in vitro resemble undifferentiated stem cells with respect to differentiation capacity and cell cycle distribution. *Cell Biol. Int.* **20**: 579–587.

Romanoff, A.L. (1960) *The Avian Embryo: Structural and Functional Development*. New York: Macmillan.

Rossant, J. (1975) Investigation of the determinative state of the mouse inner cell mass. II. The fate of isolated inner cell masses transferred to the oviduct. *J. Embryol. Exp. Morphol.* **33**: 991–1001.

Rossant, J. and Papaioannou, V.E. (1977) The biology of embryogenesis. In: *Concepts in Mammalian Embryogenesis*, ed. M.I. Sherman, pp. 1–36. Cambridge, MA: MIT Press.

Rugh, R. (1968) *The Mouse. Its Reproduction and Development*. Minneapolis: Burgess Publishing Company. (Reprinted 1990; Oxford: Oxford University Press.)

Rychter, Z. (1978) Analysis of relations between aortic arches and aortico-pulmonary septation. In: *Morphogenesis and Malformation of the Cardiovascular System*, eds G. Rosenquist and D. Bergsma, pp. 443–448. New York: Alan R Liss.

Sainio, K., Hellstedt, P., Kreidbeurg, J.A., Saxén, L. and Sariola, H. (1997) Differential regulation of two sets of mesonephric tubules by WT-1. *Development* **124**: 1293–1299.

Sakamoto, M.K., Nakamura, K., Handa, J., Kihara, T. and Tanimura, T. (1989) Morphogenesis of the secondary palate in mouse embryos with special reference to the development of rugae. *Anat. Rec.* **223**: 299–310.

Sariola, H., Holm, K. and Henke-Fahle, S. (1988) Early innervation of the metanephric kidney. *Development* **104**: 589–599.

Saunders. J.W. (1977) The experimental analysis of chick limb bud development. In: *Vertebrate Limb and Somite Morphogenesis*, eds D.A. Ede, J.R. Hinchliffe and M. Balls, pp. 1–24. Cambridge: Cambridge University Press.

Saunders, P. and Rhodes, P. (1973) The origin and circulation of the amniotic fluid. In: *Amniotic Fluid. Research and Clinical Application*, eds D.V.I. Fairweather and T.K.A.B. Eskes, pp. 1–18. Amsterdam: Excerpta Medica.

Scammon, R.E. and Kittelson, J.A. (1926) The growth of the gastro-intestinal tract of the human fetus. *Proc. Soc. Exp. Biol. Med.* **24**: 303–307.

Schambra, U. (1992) *Atlas of the Prenatal Mouse Brain*. San Diego: Academic Press.

Schoenwolf, G.C. (1984) Histological and ultrastructural studies of secondary neurulation in mouse embryos. *Am. J. Anat.* **169**: 361–376.

Schoenwolf, G.C. and Delongo, J. (1980) Ultrastructure of secondary neurulation in the chick embryo. *Am. J. Anat.* **158**: 43–63.

Schour, J. and Massler, M. (1949) The teeth. In: *The Rat in Laboratory Investigation*, 2nd edn, eds E.J. Farris and J.Q. Griffith. Philadelphia: J.B. Lippincott.

Seifert, R. and Christ, B. (1990) On the differentiation and origin of myoid cells in the avian thymus. *Anat. Embryol.* **181**: 287–298.

Seller, M.J. and Perkins-Cole, K.J. (1987) Sex difference in mouse embryonic development at neurulation. *J. Reprod. Fertil.* **79**: 159–161.

Senior, H.D. (1919) The development of the arteries of the human lower extremity. *Am. J. Anat.* **25**: 55–96.

Senior, H.D. (1920) The development of the human femoral artery, a correction. *Anat. Rec.* **17**: 271–280.

Setchell, B.P. (1978) *The Mammalian Testis.* London: Elek.

Sevel, D. (1988) A reappraisal of development of the eyelids. *Eye* **2**: 123–129.

Sevel, D. and Isaacs, R. (1989) A re-evaluation of corneal development. *Trans. Am. Ophthalmol. Soc.* **86**: 178–204 (discussion 205–207).

Severn, C.B. (1971) A morphological study of the development of the human liver. I. Development of the hepatic diverticulum. *Am. J. Anat.* **131**: 133–158.

Severn, C.B. (1972) A morphological study of the development of the human liver. II. Establishment of liver parenchyma, extrahepatic ducts and associated venous channels. *Am. J. Anat.* **133**: 85–108.

Sgalitzer, K.E. (1941) Contribution to the study of the morphogenesis of the thyroid gland. *J. Anat.* **75**: 389–405.

Shimamura, K., Hartigan, D.J., Martinez S., Puelles, L. and Rubenstein, J.L.R. (1995) Longitudinal organization of the anterior neural plate and neural tube. *Development* **121**: 3923–3933.

Shore, L.S., Shemesh, M. and Mileguir, F. (1984) Foetal testicular steroidogenesis and responsiveness to LH in freemartins and their male co-twins. *Int. J. Androl.* **7**: 87–93.

Short, R.V. (1970) The bovine freemartin: a new look at an old problem. *Philos. Trans. R. Soc. Lond. [Biol.]* **259**: 141–147.

Smith C. (1965) Studies on the thymus of the mammal. XIV. Histology and histochemistry of embryonic and early postnatal thymuses of C57BL/6 and AKR strain mice. *Am. J. Anat.* **116**: 611–630.

Smith, C. and Clifford, C.P. (1962) Histochemical study of aberrant thyroid glands associated with the thymus of the mouse. *Anat. Rec.* **143**: 229–238.

Smith, F.G., Adams, F.H., Borden, M. and Hilburn, J. (1966) Studies of renal function in the intact fetal lamb. *Am. J. Obstet. Gynecol.* **96**: 240–246.

Smith, R. and McLaren, A. (1977) Factors affecting the time of formation of the mouse blastocoele. *J. Embryol Exp. Morphol.* **41**: 73–92.

Smith, R.B. and Taylor, I.M. (1972) Observations on the intrinsic innervation of the human foetal oesophagus between the 10-mm and 140-mm crown–rump length stages. *Acta Anat.* **81**: 127–138.

Smith, R.K.W. and Johnson, M.H. (1985) DNA replication and compaction in the cleaving embryo of the mouse. *J. Embryol. Exp. Morphol.* **89**: 133–148.

Snell, G.D. and Stevens, L.C. (1966) Early embryology. In: *Biology of the Laboratory Mouse*, 2nd edn, pp. 205–245. New York: McGraw-Hill.

Snow, M.H.L. and Monk, M. (1983) Emergence and migration of mouse primordial germ cells. In: *Current Problems in Germ Cell Differentiation*, eds A. McLaren and C.C. Wylie. *Br. Soc. Dev. Biol. Symp.* **7**: 115–135.

Spark, C. and Dawson, A.B. (1928) The order and time of appearance of centers of ossification in the fore and hind limbs of the albino rat, with special reference to the possible influence of the sex factor. *Am. J. Anat.* **41**: 411–445.

Spaulding, M.H. (1921) The development of the external genitalia in the human embryo. *Contrib. Embryol. Carnegie Inst.* **13**: 67–88.

Spiegelman, H. and Bennett, D. (1973) A light- and electron-microscope study of primordial germ cells in the early mouse embryo. *J. Embryol. Exp. Morphol.* **30**: 97–118.

Spivack, M. (1943) The presence or absence of nerves in the umbilical blood vessels of man and guinea pig. *Anat. Rec.* **85**: 85–109.

Steinberger, E. and Steinberger, A. (1975) Spermatogenic function of the testis. In: *Handbook of Physiology. Endocrinology*, eds R.O. Greep and E.B. Astwood, pp. 1–19. Washington, D.C.: American Philosophical Society.

Steinhardt, R.A., Epel, D., Carroll, E.J. and Yanagimachi, R. (1974) Is calcium ionophore a universal activator for unfertilized eggs. *Nature* **252**: 41–43.

Sterin-Speziale, N., Gimeno, M.F., Zapata, C., Bagnati, P.E. and Gimeno, A.L. (1978) The effect of neurotransmitters, bradikinin, prostaglandins, and follicular fluid on spontaneous contractile characteristics of human fimbriae and tubo-ovarian ligaments isolated during different stages of the sexual cycle. *Int. J. Fertil.* **23**: 1–11.

Steven, D.H. and Morriss, G. (1975) Development of the foetal membranes. In: *Comparative Placentation. Essays in Structure and Function*, ed. D.H. Steven, pp. 58–86. London: Academic Press.

Stevens, L.C. (1975) Comparative development of normal and parthenogenic mouse embryos, early testicular and ovarian teratomas and embryoid bodies. In: *Teratomas and Differentiation*, eds M. Sherman and D Solter, pp. 17–32. New York: Academic Press.

Streeter, G.L. (1920) A human embryo (Mateer) of the presomite period. *Contr. Embryol. Carnegie Inst.* **9**: 389–424.

Strong, R.M. (1925) The order, time, and rate of ossification of the albino rat (*Mus norvegicus albinus*). *Am. J. Anat.* **36**: 313–356.

Sutherland, S.D. (1965) The intrinsic innervation of the liver. *Rev. Int. Hepat.* **15**: 569–578 (cited by Williams, 1995).

Sutherland, S.D. (1966) The intrinsic innervation of the gall bladder in *Macaca rhesus* and *Cavia Porcellus*. *J. Anat.* **100**: 261–268.

Sutherland, S.D. (1967) The neurons of the gall bladder and gut. *J. Anat.* **101**: 701–709.

Swain, A., Narvaez, V., Burgoyne, P., Camerino, G. and Lovell-Badge R. (1998) Dax1 antagonizes Sry action in mammalian sex determination. *Nature* **391**: 761–767.

Sweeny, L.J. (1988) A molecular view of cardiogenesis. *Experientia* **44**: 930–936.

Szöllösi, D. (1967) Development of cortical granules and the cortical reaction in rat and hamster eggs. *Anat. Rec.* **159**: 431–446.

Szöllösi, D. (1971) Morphological changes in mouse eggs due to aging in the fallopian tube. *Am. J. Anat.* **130**: 209–226.

Szöllösi, D. (1973) Mammalian eggs aging in the follopian tubes. In: *Aging Gametes*, ed. R.J. Blandau, pp. 98–121. Basel: Karger.

Takamura, K., Okishima, T., Ohdo, S. and Hayakawa, K. (1990) Association of cephalic neural crest cells with cardiovascular development, particularly that of the semilunar valves. *Anat. Embryol.* **182**: 263–272.

Tam, P.P.L. (1981) The control of somitogenesis in mouse embryos. *J. Embryol. Exp. Morphol.* **65** Suppl. 103–128.

Tam, P.P.L. (1986) A study on the pattern of prospective somites in the presomitic mesoderm of mouse embryos. *J. Embryol Exp. Morphol.* **92**: 269–285.

Tam, P.P.L. and Beddington, R.S.P. (1992) Establishment and organization of germ layers in the gastrulating mouse embryo. In: *Postimplantation Development in the Mouse.*

REFERENCES

CIBA Symposium no. 165, pp. 27–49. Chichester: John Wiley.

Tam, P.P.L. and Kwong, W.H. (1987) A study on the pattern of alkaline phosphatase activity correlated with observations on silver-impregnated structures in the developing mouse brain. *J. Anat.* **150**: 169–180.

Tam, P.P.L. and Meier, S. (1982) The establishment of a somitomeric pattern in the mesoderm of the gastrulating mouse embryo. *Am. J. Anat.* **164**: 209–225.

Tam, P.P.L. and Snow, M.H.L. (1981) Proliferation and migration of primordial germ cells during compensatory growth in mouse embryos. *J. Embryol. Exp. Morphol.* **64**: 133–147.

Tam, P.P.L., Meier, S. and Jacobson, A.G. (1982) Differentiation of the metameric pattern in the embryonic axis of the mouse. II. Somitomeric organization of the presomitic mesoderm. *Differentiation* **21**: 109–122.

Tam, P.P.L., Williams, E.A. and Chan, W.Y. (1993) Gastrulation in the mouse embryo: ultrastructural and molecular aspects of germ layer morphogenesis. *Microsc. Res. Techn.* **26**: 301–328.

Tan, S.S. and Morriss-Kay, G.M. (1986) Analysis of cranial neural crest cell migration and early fates in postimplantation rat chimaeras. *J. Embryol. Exp. Morphol.* **98**: 21–58.

Tarkowski, A.K. (1959) Experiments on the development of isolated blastomeres of mouse eggs. *Nature* **184**: 1286–1287.

Tarkowski, A.K. (1961) Mouse chimaeras developed from fused eggs. *Nature* **190**: 857–860.

Tarkowski, A.K. and Wróblewska, J. (1967) Development of blastomeres of mouse eggs isolated at the 4- and 8-cell stage. *J. Embryol. Exp. Morphol.* **18**: 155–180.

Teillet, M.A. and Le Douarin, N.M. (1983) Consequences of neural tube and notochord excision on the development of the peripheral nervous system in the chick embryo. *Dev. Biol.* **98**: 192–211.

Tessier-Lavigne, M. and Goodman, C.S. (1996) The molecular biology of axon guidance. *Science* **274**: 1123–1133.

Theiler, K. (1972) *The House Mouse: Development and Normal Stages from Fertilization to 4 weeks of Age*. Berlin: Springer-Verlag.

Theiler, K. (1989) *The House Mouse: Atlas of Embryonic Development*. New York: Springer-Verlag.

Thorogood, P. (1981) Neural crest cells and skeletogenesis in vertebrate embryos. *Histochem. J.* **13**: 631–642.

Thorogood, P. (1988) The developmental specification of the vertebrate skull. *Development* **103**, Suppl. 141–153.

Tickle, C., Summerbell, D. and Wolpert, L. (1975) Positional signalling and specification of digits in chick limb morphogenesis. *Nature* **254**: 199–202.

Tokioka, T. (1973) The arterial system of the spinal cord in the rat. *Okajimas Folia Anat. Jap.* **50**: 133–182 (cited by Hebel and Stromberg, 1986).

Tosney, K.W., Hotary, K.B. and Lance-Jones, C. (1995) Specifying the target identity of motoneurons. *BioEssays* **17**: 379–382.

Trainor, P.A. and Tam, P.P.L. (1994) Cranial paraxial mesoderm and neural crest cells of the mouse embryo: co-distribution in the craniofacial mesenchyme but distinct segregation in branchial arches. *Development* **121**: 2569–2582.

Trainor, P.A., Tan, S.-S. and Tam, P.P.L. (1994) Cranial paraxial mesoderm: regionalisation of cell fate and impact on craniofacial development in mouse embryos. *Development* **120**: 2397–2408.

Tsunoda, Y., Tokunaga, T. and Sugie, T. (1985) Altered sex ratio of live young after transfer of fast- and slow-developing mouse embryos. *Gamete Res.* **12**: 301–304.

Tyler, M.S. (1983) Development of the frontal bone and cranial meninges in the embryonic chick: an experimental study of tissue interactions. *Anat. Rec.* **206**: 61–70.

Tyler, M.S. and Hall, B.K. (1977) Epithelial influences on skeletogenesis in the mandible of the embryonic chick. *Anat. Rec.* **188**: 229–240.

Van Blerkom, J. and Motta, P. (1979) *The Cellular Basis of Mammalian Reproduction*. Baltimore: Urban and Schwarzenberg.

Van Blerkom, J. and Motta, P. (eds) (1984) *Ultrastructure of Reproduction. Gametogenesis, Fertilization, and Embryogenesis*. Boston: Martinus Nijhoff.

Van Heyningen, H.E. (1961) The initiation of thyroid function in the mouse. *Endocrinology* **69**: 720–727.

Van Mierop, L.H. (1969) Embryology of the heart. In: *The Ciba Collection of Medical Illustrations. The Heart*, ed. F.H. Netter, pp. 112–130. Summit, NJ: The Ciba Pharmaceutical Company (cited by Clark, 1990).

Vellguth, S., von Gaudecker, B. and Müller-Hermelink, H.K. (1985) The development of the human spleen. Ultrastructural studies in fetuses from the 14th to the 24th week of gestation. *Cell Tissue Res.* **242**: 579–592.

Vernall, D.G. (1962) The human embryonic heart in the seventh week. *Am. J. Anat.* **111**: 17–24.

Virágh, S. and Challice, C.E. (1973) Origin and differentiation of cardiac muscle cells in the mouse. *J. Ultrastruct. Res.* **41**: 1–24.

Virágh, S and Challice, C.E. (1977) The development of the conduction system in the mouse embryo heart. II. Histogenesis of the atrioventricular node and bundle. *Dev. Biol.* **56**: 397–411.

Vögler, H. (1987) *Human Blastogenesis. Formation of the Extraembryonic Cavities*. Basel: Karger.

Vosburgh, G.J., Flexner, L.B., Cowie, D.B., Hellman, L.M., Proctor, N.K. and Wilde, W.S. (1948) The rate of renewal in woman of the water and sodium of the amniotic fluid as determined by tracer techniques. *Am. J. Obstet. Gynecol.* **56**: 1156–1159.

Vuillemin, M. and Pexieder, T. (1989a) Normal stages of cardiac organogenesis in the mouse. I. Development of the external shape of the heart. *Am. J. Anat.* **184**: 101–113.

Vuillemin, M. and Pexieder, T. (1989b) Normal stages of cardiac organogenesis in the mouse. II. Development of the internal relief of the heart. *Am. J. Anat.* **184**: 114–128.

Walls, E.W. (1947) The development of the specialized conducting tissue of the human heart. *J. Anat.* **81**: 93–110.

Warkany, J. (1971) *Congenital Malformations: Notes and Comments*. Chicago: Year Book Medical.

Warkany, J. and Takacs, E. (1965) Congenital malformations in rats from streptonigrin. *Arch. Pathol.* **79**: 65–79.

Wassarman, P.M. (1988) Zona pellucida glycoproteins. *Annu. Rev. Biochem.* **57**: 415–442.

Wassarman, P.M. (1990) Regulation of mammalian fertilization by zona pellucida glycoproteins. *J. Reprod. Fertil.* Suppl. **42**: 79–87.

Webb, S., Brown, N.A. and Anderson, R.H. (1998) Formation of the atrioventricular septal structures in the normal mouse. *Circ. Res.* **82**: 645–656.

Wedden, S.E., Ralphs, J.R. and Tickle, C. (1988) Pattern formation in the facial primordia. *Development* **103**: Suppl. 31–40.

Weller, G.L. (1933) Development of the thyroid, parathyroid and thymus glands in man. *Contrib. Embryol. Carnegie Inst.* **24**: 93–140.

Wells, L.J. (1954) Development of the human diaphragm and pleural sacs. *Contrib. Embryol. Carnegie Inst.* **35**: 107–134.

Wenink, A.C.G. (1976) Development of the human cardiac conducting system. *J. Anat.* **121**: 617–631.

Whitaker, M.J. and Steinhardt, R.A. (1982) Ionic regulation of egg activation. *Q. Rev. Biophys.* **15**: 593–666.

Williams, P.L. (1995) *Gray's Anatomy. The Anatomical Basis of Medicine and Surgery*, 38th edn. New York: Churchill Livingstone.

Willis, W.D. (1995) The visual system. In: *Physiology* 4th edition, eds R.M. Berne and M.N. Levy, pp. 129–153. St Louis: Mosby.

\Wilson, A.S. (1968) Investigations on the innervation of the diaphram in cats and rodents. *Anat. Rec.* **162**: 425–432

Wirtschafter, Z.T. (1960) *The Genesis of the Mouse Skeleton: A Laboratory Atlas*. Springfield, IL: C.C. Thomas.

Wirtschafter, Z.T. and Williams, D.W. (1957) Dynamics of the amniotic fluid as measured by changes in protein patterns. *Am. J. Obstet. Gynecol.* **74**: 309–313.

Witschi, E. (1951) Embryogenesis of the adrenal and the reproductive glands. *Recent Prog. Horm. Res.* **6**: 1–23.

Witschi, E. (1962) Development: rat. In: *Growth Including Reproduction and Morphological Development*, eds P.L. Altman and D.S. Dittmer, pp. 304–314. Washington, DC: Biological Handbooks of the Federation of American Societies for Experimental Biology.

Wolfson, R.J. (1966) *The Vestibular System and Its Diseases*. Philadelphia: University of Pennsylvania Press.

Wollman, S.H. and Hilfer, S.R. (1978) Embryologic origin of the various epithelial cell types in the second kind of thyroid follicle in the C_3H mouse. *Anat. Rec.* **191**: 111–122.

Wolpert, L. (1998) *Principles of Development*. London: Current Biology Ltd; Oxford: Oxford University Press.

Wong. G.B., Weinberg, S. and Symington, J.M. (1985) Morphology of the developing articular disc of the human temporomandibular joint. *J. Oral Maxillofac. Surg.* **43**: 565–569.

Woodruff, J.D. and Pauerstein, C.J. (1969) *The Fallopian Tube: Structure, Function, Pathology, and Management*. Baltimore: Williams and Wilkins.

Wrenn, J.T. and Wessells, N.K. (1969) An ultrastructural study of lens invagination in the mouse. *J. Exp. Zool.* **171**: 359–368.

Wurst, W., Auerbach, A.B. and Joyner, A.L. (1994) Multiple developmental defects in *Engrailed-1* mutant mice: an early mid-hindbrain deletion and patterning defect in forelimbs and sternum. *Development* **120**: 2065–2075.

Xu, K.P., Yadav, B.R., King, W.A. and Betteridge, K.J. (1992) Sex-related differences in developmental rates of bovine embryos produced and cultured in vitro. *Mol. Reprod. Dev.* **31**: 249–252.

Yamauchi, A. (1965) Electron microscopic observations on the developoment of S-A and A-V nodal tissues in the human embryonic heart. *Z. Anat. Entwickl.* **124**: 562–587.

Yanagimachi, R. (1994) Mammalian fertilization. In: *The Physiology of Reproduction*, 2nd edn, eds E. Knobil and D. Neill, pp. 189–317. New York: Raven Press.

Yeom, Y.I., Fuhrmann, G., Ovitt, C.E., Brehm, A., Ohbo, K., Gross, M., Hubner, K. and Scholer, H.R. (1996) Germline regulatory element of Oct-4 specific for the totipotent cycle of embryonal cells. *Development* **122**: 881–894.

Yokoh, Y. (1970) Differentiation of the dorsal mesentery in man. *Acta Anat.* **76**: 56–67.

Young, R.W. (1985) Cell differentiation in the retina of the mouse. *Anat. Rec.* **212**: 199–205.

Zamboni, L. and Merchant, H. (1973) The fine morphology of mouse primordial germ cells in extragonadal locations. *Am. J. Anat.* **137**: 299–335.

Zamboni, L., Upadhyay, S., Bézard, J. and Mauléon, P. (1981) The role of the mesonephros in the development of the sheep testis and its excurrent pathways. In: *Development and Function of Reproductive Organs*, eds A.G. Byskov and H. Peters, pp. 31–40. Amsterdam: Excerpta Medica.

Zybina, E.V. (1970) Anomalies of polyploidisation of the cells of the trophoblast. *Tsitologiya* **12**: 1081–1093 (in Russian) (cited by Ansell, 1975).

5.7 Author index

Abramovich, D.R. 30
Adams, E.C. 118
Adams, F.H. 30, 116
Adelmann, H.B. 36, 191
Adesanya, T. 154
AhPin, P. 163
Aikawa, E. 86, 87
Aiton, J.F. 55
Alberch, P. 211
Alden R.H. 12
Aldridge, R.T. 143
Alexander, D.P. 116
Allen, B. 119
Altman, J. 186, 188, 191
Alvarez-Bolado, G. 188
Amoroso, E.C. 13
Anderson, R.H. 84
Ansell, J.D. 12, 13
Arey, L.B 143, 166
Arnott, C. 163
Auerbach, A.B. 59
Austin, C.R. 9
Avery, B. 15
Azzam, M.E.A. 5

Backhouse, K.M. 125
Baden, J.M. 15
Bagnati, P.E. 5
Bak, A. 15
Baker, T.G. 121
Baldock, R.A. 59
Bannerjee, S.D. 166
Barald, K.F. 205
Barber, A.N 195
Bard, J.B.L 3, 111, 114, 197
Barnett, J.C. 97
Barr. M.L. 91, 147, 192, 198
Barratt, E. 7
Barron, D.H 91
Bartelings, M.M. 83
Bartelmez , G.W. 27
Barton, S. 38, 80
Bates, M.N. 165
Bayer, S.A. 186, 188, 191, 192
Beatty, R.A. 9
Beck, F. 13
Beddington, R.S.P. 19
Bee, J. 47
Bell, L. 138
Bell, S.C. 11
Bellairs, R. 3, 52
Bennett, D. 118
Beringer, T. 54
Bernfield, M.R. 166
Berry, J.M. 137
Betteridge, K.J. 15
Bezard, J. 123
Biggers, J.D 13
Billington, W.D. 23
Blandau, R.J. 121
Bloom, W. 27
Bolli, V.P. 190
Bonneville, K. 17
Borden, M. 116
Borysenko, M. 54
Bourne, G 24
Boyd, J.D. 32, 156, 167
Boyden, E.A. 146
Bradbury, S. 23, 32
Brambell, F.W.R. 119
Brand-Saberi, B 52
Bray, J.J. 208
Brehm, A. 118
Bremner, C.G. 133

Brent, R. 3
Brewer, H.B. 76
Briggs, D. 6
Brinkley 159
Brökelmann, J 13
Bronner-Fraser, M. 48
Brookhart, J.M. 208
Brown J.W. 146
Brown, N.A. 15, 84
Brune, R.M. 59
Bucher, U. 146
Buckingham, M. 102
Bulfone, A. 179
Bulmer, D. 116
Burgess, D.R. 138
Burgoyne, P. 15, 118
Burke, A.C. 57
Burnside, B. 80
Burnstock, G. 143
Burr, H.S. 184
Butler, H. 3, 125
Byskov, A.G. 117, 119, 123

Camerino, G. 118
Campbell, P.E. 143
Carlson, B.M. 54, 104
Carroll, E.J. 5
Chacko, A.W. 30
Challice, C.E. 38
Chan, S.T.H. 118
Chan, W.Y 18, 22
Chang, H.H. 59, 199
Chen, J.M. 58
Chiquoine, A.D 27, 38, 118, 121
Christ, B. 52, 57, 74
Christie, G.A. 3
Clark, E.B. 77
Clark, J.M. 118
Clarke, J.T. 139
Clatsworthy, H.W. 31, 116
Clermont, Y. 123
Clifford, C.P. 74
Cobbold, P.H. 6
Cohen, J. 199
Cole, R.E. 118
Cole, R.J. 11
Collardo, J.J. 79
Coltey, P.M. 212
Comline, R.S. 91
Congdon, E.D. 64, 66, 77
Conklin, J.L. 154
Conrad, J.T. 5
Constantini, F. 49, 141
Cooper, C.W. 76
Copp, A.J. 12
Cordier, A.C. 74, 76
Coulombre, A.J. 138, 179
Coulombre, J.L. 138
Couly, G.F. 212
Coupland, R.E. 154
Cowie, D.B. 30
Cragg, P.A. 208
Crockard, H.A. 55
Cross, K.W. 91
Cunha, G.R. 122
Cuthbertson, K.S.R. 6

Dalcq, A. 12
Danziger, S. 118
David, K.M. 55
Davidson, D.D. 59
Davies, J.A. 111, 114
Davies, T 15, 17, 18, 24
Davis, C.L. 77, 78, 79, 80

Dawes, G.S. 91
Dawson, G.S 97
de Beer, G.R. 47
De Haan, R.L. 78, 87
De Martino, C. 113, 116
De Vries, P.A. 83
Delongo, J. 190
Desai, R. 146
Desmond, M.E. 179
Deuchar, E.M 165
Dickson, A.D. 11, 12
Didier, E. 121
Downs, K.M. 15, 17, 18, 24, 25
Drachman, D.B. 105
Dressler, G.R. 109
Ducibella, T. 11
Durbec, P.L. 49, 141
Duthie, G.M. 122, 123
Duval, M. 12
Dziadek, M. 12

Ebers, D.W. 143
Eddy, E.M. 118
Ede, D.A. 52
Edwards, R.G. 4, 5, 123
Eggermont,, E. 139
Eik-Nes, K.B. 123
Ellington, S.K.L. 25
Ellis, S. 163
Emery, J. 144
Enders, A.C 10, 13, 23, 32
Epel, D. 5
Erickson, R.A. 114
Eskes, T.K.A.B. 30
Espän-Ferra, J. 75
Evans, M.J. 12, 13
Everett, N.B. 118

Fairweather, D.V.I. 30
Fananapazir, K. 82
Fawcett, D.W. 123
Fekete, E. 130
Ferguson, M.W.J. 159
Ferrans, V.J. 80
Findlater, G.S. 198
Fishel, S.B. 11
Fisher, G. 199
Fitzgerald, M.J. 140
Fitzharris, T.P. 82
Fitzpatrick, L 80
Flexner, L.B. 30
Fontaine, J. 76
Forest, M.G. 118
Forsberg, J-G. 122
Fowler, R.E. 7
Franchi, L.L. 119, 121
Franklin, K.B.J. 188
Fraser, A,W 109
Fraser, E.A. 115, 165, 167
Fraser, J.E 3, 126, 137
Fredericks, C.M. 5
Friedberg, J. 76
Fritsch, B. 205
Frohman, M.A. 179
Fuhrmann, G. 118
Fujinaga, M. 15
Fujiwara, T. 30
Fukiishi, Y. 82
Furley, B. 80
Fyfe, D.M 48

Gadd, R.L. 31
Gale, T.F. 82
Gans, C. 210, 211

García-García, J.D. 75
Gardner, R.L. 10, 12, 13, 24
Gartler, S.M. 118
Gasser, R.F. 3
Gates, A.H. 4
Geelan, J.A.G. 178
Gersh, I. 113
Gershon, M.D. 49, 141
Geubelle, F. 91, 146
Gibbs, G.E. 143
Gil, D.R. 79
Gilbert, S. 3, 10, 48, 52, 95, 117, 123, 172
Gilula, N.B. 10
Gimeno, A.L. 5
Gimeno, M.F. 5
Ginsburg, M. 118
Gittenberger-de Groot, A.C. 83
Gladstone, R.J. 27
Glenister, T.W. 122
Glucksmann, A. 130
Goldstein, D.J. 23
Gondos, B. 123
Goodman, C.S. 66, 105
Gosden, R. 6
Goss, C.M. 79
Grainge, C. 5
Grand, R.J. 139
Green, E.L. 3
Green, M.N. 139
Grillo, I. 154
Gross, M. 118
Gruenwald, P. 3, 123
Grüneberg, H. 59
Gulamhusein, A.P. 13
Gupta, M. 13
Guttridge, K.M. 38, 80

Hafez, E.S.E. 5
Halbert, S.A. 5
Hall, B.K. 36, 47, 48, 49, 212
Hamburger, V. 3
Hamilton, H.L. 3
Hamilton, W.J. 3, 27, 32, 38, 53, 54, 55, 68,109, 115, 161, 198, 199, 204, 213,
Handa, J. 159
Harary, I. 80
Harmann, C. 25
Harris, M.J. 198
Hartigan, D.J. 179
Harvey, S.C. 184
Haumoont, S.M. 74, 76
Haustein, J. 198
Hay, D.A. 79
Hayakawa, K. 82
Hebel, R. 3, 105, 191, 192, 198
Hellman, L.M. 30
Hellstedt, P. 113, 123
Henke-Fahle, S. 117
Hertig, A.T. 118
Hess, A.F. 138
Hetherington, C.M. 199
Hewer, E.E. 139
Heuser, C.H. 3
Hilburn, J. 116
Hilfer, S.R. 146, 165
Hillman, N. 9, 10
Hirsch, P.F. 76
Hislop, A.A. 146
Ho, E. 83
Hogan, B.L.M. 52, 199
Holland, P.W.H. 52
Holme, K. 117
Horsburgh, G. 199
Horstädius, S. 47
Hoshino, K. 97
Hotary, K.B. 105
Huang, R. 52, 57
Hubner, K. 118
Huckins 123
Hummel, K.P. 130
Humphrey, T. 159

Hunter, C.B. 30
Hunter, R.H.F. 9
Hutchins, G.M. 83, 86
Hutchinson, D.L. 30

Igusa, Y. 6
Ingalls, N.W. 27
Isaacs, R. 198

Jacobson, A.G. 52, 178, 179, 211
Jeffcoate, T.N.A. 31
Jenkinson, J.W. 12
Jeon, U.W. 118
Johnson, F.P. 138
Johnson, M H 10
Johnson, M.L. 97
Jollie, W.P. 23, 32
Jones, H.W. 125
Jost, A. 118, 123
Joyner, A.L. 59
Juurlink, B.H.J. 3

Kalia, M. 91
Kanagasuntheram, R. 136, 138
Karlberg, P. 91, 146
Kaufman, M.H. 3, 4, 6, 7, 8, 9, 19, 27, 28, 35, 38, 59, 68, 78, 80, 82, 96, 97, 101, 122, 123, 159, 163, 178, 179, 194, 198, 199
Kawano, J. 86, 87
Keene, M.F.L. 139
Keibel, F. 3
Keiffer, H. 24
Kelly, S.J. 9
Kennedy, J.R. 118
Kessler-Hanna, A. 86
Kihara, T. 159
King, W.A. 15
Kingery, H.M. 119
Kirby D.R.S. 23, 32
Kirby, M.L. 77, 82
Kittelson, J.A. 139
Kline, D. 6
Kline J T 6
Koch, G. 91, 146
Koda, J.E. 166
Koff, A.K. 122, 123
Kollar, E. 211
Kramer, T.C. 83
Krapez, J. 6
Kratochwil, K. 197
Kreidbeurg, J.A. 113, 123
Kulikowski, R.R. 80
Kupperman, H.S. 97
Kwong, W.H. 191

Lai, F.A. 5, 6
Lance-Jones, C. 105
Langman, J. 178
Larsen, W.J. 3,20, 48, 49, 64, 74, 75, 76, 87, 163, 199,
Larsson-Blomberg, L.B. 49, 141
Lash, J.W. 52
Lawson, K, A 16, 18, 21, 22
Layton, W.M. 80
Le Douarin, N. 47, 49, 184, 212
Le Lièvre, C. 47, 184
Leeson, C.R. 30
Leeson, T.S. 30
Lemire, R.J. 190
Lewis, O.J. 152
Lieberman, M. 87
Lillie, F.R. 118
Linares, J. 75
Lind, J. 91, 92, 146
Lintern-Moore 123
Liu, H.M. 154
Lloyd, J.R. 31, 116
Lo, C.W. 10
Lomax, M.I. 205
Lovell-Badge R. 118
Lyon, M.F. 199

Mackay, S. 123
Macknight, A.D.C. 208
McBride, R.E. 83
McDonald, S.W. 123
McDougall, R.D. 7, 198
McIlhinney, R.A.J. 119
McKay, D.C. 118
McLachlan, J.C. 55
McLaren, A. 10, 118
McLeod, M.J. 198
McMinn, R.M.H. 3, 105, 106, 152
Magre, S. 123
Mall, F.P. 3
Manasek, F.J. 80
Mandl, A.M. 119, 121
Mann, I. 194
Marieb, E.N. 123
Markwald, R.R. 79
Martin, G.R. 13, 179
Martinez, S. 179
Massion, J. 185
Mathan, M. 138
Mauleon, P. 123
Meier, S. 51, 52, 211
Meister, A. 118
Melloni, B.J. 208
Meneses, J.J. 16, 18, 22
Merchant, H. 118
Mérida-Velasco, J.A. 75
Merklin, R.J. 115
Meyer, D.B. 56, 122
Meyer, R. 122
Michels, N.A. 115
Mills, R.G. 208
Miura-Yasugi, E. 130
Miyazaki, S. 6
Mizuno, T. 130
Moffat, D.B. 64
Monk, M. 118
Monroe, R.G. 80
Moore, G.W. 83, 86
Moore, K.L. 3
Morgan, B.A. 57
Morriss, G.M. 24, 211
Morriss-Kay, G.M. 47, 82
Morris-Wiman, J. 159
Mossman, H.W. 3, 32, 38, 53, 54, 55, 68, 109, 115, 161, 198, 199, 204, 213
Motta, P. 9
Mountcastle, V.B. 208
Moxey, P.C. 138
Mullen, R.J. 118
Muller, F 3, 56, 131, 188
Muller, T.S. 52
Muller-Hermelink, H-K. 152
Munson, P.L. 76
Murakami, R. 127, 130

Nadijcka, M. 10
Nakamura, K. 159
Narvaez, V. 118
Nathanielz, P.W. 91
Navaratnam, V. 38, 78, 79, 80, 87
Nelson, C.E. 57
Neslen, E.D. 30
Neubuser, A. 52
Niswander, L. 95
Nixon, D.A. 116
Noback, C.R. 97
Noden, D.M. 47, 68, 76, 211
Nolan, J.P. 140
Norris, E.H. 74

O'Neill, M.N. 140
O'Rahilly, R. 3, 56, 77, 122, 131, 144, 146, 188
O'Shea, S. 190
Odgers, P.N.B. 83
Odor, D.L. 121
Ohbo, K. 118
Ohdo, S. 82
Ohno, S. 118

AUTHOR INDEX

Okishima, T. 82
Ooka-Souda, S. 130
Orimo, H. 76
Orts Llorca, F. 79
Osathanondh, V. 114
Osmond, M. 3
Otis, E.M. 3
Ovitt, C.E. 118
Ozdzénski, W. 27, 38, 118

Pachnis, V. 49, 141
Padget, D.H. 189
Papaioannou, V.E. 10, 12
Parrington, J. 5, 6
Patten, B.M. 77, 96, 97, 101
Pattle, R.E. 146
Pauerstein, C.J. 5
Paxinos, G. 188
Peacock, A. 57
Pearce, R.M. 154
Pearson, A.A. 198
Pedersen, R.A. 16, 18, 21, 22
Pei Y.F. 198
Pelliniemi L.J. 123
Penefsky, Z.J. 80
Perchellet, J.P. 123
Perkins-Cole, K.J. 15
Persaud, T.V.N. 3
Peters, H. 121
Pexieder, T. 81, 82
Plentl, A.A. 30
Poelmann, R.E. 17, 18
Polani, P.E. 118
Pomeranz, H.D. 49, 141
Porteus, M.H. 179
Potter, E.L. 113, 114, 154
Potts, J.T. 76
Prader, A. 57
Prépin, J. 123
Pritchard, J.A. 30
Proctor, N.K. 30
Puelles, L. 179
Puschel, A.W. 109

Ralphs, J.R. 156
Rapraeger, A.C. 166
Rayner, R.M. 146
Read, J.B. 143
Reid, L. 146
Reinius, S. 17
Reisfeld, R.A. 76
Reynolds, S.R.M. 30
Rhodes, P. 83, 95
Rhodin, J.A.G. 198
Richardson, F.L. 130
Richter, D. 91
Rivest, M. 118
Robbins, R.H. 137
Rodriguez, E.R. 80
Rohwedel, J. 118
Romanoff, A.L. 66
Rossant, J. 13
Rothman, T.P. 49, 141
Rowshan, G. 30
Rubenstein, J.L.R. 179
Rugh, R. 3
Rychter, Z. 81

Sainio, K. 113, 123
Sakamoto, M.K. 159
Sariola, H. 113, 117, 123
Saunders, J.W. 83, 95
Saxén, L. 113, 123
Scammon, R.E. 139
Schambra, U. 188
Schlafke, S.J. 13
Schmidt, M. 15
Schoenwolf, G.C. 190

Scholer, H.R. 118
Schuchardt, A. 49, 141
Schwartz, Z. 59
Scott, J.S. 31
Scott, W.W. 125
Scudder, A. 30
Sehlmeyer, U. 118
Seifert, R. 74
Seller, M.J. 15
Sellman, S. 47
Senior, H.D. 104
Setchell, B.P. 123
Sevel, D. 198
Severn, C.B. 151
Sgalitzer, K.E. 165
Shan, J. 118
Shaw, J.P. 199
Shima, K. 154
Shimada, Y. 83
Shimamura, K. 179
Short. R.V. 118
Shwachman, H. 139
Skepper, J.N. 38, 80
Smart, S.D. 55
Smith, C. 74
Smith, D.I. 143
Smith, F.G. 116
Smith, R.B. 143
Smith, R.K.W. 10
Snell, G.D. 3, 12, 13, 16, 24
Snow, M.H.L. 12, 118
Sokoloff, L. 105
Spark, C. 97
Spaulding, M.H. 127
Spivack, M. 28
Spiegelman, H. 118
Steinberger, E. 121
Steinhardt, R.A. 5, 6
Steptoe, P.C. 5
Sterin-Speziale, N. 5
Steven, D.H. 24
Stevens, L.C. 3, 12, 13, 16, 24, 119
Stewart, D.E. 82
Strachan, L. 123
Streeter, G.L. 3, 27
Stromberg, M.W. 3, 105, 191, 198
Strong, R.M. 97
Sugie, T. 15
Summerbell, D. 95, 156
Sutherland, S.D. 152
Swain, A. 118
Swann, K. 5, 6
Swanson L. W. 188
Sweeny, L.J. 77
Symington, J.M. 216
Szöllösi, D. 8

Tabin, C. 57
Takacs, E. 31
Takamura, K. 82
Tam, P.P.L. 18, 19, 22, 36, 51, 52, 69, 76, 118, 191
Tan, S.-S. 47, 76
Tanimura, T. 159
Tarkowski, A.K. 9, 10, 118
Tasca, R. 9
Taylor, D.W. 208
Taylor, I.M. 143
Tecott, L.H. 188
Teillet, M.A. 49
Tessier Lavigne, M. 105
Theiler, K. 3, 15, 19, 59, 113
Thorogood, P.V. 47, 55, 209, 211
Tickle, C. 95, 156
Tokioka, T. 192
Tokunaga, T. 15
Torti, F.M. 139
Tosney, K.W. 105

Trainor, P.A. 36, 69, 76
Trier, J.S. 138
Tsunoda, Y. 15
Tucker, J.A. 144
Tyler, M.S. 212

Upadhyay, S. 123

Valentino, K.L. 188
Van Blerkom, J. 9
Van Campenhout, E. 184
Van Mierop, L.H. 79
Vellguth, S. 152
Vernall, D.G. 83
Vigier, B. 123
Virágh, S. 38
Vogler, H. S. 20
von Gaudecker, B. 152
Vosburgh, G.J. 30
Vuillemin, M. 7

Waldo, K.L. 82
Wallgren, G. 91, 146
Walls, E.W. 87
Warkany, J. 31, 46
Wassarman, P.M. 6, 121
Wasthed, B. 76
Waterman, R.E 80
Watkins, J.B. 139
Webb, S. 84, 199
Wedden, S.E. 156
Wegelius, C. 91, 146
Weinberg, S. 216
Weller, G.L. 74
Wells, L.J. 42
Wenink, A.C.G. 87
Wessells, N.K. 195
Westerfield, M. 109
Whitaker, M.J. 6
Whiten, S.C. 55
Whitten, S.C. 118
Whittingham, D.G. 6
Wigglesworth, J.S. 146
Wilde, W.S. 30
Williams, D.W. 30
Williams, E.A. 18, 22
Williams, L. 138
Williams, P.L. 3, 96, 101, 115, 123, 147, 161, 163, 165, 180, 192, 207, 208
Wilson, A.S. 105
Wilting, J. 52, 57
Wirtschafter, Z.T. 30, 97
Witschi, E. 3, 109
Wobus, A.M. 118
Wolfson, R.J. 208
Wollman, S.H. 165
Wolpert, L. 105
Wong, G.B. 216
Woodruff, J.D. 5
Wrenn, J.T. 195
Wróblewska, J. 9, 10
Wurst, W. 59

Xu, K.P. 15

Yadav, B.R. 15
Yamauchi, A. 87
Yanagimachi, R. 5, 6
Yeom, Y.I. 118
Yokoh, Y. 132
Young, R.W. 197

Zamboni, L. 113, 116, 118,123
Zapata, C. 5
Zhi, Q. 52, 57
Ziomek, C.A. 10
Zuckerman, S. 119
Zybina, E.V. 13

5.8 Index of mouse tissues

Introduction

This index is to help the reader navigate the text, but it is not complete! It documents most of the tissues mentioned, but only their first occurrence in a particular section. Note that:

- Entries are mainly to the normal names of tissues rather than their parts (e.g. temporal bone, squamous part or amniotic fold, anterior) or their system (e.g. there are no entries under "heart") and to text and footnotes rather than drawings. If a particular tissue is not mentioned under its tissue (e.g. pituitary), it may well be under its own name (e.g. pars nervosa).
- Tissues are designated by their main name rather than by modifiers: thus the *ascending aorta* will be found under *aorta ascending*.
- There are no general groups of tissues in this index (e.g. each gland will be found under its own name rather than under *glands*)
- References are to branchial arch tissues rather than pharyngeal, aortic or visceral arches. Entries are also to the number of the arch

Most tissues are not arranged in a hierarchy but are simply itemised under their most common name although a few component tissues are included under their main organ name (gut - foregut etc.)

abdominal paraganglia (of Zuckerkandl) 143
abducens (VI) nerve 176, 186, 199, 208
accesory (XI) nerve 70, 186
acetabular cup (fossa) 100
acoustic meatus, external 70, 205
acoustico-facial [VII–VIII] pre-ganglion complex 205
acromion 101
adenohypophysis 158, 181, 193
adrenal gland
　cortex 143
　medulla 50, 143
alar plate, rhombic lip 186
allantoic bud 18, 19
allantois 24
amnion 24, 26, 136
amniotic fluid 30, 116
amniotic fold
　anterior 23
　posterior 23
anal region
　canal 138
　membrane 39, 127, 138
　pit 138, 157
　sphincter, external 127, 139
anconeus muscle 106
annular ligament 95
anulus fibrosus 57
aorta 91
　ascending (thoracic) 63, 81, 83
　dorsal 63, 79
　ventral 63
aortic arch 47, 64, 79
aortic sac 63, 79
aortico-pulmonary spiral septum 63, 65, 81, 83
apical ectodermal ridge 93
appendix
　epididymis, 125
　testis 125
aqueous chamber of eye 198
arachnoid
　granulations 185

meningeal layer 177
　(subarachnoid) space 184
archicerebellum 176
arrectores pilorum 55
ary-epiglottic folds 144
arytenoid
　cartilages 68
　swellings 144, 147
atlanto-axial joint 57
atlanto-occipital joint 57
atlas vertebra 55
atrium (atrial chamber)
　left, right 79, 91
　common (primitive) 79
　septum 82
atrio-ventricular
　groove 80
　bulbar cushion 82
　canal 83
auditory region
　(ext. acoustic) meatus (canal) 202
　(vestibulocochlear) (VIII) nerve 186
　(VIII) nerve 207
　[VIII] nerve, vestibular & cochlear divisions 205
　cortex 207
　hillocks (tubercles) 71, 202, 205
　ossicles 210
Auerbach's plexus 141
autonomic nerve supply
　pancreas 154
　gonads 128
axial arteries to limbs 103
axillary veins 104
axis vertebra 57
azygos vein 104, 145

basilar
　artery 188
　membrane 207
basilic vein 106
basioccipital bone 209, 214
basis pedunculi 185
basi-sphenoid bones 209
bile duct 149, 151
biliary system 152
　diverticulum (bud) 39, 149, 152
　intra- and extra-hepatic components 149, 152
bladder 109, 116, 127
blastocyst 12
blood 32, 38
　islands 37
Bochdalek's foramina 46
body cavities 33
bony labyrinth 207
brachial
　artery 104
　plexus 105, 106
brachiocephalic
　vein 87, 91
　　left 104
　trunk 64, 66
brachium pontis 188
brain stem 176, 188
branchial afferent nerves 69
branchial arch system
　arteries 63, 79
　　first 64
　　fourth 64
　　second 64
　　sixth 64, 144
　　third 64
　cartilages 202
　　first 68, 203, 209

second 68, 203
　fourth 68
　sixth 68
　third 68
cleft, first 70
cyst 72
efferent nerves 69
first arch 69, 204
　mandibular component 60
　maxillary component 60
fistulae 72
fourth arch 143
groove (cleft), first 204
membranes 72
muscles 76
pouches 72, 158
　first 132, 203, 205
　fourth 74
　second (hyoid) 60
　third 74
second arch 70, 204
summary 60
third arch 70
branchial (cervical) sinus 72
broad ligament 123
bronchi 144
　main 132
　segmental 145
bronchial arteries 91, 144
broncho-pulmonary segments 146
buccopharyngeal membrane 18, 39, 132, 157, 192
bulbar cushion 82
bulbar ridges 83
bulbo-ventricular grooves 80
bulbus cordis 79

caecum 138, 139
calcaneum 96
calcarine fissure 181
calvarium (frontal, parietal, squamous parts) 209
cardiac jelly 37, 79
cardinal vein system
　anterior 82, 104
　common 46, 82, 88
　posterior 82, 104
　right 87
cardiogenic plate 19, 37, 77, 79
carotid arteries 188, 189
　common 64, 66
　external 64, 66
　internal 64, 66, 79, 193
carotid body 64
carotid sinus 64
carpal bones 96
carpal region (forelimb) 95
cauda equina 184
caudal eminence 189
caudal neuropore 178
caudate nucleus 183
cavities of early embryo 33
cavities, intraembryonic 42
central canal (spinal cord) 173, 179
central tendon of diaphragm 40, 46
cephalic neural crest 48
cerebellar artery 189
cerebellomedullary cistern 185
cerebellum 176, 188, 189
　peduncle, juxtarestiform body 188
　peduncles (middle, superior, inferior) 188, 176
　plate 188
　primordium 188
　interventricular portion 186
　restiform body 188

282

INDEX OF MOUSE TISSUES

vermis 188
cerebral aqueduct 176, 178
cerebral
   arteries 64, 189
   cortex 173, 183
   peduncle 185
   vein 189
cerebrospinal fluid 176
cervical
   [C2] nerve, second 206
   flexure 179
   ganglion, superior 49
   sinus 71
cervico-thoracic ganglion 50
cervix uteri 122
choana,
   definitive secondary 159
   primary 159
chondrocranium 209, 210
chorio-allantoic placenta 24, 28
chorion 24
chorionic plate 13, 32
choroid
   fissure 195
   layer of eye 47, 49, 198
   plexus 183, 186, 189
ciliary
   body 197
   muscle 47, 198, 199
circle of Willis 188
clavicle 100, 101
clitoris 127
cloaca 39, 127
cloacal membrane 18, 39, 127, 138
   sphincter 127
coagulating gland 127
coccygeal nerve 108
cochlea 207
cochlear duct 202, 207
coeliac
   ganglion 143, 154
   plexus 117, 143
   trunk 140, 152, 154
coelomic cavity, extra-embryonic 23
   intraembryonic 19, 77, 79
colon
   ascending 138
   descending 138
   sigmoid 138
   transverse 138
commissure, anterior 181
conchae 161
condylar process of mandible 216
conus medullaris 190
copula of tongue 163
coracobrachialis muscle 106
coracoid process 101
cornea 47, 197
corniculate cartilages 68, 144
corona radiata 4
coronary
   arteries 86
   ligaments 40
   sinus 46, 87, 90, 104,149
coronoid processes of mandible 216
corpus
   callosum 189
   luteum 5
   striatum 181
costal cartilages 57
cranial nerves (table) 171
cranial outflow 191
cribriform plate of the ethmoid bone 161, 162, 181
cricoid cartilages 68, 147
crista
   dividens 84
   terminalis 82
crura 41
crus commune 202
cumulus mass, 5
cuneiform cartilages 68

cystic system
   artery 155
   duct 151
   gall bladder primordium 149

decidual reaction 13
deltoid tuberosity 96
dental
   lamina 169
   papilla 169
dermatome 51, 55
dermatome pattern 106
dermomyotome 51
Descemet's membrane 197
detrusor muscle 116
diaphragm 40, 46, 146
diaphragma sellae 193
diencephalon 173, 181, 192
digital
   interzones 95
   pads 95
   rays 95
diploic vein 217
distal convoluted tubule of kidney 114
dorsal
   aorta 117, 189
      left 64
   mesocardium 38, 79
   mesogastrium 152
   root ganglia 49, 191
ducts of Cuvier 82
ductus arteriosus 65, 66, 83, 90, 92
   patent 92
   caroticus 66, 189
   deferens 112, 125, 128
   venosus 40, 88, 91, 149
duodenum 138
dura mater 184
dural layer 177
   venous sinuses 185
      superior sagittal 185

ear 202
   drum 72
   external 204
   external acoustic meatus 167
   middle 73, 202
      ossicles 49, 73
ectoderm, extra-embryonic 13
ectodermal ridge of limb, apical 93
ectoplacental cavity 24
   cone 10, 24
egg cylinder 12, 24
   late 16
egg, cortical granules 6
ejaculatory ducts 125
elbow joint 99
embryonic ectoderm 13
emissary veins 185
enamel epithelium 169
endocardium 37, 79
endoderm, definitive 18
endolymphatic
   appendage (diverticulum) 202
   duct, sac 208
enteric nervous system 49
epaxial muscles 53
ependymal layer of forebrain 176
epiblast 11, 13
epibranchial placode 74
epicardium 79, 83
epididymis 112, 123, 125,128
epigastric veins 150
epiglottis 143, 147, 163, 183
epiphysis 183
epiploic foramen 136
epithalamic sulcus 181
epithalamus 173, 181
ethmoid bone 161, 216, 173
   cribriform plate 162, 181
Eustachian tube 73, 203, 206

exoccipital bone 214
exocoelomic cavity 24
external acoustic meatus 206
external carotid artery 64
eye 194
   nerve supply 198
   posterior chamber 199
eyelids 198

fabella 96
facial (VII) nerve 70, 163, 176, 186, 206
   chorda tympani branch 70, 163
   infraorbital branch 169
facio-acoustic (VII-VIII) ganglion complex 186
   neural crest 202
falciform ligament 40, 88, 132, 149
Fallopian tubes (oviducts) 122
falx
   cerebelli 185
   cerebri 185
femoral artery 104
fenestra vestibuli 203
filum terminale 191
fontanelles, anterior, posterior. sphenoidal mastoid 209
footplate of stapes bone 95
foramen
   caecum 39, 132, 165
   magnum 188, 214
   ovale (of cranial cavity) 185
   ovale (of heart) 84, 91
   spinosum 185
foregut diverticulum 39, 129
   enteric nervous system 49
   pocket 19, 129
   portal 132
foregut-midgut junction 138
fourth ventricle 173
   lateral recess (Luschka's foramen) 186
   lumen 178
   median recess (Magendie's foramen) 184
frontal bone 215
frontal suture 215

gall bladder 152, 154
   primordium 149
gastric artery (left), oesophageal branch 140
gastroduodenal artery 154, 155
gastro-splenic ligament 136,152
gastrulation 18
geniculo-calcarine cortex 181, 183
genital
   canal, common 122
   folds 127
   papilla 127
   tubercle 127
giant cells
   primary 13, 29
   secondary 13, 28
   trophoblastic 31
glans clitoridis 127
   penis 127
gleno-humeral joint 99
glenoid fossa of scapula 101
globus pallidus (pallidum) 183
glossopharyngeal (IX) nerve 70, 147, 163, 186
gonad primordium 119
   cortical cords of ovary 119
great
   auricular nerve 206
   cerebral vein (of Galen) 189
gubernaculum 122
gum 167

hair
   follicles 169
   muscle 55
handplate 95
head mesoderm 173
head-folds 19
hemiazygos vein 145

hepatic artery 151, 155
　ducts, right, left and common 151
　sinusoids 40
hepato-cardiac channel 149
hernia, physiological umbilical 28
hindbrain 186
hindgut
　diverticulum 39
　dorsal mesentery 112
　portal 129
　primitive 39
hip bone 99, 100
hippocampus 189
humerus 96
hyaloid system
　cavity 199
　membrane 198
　vessels 199
hymen 123
hyoglossus muscle 214
hyoid
　arch 60
　artery 64
　bone 214
　　greater horns 68
　　lesser horn 68, 210
hypaxial muscles 53
hypobranchial eminence 132, 147, 163
hypochordal bow 55
hypogastric nerve 117
hypogastric plexus 128
hypoglossal (XII) nerve 165, 173, 186
hypophyseal
　artery 189
　cleft 193
　fossa 193
hypothalamic sulcus 181
hypothalamus 173, 181, 189

iliac
　arteries, external 104
　crest 100
　veins, external 104
incisive
　canals 161
　fossa 161
incisor 169
incus 68, 202, 207, 210
inferior
　colliculi 207
　gluteal artery 104
　vena cava 46, 88, 89, 91, 149
infundibular recess 181, 192, 193
inguinal canal 126
inner cell mass 10
innominate bone 100
interatrial septum 79
intermediate plate mesoderm 109
intermolar eminence 163
internal carotid artery 199
interosseous artery 104
interparietal bone 214, 216
inter-phalangeal joints 99
intersegmental artery
　fifth lumbar 103
　seventh cervical 103
inter-thalamic adhesion 181
interventricular
　foramina 173, 178
　septum 83
　　membranous component 83
　　muscular component 83
intestine 139
intraembryonic coelom 42
intraocular muscle 49
iris 197
iris, smooth muscle 198
ischiadic artery 104
ischial bones 100

jejunum 138

jugular veins 185, 189

kidney
　calyces, major & minor 114
　collecting duct 114
　cortex 114, 117
　definitive 109, 137
　glomerulus 114
　hilar region 116
　loop of Henle 114
　medulla 114, 117
　mesonephros 112
　nephron, 114
　pelvis 116
　primitive renal pelvis 114
　pronephros 112
　proximal convoluted tubule 114
　stem cells 115
knee joint 99

labia majora, minora 127
labio-gingival sulcus 167
labio-scrotal folds 127
lacrimal bone 216
lamina terminalis 179, 181
lanugo 169
laryngeal
　aditus 143
　cartilages 68, 214
　primitive 165
laryngo-tracheal groove 132, 143
larynx 147
late primitive streak stage embryo 24
lateral
　geniculate body 183, 198
　plate mesoderm 37
　　somatopleuric component 37
　　spanchnopleuric component 37
　ventricle 178, 189
lens 199
　placode 195, 199
　vesicle 195
lentiform nucleus 183
leptomeninges 184
lesser omentum 40
　lesser sac of peritoneal cavity 133
lieno-renal ligament 136, 152
ligamentum
　arteriosum 66, 70, 92
　teres 40, 91, 149
　venosum 40, 91
limb bone
　diaphysis 96
　epiphysis 96
lingual
　mucous glands, anterior 166
　nerve 163
　swellings, lateral 162
lingula 68
linguo-gingival sulcus 167
lips 167
liver 39, 88, 129, 132, 149, 151
　coronary ligament 150
　bare area 40, 150
　parenchyma 149
lumbo-sacral plexus 105, 106
lung 140, 147
　blood flow 83
　buds 46, 132, 144
　lingula 145
　lobes 145
　alveoli 146
Luschka's foramen (lateral apertura of fourth
　　ventricle) 184
lymph glands 146
lymphoid nodules 74

Magendie's foramen (median apertura of fourth
　　ventricle) 184
malleus 68, 202, 207, 209
mammillary bodies 192

mandible 68, 209, 216
mandibular foramen 68
mantle layer of forebrain 176
manubrium 58
marginal
　layer of forebrain 176
　vein 104
massa intermedia 181
masseter muscle 169
mastoid
　air cells 204
　antrum 73
　process 73, 202
maxilla 216
　bone 169, 216
　intermaxillary segment 159
　premaxilla 159
maxillary artery 64, 185
Meckel's cartilage 68, 156, 202, 209, 216,
medial geniculate body 208
median nerve 105, 108
　median & medial umbilical ligaments 30
medulla oblongata 186
Meissner's plexus 141
membranous labyrinth 202, 207
meningeal
　arteries, 185
　veins 185
meninges 177, 184
mesencephalon 49, 171
mesenchyme of otocyst 202
mesenteric
　artery, superior and inferior 140
　ganglia, superior and inferior 143
　plexus, superior and inferior 143
mesentery of foregut 132
mesocardium, dorsal 38, 42, 79
mesoderm
　head 173
　intra-embryonic 18
　paraxial 52
mesogastrium, dorsal 136
mesometrium 123
mesonephric
　(Wolffian) duct 115, 121, 123, 126
　tubules 113
mesonephros 109, 137
mesorchium 125
mesothelial lining of body cavity 43
mesovarium 123
metacarpal bones 96
metacarpo-phalangeal joint 99
metanephric blastema 111
metanephros 109, 137
metatarsal pads 95
metencephalon 176, 196
metopic suture 215
metrial gland 32
midbrain 185
　flexure 179
midgut 136
　loop 28
　mesentery 138
　region 129
molar tooth 169
morula 10
mouth pit 39, 132, 157
Müllerian (paramesonephric) duct 115, 121, 123
muscles of facial expression 70, 76
　mastication 70, 76
musculature, non-somite-derived
myelencephalon 176, 186
myenteric (Auerbach's) plexus 141
myocardial
　mantle 79
　muscle 79
　plate 19, 37
myocoel 55
myotome 51, 93

nail primordia 95

INDEX OF MOUSE TISSUES

nares
  external 159
  posterior 159
nasal
  bone 216
  capsule 162, 210
  cavity 181
  placodes 156
  process, lateral and medial 158
  septum 156, 154, 180, 214
naso-lacrimal
  duct 162
  groove 162
nasopharynx 158, 159
neocerebellum 176
neopallial cortex 181, 183
nephric duct 109
nephrogenic cord 37, 109
neural crest 47, 50, 117
  cranial 209
  occipito-cervical 141
  sacral 141
  thoraco-lumbar 141
  trigeminal (V) 36, 48
neural folds 33, 173, 193
  plate 19
neurocranium 209
  cartiliginous 69
  membranous 69
neurohypophysis 181, 193
node formation 18
nose
  (nasal) cavities 161
  vomeronasal (Jacobson's) organ 156, 161
nostrils 159
notochord 129
notochordal
  plate 19, 132
  process 18
nucleus pulposus 57

occipital
  bone (occiput) 55, 214, 216
  myotome 164, 214
  somites 173
ocular muscles, extrinsic, intrinsic 199
oculomotor (III) nerve 49, 176, 185, 199, 208
odontoid process of axis vertebra 55
oesophageal
  mesentery 46
  region 132
oesophagus 141
olfactory (I)
  bulb, accessory 161, 181
  cortex 181
  epithelium 181
  lobe 161, 181
  nerve 162, 173, 181
  pits 158
  placodes 158, 179
olivary nucleus, inferior 188
omental bursa 136
omentum
  greater 136
  lesser 132, 149
omphalomesenteric veins 88
ophthalmic artery 64, 189, 199
optic (II)
  chiasma 179, 183, 188, 192, 194, 195
  cup 195
  eminence 194
  nerve 181, 183, 108
  pit 36, 194
  placode 194
  stalk 179, 194
  sulcus 179, 194
  vesicle 179, 194
oral pit 139
orbital fissure, superior 214
organ of Corti 207
oropharynx 158, 158, 203

os clitoridis 130
os penis 128
ossicles, middle ear 68
ostium primum 83
ostium secundum 83
otic (auditory)
  capsule 68, 202, 210, 214
  ganglion 70
  pit 202
  placode 36, 202
  vesicle 202
otocyst 202, 206
  mesenchyme 202
  oval window 203, 207
otolith membrane 208
outflow tract of heart 79
oval window 203, 207
ovarian
  follicles 4
  ligament 122
ovary 121, 139
  rete ovarii 123
oviduct 122, 128
  ampullary region 5
  fimbriated os 5

pacchionian granulations 185
pacemaker of heart 87
palate
  crypts 74
  hard 159, 161
  palatal (palatine) shelves 156, 158, 159
  palatine bones 161, 216
  primary 158
  rugae 159
  secondary (definitive) 158, 159
  soft 159
palatoquadrate bone, cartilage 68, 210
paleocerebellum 176
paleocortex 181
pallium 181
pancreas 136, 152, 154
  dorsal and ventral buds 152
  tail region 154
  diverticulae 39
papilla
  circumvallate 70
  median vallate 70
parachordal (basal plate) of chondrocranium 214
paramesonephric (Müllerian) duct 39, 115, 121, 123
parasympathetic
  fibres, vagal 191
  ganglia 49, 191
  system 147, 152, 154
parathyroid glands 47, 74
paravertebral muscle masses 57
paraxial mesoderm 52
parietal
  bone 215
  lobe of brain 147
parotid gland 167
pars
  anterior (of pituitary) 193
  distalis 193
  intermedia 193
  nervosa 193
  posterior 193
  tuberalis 193
pathfinder nerves of limb 105
pectinate line (anal canal) 139, 157
pectoral girdle 100
pelvic girdle 100
pelvis, splanchnic nerves 143, 191
penile urethra 125
penis 125, 127
pericardial cavity 46
pericardial sinus (transverse, oblique) 79, 80
pericardio-peritoneal canals 46, 79, 144
pericardium 149

fibrous layer 46
  parietal 43
  serous 79
  visceral 43, 83
perineal body 39, 127, 138
perineum 138
  central tendon 138
periosteum 185
peritoneal
  cavity 40
  greater sac 129, 136
  lesser sac 129, 133
peroneal artery 104
pharyngo-tympanic (Eustachian) tube 73, 203, 206
pharynx 143, 147, 149
philtrum 156, 167
phrenic nerves 46
physiological umbilical hernia 136
pia mater 177, 184, 186
pia-arachnoid 184
pigment layer of retina 49
pineal
  gland 181
  primordium 183
  recess 183
pinna 205, 206
pituitary
  gland 30, 47, 181, 188, 192, 193, 214
  fossa 214
placenta 30
  interplacental space 29
  labyrinthine part 31
  spongy zone 32
placode
  lens 199
  olfactory 179
  otic 36, 202
pleural cavities 40, 146
pleuro-pericardial
  canals (channels) 46
  membranes (folds) 40, 43, 46
pons 147, 176, 189
pontine flexure 179
pontocerebellum 176
popliteal artery 104
portal
  triad 152
  vein 40, 140, 149, 152, 154
pre-allantoic bud stage 19
primary trophoblast giant cells 11
primitive
  meninx 184
  oropharynx 192
  streak region 19
    early streak stage 18
    pre-streak stage 16
primordial germ cells 27, 118, 123
proamniotic
  canal 13, 16, 24
  cavity 12, 13, 16, 23
proctodaeum 18, 127, 139, 157
pronephros 109
pronucleus, male, female 9
prosencephalon 49, 171
prosomeres 179
prostate
  gland 125, 127
  primordium 125
prostatic
  region of the urethra 125
  utricle 116, 125
pterygoid venous plexus 185
pterygoquadrate
  bone 210
  cartilage 68
pubic bones 100
pudendal nerve 127
pulmonary
  arteries 64, 91, 144
  trunk 64, 65, 81, 83, 91
pupil 197

pupillary
  membrane 198
  muscles 47
putamen 183
pyloric sphincter 136

radial nerve 105
Rathke's pouch 39, 132, 158, 181, 192
rectal arteries, superior and inferior 141
rectum 138
red nucleus 185, 188
Reichert's
  cartilage 68, 202, 203, 207
  membrane 13, 25
Reissner's membrane 202, 207
renal: see kidney
respiratory tract, lower 143
retina 195
  central artery 199
  neural retina 195
  pigment layer 195
rhombencephalon 171
rhombomeres 36, 186
rib 52, 57
  costal cartilage 55
round window 207

saccule 202, 207, 208
sacral nerve 108, 141
sacro-iliac joints 100
sacrum 100
scala tympani, vestibuli 202, 207
scapula 100, 101
sciatic nerve, artery to 104
sclera 47, 49, 198, 210
sclerotomes 51
segmental spinal nerves 55, 191
sella turcica 193, 214
semicircular canals 202, 208
seminal
  colliculus 127
  vesicles 112
seminiferous tubules 119
septum of heart
  primum 82
  secundum 84, 91
septum transversum 33, 38, 39, 129, 147
sex cords, primary 119
shoulder joint 99
sinus
  venarum 82
  venosus 37, 46, 79, 82, 149
    right horn 87
sinusal tubercle 122
sinuvaginal (sino-vaginal) bulb 122
sixth branchial arch 70
skull 216
  membrane bones 184
somatic motor cells 190
somatopleure 37
somites 33, 51, 93
somitomere 49, 51, 76, 173
spermatogonia 123
sphenoid bone 214
  greater wing 68, 207, 210, 214
  lesser wing 214
  spine 68, 209
sphenomandibular ligament 68, 210
sphincter pupillae 198
spinal
  artery 192
  cord, 173, 176, 179, 189, 191
    alar columns 190
    basal columns 190
    floor plate 190
    mantle layer 190
    marginal layer 190
    roof plate 190
    ventral roots 191
    ventricular layer 190
  ganglia 47

vein 192
spinocerebellum 176
spiral
  organ of Corti 202
  septum of heart 47
splanchnopleure
spleen 129, 136, 152
splenic artery 154
splenunculi 152
stapedial
  artery 64
  ring 203
stapedius muscle 203, 205, 207
stapes 68, 202, 207, 210
stellate
  ganglion 50
  reticulum 169
stem cells
  pluripotential 12
  kidney 115
sternebrae 58, 100
sternum 58, 100
stomach 132
  cutaneous region 136
  fundic region 136
  proventricular region 136
  pyloric region 136
stomatodaeum 18, 39, 132, 156, 157
striate cortex. 183
striatum 181
stylohyoid ligament, 68, 210
styloid process 210
subarachnoid
  cisterns 184
  space 177, 184, 188
subclavian
  artery 64, 104, 188
  veins 104
subdural space 184
sublingual
  ducts 166
  salivary gland 166
submandibular salivary gland 166
submucosal (Meissner's) plexus 141
substantia nigra 185
sulcus
  limitans 176, 190
  terminalis 163
supraoccipital bone 214
sutures 209
sympathetic
  system 152
  trunks 191
symphysis menti 209, 216

tactile pads 95
tail 190
talus 96
tarsal pads 95
taste buds 70
tectum 185
tegmentum 185
tela choroidea 183, 188
telencephalic vesicles 173, 178, 179
temporal bone 216
  frontal, parietal parts 209
  petrous part 186, 202, 210
  squamous part 68, 209, 216
  styloid process 68, 210
temporo-mandibular joint 68, 216
tensor
  tympani muscle 203, 207, 205
  veli palatini muscle 70
tentorium cerebelli 185, 189
testicular cords 119
testis
  appendix testis 125
  efferent ductules 123
  gubernaculum testis 125, 126
  mediastinum (hilar region) 123
  rete testis 123

seminal vesicles 125
seminiferous tubules 123
tunica albuginea 125
tunica vaginalis 125
thalamus 173, 181, 188, 189
third arch artery 189
third ventricle of brain 173, 178, 189, 192
  infundibular recess 192
  pineal recess 183
thymus gland 47, 74
thyroglossal duct 165
thyroid gland 47, 147, 165
  C-(parafollicular) cells 47
  cartilages 68
  diverticulum 132
  primordium 39
tongue 143, 147
  copula 163
  extrinsic musculature 164
  filiform papillae 163
  frenulum 163
  fungiform papillae 163
  intrinsic musculature 164
  lingual swellings 159
  median circumvallate papilla 163
  sublingual (caruncles) papillae 163
tonsil
  cleft 74
  lingual 74
  pharyngeal 74
  tonsillar bed 74
  tubal 74
tooth 49, 169
  germ 169
  papilla, dentine 47
trabecular cords 32
trachea 132, 144
  diverticulum 144
  rings 68, 146
tracheo-bronchial groove 143
tractus solitarius 147
triceps muscle 106
trigeminal (V) nerve 163, 169, 176, 185, 186, 188, 198
  auriculo-temporal branch 206
  mandibular division 69, 203, 206
  maxillary division 69
  neural crest 48
trigone 116
triradiate ligament, ossified central part (hypochordal bow) 55
trochlear (IV) nerve 176, 185, 186, 199, 208
trophectoderm 10
trophoblast cells 10
  mural 10, 12
  primary giant cells 10
  secondary giant cells 10
truncus arteriosus 79, 82
tuber cinereum 192
tuberculum impar 162, 165
tubo-tympanic recess 73, 203, 202, 204
tunica vasculosa lentis 199
turbinate bone 161, 216
tympanic
  antrum 204
  cavity 73, 202, 204
  membrane 72, 73, 202, 203, 204, 206
  plate 68
  plexus 206
  ring 214

ulnar nerve 105
ultimobranchial bodies 47, 49, 74
umbilical
  arteries, 91
  artery, common 28
  cord 25, 26, 27, 136
  ligament, median 91, 116
  ring 26, 27, 33, 136
  vein, common 28, 40, 88
    left 91, 149

urachus 30, 116, 127
ureteric bud 111
ureters 109
urethra 109, 125, 127
   crest 115
   groove 127
   plate 125
   prostatic region 125
urogenital diaphragm 127
   membrane 39, 127, 138
   mesentery 112
   ridges 112
   sinus 39, 111, 115, 122, 125, 138, 141
urorectal septum 39, 127, 138
uterine
   artery 32
   horns 122
uterovaginal canal 122
uterus masculinus 125
uterus, round ligament 122
utricle 202, 207, 208

vagal parasympathetic fibres 191
vagina 122
vaginal plate 122
vagus (X) nerve 70, 141, 147, 163, 186
   auricular branch 206
   recurrent laryngeal branches 70
   superior laryngeal branches 70
vault of skull 209, 215
vena cava
   inferior 46, 87, 104, 149
   superior 87, 86
ventricle
   left, right 79, 91
   primitive 79
vertebra 55, 188
   centrum 57
   dorsal arch 57
   intervertebral discs 57
vertebral
   artery 188
   body 55, 57
verumontanum 115
vestibular membrane 202
vestibulocerebellum 176
vestibulocochlear (VIII) nerve 176, 186
   ganglia 206
vibrissae 169
viscerocranium
   bones 49, 50, 209, 216
   membranous bone 68

visual cortex 181, 183
vitelline
   duct 27
   system 40
   vein 40, 88
   venous plexus 140
vitello-intestinal (vitelline) duct 136
vitreous body 199
vomer 216
vomeronasal (Jacobson's)
   nerve 180
   organ 161, 180, 216

Wharton's jelly 28
Wolffian (mesonephric) duct 115, 121, 123, 126
wrist joint 25, 38, 99

xiphoid process 58

yolk sac 136
   cavity 12, 13, 23
   vasculature 27, 37

zona pellucida 9
Zuckerkandl, paraganglia of 143
zygomatic bone 216

5.9 Subject index

Note that tissues are mainly cited in Index 5.8

acrosome reaction 6
allantois formation and fate 24
amnion formation and fate 24
amniotic cavity
    formation 23
    fluid 26
        production 116
        source 30
        volume 30
amniotic sac puncture 59
androgenesis 8
anophthalmia 199
anteater tongue papillae 164
anti-Müllerian Hormone (AMH) 117, 123
antral follicle 5
apoptosis
    heart development 82
    neural crest 186
    mesonephros 113
Aristotle, naming of meconium 139
arterial system
    early development 87
    heart 87
ascent of tissues
    forelimb 63, 93
    hindbrain cavity (fourth ventricle) 95
    kidney 115, 125
auditory system early formation 36
autonomic nervous system *see* innervation

biliary system
    development 40, 149
    innervation (autonomic) 152
bladder development 116
blastocoel cavity
    formation 10
    fluid 10
blood cell formation 38
blood supply to brain 188
bone formation 213
brain
    blood supply 188
    cerebrospinal fluid formation 176
    choroid plexus 184
    flexures 179
    forebrain
        cavity (third ventricle) 172
        cerebral cortex formation 183
        diencephalon derivatives 181
        early development 179
        olfactory development 173
        rostral-caudal organization 179
        telencephalic enlargement 173,179
    hindbrain
        ascent of cavity (fourth ventricle) 95
        cerebellar peduncles 188
        cerebellum formation 186
        development 176, 186
    meninges, origin, nomenclature and function 184
    midbrain development 176, 185
    morphology 184
    prosomere formation 179
    stem 176, 188
    visual input 181
branchial arch
    arteries and derivatives 63
        congenital abnormalities 63, 68
    cartilage and derivatives 68
    components 63
    components of thyroid 165
    congenital abnormalities 63, 76

contribution to skull 209
    evolution 60
    grooves (clefts) and derivatives 70
    mesoderm and derivatives 76
    nerves and derivatives 69
    organization 63
    pouches and derivatives 72
    table of derivatives 64
    timing of development
    ultimobranchial bodies, origin 76 63
buccopharyngeal membrane
    break down 39, 158
    formation 18, 39

C57BLxCBA F1 mouse 19
C57BLxCBA F1 v. PO mouse differences (table) 16
cardia bifida 78
cardiovascular system early stages 37
cat: oogenesis 5
cattle: freemartin and sex determination in 117
cavities
    ascent of hindbrain cavity (fourth ventricle) 95
    ectoplacental 24
    exocoelomic 23
    extraembryonic 23
    formation in early embryos 23
    intraembryonic (coelomic) 42
    partitioning of peritoneal cavity 136
C- (parafollicular) cells 74, 165
cell lines, pluripotential 12
cerebellar peduncles 188
cerebellum formation 186
cerebral cortex formation 183
cerebrospinal fluid
    cysterns 184
    formation 176
    location 184
cervical muscle origin (table) 54
chicken
    branchial arch arteries 66
    branchial arch cartilage lineage 68
    heart development 78
    Henson's node 18
    hindgut development 138
    limb experimentation 93
    limb innervation experiments 105
    neural crest 49
    placodal origin of nerves 171
    skull lineages 212
chimaeras
    chicken-quail
        thyroid lineage 74
        gut (autonomic) innervation 143
        neural crest 47
        usage 49
    use in sex determination studies 117
chondrocranium development 214
    ossification 214
chorio-allantoic placenta 24, 28
chorion formation and fate 24
chorionic gonadotrophin (HCG) 10
chorionic plate formation 13
choroid plexus formation 181
chromaffin cells 50, 143
chromosomal abnormality
    androgenesis 8
    gynogenesis 8
    parthenogenetic development 8
    tetraploid mice 199
    triploidy 8
cisterns for cerebrospinal fluid 184
coloboma of iris 199
compaction 10

conduction system of heart 87
congenital abnormalities
    annular pancreas 154
    anophthalmia 199
    branchial arch grooves 71
    branchial arch arteries 68
    branchial arch system 63, 76
    cardia bifida 78
    coloboma of iris 199
    enlargement of foramen magnum 31
    haematocolpos (haematocolpometra) 122
    heart 90
    hydramnios (polyhydramnios) 31, 116
    oesophagus 146
    oligohydramnios 116
    recto-vesical (rectovaginal) fistula 138
    ribs 59
    skull 214
    sternum 59
    thyroid 165
    urachus (closure) 116
    vascular "rings" 104
    vitelline duct 30
cornea development 197
coronary arteries 86
corpus luteum 5
cortical reaction to fertilisation 6
cranial muscle origin (table) 54
cranial nerve formation 186
    origin and functions (table) 171

decidual reaction 11, 13
dermal bones 213
dermotome differentiation 55
descent of testes 125
diaphragm differentiation 40
diencephalon derivatives 181
dormouse: tongue papillae 163
ductus arteriosus closure 92
ductus venosus early function and post-birth loss 149

ear
    endolymph 207
    evolution of ossicles 68
    external ear development 204
    function 206
    inner ear development 202
    innervation 205
    middle ear development 202
    middle ear musculature 203
    origin 202
    perilymph 207
    post-birth changes 203
    semicircular canal function 207
    vascular system 206
    vestibular function 207
ectoplacental cavity formation 13, 23
egg cylinder
    formation 12, 26
    stage 16
egg maturation 4
egg-sperm interactions 6
embryonic stem cells 12
endolymph 207
endometrium 13
endoreduplication 10
engrailed gene 59
epaxial muscles 52
erythropoiesis 38
evolution
    branchial arch system 60
    jaw 60
    neck arteries 63

SUBJECT INDEX

skull 209
spleen 152
external development
  ear 204
  genitalia 127
extraembryonic development
  cavity formation 23
  membranes 23
extrinsic ocular muscles 199
eye
  abnormalities 199
  coloboma of iris 199
  cornea development 197
  early development and origins 194
  extrinsic ocular muscles 199
  eyelid closure and embryo ageing 198
  innervation 198
  intrinsic ocular muscles 55, 199
  iris development 197
  lens differentiation 195
  optic nerve development 197, 198
  pupil development 197
  retina development 195
  vascular system to the 199
eyelid
  closure and embryo ageing 198
  formation 198

facial skeleton development 216
fast (primary) block to polyspermy 5, 6, 8
fertilization
  block to polyspermy
    fast (primary) 5, 6, 8
    slow (secondary) 6, 7, 8
  cortical reaction 8
  immediate effect of 5
filum terminale of spinal cord 190
fish
  middle ear apparatus 63
  sex determination 117
fistula, rectovesical (rectovaginal) 138
flexures of brain 179
follicle-stimulating hormone (FSH) 4
  effect on males 123
fontanelles 213
foramen ovale closure 91
freemartin and sex determination in cattle 117

gall bladder
  absence in rat 152
  formation 152
  vascular supply 152
gastrulation
  human 21
  mouse 17, 18
gene engrailed 59
  testis determining factor 123
genetic mutation *small eye* mice 199
genital system
  early development 117
  external development 127
  females 121, 122
  males 123
  sex determination 117
genitalia external 127
giant cell formation 11, 12
glycoproteins ZP1, ZP2 and ZP3 6
gonad development
  autonomic innervation 128
  female 121
  indifferent stage 118
  male 123
gonadotrophins 4, 181
guinea pig
  oogenesis 5
  turning of embryo 34
gut
  autonomic nervous system 143
  blood supply 140
  early development 129
  early mesentery development 132

foregut
  development 39, 132
  ligaments 136
  mesenteries 132
  stomach 132
hindgut
  development 138
  parasympathetic supply 143
  initiation of function 139
  innervation 141, 143
  meconium 139
midgut
  development 136
  extension and rotation 136
  mesenteries 138
  physiological umbilical hernia 28, 136
  neural crest cells 141
  parasympathetic system 141, 143
  smooth muscles 143
  vascular system 140
gynogenesis 8

haematopoiesis 38
  stem cells 26, 149
hair
  follicles 169
  formation 55
  innervation 169
  lanugo 169
  vibrissae 169
hearing anatomical basis 206
heart
  abnormalities 90
  associated vascular system 87
  blood flow before birth 83
  blood vessels redundant after birth 91
  cardiac muscle 54
  conduction system 87
  coronary arteries 86
  development in chicken 78
  differences between human and mouse 86
  early development 77
  early location 42
  first contractions 87
  inflow 87
  initial heartbeat 78
  innervation 83
  looping stage 80
  outflow tract nomenclature 78
  pacemaker 80, 87
  post-birth changes 91
  pumping 79
  remodelling 81
  role of neural crest 82
  role of septum secundum 91
  septation 81
  septum transversum contribution to heart 149
  trabeculation of atrial chamber 82, 84
  vascular supply 87
  venous system 87
Henson's node 18
Hilton's law 105
Hilton's white (pectinate) line 139
hormone
  anti-Müllerian Hormone (AMH) 117, 123
  chorionic gonadotrophin (HCG) 10
  follicle stimulating hormone (FSH) 4
    effect on males 123
  gonadotrophin 4, 181
  progesterone 5
horseshoe kidneys 116
human development
  abnormal kidney development 115
  amnion 26
  amniotic fluid 30
  caput medusae 150
  chondrocranium development 214
  cloacal development 139
  colon development 138
  dorsal mesocardium development 78
  early embryonic development 19

  eyelid formation 198
  failure of testis descent 125
  gastrulation 21
  growth of intestine 139
  heart inflow 86
  limb innervation 105
    nerve roots in 108
  liver development 150
  lung development 146
  mesonephros development 113
  midgut villus formation 137
  muscle formation 53
  oesophagus differentiation 143
  oogenesis 5
  palate formation 159
  pancreas development 154
  parathyroid development 74
  parotid gland 167
  pituitary 193
  spinal cord development 191
  stomach muscle development 143
  teratomas 119
  tongue papillae 163
  vascular supply to limbs 103
  venous drainage of head 46
  vomeronasal (Jacobson's) organ 161
  yolk sac 26
hydramnios (polyhydramnios) 31, 116
hymen formation and persistence 122
hypaxial muscles 52

immunological privilege of trophoblast 26
implantation 11
inner cell mass 10
inner ear development 202
innervation
  autonomic
    biliary system 152
    enteric nervous system 49
    eye 198
    gonads 128
    gut (enteric) 49, 141, 143
    hair 83
    heart 83, 117
    kidney 117
    liver 152
    lung 141, 147
    pancreas 154
    vibrissae 169
  non-autonomic
    branchial arch system 69
    ear 205
    eye 198
    limb 105
    limb joints 105
    tongue 163, 165
    vomeronasal (Jacobson's) organ 161
inside-outside hypothesis 10
intermediate mesoderm development 112
intersegmental vertebral vascular system 59
intraembryonic coelomic cavity
  coelom fate 42
  coelom formation 40, 54
intrinsic ocular muscles 199
inversion of germ layers, *see* turning of embryo 33
iris development 197

jaw
  development 216
  evolution 60
joint differentiation in limbs 99

kidney
  abnormal development in human 115
  ascent 115, 125
  cysts 114
  horseshoe 116
  initiation of function 116
  innervation 117
  mesonephros 113

kidney – continued
    metanephros 109, 114, 115
    vascular system 115
    Wolffian duct differentiation in male 112

lanugo 169
lens
    differentiation 195
    vascular system 199
Leydig (interstitial cells) 117, 126
limb
    early development 93
    forelimb ascent 63
    innervation (table) 106
    joint differentiation 99
        innervation 105
    muscle development 102
    muscles 105
    nerve roots (table) 106
    ossification 96
    ossification centres timing (table) 97
    pathfinder nerves 105
    paw development 95
    rostral ascent 93
    rotation of forelimb (pronation) 95
    synovial joint formation 100
    vascular system 103
lip
    development 167
    morphology in mammals 158
liver
    development 149
    early development 40
    innervation 152
    vascular supply 151
lung
    changes at birth 146
    development 43, 144
    innervation 141, 147
    lymphatic drainage 146
    stages of development 146
    vascular supply 146
luteinizing hormone (LH) 4

male gonad
    development 123
    ducts 123
mandible development 216
maxilla, premaxilla and tooth formation 158
meconium 139
meiosis 5
melanocytes (pigment cells) 50
membrane bone
    clitoris 128
    mandible 68
    skull 213
meninges 177, 184
mesoderm
    associated with brain (somitomeres) 51
    branchial arch system 76
    development 37
    paraxial 51
    skull derivatives 209
mesothelium differentiation 43
metanephros, see kidney
middle ear development 202
morula polarization 10
motor cell formation 191
mouse embryos
    gastrulation 18
    oogenesis 5
    staging 15
mouth development
    early development 158
    palate formation primary 159
    palate formation secondary (definitive) 159
Müllerian duct derivatives 122
muscle
    body wall 52
    branchial arch system mesoderm 76
    cardiac 54

cervical 54
cranial 54
cranial and cervical origin (table) 54
epaxial 52, 59
extrinsic ocular 199
hair (arrectores pilorum) 55
hypaxial 52, 59
intrinsic ocular 55, 199
limb 102
middle ear 203
myotome-derived 52
non-myotome-derived 54
smooth muscles of gut 143
tongue, extrinsic and intrinsic 163, 164
myotome derivatives 52

nasal bones origin and ossification 162
nerve roots in human limb 108
nerve supply see innervation
nerves segmental spinal 191
nervous system
    basic subdivisions 173
    layering 176
    peripheral 191
neural crest
    cephalic 48
    chicken-quail chimaeras 47
    derivatives (table) 47
    dorsal roots 191
    experimental studies 49
    formation 47
    gut 143
    migration 36, 48
    occipital 49
    rhombomere origin 186
    role in heart development 82
neural fold closure 178
neural tube development
    early stages 178
    layering 176
    late stages
        filum terminale 190
        retrogressive differentiation 189
neurulation
    development 18, 36
    pathfinder nerves 105
    secondary, in spinal cord 191
node formation 18
nose development
    external 162
    internal 161
    early development 158
    internal cavity 161
    nasal bones 162
    vomeronasal (Jacobson's) organ 180

olfactory system
    development 179
    placode formation 158
oligohydramnios 116
oocyte maturation 4
oogenesis
    cat 5
    guinea pig 5
    human 5
    mouse 5
    rabbit 5
oropharynx early development 157
ossification
    chondrocranium 214
    hyoid bone 215
    jaw 216
    limb 96
    mandible 170
    maxilla 170
    nasal bones 162
    palate 159
    pectoral girdle 100
    pelvic girdle 100
    rib 59
    sternum 59

vault of skull 213, 216
vertebral axis 57
viscerocranium 216
otic (auditory) placode 202
ovary
    descent 122
    development 121
ovulation, superovulation 10

pacemaker of heart 80, 87
palate formation
    primary 158
    secondary (definitive) 159
pancreas
    annular 154
    formation 152
    innervation 154
    vascular supply 154
parafollicular (C-) cells 74, 165
paramesonephric (Müllerian) duct formation 121
parasympathetic system
    ganglia 191
    gut 141, 143
    heart 82
    neural crest 49
paraxial mesoderm segmentation 51
parotid gland formation 167
parthenogenetic development 8
pathfinder nerves 105
paw development 95
pectoral girdle formation 99
pelvic girdle formation 100
perilymph 207
peripheral nervous system 191
pes anserinus 167
pheromone receptors 179
physiological umbilical hernia 28, 136
pituitary
    absence of induction in its formation 193
    gland 158, 181, 192
placenta
    chorio-allantoic 24, 28
    formation 32
placode
    epibranchial 74
    nasal 158
    olfactory 158, 179
    optic 36, 194
pluripotential embryonic stem cells 12
PO strain mouse 15, 19
PO/C57BLxCBA F1 mouse differences (table) 16
polyhydramnios (hydramnios) 31, 116
polyspermy
    fast (primary) block 5, 6, 8
    slow (secondary) block 6, 7, 8
porcupine: tongue papillae 163
post-birth loss of ductus venosus 149
pre-primitive streak stage 16
primary sperm receptor 6
primitive streak formation 17, 18
primordial germ cells
    formation 27, 117, 118
    migration 117
    numbers 121
progesterone 5
pronation of limb 95
prosomere formation in brain 179
prostaglandins 92
prostate formation 127
pupil development 197
purkinje cells 87

rabbit
    oogenesis 5
    sex determination 117
    spleen vascular supply 152
    turning of embryo 34
rat
    spinal cord 191
    tongue papillae 163

SUBJECT INDEX

renal system
   cloacal origins 116
   see kidney
reproductive system
   duct system (internal)
      female 121
      male 123
   ejaculatory fluid 126
   external genitalia 127
   gonad formation, early stages 118
   gonad, female 121
   gonad, male 123
   initial development 117
   primordial germ cell migration 117
   prostate formation 127
   sex differences at birth for males and females 127
   sperm formation and maturation 123
   testis development (TEM) 123
   vaginal plug 127
reptile: branchial arch system arteries 63
respiration, control of 147
respiratory tract, lower 143
retina development 195
retrogressive differentiation of neural tube 189
rib
   congenital abnormality 59
   formation 52, 57
   number abnormalities 59
rotation of forelimb (pronation) 95

salivary gland formation 167
Schwann cells 50
sclerotome differentiation 55
segmental spinal nerves 191
semicircular canals function 207
septation in heart 82
septum secundum function of 83
septum transversum
   derivatives 147
   formation 39
Sertoli (sustentacular) cells 117, 126
sesamoid bones 96
sex determination 117
   differences at birth for males and females 127
   rabbit 117
skull
   bones, dermal and membrane 213
   branchial arch contribution 209
   branchial arch system contribution 209
   chondrocranium 214
   embryological origin 211
   evolutional constraints 209
   lineage analysis 211
   neural crest contribution 209
   sutures 209, 213
   terminology 209
   vascular supply 217
   vault of skull 215
   viscerocranium (facial skeleton) 216
slow (secondary) block to polyspermy 6, 7
small eye mice 199
somatopleure formation 54
somite
   breakdown 51
   formation 33, 37, 51
   numbers (table) 52

somitomere 51
   origin of branchial mesoderm 76
sperm formation and maturation 123
   primary receptor 6
spermatogonial cells 126
sperm-egg interactions 6
spinal cord
   cellular differentiation 190
   development 189
   dorsal and ventral roots 191
   filum terminale 190
   formation of caudal region 189
   link to peripheral nervous system 191
   rat 191
   retrogressive differentiation 189
   vascular supply 191
splanchnopleure formation 54
spleen
   evolution 152
   formation 152
   function 152
staging mouse embryos 15, 18, 26
sterility in mutant mice 127
   transgenic mice 128
sternum
   congenital abnormality 59
   formation 58
stomach
   development 132
   rotation 132
sublingual gland formation 166
submandibular gland formation 166
superovulation 9
sympathetic system
   gut 143
   neural crest 49
   thoraco-lumbar outflow 50
   trunks 191

telencephalic enlargement 179
testes
   descent 125
   determining factor 123
   development 123
tetraploid mice 199
Theiler stages 15, 220 (table of features)
thoraco-lumbar outflow 50
thyroid
   abnormalities 63
   C- (parafollicular) cells 165
   gland 165
   initiation of function 165
   lineage chicken-quail chimaera data 74
tongue
   branchial arch system origin 162
   development 162
   innervation 163, 165
   musculature 164
tooth
   differentiation 169
   eruption times 170
   incisor origin 158
trabeculation of atrial chamber 84
transgenic mice sterility 128
triploidy 8
trophectoderm formation 12
trophoblast immunological privilege 26

turning
   mouse embryo 16, 26, 33
   guinea pig embryo 34
   rabbit embryo 34

ultimobranchial bodies origin 76
umbilical cord
   formation 26, 27
   length 29
umbilical ring 26, 27, 33
urachus, congenital abnormality of closure 116
urethra, formation in males 126
urorectal septum growth 138
uterolysins 11

vagina development 122
vascular supply
   ear 206
   eye 199
   gall bladder 155
   gut 140
   heart 87
   intersegmental (of vertebrae) 59
   kidney 115
   limbs 103
   liver 149, 151
   lung 146
   pancreas 154
   post-birth changes 92, 149
   septum transversum 149
   skull 217
   spinal cord 191
   venous drainage of head 46
vasculature yolk sac 37
vault of skull development 215
venous system
   early development 87
   heart 87
vertebra organization 55
vertebral axis
   blood supply 59
   early development 55
   ossification 57
   post-cranial 57
   upper cervical 55
vestibular function of ear 207
vibrissa formation 169
viscerocranium
   ossification 216
   development 213, 216
visual cortex development 181
visual input to brain 181
vitelline duct congenital abnormalities 30
vomeronasal (Jacobson's) organ 161, 180

Wolffian duct
   differentiation of ducts in male 123
   formation 109
   reproductive system 112

yolk sac
   circulation 26
   vasculature 26, 37

ZP1, ZP2 and ZP3 glycoproteins 6